Science Underground
(Los Alamos, 1982)

To Henry Primakoff

Mentor, colleague, and friend to so many of us

An *hidalgo*, in the finest sense of the word

AIP Conference Proceedings
Series Editor: Hugh C. Wolfe
Number 96

Science Underground
(Los Alamos, 1982)

Edited by
Michael Martin Nieto, W. C. Haxton, C. M. Hoffman,
E. W. Kolb, V. D. Sandberg, J. W. Toevs
Los Alamos National Laboratory

American Institute of Physics
New York   1983

Copying fees: The code at the bottom of the first page of each article in this volume gives the fee for each copy of the article made beyond the free copying permitted under the 1978 US Copyright Law. (See also the statement following "Copyright" below). This fee can be paid to the American Institute of Physics through the Copyright Clearance Center, Inc., Box 765, Schenectady, N.Y. 12301.

Copyright © 1983 American Institute of Physics

Individual readers of this volume and non-profit libraries, acting for them, are permitted to make fair use of the material in it, such as copying an article for use in teaching or research. Permission is granted to quote from this volume in scientific work with the customary acknowledgment of the source. To reprint a figure, table or other excerpt requires the consent of one of the original authors and notification to AIP. Republication or systematic or multiple reproduction of any material in this volume is permitted only under license from AIP. Address inquiries to Series Editor, AIP Conference Proceedings, AIP, 335 E. 45th St., New York, N. Y. 10017

L.C. Catalog Card No. 83-70377
ISBN 0-88318-195-9
DOE CONF- 820989

TABLE OF CONTENTS

Workshop Staff xi

Workshop Participants xii

Foreword xv

I. SCIENCE UNDERGROUND. 1
    Underground Science at Homestake
        R. Davis Jr., et al. 2
    Proposal for a National Underground Science Facility
        A. K. Mann 16
    Perspectives of Fundamental Research in the Gran
    Sasso Underground Laboratory
        M. Conversi 37
    The Gran Sasso Project
        A. Zichichi 52

II. SOLAR NEUTRINOS. 65
    The Analysis of Solar Models--Neutrinos and
    Oscillations
        R. K. Ulrich, et al. 66
    The Case of the Missing Solar Neutrinos
        W. A. Fowler 80
    The Gallium Solar Neutrino Detector
        W. Hampel, et al. 88
    A Radiochemical Solar Neutrino Experiment Using
    $^{81}Br(\nu,e-)^{81}Kr$
        G. S. Hurst, et al. 96
    A Proposed Geological Solar Neutrino Measurement
        G. A. Cowan and W. C. Haxton 105

III. PROTON DECAY 109
    Theoretical Prediction for Baryon Number Violation
        P. Langacker 110
    The NUSEX Experiment and the Future of the Mont
    Blanc Laboratory
        G. Battistoni, et al., P. Picchi, speaker 124
    IMB Detector-the First 30 Days
        R. M. Bionta, et al., D. Sinclair, speaker. . . . 138
    The Harvard-Purdue-Wisconsin Baryon Decay Experiment
    at Park City, Utah
        D. B. Cline 143
    The Soudan Nucleon Decay Experiments
        L. E. Price 161
    Proton Decay Experiment in the Kolar Gold Fields
        M. R. Krishnaswamy, et al., S. Miyake, speaker . 168
    The Nucleon Decay Experiment in the Frejus Tunnel
        J. Ernwein 175

PROTON DECAY (cont.)
    A Proton Decay and Solar Neutrino Experiment with a
    Liquid Argon Time Projection Chamber
        H. H. Chen, et al. 182
    Magnetic Monopoles, Nucleon Decay and DUMAND
        P. C. Bosetti 186

IV. COSMIC RAYS. 190
    The Fly's Eye
        R. Cady, et al., J. Elbert, speaker 191
    Physics with Underground Leptons of Atmospheric
    Origin
        T. K. Gaisser 203
    Detection of Gravitational Collapse
        J. C. Wheeler and J. A. Wheeler 214
    Signatures for Underground Neutrinos
        M. Crouch 225
    Particle Physics Below the Earth's Surface, An Overview of Possibilities
        J. G. Learned 236
    Beyond Proton Decay: Other Physics Possibilities
    with the Homestake Nucleon Decay Detector
        M. L. Cherry, et al., K. Lande, speaker 248
    Cosmic Ray Physics Underground: Some Puzzles
        M. R. Krishnaswamy, et al., S. Miyake, speaker . 265
    A Solar Breeder to Explain the Lack of Neutrinos and
    Constant Luminosity
        C. Alexander, et al., L. M. Libby, speaker. . . . 273

V. GEOPHYSICS 277
    Siting, Constructing, and Maintaining a Deep
    Underground Science Laboratory
        R. R. Sharp, Jr. 278
    Subterranean Gravity and Other Deep Hole Geophysics
        F. D. Stacey 285

VI. POSSIBLE DIRECTIONS 298
    Monopoles Underground
        J. A. Harvey 299
    Non-LTE Astrophysics and the Origin of High Energy
    Cosmic Rays
        S. A. Colgate 306
    Thoughts on Family Symmetries
        F. Wilczek 313
    Underground Neutrino Astronomy
        D. Schramm 318

VII.	GRAVITY WAVES.	324
	An Heuristic Introduction to Gravitational Waves	
	V. D. Sandberg	325
	Review of Resonant Bar Gravity Wave Experiments	
	W. C. Oelfke	331
	Laser Interferometer Gravitational Radiation Detectors	
	R. W. P. Drever	336
	Prospects for Ground Based Detectors of Low Frequency Gravitational Radiation	
	R. Spero	347
VIII.	DOUBLE BETA DECAY	351
	Double Beta Decay, Massive Neutrinos, and Lepton Number Conservation	
	S. P. Rosen	352
	Double Beta Decay - An Experimental Review	
	C.-S. Wu	374
	Geochemical Double Beta Decay Experiments	
	T. Kirsten	396
	The Irvine ^{82}Se Experiments	
	M. K. Moe and A. A. Hahn	411
	Early Results from the Batelle-Carolina ^{76}Ge Double Beta Decay Project	
	F. T. Avignone, III, et al.	419
	The Status of the UCSB-LBL ^{76}Ge Double Beta Decay Experiment	
	M. S. Witherell	427
	Recent Results Obtained in the Mt. Blanc Experiment on $\beta\beta$ Decay of ^{76}Ge	
	E. Bellotti, et al., C. Liguori, speaker	430
IX.	DISCUSSION ON A NATIONAL UNDERGROUND SCIENCE FACILITY . .	434
	Panel: W. A. Fowler, chairman, R. Davis Jr., T. Gaisser, A. K. Mann, S. P. Rosen, F. D. Stacey, and W. Wenzel.	435

MANKIND AND MOTHER EARTH

by Gilbert Atencio

At the beginning of life man was created as a good being, but as time went by man also became evil. Man can still be good, as the figure holding vegetation portrays growth and prosperity. But mankind can also overstep what is his due and destroy what is good, as the figure with lightning in his hand represents.

The whole of nature is greater than man. In the end it is those who have respect for Mother Earth and all of nature, and use them wisely, who will survive.

Below the sun and above the moon is the earth, which is on the back of the great sea turtle. Surrounding the earth is the sea serpent, which provides moisture. On the earth are the signs of the six important creations: water, minerals, plant life, sea life, animal life, and man. All of these creations are important to one another.

Mankind and Mother Earth — by Gilbert Atencio

WORKSHOP STAFF

WORKSHOP HOST:

Los Alamos National Laboratory of the University of California, Associate Directorship for Physics and Mathematics

FUNDING:

United States Department of Energy

WORKSHOP ADVISORY COMMITTEE:

L. W. Alvarez, UC, Berkeley
W. A. Fowler, Caltech
T. Gaisser, Bartol
S. L. Glashow, Harvard
M. Goldhaber, Brookhaven
A. K. Mann, Pennsylvania
H. Primakoff, Pennsylvania

F. Reines, UC, Irvine
S. P. Rosen, Purdue and NSF
F. D. Stacey, Queensland
A. L. Turkevich, Chicago
J. A. Wheeler, Texas
C. S. Wu, Columbia

WORKSHOP ORGANIZING COMMITTEE:

W. C. Haxton, Chairman
S. A. Colgate
C. M. Hoffman
E. W. Kolb

M. M. Nieto
V. D. Sandberg
L. M. Simmons, Jr.
J. W. Toevs

WORKSHOP ADMINISTRATORS:

F. Gomez
L. Lauer
M. Martinez

ADMINISTRATIVE ASSISTANCE:

F. Archuleta
E. Morris
L. Olson
L. Woodwell
M. S. Wooten

SESSION CHAIRMEN:

B. Butcher, Sandia
P. A. Carruthers, Los Alamos
A. Cox, Los Alamos
R. L. Forward, Hughes
T. Goldman, Los Alamos

W. C. Haxton, Los Alamos
E. W. Kolb, Los Alamos
N. F. Ramsey, Harvard
G. J. Stephenson Jr., Los Alamos
A. L. Turkevich, Chicago
J. D. Ullman, CUNY

WORKSHOP PARTICIPANTS

ANDER, M., Los Alamos National Laboratory
ANDERSON, H. L., Los Alamos National Laboratory
AVIGNONE, F. T., III, University of South Carolina, Columbia, SC
BARR, D., Los Alamos National Laboratory
BECKER, R. L., Oak Ridge National Laboratory, Oak Ridge, TN
BELL, G. I., Los Alamos National Laboratory
BISHOP, A., Los Alamos National Laboratory
BOSETTI, P. C., III Phys. Institut, Aachen, West Germany
BOWEN, T., University of Arizona, Tucson, AZ
BOWLES, T., Los Alamos National Laboratory
BOWMAN, D., Los Alamos National Laboratory
BOYD, R., Ohio State University, Columbus, OH
BRATTON, C. B., Cleveland State University, Cleveland, OH
BRODZINSKI, R. L., Battelle Northwest Labs, Richland, WA
BROLLEY, J., Los Alamos National Laboratory
BROWN, R. E., Los Alamos National Laboratory
BROWNE, J., Los Alamos National Laboratory
BURMAN, R., Los Alamos National Laboratory
BUTCHER, B., Sandia National Laboratory, Albuquerque, NM
CAMPBELL, D. K., Los Alamos National Laboratory
CARRUTHERS, P. A., Los Alamos National Laboratory
CHAPLINE, G. C., Jr., Lawrence Livermore Laboratory, Livermore, CA
CHEN, H. H., University of California, Irvine, CA
CLEVELAND, B. T., Brookhaven National Laboratory, Upton, NY
CLINE, D., University of Wisconsin, Madison, WI
COGBILL, A. H., Los Alamos National Laboratory
COLEMAN, P. J., Jr., Los Alamos National Laboratory
COLGATE, S. A., Los Alamos National Laboratory
CONVERSI, M., University of Rome, Italy
COWAN, G. A., Los Alamos National Laboratory
COX, A., Los Alamos National Laboratory
CROUCH, M. F., Case-Western Reserve University, Cleveland, OH
DAVIS, J. F., Los Alamos National Laboratory
DAVIS Jr., R., Brookhaven National Laboratory, Upton, NY
DETWEILER, S., University of Florida, Gainesville, FL
DeVRIES, R., Los Alamos National Laboratory
DEY, T., Los Alamos National Laboratory
DOE, P., University of California, Irvine, CA
DOMBECK, T., Los Alamos National Laboratory
DONOGHUE, T., Ohio State University, Columbus, OH
DOOLEN, G., Los Alamos National Laboratory
DREVER, R. W. P., California Institute of Technology, Pasadena, CA
DUONG-VAN, M., Rice University, Houston, TX
EISBERG, R., University of California, Santa Barbara, CA
ELBERT, J. W., University of Utah, Salt Lake City, UT
ERDAL, B. R., Los Alamos National Laboratory
ERNWEIN, J., CEN-Saclay, France
EVANS, J. C., Battelle Northwest Labs., Richland, WA
FENYVES, E. J., University of Texas, Dallas, TX
FORWARD, R. L., Hughes Research Laboratories, Malibu, CA

FOWLER, W. A., California Institute of Technology, Pasadena, CA
FRAZER, W. R., University of California, Berkeley, CA
GABBARD, F., University of Kentucky, Lexington, KY
GAISSER, T., Bartol Research Foundation, Newark, DE
GLASER, H., University of California, Berkeley, CA
GOLDMAN, T., Los Alamos National Laboratory
GOULDING, C., EG&G, Los Alamos, NM
HAMPEL, W., Max Planck Institut fur Kernphysik, West Germany
HARRIS, R., Los Alamos National Laboratory
HARVEY, J. A., Princeton University, Princeton, NJ
HAXTON, W. C., Los Alamos National Laboratory
HENDERSON, D. B., Los Alamos National Laboratory
HENSLEY, W. K., Battelle Northwest Labs, Richland, WA
HERCZEG, P., Los Alamos National Laboratory
HEYDEGGER, H. R., Purdue University Calumet, Hammond, IN
HILLS, J. G., Los Alamos National Laboratory
HOFFMAN, C. M., Los Alamos National Laboratory
HOFFMAN, D., Los Alamos National Laboratory
HURST, G. S., Oak Ridge National Laboratory, Oak Ridge, TN
HYNES, M. V., Los Alamos National Laboratory
JARMIE, N., Los Alamos National Laboratory
JOHNSON, M., Los Alamos National Laboratory
KIRSTEN, T., Max Planck Institut fur Kernphysik, West Germany
KOLB, E. W., Los Alamos National Laboratory
KOTZER, P., Western Washington University, Bellingham, WA
KROPP, W., University of California, Irvine, CA
LANDE, K., University of Pennsylvania, Philadelphia, PA
LANGACKER, P., University of Pennsylvania, Philadelphia, PA
LEARNED, J., University of Hawaii, Honolulu, HI
LIBBY, L. M., University of California, Los Angeles, CA
LIGUORI, C., I.N.F.N.-Sef. di Milano, Italy
LYONS, P. B., Los Alamos National Laboratory
MacCALLUM, C., Sandia Laboratories, Albuquerque, NM
MALIK, J., Los Alamos National Laboratory
MANN, A. K., University of Pennsylvania, Philadelphia, PA
MANNOCCHI, G., Laboratori Nazionali di Frascati, Italy
MATHEWS, M., Los Alamos National Laboratory
MATTHEWS, J., University of Wisconsin, Madison, WI
McHENRY, R., Purdue University, West Lafayette, IN
McKELLAR, B., University of Melbourne, Australia
METROPOLIS, N., Los Alamos National Laboratory
MISCHKE, R., Los Alamos National Laboratory
MIYAKE, S., University of Tokyo, Japan
MOE, M. K., University of California, Irvine, CA
MUTCHLER, G. S., Rice University, Houston, TX
MYERS, C. W., Los Alamos National Laboratory
NAGLE, D., Los Alamos National Laboratory
NEZRICK, F., Department of Energy, Washington, DC
NIETO, M. M., Los Alamos National Laboratory
NORMAN, E. B., University of Washington, Seattle, WA
NORMAN, J., Los Alamos National Laboratory
OELFKE, W. C., Louisiana State University, Baton Rouge, LA
OKA, T., Los Alamos National Laboratory

PEAK, L. S., University of Sydney, Australia
PEHL, R., Lawrence Berkeley Laboratory, Berkeley, CA
PICCHI, P., Laboratori Nazionali di Frascati, Italy
PRICE, L., Argonne National Laboratory, Argonne, IL
RAGAN, C., Los Alamos National Laboratory
RAMSEY, N. F., Harvard University, Cambridge, MA
ROBERTSON, H., Los Alamos National Laboratory
ROSEN, S. P., National Science Foundation, Washington, DC
ROSEN, L., Los Alamos National Laboratory
SANDBERG, V. D., Los Alamos National Laboratory
SCHRAMM, D., University of Chicago, Chicago, IL
SHARP, R. R., Jr., Los Alamos National Laboratory
SHUPE, M., University of Minnesota, Minneapolis, MN
SIMMONS, L. M., Jr., Los Alamos National Laboratory
SINCLAIR, D., University of Michigan, Ann Arbor, MI
SLANSKY, R., Los Alamos National Laboratory
SOLEM, J., Los Alamos National Laboratory
SPERO, R., California Institute of Technology, Pasadena, CA
STACEY, F. D., University of Queensland, Australia
STEIN, N., Los Alamos National Laboratory
STEPHENSON Jr., G. J., Los Alamos National Laboratory
STEVENSON, M. L., Lawrence Berkeley Laboratory, Berkeley, CA
SUMNER, J., University of Arizona, Tucson, AZ
TAKASUGI, E., Osaka University, Japan
TALBERT, W., Los Alamos National Laboratory
THIESSEN, H. A., Los Alamos National Laboratory
TOEVS, J. W., Los Alamos National Laboratory
TRIPP, R., Lawrence Berkeley Laboratory, Berkeley, CA
TURKEVICH, A. L., University of Chicago, Chicago, IL
ULLMAN, J. D., Herbert H. Lehman College of the CUNY, New York, NY
ULRICH, R., University of California, Los Angeles, CA
VESSER, L. R., Los Alamos National Laboratory
WACHSMUTH, H., CERN, Switzerland
WALLACE, R., Los Alamos National Laboratory
WENZEL, W. A., Lawrence Berkeley Laboratory, Berkeley, CA
WEST, G. B., Los Alamos National Laboratory
WHEELER, J. C., University of Texas, Austin, TX
WHITE, D. H., Brookhaven National Laboratory, Upton, NY
WILCZEK, F., University of California, Santa Barbara, CA
WILLIAMS, W. S. C., Los Alamos National Laboratory
WITHERELL, M., University of California, Santa Barbara, CA
WU, C. S., Columbia University, New York, NY
ZICHICHI, A., I.N.F.N., University di Roma, Italy
ZIOCK, H., Los Alamos National Laboratory

FOREWORD

In the summer of 1981, I encountered Al Mann at Los Alamos and at the Aspen Center for Physics. Al wanted to discuss with one and all his ideas for a national underground physics laboratory. The conversations interested me, especially when i) the talk began to focus on a possible Los Alamos participation in the project, and ii) the project itself began to enlarge from a specialized particle physics facility, geared to studying proton decay, to one which would be a general underground science facility.

In fact, I remember the point at which my involvement became active. It was as I was walking back to my office after lunch during a snow storm. Mike Simmons, who was coordinating the idea in the Associate Directorship for Physics and Mathematics, asked me if I could think of any gravity or geophysics experiment that could use a hole a mile deep. One day later I answered, "Yeah, I know this guy in Australia," and I was in.

In their own individual ways, each member of the local organizing committee became involved. It was a matter of obvious chemistry that Wick Haxton became chairman; a matter of my having piped up with the suggestion that we publish through the AIP that I became editor; and a matter of either permission being given for Frankie Gomez, Laurie Lauer, and Marian Martinez to be the workshop administrators or else someone could find a new organizing committee.

There are many people who, in large and small ways, contributed to the success of both the Workshop and, hopefully, these Proceedings. Some, but not all of them, are listed on the "Workshop Staff" page. I, of course, want to especially thank the five who joined me in the editorshop of this volume and the Workshop administration. The help of Hugh Wolfe, and the staff at the AIP, and Robin Odette, and the staff at Publication Press, Inc., is greatfully acknowledged. Gilbert Atencio, who, complaining to the end, agreed to paint our frontspiece, added the right touch. Lastly there were the lively talks, the spirited discussions by the participants, and the authors who managed, some of them kicking and screaming, to send in their articles.

We here at Los Alamos hope that our efforts will be judged to have been enlightened self-interest, well done. From the responses we received during, at, and after the Workshop, we think that part was. Similarly, I hope that the printed volume will give me cause to open the bottle of Martell Cordon Bleu cognac that I have been saving. If so, I must thank all those I have mentioned once more; this time for my pleasure. In no way can I claim, as did my namesake, that

Here arrived the Señor and Governor

Don Francisco Manuel de Silva Nieto

Whose indubitable arm and valor

Have overcome the impossible ...

A thing which he alone put into this effect ...

    Translation of the Spanish
  inscription dated August 5, 1629,
on El Morro Rock (National Monument),
          New Mexico

                                      Michael Martin Nieto
                                      Los Alamos, New Mexico
                                      January, 1983

Chapter I

SCIENCE UNDERGROUND

In a hole in the ground there lived a hobbit. Not a nasty, dirty, wet hole, filled with the ends of worms and an oozy smell, nor yet a dry, bare, sandy hole with nothing in it to sit down on or to eat: it was a hobbit-hole, and that means comfort.

J. R. R. Tolkien,
The Hobbit, or There and Back Again,
Chapter I

UNDERGROUND SCIENCE AT HOMESTAKE

R. Davis Jr., B. T. Cleveland, and J. K. Rowley
Brookhaven National Laboratory, Upton, NY 11973

ABSTRACT

A brief overview is given of some of the scientific work that has been done in the Homestake mine. The problems and advantages of working in active mines are discussed. Some details on the construction of the chlorine solar neutrino experiment are presented and the current results of this experiment are given. The report concludes with a discussion of the importance and feasibility of a much larger chlorine experiment.

INTRODUCTION

I gather from the organizers of this workshop that I am expected to discuss our experiences in carrying out our work in the Homestake mine. I will try to give an impression of how well our arrangements with Homestake have worked over the past 18 years, and how we have accommodated other experimentalists that wished to use our facilities. A number of other groups that wanted to carry out experimental work underground have solved their problem by working in active privately-owned mines. Fred Reines' group has had a long experience with the Morton Salt Company mine in Painesville, Ohio, and in the East Rand Proprietary gold mine at Johannesburg, South Africa. S. Miyake has used a mine in the Kolar gold field in India for many years. To my knowledge these groups have had very productive associations with these private mining companies, and have carried out their work with minimum difficulty and effort.

It is interesting that in Europe, experimenters have taken advantage of the fact that the deepest underground sites are automobile tunnels in the Alps. Some groups have arranged to use the tunnels, apparently without much difficulty, and at times space has even been offered to outside experimental groups. The Soviet group under G. Zatsepin has invested great effort in building an underground site for neutrino astronomy and cosmic ray physics.[1] They chose to drive a horizontal tunnel and will have at least four very large rooms that will house gallium and chlorine solar neutrino detectors, and a 1000-ton liquid scintillator. The greatest depth available in this facility will be 4000-4300 m.w.e.

After discussing our experiences and arrangements at Homestake I will give a brief statement of the results of our chlorine solar neutrino experiment. W. A. Fowler and others at this workshop will give the theoretical and nuclear physics background important to understanding these results. We will then discuss the importance of building a much larger chlorine experiment. In our view a project of this magnitude could be a commendable goal for the underground science facility.

EXPERIMENTS IN THE HOMESTAKE MINE

Building the chlorine detector.--When planning the Brookhaven chlorine solar neutrino experiment in the early 1960's we needed to locate a mine with a depth of at least 4000 m.w.e.[2] This depth was estimated to be sufficiently great to reduce the background from cosmic ray muons to less than one percent of the expected signal from solar neutrinos. Armed with the advice of the Bureau of Mines we examined three mines: Homestake (4200 m.w.e), the Sunshine silver mine, Wallace, Idaho (4500 m.w.e), and the Anaconda copper mine at Butte, Montana (∼4000 m.w.e). After considering the rock structure, depth, size of the hoist, and contractual arrangements, we chose the Homestake mine as the most suitable for our purpose. I might add that the Anaconda copper mine was disappointed that we did not use their mine. The main difficulty in the Butte mine was that the rock structure was not sufficiently strong to allow opening a large cavity to contain our tank.

Our experiment required a 390,000-liter tank that needed a room 10 m. wide and 20 m. long with an 11 m. high ceiling. The steel tank had to be built in place, and all of the parts were sized to fit the Homestake hoist (1.5 m. x 3.5 m. x 2 m.). The cavity was designed and built by the Homestake Mining Company on a fixed cost contract. The walls of the cavity were covered with chain-link fencing held in place with many rock bolts. This arrangement served to stabilize the rock wall. The excavation included adjacent rooms to serve as a pump room (5 m. x 5 m. x 3 m.) and a space to erect a prefabricated metal building for the processing equipment and instrument panel. These areas had concrete floors, and a rail track so that heavy equipment, tank parts, etc. could be brought in. Two steel bulkhead doors were provided at the access drifts that could be closed in the event of a liquid spill or vapor release to prevent perchloroethylene from entering the mine air. The cavity as shown in Figure 1 was built in three months in the spring of 1965 at a cost of $125,000.

The tank was built by the Chicago Bridge and Iron (CBI) Company on a separate contract. The Homestake company assisted CBI in bringing the tank parts underground and provided electrical services. In addition Homestake provided an engineer to oversee the tank construction on a daily basis. The final operation was to fill the tank with 380,000 liters of perchloroethylene (10 railroad tank cars). We chose to accomplish this task by using three 2500-liter tank cars that would fit the hoist and mine rail system.

Although building this experiment was at times a considerable load on the Homestake company, there were no serious problems and the entire operation was carried out as an additional work load to their normal mining activities. The shaft used for this work is the main one used for transportation of supplies and personnel in and out of the mine. This shaft also contains the skip for

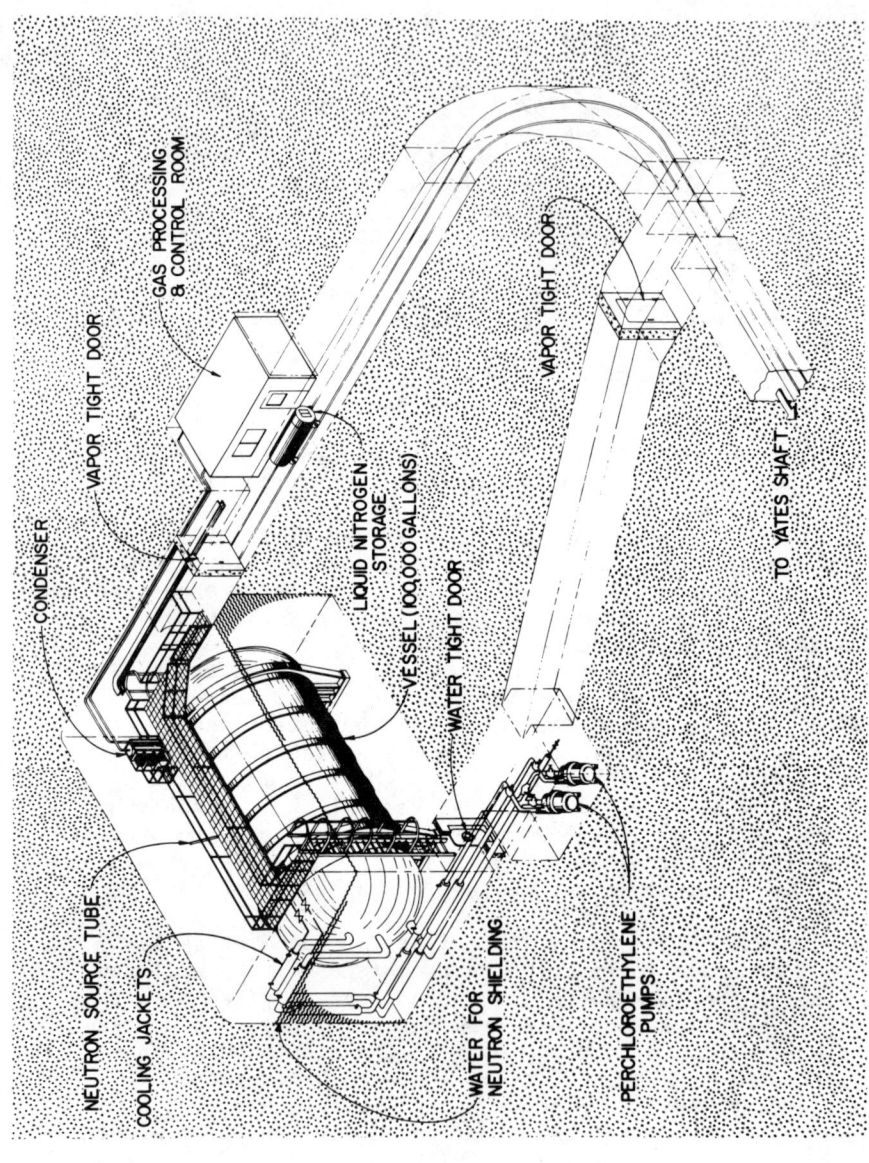

Figure 1. The chlorine solar neutrino experiment.

bringing the ore out of the mine. I mention the magnitude of the work load because it is actually a smaller and simpler task than building a large proton decay experiment like the one built by the Irvine-Michigan-Brookhaven collaboration, or the second generation experiments now being proposed.

Safety Procedures.--When carrying out a scientific experiment in an operating mine, one must accept the fact that there are hundreds of men working underground and their safety could be affected. In our case we had a very large volume of potentially hazardous liquid and we had to insure that large quantities of the vapors did not get into the mine air. Perchloroethylene was chosen instead of carbon tetrachloride because its vapor pressure was lower and it was eight times less toxic. A monitor for halogenated hydrocarbons is continuously operated that automatically turns off the ventilation fans if fumes are detected and turns on a warning light at the hoist. As mentioned earlier, steel doors were provided that were designed to completely isolate the experiment. At a later time we built a set of liquid scintillation counters, and we were careful to explain the consequences of having approximately 5000 liters of this liquid underground. Since our liquid has a high flash-point we were certain it was perfectly safe.

In the early stages of developing our chlorine detector we engaged the Bureau of Mines to examine our plans and insure that we were following normal safety procedures. I believe the Bureau of Mines or a qualified consultnt should make some check on procedures and materials brought underground. This is a simple and inexpensive precaution that should be taken by the individual groups. At the present time all underground work is monitored by the Mine Safety and Health Administration. They are very observant in normal safety matters, but in general are not very experienced in evaluating complex scientific apparatus. It would be well to have the funding agencies set up a review panel to approve the apparatus and procedures in the planning stage, and periodically examine the facility while in operation. A number of the experiments being planned in the U.S. have potentially hazardous features, e.g., large volumes of liquid scintillator, dewars of liquid argon, and electronic circuitry. These problems can be solved if carefully considered and if emergency procedures are worked out. Safety considerations are taken into account at accelerator laboratories, but not in present underground operations in an active mine where the hazards and the number of people involved are much greater. More care should be exercised in these matters.

These considerations alone will introduce a limitation on the sort of experiments that can be carried out in a private mine, because many workers are underground, and the scientist must secure approval from the mine management. If, on the other hand, a national facility dedicated to underground science is established, then it would be possible to carry out experiments that would be unacceptable to a mining company.

Other Experiments.-- There have been a number of scientists interested in using our facilities. They have contacted me to explore the possibilities and, if they are interested, to visit the mine. In all cases we have been able to accommodate them in our space. If they would like to use the facilities, they explain their experiment to the Homestake engineer and write a letter asking for permission. It is especially easy to make arrangements if their work can be done in conjunction with our periodic visits. Brookhaven pays a fee of $1000 a year for access and hoist services. Guest scientists are included and no other fees are necessary. If Homestake personnel are needed for assistance, their services are charged to our account or the guest scientist's account at an agreed upon rate.

The following experiments have been accommodated in our space in the mine. To our knowledge all proposals for new experiments have been given permission by Homestake management.

University of Pennsylvania.--Ken Lande's group has built a series of water Cerenkov detectors. Their present system and their latest proposal is described at this workshop. This is the largest group using our space and they have an independent and growing program. This work started in 1972.

Smithsonian Center for Astrophysics.--Edward Fireman has a tank containing 1.7 tons of potassium acetate powder designed to measure muon production of ^{37}Ar by the photonuclear process $^{39}K(\mu^\pm,\mu^\pm np)^{37}Ar$. Measurements have been performed at various levels in the mine to study this process as a function of average muon energy.

University of Washington.--Jerry Lord and Peter Kotzer have performed several photoplate measurements of muon interactions. The emulsions were prepared underground and exposed for long periods. This work is now completed.

Washington University, St. Louis.--George Flynn has made a search for superheavy elements in chemical fractions of the Allende meteorite using plastic track detectors. This work is now completed.

Additional space in the Homestake mine is needed to build the gallium solar neutrino experiment and the 1400-ton tracking spectrometer described by Ken Lande at this workshop. We believe it is most economical to have a single large room to contain both experiments. With this in view we have asked the Homestake engineering staff to estimate the cost of excavating a chamber 32 ft wide by 160 ft long with 30 ft high ceiling near our present room, and to consider how to carry out the work with minimum interference with their gold mining operation. The mining engineering has been sketched out, and the costs have been estimated. We are now ready to explore the views of the management towards carrying out this work for us.

Working in the Homestake mine during these last 15 years has been extremely pleasant. The management and the people that have assisted us have been very skillful and have always been very generous and cooperative. We have asked them to accomplish numerous tasks that are far outside the interest of the company, and they have always responded with a ready solution to our problem. We feel very much at home there.

RESULTS OF THE CHLORINE SOLAR NEUTRINO EXPERIMENT

The chlorine solar neutrino experiment is based upon the neutrino capture reaction $^{37}Cl(\nu,e^-)^{37}Ar$. The experimental procedure is simply to remove ^{37}Ar from a 380,000-liter tank of perchloroethylene by a helium purge. The argon is collected on a charcoal adsorber, purified and placed in a small low-level proportional counter to observe the x-ray at 2.82 keV from the electron-capture decay of ^{37}Ar. The design, operation, tests of efficiency and results have been given in earlier reports.[3] The neutrino detection sensitivity is based upon theoretically calculated neutrino capture cross sections,[4] and the expected neutrino capture rate from the sun has been recently reviewed by J. N. Bahcall and his associates.[5] The experiment has been operating since 1967. In 1970 the sensitivity for detecting ^{37}Ar was greatly improved by characterizing ^{37}Ar decay events in the proportional counter by measuring both the pulse rise-time and the pulse height. In addition to this change in the counting electronics, there have been gradual improvements in the counter shielding and changes in the counter materials so as to reduce the number of events due to radioactive impurities. At present the total frequency of counts in the energy range from 1 to 5 keV that are not in coincidence with a surrounding NaI detector is less than one event every two days. The individual events from the proportional counter are recorded to give the pulse height, pulse rise-time, and the time of the event. The events having a pulse rise-time and pulse height correct for ^{37}Ar decay are selected. The ^{37}Ar-like events are treated by a maximum likelihood statistical treatment and resolved into a decaying component with a 35-day half life and a constant background component. The statistical treatment has been specifically designed for treating the data from this experiment.[6] The resulting ^{37}Ar production rates for individual experimental runs are given in Figure 2. The combined ^{37}Ar production rate for 615 metric tons of perchloroethylene is 0.42 ± 0.05 ^{37}Ar atoms/day. There is a cosmic ray background production of ^{37}Ar in the detector that must be subtracted from this rate to obtain the ^{37}Ar production rate presumably from solar neutrinos. The background ^{37}Ar production rate is 0.08 ± 0.03 ^{37}Ar atoms per day, and it arises from photonuclear interactions of muons.[7] The net ^{37}Ar rate that could be ascribed to solar neutrinos is then 0.34 ± 0.06 ^{37}Ar atoms/day in the detector; the error expressed corresponds to 1σ. The rate expressed in SNU (Solar Neutrino Units = SNU = 10^{-36} captures/second·atom) is:

Neutrino capture rate, SNU = $5.31 \times (0.34 \pm 0.06) = 1.8 \pm 0.3$ (1σ)

$$= 1.8 \,^{-0.7}_{+0.8} \quad (3\sigma)$$

Figure 2. Argon-37 production rates observed in the chlorine solar neutrino experiment 1970-1981.

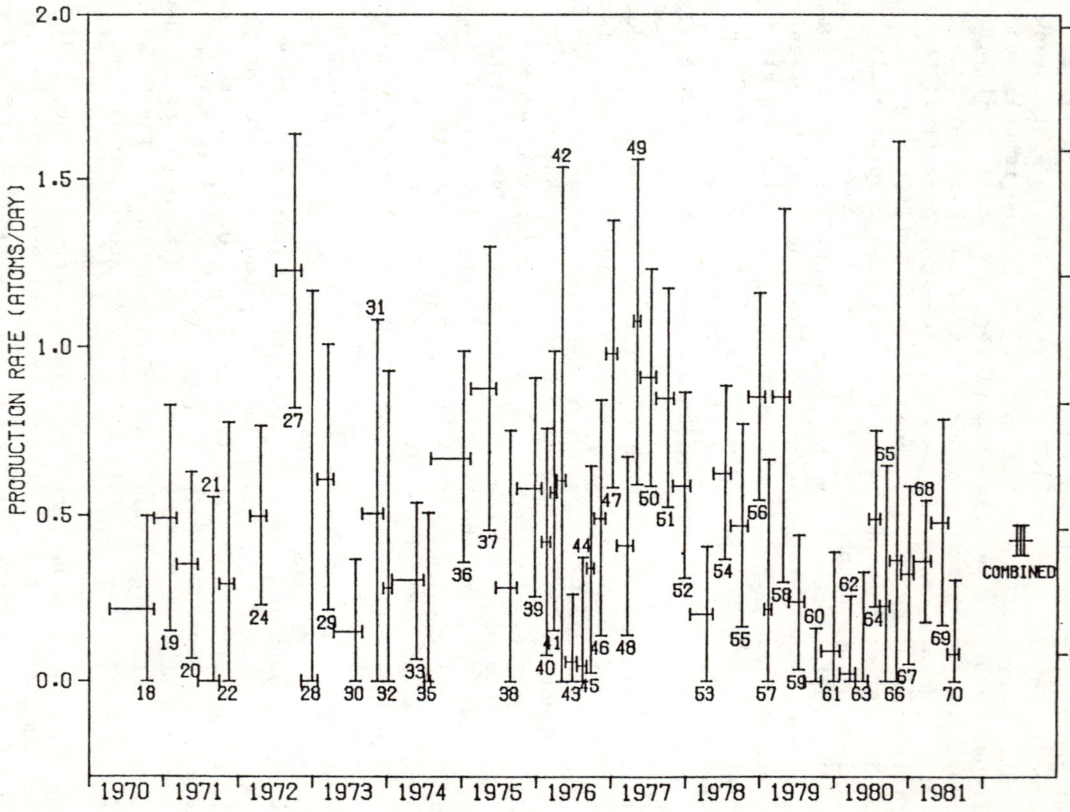

The current theoretical analysis of the expected solar neutrino capture rate in ^{37}Cl is 7.6 ± 3.3 SNU in which the error expressed is an effective 3σ.[5] The error is derived from errors in the input physical measurements that are used in the solar model (nuclear reaction rates, solar constant, surface element abundances, and solar age), and estimated calculational errors for parameters not amenable to physical measurements (opacities, neutrino capture cross sections).

The error stated does not include the fact there is at the present time a discrepancy in the measurements of the important $^3He(\alpha,\gamma)^8Be$ reaction.[8] There are some basic assumptions in the standard solar model that may well be incorrect and could account for the disagreement with the chlorine experiment. For example, the sun is assumed to be initially uniform in composition with a known equation of state, to contain no significant magnetic fields and to be rotating too slowly to affect the structure.

It is interesting to see how Figure 2 would appear if indeed the sun were producing a neutrino spectrum as forecasted by the standard solar model with the total neutrino capture rate of 7.6 SNU. A Monte Carlo calculation was performed using the exact exposure times, counting periods, counting and extraction efficiencies as were measured in the actual experimental runs. The result is shown in Figure 3. One can note from this plot that there is one run with a zero ^{37}Ar production (run 28) and the highest runs are 13 SNU (nos. 29 and 70). It is interesting to note that if the solar neutrino capture rate in ^{37}Cl were 7.6 SNU then we would have measured the ^{37}Ar production rate with a 5 percent statistical error (1σ).

The chlorine solar neutrino experiment has given us the only direct measure of the internal processes that are furnishing the sun's energy. We find after many years that the observed neutrino capture rate is below expectation and there is no accepted explanation for this low result. One can raise the question of whether or not the Homestake chlorine experiment is indeed observing solar neutrinos. The background from cosmic ray muons is presently estimated to be 0.08 ^{37}Ar atoms/day, a rate that is approximately 20 percent of the total signal observed. We have plans to redetermine the muon background using the technique of E. L. Fireman[9] by measuring the variation with depth of the photonuclear cross section of the process $^{39}K(\mu^{\pm},\mu^{\pm}np)^{37}Ar$. A possible way of testing whether or not the small signal observed by the chlorine experiment is related to the sun is to observe a correlation with either the earth-sun distance or with solar activity. The signal observed in the Homestake experiment is too small to demonstrate a variation with the earth-sun distance. In fact Ehrlich has made a careful statistical analysis of our data and found no evidence for such a variation.[10] Several investigators have examined our data for a correlation of the ^{37}Ar production rates with solar activity.[11] Three investigators, Subramanian, Sakurai and Basu, have concluded that a correlation does exist between our ^{37}Ar production rates and solar activity as measured by sunspot occurrences and solar flares. However,

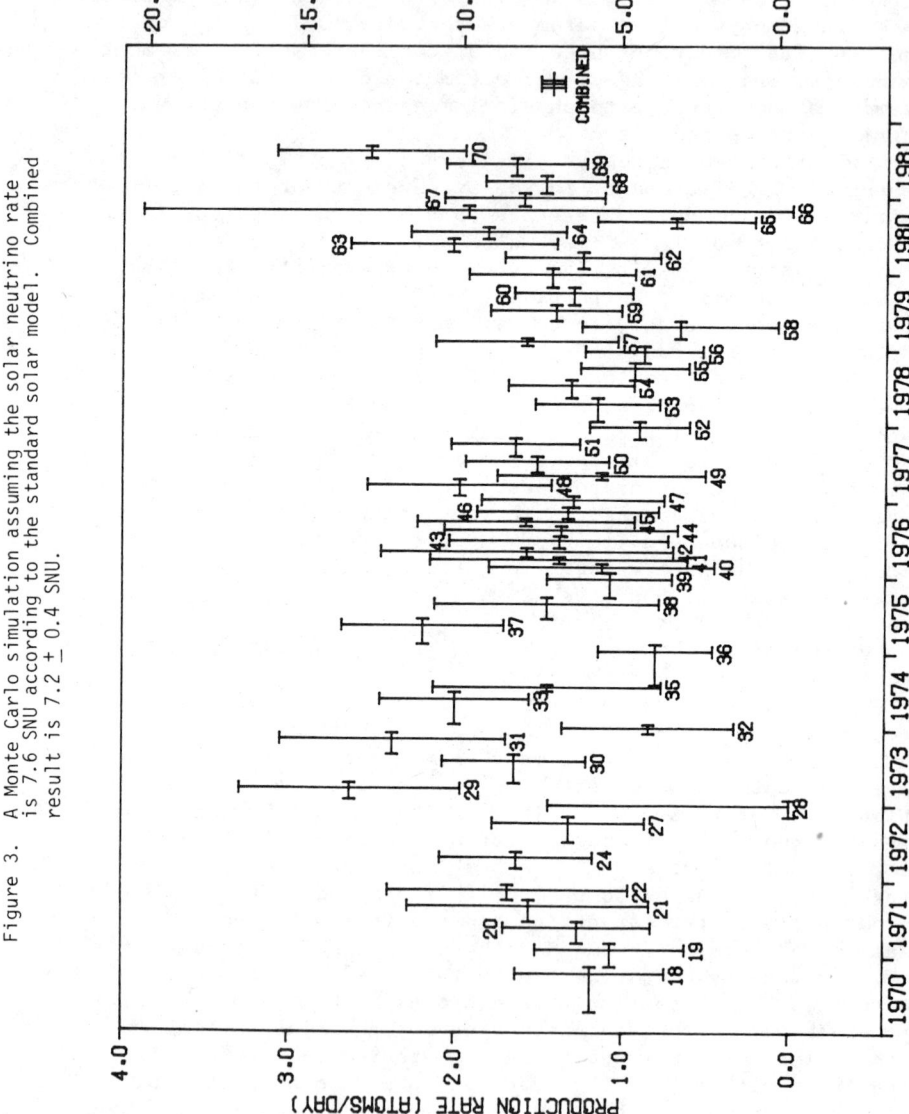

Figure 3. A Monte Carlo simulation assuming the solar neutrino rate is 7.6 SNU according to the standard solar model. Combined result is 7.2 ± 0.4 SNU.

Lanzerotti and Raghavan say that they do not find evidence for a correlation. A problem in searching for such correlations is that the rate observed in the individual experiments is very low and therefore the statistical errors are large. This situation can be improved in searching for long term changes associated with the solar cycle by taking yearly averages. Figure 4 shows a plot of the yearly averages, and one can see a change at the 1σ level that appears to be anticorrelated with solar activity! The data, especially in the early seventies, are less accurate because fewer runs per year were recorded, and the 1972 value based on 3 measurements contains our highest run, no. 27 at 6.1 ± 2.2 SNU. Searching for possible variations is important and we plan to continue making observations at the rate of six per year. We would like to point out that the standard viewpoint is that the energy production process occurring in the solar core should only increase slowly with time, 5 percent per billion years, and could not possibly vary with periods less than 10^4 to 10^6 years. Changes on this time scale can in principle be tested using a geologically old sample.[12] An experiment using this approach will be discussed by George Cowan at this workshop.[13]

A LARGER CHLORINE SOLAR NEUTRINO EXPERIMENT
AND THE UNDERGROUND SCIENCE FACILITY

In the spirit of this workshop we would like to propose building another chlorine experiment that is 5 or 10 times larger than the Homestake experiment.[14] We believe a project of this magnitude would be a worthy goal for an underground science facility. Building an experiment of this magnitude would indeed be on a scale befitting a national facility. It would be operated for a long period of time monitoring the solar neutrino flux and sporadic neutrino sources in our galaxy.

Scientific Goals

The first goal of a larger chlorine experiment at a new and deeper location would be to check the results of the Homestake detector. A result in agreement with the Homestake detector would strongly indicate that the signal observed is from neutrinos. An accurate and verified rate for the chlorine detector is crucial to understanding the solar interior and would serve as a basic measurement for comparison with other solar neutrino experiments directed at understanding the solar neutrino spectrum. Secondly, such a chlorine experiment would be large enough to search for variations in the neutrino luminosity of the sun. We will argue later that the chlorine neutrino capture reaction is unique, and superior to all other reactions for this purpose. It is important to search for these variations. If it is clearly shown that the sun's neutrino luminosity is constant, an accurate base line would be established for comparison of sporadic sources in our galaxy.[15] Finally, monitoring the galaxy for neutrino sources is extremely important and must be continued for many years by a reliable

Figure 4. Yearly averages of the ^{37}Ar production rate obtained from the chlorine solar neutrino experiment.

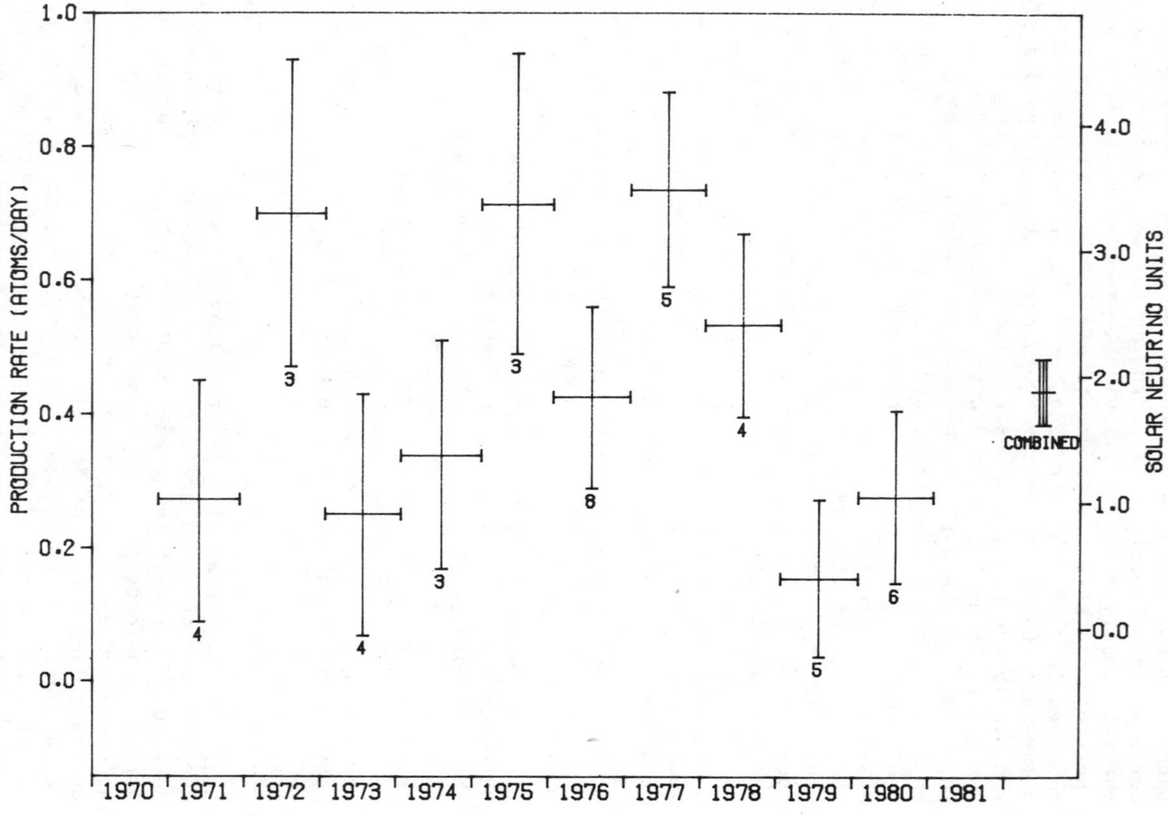

detector with an accurately known sensitivity. Although this detector would be incapable of giving information on the neutrino pulse shape and on the energy spectrum of the neutrinos from collapsing stars, it would give the integrated flux-cross section product for these neutrinos.

Advantages of the Chlorine Detector

The <u>chemical extraction</u> of ^{37}Ar from large volumes of perchloroethylene is a very simple, reliable chemical procedure that has been thoroughly tested. This chemical procedure could be easily engineered for a detector ten times larger. The cost of building a chlorine detector would be far less than that of any other radiochemical neutrino detector of comparable sensitivity. A radiochemical system has a lower inherent background than a direct counting system because it is insensitive to gamma radiation. Therefore, building a very large detector with a low neutrino energy threshold is possible following the radiochemical approach. Building a direct counting detector with high sensitivity and low energy threshold is very difficult.

The <u>neutrino capture cross section</u> of ^{37}Cl to produce ^{37}Ar in the ground state and in excited nuclear states is well known, a situation nearly unique in nuclear physics.[4] With neutrinos of energy greater than 6 MeV, the isobaric analog state in ^{37}Ar is formed by a superallowed transition. Because of this fact, the chlorine neutrino detector is primarily sensitive to the ^{8}B decay neutrino component of the solar neutrino spectrum, even though the branch in the proton-proton chain that produces ^{8}B is a very minor one. Because the production of ^{8}B in the sun increases dramatically with central temperature, ^{8}B is produced only in the solar core. Observing the ^{8}B neutrino flux is the most sensitive indicator of variations in the core temperature of the sun. We point out that in the presently conceived proton-proton chain there is competition between the p-p I_3 chain terminating reaction ^{3}He + ^{3}He → ^{4}He + 2H and the reaction ^{3}He + ^{4}He → ^{7}Be + γ, and it is the latter reaction that leads ultimately to ^{8}B. Even if the signal we are observing arises from nuclear processes on the surface of the sun associated with solar flare phenomena, the chlorine detector would also be best suited for searching for any fluctuations.

The <u>depth requirement</u> for a larger chlorine detector is a very important consideration. From our analysis of the muon background for the Homestake experiment 20 percent of the signal observed is from energetic muons. It would be important to build a larger chlorine experiment at a greater depth than the present Homestake detector. If we use as a guide the detailed calculations of Zatsepin, Kopylov and Shirokova,[7] the background could be reduced by a factor of 2, 5 or 10 by increasing the depth by 500, 1100 or 1600 m.w.e. We conclude from this information that a depth of 5000 m.w.e. would be suitable and consistent with depths that could be achieved from an engineering viewpoint. Needless to say, the deeper the better!

The engineering requirements other than the depth are relatively easy to achieve. First the size of the room needed could be similar to the present Homestake room, only longer: 10 m. by 10 m. and 75 m. long, for example, for a five times larger detector. The tank, pumps, and piping could be built underground using the construction procedures and a hoist size identical to those used at Homestake. The liquid could be transported underground by small tank cars, or by a pipe down the shaft. It would be best to have a rail line near the site of the shaft since 50 railroad tank cars would be needed to carry the liquid. The entire operation is consistent with normal industrial practices, and could be carried out without great difficulty.

ACKNOWLEDGMENT

This research was performed at Brookhaven National Laboratory under contract with the U.S. Department of Energy and supported by its Division of Basic Energy Sciences.

REFERENCES

1. A. A. Pomanskii, Soviet Atomic Energy 44, 433 (1978).
2. J. N. Bahcall and R. Davis Jr. Essays in Nuclear Astrophysics, C. A. Barnes, D. D. Clayton and D. N. Schramm, Editors, Cambridge Univ. Press, 1982.
3. R. Davis Jr., D. S. Harmer and K. C. Hoffman, Phys. Rev. Lett. 21, 1205 (1968); Brookhaven Solar Neutrino Conference, p. 1 (1978), G. Friedlander, Editor, BNL 50879; J. N. Bahcall and R. Davis Jr., Science 191, 264 (1976); B. T. Cleveland, R. Davis Jr. and J. K. Rowley, AIP Conf. Proc. No. 72, G. B. Collins, L. N., Chang and J. R. Ficenec, Editors (1980); R. Davis, Proc. Proton Decay Workshop, Argonne National Laboratory, September 1980.
4. J. N. Bahcall, Rev. Mod. Phys. 50, 881 (1978)
5. J. N. Bahcall, W. F. Huebner, S. H. Lebow, P. D. Parker and R. K. Ulrich, Rev. Mod. Phys. 54, 767 (1982).
6. B. T. Cleveland, paper to be submitted to Nuclear Instruments and Methods.
7. A. W. Wolfendale, E.C.M. Young and R. Davis Jr., Nature Phys. Sci. 238, 130 (1972); G. L. Cassiday, Proc. Int. Cosmic Ray Conf., 13th (Denver) 3, 1958 (1973); G. T. Zatsepin, A. V. Kopylov and E. K. Shirokova, Sov. J. Nucl. Phys. 33, 200 (1981).
8. Discussed in W. A. Fowler's report at this workshop.
9. E. L. Fireman, 16th Int. Cosmic Ray Conf., Kyoto, Japan, Paper Code 5-3, Vol. 12 (1979).
10. R. Ehrlich, Phys. Rev. D 25, 2282 (1982).
11. K. Sakurai, Nature 269, 401 (1977), Solar Physics 74, 35 (1981); A. Subramanian, Current Science 48, 705 (1979) and preprint TIFR-BC-82-4; L. J. Lanzerotti and R. S. Raghavan, Nature 293, 122 (1981); D. Basu, private communication 1982, to be published in Solar Physics.

12. J. K. Rowley, B. T. Cleveland, R. Davis Jr., W. Hampel, and T. Kirsten, Proc. Conf. Ancient Sun, pp. 45-62, R. O. Pepin, J. A. Eddy and R. B. Merrill, Editors (1980) and references therein.
13. G. A. Cowan and W. C. Haxton, Science <u>216</u>, 51 (1982); G. A. Cowan's report this workshop.
14. Only a brief mention of this topic was made in our verbal report. We have decided to elaborate on this possibility in our written report.
15. J. N. Bahcall, submitted to Ap. J. Letters, Nov. 1982.

PROPOSAL FOR A NATIONAL UNDERGROUND
SCIENCE FACILITY

Alfred K. Mann[*]

Department of Physics
University of Pennsylvania
Philadelphia, Pennsylvania 19104

ABSTRACT

The idea is explored of establishing a laboratory complex shielded from the cosmic ray flux at the earth's surface for the purpose of housing and providing technical support for experiments in particle physics, astrophysics, and other scientific disciplines. The scientific motivation for such an underground science facility is described, and the questions of location and desired properties of the facility are discussed.

INTRODUCTION

In this talk I wish to explore the idea of establishing a laboratory complex or science facility that would be shielded from the cosmic ray flux at the earth's surface, and also in large part from seismic and other disturbances present at the earth's surface. Such a facility would be deep underground with an appropriate rock overburden to provide the shielding. The purpose of the facility would be to house and provide technical support for experiments in particle physics and astrophysics, in gravitation and geophysics, in applied physics and engineering, and in such other scientific disciplines as would benefit from the features of an underground laboratory complex.

Let me begin by discussing the question of scientific motivation which must provide the justification for establishing any new basic science facility. Following that, I will turn to the questions of where these motivating experiments might be carried out, and what are the desired characteristics of an appropriate facility. These lead in a natural way to a brief discussion of the Los Alamos National Laboratory proposal for a national underground science facility, and thence to my conclusions.

SCIENTIFIC MOTIVATION

The urge to explore higher regions of energy and new regions of distance--both very small and very large--in the search for new phenomena has motivated many experiments in the past. When successful, these experiments have, in turn, stimulated new theories which occasionally reach far beyond the boundaries of the initial experiments. For example, the discovery of the relic black body radiation,

---

[*] John Simon Guggenheim Fellow, 1981-82.

in conjunction with the implications of the Hubble constant and the abundance of helium in the universe, has given additional strong support to the idea of an expanding universe and, particularly, to the "hot big bang" cosmology. In particle physics, the discoveries of the weak neutral current and of new families of elementary fermions have consolidated unification of two of the fundamental physical forces--the weak and electromagnetic forces--and led to the present "electroweak" theory.

Taken together, these developments of the past two decades provide a new perspective from which to view the future of particle physics and astrophysics. We think now of unification on a grander scale, one in which the strong as well as the weak and electromagnetic forces would flow naturally from a single general theory. But, of equal importance, grand unification is intimately related to the hot big bang cosmology, and events which occurred during the early moments of the big bang and were subsequently subjected to expansion are thought to determine the nature of certain phenomena that are observed now, viz., the relic black body radiation.

These ideas and the theories obtained from them present a new challenge to experimental physics, which lies not only in experiments to test specific predictions where available, but also to explore as widely as possible beyond the boundaries of the standard models for phenomena not even anticipated. For the moment, however, success in formulating grand unified theories (GUTS)[1] has led to certain quantitative predictions that clearly need experimental verification or repudiation.

Prediction of the Weinberg Angle.[2] One such prediction that follows directly from the grand unified theory denoted by "minimal SU(5)" has already been tested. This is the prediction of the magnitude of the single fundamental constant $\sin^2\theta_W$ of the now standard electroweak theory, where θ_W is the so-called Weinberg angle. One obtains

$$(\sin^2\theta_W)_{SU(5)} = 0.214 + 0.0006 \ln (0.16 \text{ GeV}/\Lambda_{\overline{MS}}) \qquad (1)$$

$$= 0.214^{+0.004}_{-0.003} \; ; \quad \Lambda_{\overline{MS}} = 0.16^{+0.10}_{-0.08} \text{ GeV}$$

which is to be compared with

$$(\sin^2\theta_W)_{Expt} = 0.215 \pm 0.014, \qquad (2)$$

the result of precision measurements on semileptonic weak neutral current processes. In eq. (1), $\Lambda_{\overline{MS}}$ is the measured value of the QCD mass scale. It is the remarkable agreement between eqs. (1) and (2) which lends credence to other, as yet untested, consequences of grand unified theories.

There are several consequences of potentially great significance that need to be tested by experiment. Depending on the detailed nature of the specific models, these may be in the form either of quantitative predictions or of "natural," if not yet quantitative, explanations of known and conjectured phenomena. Among such known (but not understood) phenomena are, for example, the baryon asymmetry in the universe and the violation of CP-invariance, both of which are at least amenable to treatment in grand unified theories. There are a number of conjectured phenomena but it is not appropriate here to attempt an exhaustive survey of all of them.[3] For our purpose, it is sufficient to choose two regions where current interest is especially high: possible violations of baryon number and lepton number conservation. In the former, many grand unified models require protons and bound neutrons to be unstable, i.e., nucleons to decay spontaneously, albeit with a very long lifetime. In the latter, there is room for non-zero neutrino masses and mixing among different neutrino flavors which would be manifested in neutrino oscillations and possibly in double beta-decay. I want to discuss nucleon decay experiments and cosmic ray neutrino experiments as illustrations of this physics, and which must be done in an underground laboratory.

<u>Nucleon Instability</u>. In the minimal SU(5) theory the nucleon lifetime is given as

$$\tau_N \approx \text{const } M_x^4/M_N^5 \approx 2 \times 10^{(29 \pm 2)} \text{ yr,} \qquad (3)$$

where the ± 2 in the exponent is due to uncertainties in $\Lambda_{\overline{MS}}$ and matrix element evaluation. M_x is the grand unification mass scale, of order 10^{14}-10^{15} GeV, at which it is speculated that the coupling strengths of the fundamental interactions, weak, strong and electromagnetic, would become equal.

The present limit of roughly 10^{30} yr on the mean life of nucleons is obtained from two experiments[4] carried out with detectors that were ingeniously diverted from cosmic ray studies to the search for nucleon decay. In addition, there are several experiments that have very recently begun to take data or are within about one year of doing so; their salient properties are shown in Table I. One sees that three of the detectors are very large water Cherenkov counters which are most sensitive to two and three body decay modes. The capability of these detectors for spatial reconstruction is about 70 cm in space and 5 to 10 degrees in angle. They are insensitive to particles with velocities below the threshold for producing Cherenkov light (~100 MeV pion energy), but are calorimeters for showering particles. The Cherenkov light also gives track directionality. Their advantages are relatively low cost for large mass and simplicity of construction, both of which make it possible for them to be early on the scene. Of the two other detectors in Table I, the one with small mass has just begun to take data while the larger detector has just begun construction. Both of these have relatively small grain size which, within certain limitations, allows for spatial reconstruction and perhaps particle identification for a

Table I. Properties of "large" mass nucleon decay detectors likely to take data before the end of 1983. (For details of earlier experiments see reference 4.)

GROUP	LOCATION	DEPTH (mwe)	FIDUCIAL MASS(kT)	DETECTOR TYPE
HPW	Park City, Utah, Silver Mine	1800	~0.6	Water Cerenkov; PMT in volume and on surface
IMB	Cleveland, Ohio Salt Mine	1570	3.7	Water Cerenkov; PMT on surface
CERN, Frascati, Milano, Torino	Mont Blanc tunnel, Alps	5000	~0.1	Fine-grained; iron plates (1 cm) between streamer counters
Saclay, Wuppertal	Frejus tunnel, Alps	4000	~1	Fine-grained; iron plates (0.3 cm) between flash tubes
Tokyo	Kamioka	2400	~1	Water Cerenkov; large area PMT

given event that are superior to those in the large water detectors. All of the experiments mentioned here are described at length in the sessions on proton decay in these Proceedings.

In planning for the future beyond about 1984, there appears to be a consensus[5] that one of three alternative possible results is likely to be forthcoming from the experiments listed in Table I. These are: (i) that a lower limit to τ_N is found (probably at the level of a few times 10^{31} yr) which is set by the observation of a number of events ("candidates" plus "background"), but not by the total nucleon content of the detectors; (ii) that a clear signal above background is found at a lifetime of a few x 10^{31} yr; and (iii) that zero signal is found at the limit of the total nucleon content of the detectors. Under any of these alternatives it is probable that at least one additional multi-kiloton, fine-grained detector will be necessary in the U.S. to provide either a definitive lower limit on τ_N of the order of 10^{33} yr or to study quantitatively the various nucleon decay modes.

The desired properties of that "later" generation detector can be specified even now[5] in a general way as indicated by the following numerical example: a mass of 1 metric kiloton contains 6 x 10^{32} nucleons, so that in a detector of fiducial mass 1 metric kiloton with a detection efficiency summed over all possible decay modes of 50 percent, the event rate corresponding to a nucleon mean life of 10^{32} yr would be 3 events/yr. It is unlikely that a definitive conclusion concerning nucleon decay can be reached without full reconstruction of those few events and the larger (by an order of magnitude) number of neutrino-induced background events that will attend them. This example suggests also that the fiducial mass of any "later generation" detector should be at least several metric kilotons with the upper limit set by financial and technological restrictions. If the detector construction is modular, it can be produced and brought into operation in stages. An approximate cost of (5-10) x 10^6 dollars/kiloton of fiducial mass is indicated for such detectors, the more expensive detectors providing more information per event and possessing greater redundancy. We summarize the expected results from "near" and "far" future detectors and compare them with the present limits on nucleon decay in Fig. 1, which conveys graphically the magnitude of the effort required from experiments in the next ten years.

Cosmic Ray Neutrinos and Neutrino Oscillations. Supposing that a detector for nucleon decay with properties similar to those described above were to exist, it would be possible to carry out the first intensive study of the cosmic ray neutrino flux, and thereby to make a small beginning in neutrino astronomy. Equally tempting is the possibility of performing an especially sensitive search for neutrino oscillations using cosmic ray neutrinos.

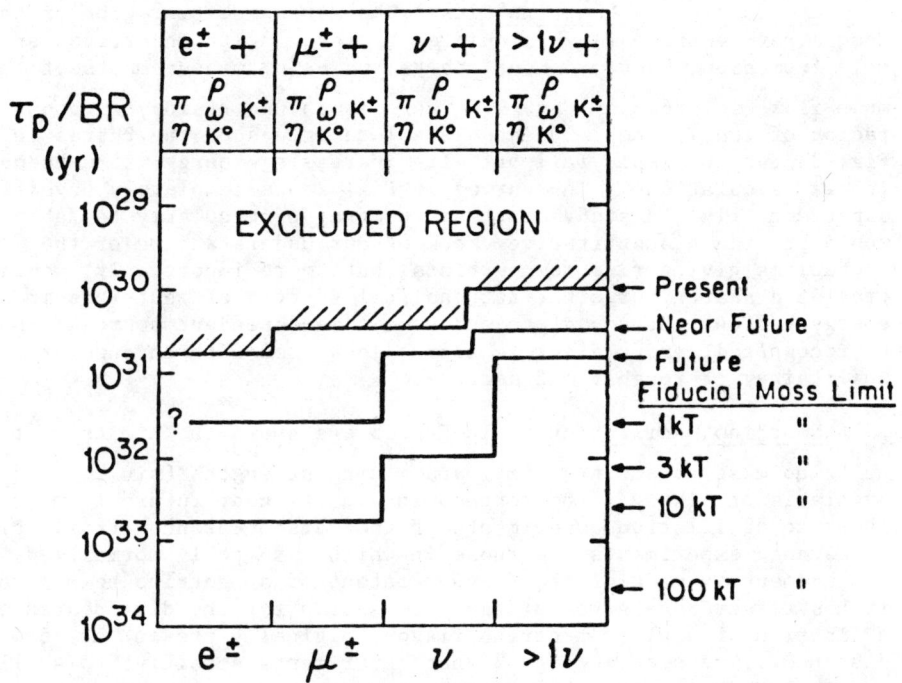

Fig. 1. Approximate expectations for present and future experimental limits on the nucleon lifetime for various decay modes. The limits shown would result from either (1) fewer than five decay events per year or (2) fewer decay events, than neutrino background events, assuming 100% of the decays go into the given channel. Detectors are assumed to have a 33% detection efficiency for decay events, after cuts to remove the neutrino background events. The curve labelled "Present" is for experiments which have been in operation for some time, and are characterized by minimal rejection of the neutrino-induced background (assumed to consist of $\nu_\mu:\nu_e$ = 2:1). The curve labelled "Near Future" refers to the expected results from the water Cerenkov experiments and the Frejus tunnel calorimeter. The "Future" curve refers to expectations from fine-grained detectors with $\gtrsim$ 10 kton fiducial mass and 100 times better background rejection for the electron and muon modes than present experiments. The fiducial-mass limits indicated show what could be achieved with a one-year exposure on the basis of the nucleon content of a detector alone. Taken from D. S. Ayres et al., Proceedings of the Division of Particles and Fields Summer Study on Particle Physics and Facilities, Snowmass, Colorado, June 28 - July 16, 1982.

The Cosmic Ray Neutrino Flux. The lower energy region of the cosmic ray neutrino flux calculated by assuming the neutrinos arise only from pion, kaon and muon decays and using measurements of the muon flux[6] is shown in Fig. 2. There is an uncertainty of about a factor of two in the calculated absolute values of the curves in Fig. 2, but the rapid fall-off with increasing energy is reproduced in all calculations. The curves in Fig. 2 are completely unverified experimentally. A study of those spectra, particularly the shapes, would provide a quantitative check on our understanding of the mechanisms giving rise to neutrinos, but, more importantly, would provide a search for other astronomical sources of neutrinos in that energy region. Steady state and some time-dependent sources might be recognized, and a discrete source localized with an angular uncertainty of roughly $\pm$ 5 degrees.

Neutrino Oscillations. In Fig. 3 are shown the limits on the neutrino mass parameters (Δm^2) and mixing strengths ($\sin^2 2\alpha$) now available or likely to be forthcoming in the near future from neutrino oscillation experiments of the "disappearance" type. Disappearance experiments are those in which a suitably normalized measurement is made of the flavor content of a neutrino beam after it has traversed a given distance to search for the disappearance of a fraction of a given neutrino flavor originally present at zero distance. One sees in Fig. 3 that experiments sensitive to small values of Δm^2 are done with $\bar{\nu}_e$ from reactors at the level $\Delta m^2 \geq 10^{-2}$ eV^2; they may also be done with from the sun at the level $\Delta m^2 \geq 10^{-11}$ eV^2. (The strength of neutrino mixing that is accessible in those disappearance experiments is of magnitude $\sin^2 2\alpha = 0.1$). It is of interest to note also the relatively high upper bound on Δm^2 that has been obtained or is likely to be obtained in the future in experiments at accelerators searching for ν_μ disappearance.

For the purpose of this talk it is useful to point out that a sensitive neutrino oscillation experiment involving the possible disappearance of ν_e and ν_μ into other flavors, e.g., ν_τ, is provided by measurement of the ratio of the total number of interactions of upward- and downward-going cosmic ray neutrinos that occur in and are contained in the massive detector.[7] The geometry of the experiment is shown in Fig. 4. The experiment has the following advantages: (1) it is the only experiment that is capable of searching for the disappearance of ν_μ and $\bar{\nu}_\mu$ at the limiting value $\Delta m^2 \geq 10^{-4}$ eV^2 (see Fig. 3); (2) because it measures the quantities $N_{tot}(up)$ and $N_{tot}(dn)$ the experiment is relatively insensitive to systematic errors; (3) the experiment is capable of observing time averaged probabilities $\langle P_{e\tau} \rangle_t$ and $\langle P_{\mu\tau} \rangle_t$ of magnitude set by mixing strengths corre-

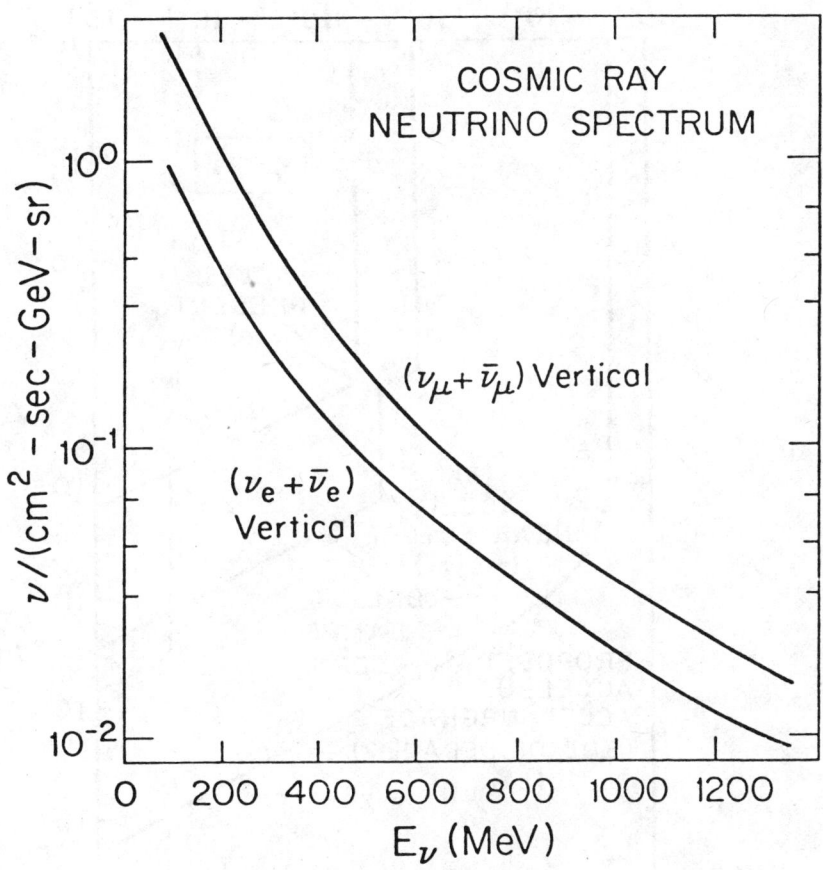

Fig. 2. Calculated cosmic ray neutrino spectra from reference 6.

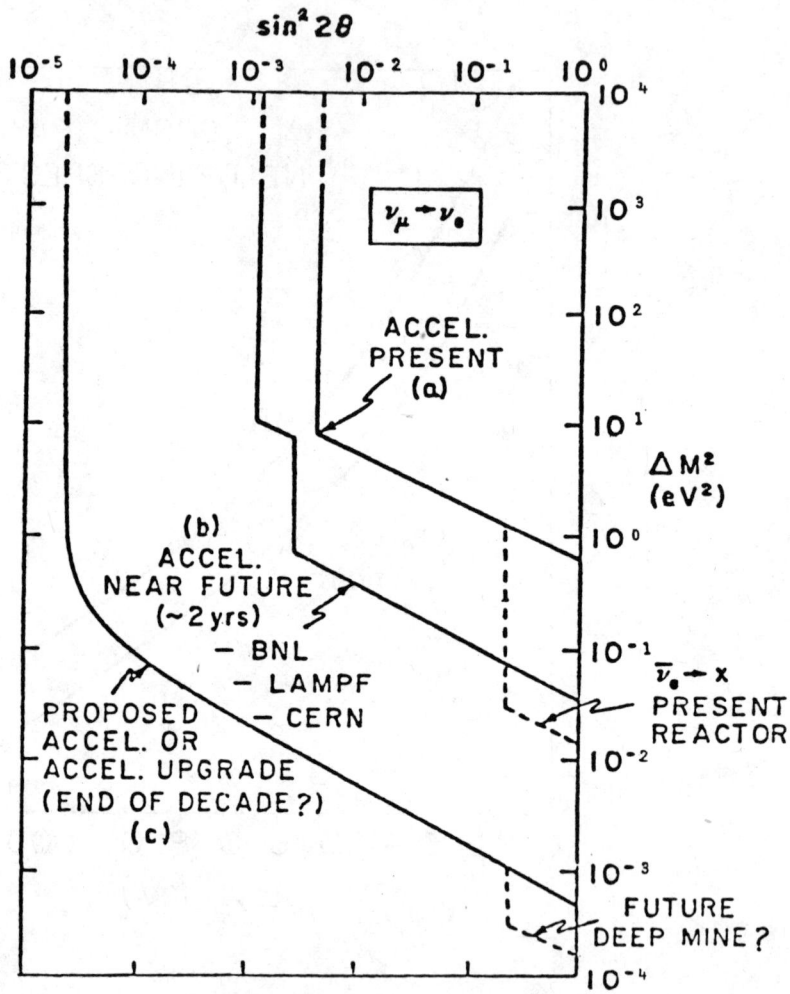

Fig. 3. Summary of neutrino oscillation experiments of the type $\nu_\mu \to \nu_e$, showing (a) the envelope of completed experiments, (b) near future accelerator experiments, and (c) proposed (end of decade) experiments. Reactor experiments are shown for reference. Taken from R. E. Lanou, Proceedings of the Division of Particles and Fields Summer Study on Particle Physics and Facilities, Snowmass, Colorado, June 28 - July 16, 1982.

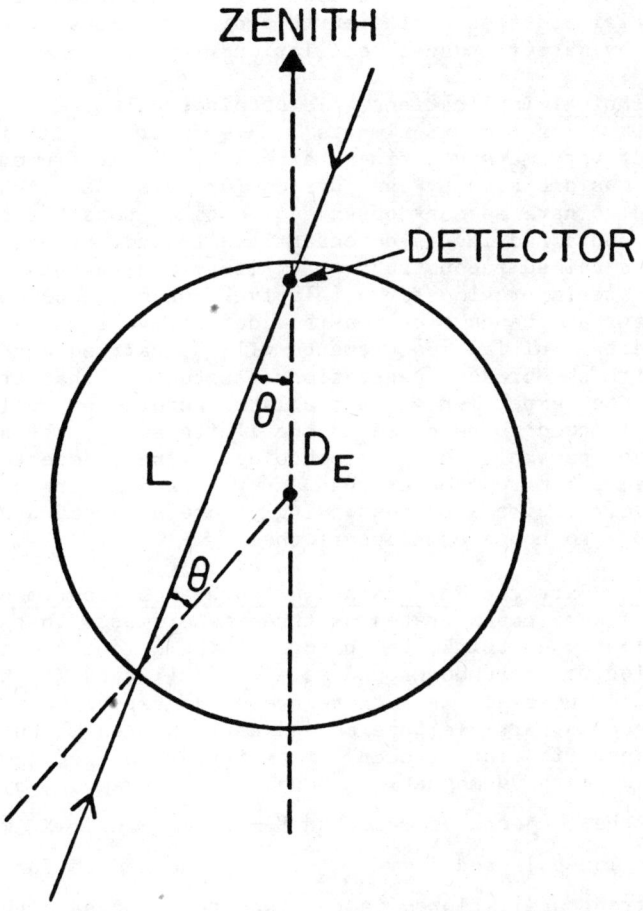

Fig. 4. Sketch of the experimental method. The neutrino detector is located as indicated roughly 1-2 km below the earth's surface. Neutrinos originate in the 10-20 km thick atmospheric shell surrounding the earth. Neutrinos from near the zenith that intersect little of the earth's matter before interacting in the detector are called down-going, N(dn). Neutrinos that have traversed a large fraction of the earth's diameter ($D_E = 1.3 \times 10^7$ m) and are observed to produce upward-going interactions in the detector are called up-going, N(up). Present accelerator limits on neutrino oscillations suggest that oscillations should have a negligible effect on the down-going atmospheric neutrino flux. Taken from reference 7.

sponding to, e.g., the d- to s-quark mixing strength; (4) although the experiment relies on the upward-going neutrinos traversing a substantial fraction of the earth's diameter, its sensitivity is not limited by matter-induced oscillations.

<u>Technical Implications</u>. To obtain conclusive results--positive or negative--in the experiments I have described, it is necessary to construct very massive, fine-grained, highly instrumented detectors. The reasons dictated by the physics for this construction are, first, the need to have as many observed events as possible occur and be totally contained in the detector itself, and, second, the requirement of detailed reconstruction of individual events. Together, these criteria provide for a relatively unambiguous interpretation of each event and hence make possible definitive conclusions from each experiment. Initial measurements will be carried out in the next few years with the present generation of detectors that are directed toward these experiments. But a later generation will be required which will need to be close to the limits set by the state of the art of present particle physics technology. These detectors will be more expensive and more complex than any yet used underground. They will, I believe, require a correspondingly more advanced underground laboratory to house and support them.

<u>Other Particle Physics and Astrophysics Experiments</u>. For want of time I have concentrated on three experiments that illustrate quite clearly, I think, the nature of the scientific and technical motivation for an underground science facility. It should be emphasized, however, that there are other experiments of special importance to particle physics and astrophysics which also contribute to the justification of such a facility. The list includes: (i) the search for massive magnetic monopoles, for example, by seeking to observe the conjectured reaction $M + p \rightarrow M + e^+ + X$, where M is the magnetic monopole and $X = \pi^o, \pi^{\pm}$; (ii) the search for neutron-antineutron oscillations via the two-step process n (bound) $\rightarrow \bar{n}$, $(\bar{n} + n)$ annihilation; (iii) the search for the formation of various hypothesized "collapsed" nuclei by study of possible energy release on their formation; (iv) precise measurements of the solar neutrino fluxes by radiochemical detection, and by detection in real time of individual $\nu_e + e^- \rightarrow \nu_e + e^-$ interactions. Most of these experiments are discussed fully in papers in these Proceedings. Each of them is a difficult, very low rate experiment, requiring a sophisticated detector (in some cases, however, the same as for the nucleon stability search) which must be in operation for a long period of time to allow a definitive conclusion to be reached. But a positive result in any one of them would stimulate great excitement and many subsequent experiments.

<u>Still Other Experiments</u>. The prospect of massive, well-instrumented detectors deep underground has suggested that new information on cosmic ray secondaries, may be obtained. Emphasis is on the

measurement of multiple coincident muons to study the composition of primary cosmic rays at energies inaccessible to direct experiments on the primaries. A fuller discussion of this possibility is given by Gaisser in a paper in these Proceedings.

Another stimulating possibility in a completely different field is that the isolation from seismic and other surface disturbances in a deep underground laboratory would allow the search for gravity waves to be conducted at frequencies lower than those permitted on the earth's surface. One interesting detector would be a very long base line (~1 km) laser interferometer which might be sensitive to frequencies in the 1 to 100 hz region. This type of experiment would place severe restrictions on the geological quietude and permissible water flow rates of the underground environment. The subject is addressed in more detail in the papers on gravity waves in these Proceedings.

In geophysics, a deep underground laboratory would permit long-term, precise measurements on rock subjected to significant strain. These are of importance for studies of earthquake prediction and waste isolation. In addition, there is a category of gravity experiments relating to the possible time dependence of the gravitational constant which might benefit from a location deep underground. Such experiments, using a delicate, rotating, magnetically levitated system would also be sensitive to variations in the gravitational gradients and require geological quietude.

Finally, it is only prudent to point out from past experience that tomorrow always brings ideas for new experiments that did not exist today. And experiments done well are also stimuli of new experiments. Accordingly, a partial justification of a science facility is provided by the not-yet-thought-of-experiments that will need to be located in and supported by the facility if they are to be done successfully.

WHERE MIGHT EXPERIMENTS OF SUCH COMPLEXITY BE DONE?

Scientific research--mainly in physics--has been carried out in underground locations for many years. The solar neutrino experiment in the Homestake Mine in South Dakota by Davis and his collaborators, and the cosmic ray neutrino studies of Reines and his associates in mines in Ohio and in South Africa are examples. More recently, as we have seen, experiments of considerable ingenuity and complexity to search for nucleon decay have been started in mines in Ohio and in Utah, and in a laboratory in the Mont Blanc tunnel under the Alps. As a consequence there exists much insight into the conditions that need to be satisfied by underground research facilities. This allows us to extrapolate to the features that would be especially desirable in a facility devoted exclusively to science underground. I want to discuss these features in some detail. They are concerned with possible sites, the depth and extent of the shielding, the size of the rooms for experiments, ease of access to the laboratories, safety, technical support on the surface and underground, and possibilities for growth and expansion.

Possible Sites: General. It should be emphasized at the outset that the number of existing underground locations which are suitable to house advanced technology experiments is quite limited. Relatively extensive searches by physicists and geologists for deep mines and tunnels in the U.S. have been made during the past decade. These have sought sites that might be useful for underground laboratories or for storage. None of them has disclosed new, available sites that are clearly superior to the few already in use.

Existing tunnels in the U.S. are not at present employed for these purposes. Our mountain ranges, such as the Rockies, do not rise precipitously from a low lying plain, but rather slowly with moderate slopes over long distances. Thus, tunnels in the U.S. tend to have termini at high elevations so that the total overburden or shielding is less high than might initially be thought. For example, the Moffat Tunnel--a railroad tunnel in northern Colorado--is under a region of the Rockies with peaks of about 12,000 ft, but the termini are at about 9,000 ft; the resultant average cover turns out to be only 2,200 ft. No major physics experiments are located in or adjacent to existing tunnels in the U.S. The situation in this regard is different in Europe where, as I noted, there is an operating physics laboratory in the Mont Blanc automobile tunnel between France and Italy, and where underground laboratories are being prepared in the Frejus tunnel in the Alps (near Modane, France) and in the Gran Sasso tunnel in the Apennines (near L'Aquila, Italy). These laboratories are all at average depths of 4,000 meters of water equivalent (mwe) or greater, and the overburden is extensive.

There are several working mines in the U.S. in use as sites of underground experiments. I mentioned the Homestake Mine which is an active gold mine, and has provided a room for the Davis radiochemical solar neutrino detector since about 1970. More recently, the remaining available space in that room has been occupied by Lande and his collaborators who have adopted their cosmic ray apparatus to carry out a relatively early search for proton decay. This symbiotic relationship between two groups working at complementary experiments suggests the need for more research space underground, and is perhaps indicative of the value of cooperation on joint facilities underground.

The laboratory in the Homestake Mine at 4400 mwe, is the only physics research facility now being utilized in the U.S. at a depth greater than 2000 mwe (see Table I). There is a huge water Cherenkov counter directed primarily toward nucleon decay that has recently begun operation in the Morton Salt Mine in Fairfield, Ohio at a depth of 1570 mwe. This working mine has over many years been the home of several physics experiments by Reines and by Wu. In Utah, there is the Silver King Mine at Park City, at a depth of 1800 mwe, which is no longer worked, but has been used for underground cosmic ray experiments by the University of Utah for a number of years, and is now occupied by another large water Cherenkov counter for nucleon decay. Lastly, in Minnesota there is the Soudan mine, formerly a working iron mine but now operated as a museum site by the State of Minnesota, which houses a small experimental physics program at a depth of about 1900 mwe. To the best of my knowledge, these are the only relatively deep mines in the U.S. now being used in part as laboratories for physics experiments of greater than minimal complexity. It is significant, I think, that some of these same mines have been used for experiments again and again over many years.

At present, the only mine abroad that is used for a major physics experiment is a very deep mine (~8000 mwe) in the Kolar Gold Field in India, where an Indian-Japanese collaboration working in cosmic ray and nucleon decay experiments has existed for a number of years.

Depth. There is no unique depth criterion for a laboratory underground. Certain experiments need less shielding than others. For example, the active detectors used in nucleon decay experiments can function well at lesser depths than the radiochemical experiments seeking solar neutrinos. Indeed, cosmic ray muons at some detectable rate are useful in the former experiments as a means of continuous testing and calibration of the apparatus. It would be advantageous to have available laboratories at two or more depths for different types of experiments but, in the absence of that luxury, most experimenters would agree that an average depth of not less than about 2500 mwe is desirable for many experiments, which for rock of density 2.8 gm/cm^3, corresponds to 2927 ft of overburden.

Number and Size of Laboratory Rooms. Similarly, the question of the size and shape of the rooms for experiments has no single answer. We can, however, anticipate that some quite large rooms will be required for the next generation of massive detectors for nucleon decay and cosmic ray neutrino studies. Rooms as large as 50 ft x 200 ft in area and 50 ft high (at the top of the arch) need to be considered. Again, very long drifts or passageways, possibly a kilometer in length, might be needed for the gravitational wave laser interferometers mentioned earlier. These criteria indicate the need for high stability of the rock in which the laboratory rooms are constructed. Incidentally, the requirement of environmental stability for very sensitive gravitational wave detectors suggests that they are unlikely to be able to function in a working mine.

Access and Safety. Convenient access to the laboratory is somewhat difficult to ensure in a working mine where emphasis is on the mining operation, although satisfactory compromises have been reached in the past. As experiments underground become more complex, however, the need for access increases, and must be an important consideration in site selection.

A concomitant requirement is safety. Here the working mine may have an advantage because it employs personnel whose primary function is to enforce safety rules and protect the miners. An underground laboratory with convenient access for scientists and technicians would also need personnel to monitor that access with great care and to maintain the highest standards of safety in other respects.

Technical Support of Experiments. In addition to the individuals charged with controlling access and maintaining safety, it would be desirable to employ personnel to provide direct technical support for experiments. Indeed, provision of a wide variety of technical support for experiments would be one of the primary functions of an underground science facility, as it is for major accelerator laboratories and astronomical observatories. This is especially necessary as experiments grow in size, mass and complexity. A complement consisting of an engineer and, say, ten technicians (of whom six would be concerned with access and safety) would appear to be a reasonable permanent staff. Other technical help would, presumably, be brought for short periods by users of the facility.

Growth and Expansion. Lastly, the site of an underground science facility needs to be chosen with the possibility in mind that a successful experimental program will require growth and expansion of the facility. A site in which future growth underground or on the surface would be seriously inhibited is of limited value.

Summary. Let me close this section with Table II which summarizes some of the physical features that appear desirable in an underground science facility. The Table is intended primarily to provoke additional thought and proposals.

A DEDICATED NATIONAL UNDERGROUND SCIENCE FACILITY (NUSF)

One means of achieving these desired properties in an underground science laboratory would be to establish a facility devoted exclusively to basic research. That facility would presumably be national in character in the same sense that the major particle accelerator laboratories and astronomical observatories in the U.S. are national in character. In such a dedicated underground science facility it would be possible to conduct experiments that are at the forefront of their disciplines, and that utilize the most advanced experimental techniques. The potential richness of the experiments that are foreseen in that facility would, I believe, justify the initial investment of time and money. It is likely that the facility would be fully occupied and profitably used for at least two decades. It is of equal importance to the justification to recognize the stimulating effect that scientific facilities often produce by bringing together scientists in different disciplines, and by providing an intellectual environment in which students are made aware of a wider horizon than that delineated by their thesis topics.

It is also important to note here that the underground laboratory in the Apennines that I mentioned earlier is intended to be a national research facility of magnitude sufficient to accommodate a variety of experiments in different disciplines. With similar intent, a laboratory has been constructed at Baksan in the Caucausus of the U.S.S.R. which is dedicated to science underground.

The form or structure that a national underground science facility in the U.S. might take is not hard to envision. It would

Table II. Summary of Some Desirable Physical Properties of an Underground Science Facility.

Property	Quantity or Quality
Geology	Especially stable
Depth	$\geq$ 2500 mwe
Adit or Shaft	$\geq$ 12 ft diameter
Roadway or cage	Moderate load capacity; Auxiliaries for safety
Access	Not heavily committed to other purposes
Laboratory Rooms	Several, at least one large room, e.g., 50 ft x 200 ft x 50 ft (high); two moderate size rooms, e.g., 30 ft x 50 ft x 25 ft (high); others as proposed for experiments
Ventilation, Temperature and Humidity	Comfortable for apparatus and human occupancy; cool and dry
Floors	Covered with low radioactivity concrete
Walls and Ceiling	Rock-bolted, wire-netted, shotcreted with low radioactivity material
Surface Support	Adequate road to major highway; sufficient power, water. Small buildings for modest machine and electronics shops, for computers and experiment command posts, for limited dormitory space

be akin in number of scientists to a small university physics department, and have a yearly budget probably one order of magnitude less than that of the lower energy particle accelerator laboratories. Much of the research would be done by university physicists and students whose experiments had been approved by the program advisory committee of the facility. Thus it would be too small in scope, in the number of scientific personnel, and in budget to be a completely independent entity. It would therefore be advantageous for the facility to be an adjunct of an established, larger, national laboratory. The overhead costs of the facility would be reduced, the occasional peak loads that would seriously strain the resources of a small independent laboratory would be smoothed, and special skills of the parent laboratory would be available.

There are good reasons that suggest Los Alamos National Laboratory (LANL) as a possible parent of a national underground science facility. Among these are: (i) LANL has a high level of technical competence in geology and geophysics, and wide experience in mining and drilling in the Nevada Test Site (NTS) and elsewhere; there is also a program in the storage of radioactive wastes which has constructed a major test facility in the Climax Stock area of the NTS and is engaged in studies for construction of another such facility; (ii) there is a tradition of sustaining large technical efforts over long periods far from Los Alamos; (iii) many of the deep mines and tunnels which might serve as possible sites for NUSF are in western U.S.; and (iv) the physics prospectus of NUSF would be an important complement to the present physics programs of LANL.

A Possible Site for NUSF. I have made the point earlier that there are very few underground sites in the U.S. which appear suitable for development as an underground laboratory. I want to reaffirm this conclusion by observing that for more than a year there has been an on-going search for such sites led by R. R. Sharp, Jr. of LANL. The principal result of this search, apart from verifying again the small number of potentially useful locations, was the determination that there are several tracts in the Nevada Test Site (NTS) which would be especially favorable for the construction of a national underground science facility ab initio. (I might add, parenthetically, that this suggestion originated in an informal conference of cartographers and mining engineers at the U.S.G.S. in Denver before Los Alamos entered the picture.) A map of the NTS showing the location of the favorable tracts is given in Fig. 5.

The reasons that recommend the NTS are the following. (i) An area--designated the Nevada Research and Development Area (NRDA)--has been set aside (see Fig. 5) for the purpose of encouraging non-classified scientific research in fields not connected with weapons development. It would be possible to establish a dedicated basic science facility there, freely accessible, and capable of growth and expansion as the circumstances would warrant. The connection between the facility and LANL would be completely natural. (ii) The geology, hydrology, and underground temperature gradients in the NTS have been intensively studied empirically for other purposes. (iii) There is a long-standing, high level of experience in and technical support for

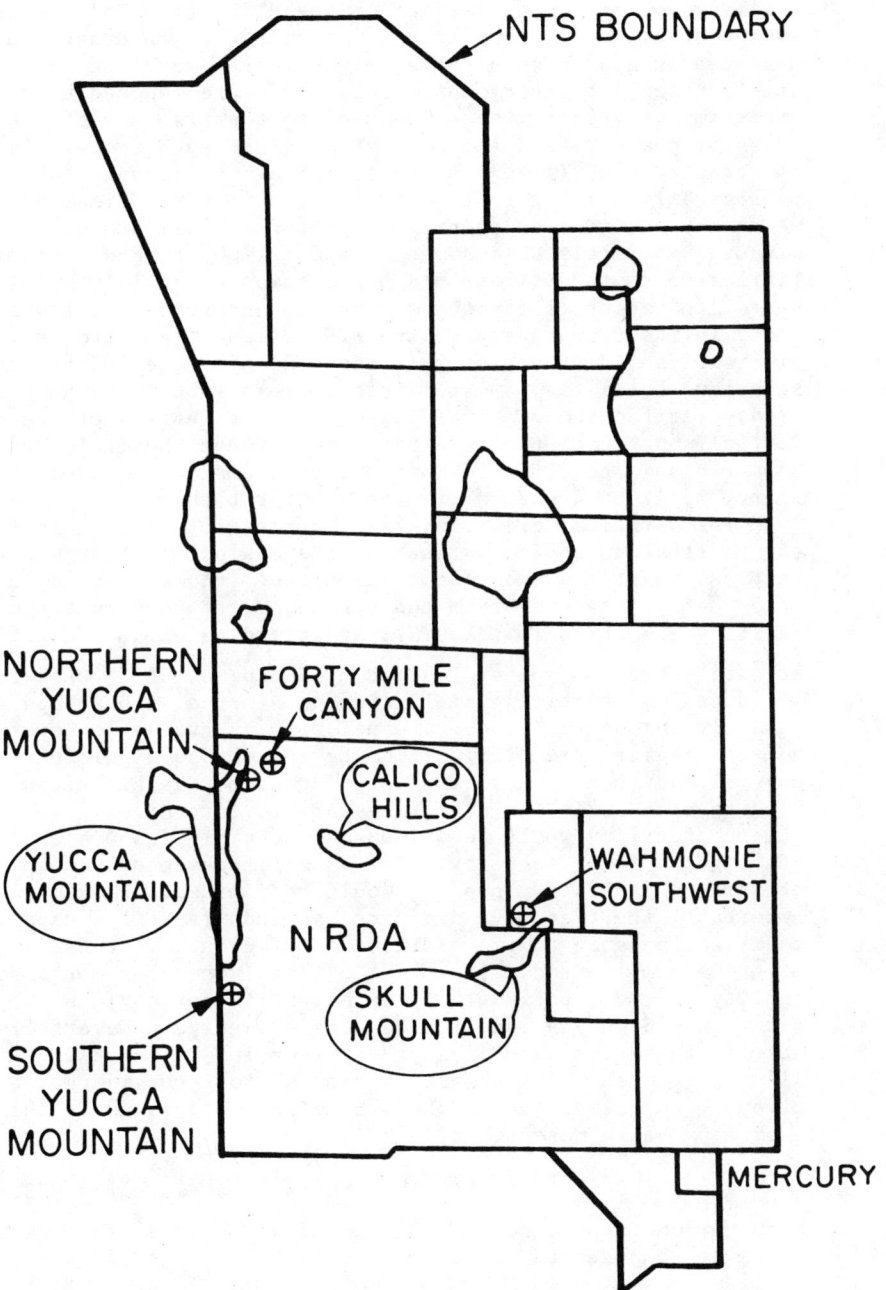

Fig. 5. Location of sites ⊕ most likely to prove adequate for a national underground physics laboratory within or near the Nevada Research and Development Area (NRDA) of the Nevada Test Site.

drilling and mining operations in the NTS. (iv) The tracts in the NRDA that are most promising geologically for an underground facility have nearby good access roads, electrical power, and existing, available surface laboratory buildings. They are reached by a major highway from Las Vegas without the need to traverse any of the restricted areas of the NTS, and they do not overlook such areas. (v) One of the tracts which is very suitable for an underground laboratory is geologically similar to a distant tract in NTS (Climax Stock area) where there exists an operating spent fuel cell storage facility constructed <u>ab initio</u> at a depth of about 1700 ft, and with many features similar to those desired in a basic science facility. (vi) Finally, there are potential advantages that an underground science facility might in its turn convey to the NTS. Among these are, briefly, the diversification of activity in the NTS, possible instrumentation and technique spin-offs, and possible improvements in methods of test ban treaty verification. These suggest that a sharing of the cost of the initial construction of the facility between the Office of Energy Research and the Office of Defense Programs of the Department of Energy might be a worthwhile venture for both.

The principal disadvantage of a site in the NRDA is the necessary initial cost of constructing and equipping a large diameter vertical shaft and underground laboratory rooms. It would be even more costly, starting from the beginning, to construct and equip a facility elsewhere in the U.S., as is indicated in a detailed study on site selection and evaluation by Sharp, <u>et al</u>.[8] Clearly, it would be less costly initially if one started with an operating (or possibly abandoned) mine or tunnel with an available working shaft or adit. However, the difficulties here are also significant. As we have already seen, there are very few such possible sites. After finding a site, it would be necessary to determine if an underground science facility would be welcomed by the site owners and to define the rights of the facility. In general, the access and possibility of surface laboratory support would be limited. Perhaps most important, the idea of a dedicated science facility, capable of modification and growth as dictated by scientific considerations, might be compromised. If this, in turn, were to compromise the quality of the experimental program, it is not obvious which alternative would in the long run be less expensive. Nevertheless, despite these disadvantages, the lesser initial cost involved in this alternative makes it attractive, and serious consideration is being given to locations (other than those in NRDA) found in the site search by Sharp referred to earlier.

<u>Present Status of the LANL Proposal to Establish NUSF</u>. LANL has been guided in part by a committee of scientists[9] which was formed to advise on matters relating to science underground and the proposed facility.

The LANL proposal to construct a basic science facility underground in the NTS has been exposed to the Offices of Energy Research and Defense Programs of the Department of Energy in a series of informal presentations. In addition, discussions have been held with individuals responsible for the various programs at the NTS to solicit their reactions to the proposal. Based on a LANL technical report,[8] a formal proposal (Schedule 44) also has been submitted which is now under review.

CONCLUSIONS

We have seen that particle physics theory and experiment are faced with the prospect that further unification of our understanding of the fundamental forces of nature may be in the offing. These ideas also relate particle physics to astrophysics and cosmology even more closely than before, and suggest a multiplicity of experiments to search for rare phenomena at the most basic level of physical science. They open our minds to the possibility of exploratory experiments which would test far beyond the present "standard models" of particle physics and astrophysics.

The mass scale that enters these speculations (roughly 10^{14} GeV) and the phenomena that follow from them, are not in general accessible to study by particle accelerators. We need experiments in addition to the collision-type experiments at particle accelerators that have been so valuable in recent years. I am of the opinion, as are many others, that experiments utilizing the techniques of particle physics but not particle accelerators themselves will become increasingly important to the progress of particle physics in the future, as they were in the past.

To encourage such a resurgence, and to provide facilities to help ensure the highest quality experimentation, it is desirable to concentrate resources for "non-accelerator" physics as has been done for "accelerator" physics. In addition to its direct technical advantages, this concentration might also foster wider participation of the particle physics community in non-accelerator physics.

The precise step that should be taken to accomplish these ends is open to debate. That contemplated here--a national underground science facility--is in keeping with steps that have taken us to higher collision energies at particle accelerators, and to far-reaching exploration at astronomical observatories. There is reason to believe that the step toward a national underground science facility in the U.S. may ultimately prove to be of comparable scientific value.

ACKNOWLEDGEMENTS

I am indebted to L. M. Simmons, Jr. for encouragement and moral support. Much of what I know about mines and mining has been learned from R. R. Sharp, Jr.

REFERENCES

1. H. Georgi and S. Glashow, Phys. Rev. Lett. 32, 438 (1974); P. Langacker, Phys. Rep. 72, 185 (1981).
2. H. Georgi, H. Quinn and S. Weinberg, Phys. Rev. Lett. 33, 457 (1974); W. Marciano and A. Sirlin, Phys. Rev. Lett. 46, 163 (1981).
3. See, however, A. K. Mann, Report of the Working Group on Low Energy and Cosmic Ray Tests of Particle Physics, in Proceedings of the Division of Particles and Fields Summer Study on Particle Physics and Facilities, Snowmass, Colorado, June 28 - July 16, 1982.
4. M. L. Cherry et al., Phys. Rev. Lett 47, 1507 (1981); M. R. Krishnaswamy et al., Phys. Lett. 106B, 339 (1981).
5. A. K. Mann, On A Possible National Underground Physics Laboratory in the USA, in Proceedings of the GUD Workshop on Physics and Astrophysics With a Multikiloton Modular Underground Track Detector, Rome, Italy, October 29-31, 1981, edited by G. Ciapetti, F. Massa, and S. Stipcich, p. 148; A. K. Mann, Report of Working Group #5: Physics in a Multipurpose Underground Laboratory, submitted to the 1982 Summer Workshop on Proton Decay Experiments, Argonne National Laboratory, Argonne, IL, June 7 - 11, 1982; A. K. Mann, Report of the Working Group on Low Energy and Cosmic Ray Tests of Particle Physics, in Proceedings of the Division of Particles and Fields Summer Study on Particle Physics and Facilities, Snowmass, Colorado, June 28 - July 16, 1982.
6. J. L. Osborne, S. S. Said and A. W. Wolfendale, Proc. Phys. Soc. (London) 86, 93 (1965); E. C. M. Young in Cosmic Rays at Ground Level, ed. by A. W. Wolfendale, (Inst. of Physics Press, London and Bristol, 1973), p. 105. L. V. Volkova, Sov. J. Nucl. Phys. 31, 1510 (1980) has recently computed the flux of muon and electron type neutrinos and anti-neutrinos above 1 GeV. See also L. V. Volkova, Proc. DUMAND Workshop 1980, p. 75. At 1 GeV the calculation of Volkova is about 70% higher than the earlier calculation.
7. D. S. Ayres, T. K. Gaisser, A. K. Mann, and R. E. Shrock, Neutrino Oscillation Search with Cosmic Ray Neutrinos, submitted to the Division of Particles and Fields Summer Study on Particle Physics and Facilities, Snowmass, Colorado, June 28 - July 16, 1982.
8. R. R. Sharp, Jr., R. G. Warren, P. L. Aamodt, and A. K. Mann, Preliminary Site Selection and Evaluation for a National Underground Physics Laboratory, Los Alamos National Laboratory Report LA-UR-82-556.
9. Members of the Advisory Committee for the National Underground Science Facility are: George A. Cowan, LANL; Raymond Davis Jr., BNL; William A. Fowler, CIT; Alfred K. Mann, Pennsylvania; Norman F. Ramsay, Harvard; and Frederick Reines, UC Irvine.

PERSPECTIVES OF FUNDAMENTAL RESEARCH IN THE GRAN SASSO UNDERGROUND LABORATORY[*]

M. Conversi
Department of Physics, University of Rome, Italy.

INTRODUCTION

The opinion is by now widespread that the installation of very massive and sophisticated particle detectors deep underground will provide a new way to attack frontier problems of subnuclear physics, astrophysics and cosmology, opening up a new field of fundamental research probably destined to expand into the year 2000. Matter instability, neutrino oscillations, extra-terrestrial neutrino sources, extremely high-energy cosmic ray events, heavy monopoles, neutrino bursts from stellar collapse,... are examples of such problems; all beyond the frontier of our present knowledge and mostly impregnable by other existing tools such as particle accelerators, reactors and the largest present-day astronomical installations.

These considerations fully justify, in my opinion, the efforts being made in different parts of the world to open up the new field of "Underground Science", to which the present meeting is dedicated. But perhaps some effort should also be made to coordinate the many initiatives on a world scale in order to avoid a dispersion of the available forces (in manpower and economics) which are necessary for these new big enterprises to become a reality.

First let me briefly recall the present situation of underground science in western Europe, where right now the only existing underground laboratory, already in operation for several years, is located under Mont Blanc, at a depth of about 5000 m of water equivalent (w.e.). Unfortunately this laboratory has a limited capacity which cannot realisticly be increased. A proton decay "calorimetric detector" of mass ~160 t, developed and installed there by the "Nusex Collaboration"[1], recently entered into operation and has already recorded one possible nucleon decay candidate, as reported later at this meeting[2].

Another underground laboratory, of useful volume about 10 times larger than that of the Mont Blanc laboratory, is now being excavated near the Fréjus tunnel, at a depth of about 4500 m w.e. A detector of mass 1.5 Kt, being constructed by a French-German Collaboration[3], again using a track calorimeter but of improved granularity and lower cost per ton with respect to "Nusex", will be installed there starting next year.

(*) Dedicated to the memory of Lorenzo Federici
(Ranzanico 2/9/39 - Roma 11/1/82).

Finally, a much larger laboratory will be constructed under the Gran Sasso d'Italia (see next Section) at a depth similar to that of the Fréjus laboratory. The lines of a project for a modular Giant Underground Detector ("GUD") of mass expandable into the 10 Kt region, to be installed in this last laboratory, were presented at the "GUD Workshop" held in Rome last year[4] following a letter of intent sent in June, 1980, by a Frascati-Milan-Rome-Turin Collaboration, to the Italian authorities.

The conclusions of the GUD Workshop confirmed the validity of the wide-range experimental program made possible by the construction under the Gran Sasso Laboratory of a modular track detector of high space-time resolution and of mass expandable beyond the 10 Kt limit.

THE GRAN SASSO LABORATORY PROJECT

The project of the Gran Sasso Laboratory (GLS) is going on under the direct responsibility of the present President of the Italian Institute for Nuclear Physics (INFN) who already presented its basic features at the GUD Workshop[5] and will report in detail its present status[6]. Here I will merely recall the most relevant characteristics and the main "figures of merit" of this new large INFN facility.

The GSL will be located about 100 miles from Rome, near the tunnel excavated under the Gran Sasso d'Italia on the highway which links up l'Aquila with Teramo. The GSL will have a total volume of $\sim 5 \cdot 10^4$ m^3. It will be under an average thickness of matter corresponding to about 4300 m w.e. In the present version of the project it will essentially consist of three holes of $(9m)^2$ useful cross-sectional area excavated over a length of nearly 100 m.

The GSL will be international in character and it will offer the possibility of installing there several large detectors to attack a wide range of problems of "Underground Science".

The financial support for the construction of the GSL and its facilities (not for the experimental set-ups) was secured early this year. The excavation of the first hole started early in September.

THE GUD PROJECT

As stated at the GUD Workshop, the project under study, a Giant Underground track-Detector (GUD) to be installed in the Gran Sasso Laboratory, is already oriented towards a "calorimetric approach", as opposed to a "water Cerenkov" detector. However the actual technical solutions are by no means frozen, even though in the general lines of the project, as presented at the GUD

Workshop and here, reference is made to definite techniques which appear particularly suitable for meeting the requirements of "high performance at a reasonable cost".

A set of "physics requirements" leads to the choice of a modular structure detector made of massive fine grained track calorimeter modules, of cost compatible with the goal of overcoming the 10 Kt mass limit. This goal is one of the main justifications for a new generation experimental program in an underground laboratory of the size of the GSL. This, jointly with a detector high performance (specifically a high space-time resolution) makes it possible to attack new areas of physics and astrophysics, such as the detailed investigation of the nucleon decay through identification of the various anticipated decay modes, study of the possible $\Delta B = 2$ transitions, sensitive searches for oscillations of atmospheric neutrinos and for extra-terrestrial neutrino sources, investigation of muon groups and other peculiar high-energy cosmic ray events, etc.

GUD can be regarded as the natural extension of the other two Western Europe experiments[1] [3], in that it will also use "hit calorimeters"[7]. These are total absorption track detectors which provide topological information on each event and allow one to derive the energy merely from the number of "counter hits", no matter which specific type of counter is utilized in the counter planes that in these calorimeters are alternated with shower sampling inert plates of dense material. But, in addition to its much larger mass, GUD should also provide precious time information, at a few ns resolution level, and thereby the possibility of deriving in many istances the direction of motion of the recorded particles from time-of-flight measurements.

The main ingredients in the proposal already presented "more as an example" at the GUD Workshop[4] are flash chambers of plastic material (PFC)[8] and resistive plate counters (RPC)[9]. The latter are used both to trigger the chambers and to provide continuous time information.

PFC's made of extruded polypropylene sheets alternated with metal plates for shower sampling ("flash calorimeters")[10] are now currently in use in large experiments for neutrino physics, in particular at Fermilab[11] and LAMPF[12]. The technique is well established and I shall only recall here the following facts:

a) The energy E of a primary electron or photon can be derived from the number N_f of recorded flashes (Fig.1, from reference 7, shows the linear relationship between N_f and E up to E ~500 MeV). The r.m.s. measurement error under the conditions of the GUD project[4] and for particles impinging <u>perpendicularly</u> to 3 mm thick Fe plates (as in GUD) is given by

$$\sigma/E = 7\%/\sqrt{E_{GeV}} \quad ;$$

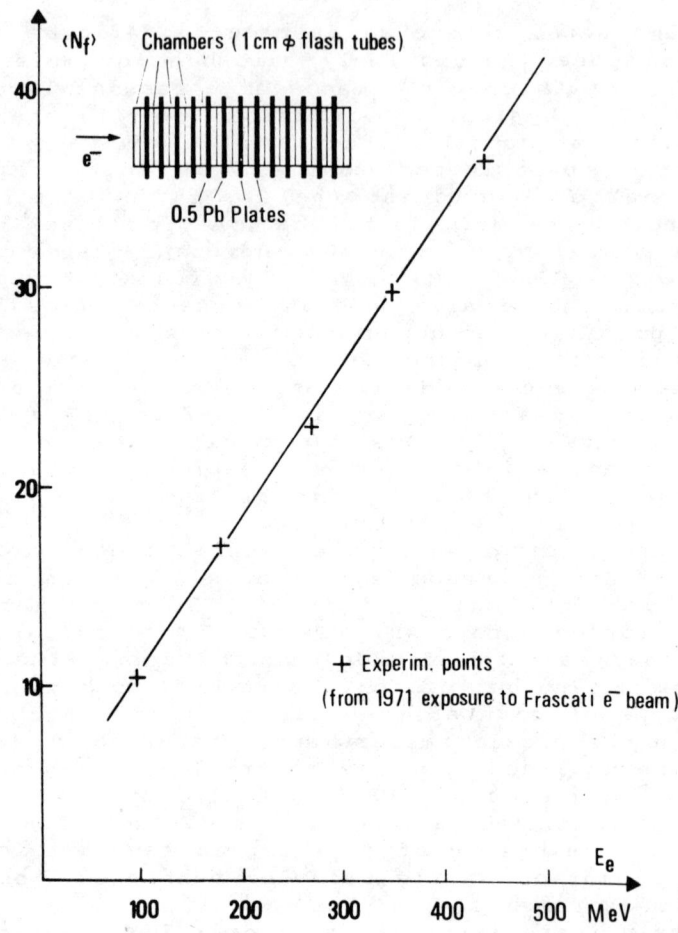

Fig. 1 - The insert at the top left shows a sketch of an "hit calorimeter", made of counter (flash-tube) planes, alternated with thin lead plates. The straight line through the experimental points shows the linear response of this calorimeter to electrons of energies from ~ 50 MeV to ~ 500 MeV.

b) The technique allows for some particle identification, as illustrated by the examples of Fig. 2 (from Ref. 10)) where the patterns of hadronic showers clearly appear different from those of electromagnetic showers.
c) Muon identification is unambiguous (straight track) and accurate energy determination is obtained from range measurements if the muon stops in one of the calorimeter plates.

d) The muon charge can be determined from the $\mu^+ \to e^+$ decay (μ^-'s being captured in iron in 98% of the cases) <u>provided that</u> the chambers are operated with a sensitive time of a few μs.

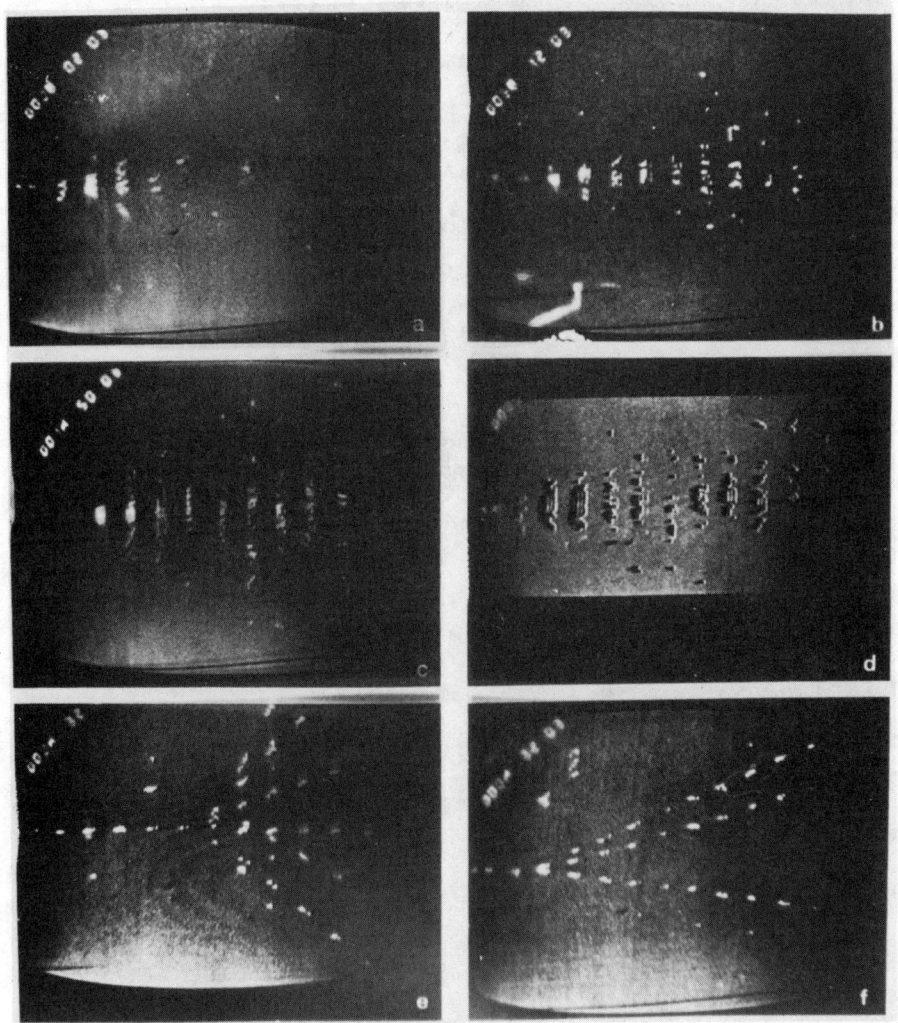

Fig. 2 - Response of the "plastic flash calorimeter" of Ref. 10 to electrons and pions of various energies. Videograms <u>a)</u> to <u>d)</u> show showers initiated by e^- of 0.5, 1, 1.5, 3, 4 GeV/c; videograms <u>e)</u> and <u>f)</u> show showers initiated by pions of 4 GeV/c and 5 GeV/c, respectively.

RPC's have been developed only recently[9] [13] and will be used for the first time in a search for neutron-antineutron oscillations at the Pavia reactor[14]. They are fast, wireless, d.c. operated counters made of parallel electrodes of high resistivity ($\sim 10^{10}$ Ω cm), between which a gas mixture, typically 50% argon and 50% butane, flows at atmospheric pressure. Their principle of operation is similar to that of the "Pestov counter"[15], but differently from the latter they can be built of large area at a low cost. At any particle traversal, which is recorded with 99% efficiency, they yield pulses of up to 1 volt, 2-3 ns risetime, on inductive pick-up aluminium-foil strip lines of 50 Ω impedence, placed on the counter walls as shown in Fig.3. Under the effect of the strong electric field ($\sim$50 kV/cm) present in the 2 mm gap between the resistive plate electrodes, the ionization electrons freed by the primary particle develop quickly into an avalanche, and then into a local streamer which, however, does not propagate via photoionization processes (as in Geiger counters or flash tubes) because of the butane photon absorption, nor does it evolve into a spark because of the high resistivity of the electrodes.

When the RPC's are used as trigger counters of very large area it is important to couple two RPC's into a single unit, as shown in Fig.3, in order to reduce the single counting rates, and therefore the accidental rate of twofold coincidences between pairs of these coupled units. One such unit, of 15 cm x 220 cm useful area and over-all thickness ~2 cm, with pick-up strip lines 3 cm wide, has recorded a "single" counting rate of less than 1/s at a position under the Gran Sasso tunnel not far from that forseen for the undeground laboratory.

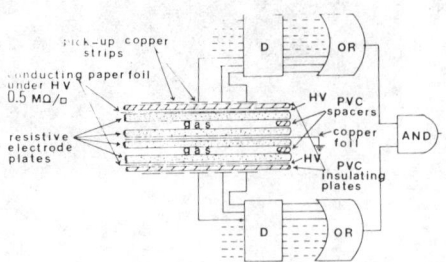

Fig. 3 - Sketch of double-layer "resistive plate counters"[9] with incorporated "cheap electronics" twofold coincidences, to provide a unit of low "single" counting rate.

The combination of the PFC and RPC techniques should provide a remarkable over all space-time resolution at such a low cost per detector-ton as to make realistic the reaching of the 10 Kt mass goal for GUD.

Fig. 4 shows a small portion of one of the GUD modules with the structure already discussed at the GUD Workshop[4]. PFC's are alternated with 3 mm thick iron plates (as in the Fréjus experiment) and RPC's are interleaved at a distance which leaves ~12 g/cm^2 of material between any pair of contiguous RPC planes. Under these conditions the energy trigger threshold is low enough to secure a high detection efficiency even for low energy events.

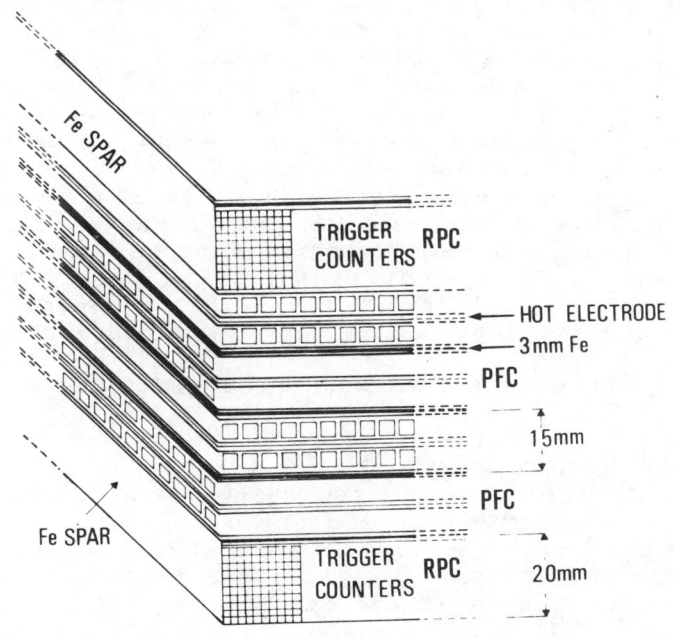

Fig. 4 - Showing a small portion of a corner of one of the (8 m)^3 GUD modules, to illustrate a possible module structure with low energy trigger threshold (~ 12 g/cm^2 of material between trigger counter planes). The trigger counters are RPC[9]; fast enough to also provide time-of-flight measurements. They are coupled in double layers (see Fig. 3) in order to drastically reduce the accidental trigger rate ("twofold" coincidences between contiguous counter planes). The chambers (PFC) are made of two extruded polypropylene plates incorporating the "hot electrode". They contain the sensitive "flash cells", of 8 m length and (4 mm)^2 cross sectional area. A (8m)^3 module (~1 Kt mass) contains more than one million sensitive cells.

Under the new conditions of the Gran Sasso Laboratory project[6] the dimensions and disposition of the GUD modules need to be changed with respect to those reported at the GUD Workshop[5]. Ten modules of the same structure, of $(8\ m)^3$ volume and $\sim 2\ g/cm^3$ density, could be aligned along one of the three GSL main holes to total 10 Kt.

For a total GUD mass of about 10 Kt the number of sensitive cells of $(4\ mm)^2$ cross-sectional area and of $\sim 8\ m$ length is still of the order of 10^7. This is indeed a huge number; in fact unrealistically large from the stand point of both cost and work required, should the wireless flash tubes be replaced by any type of wire counter(*). Thus, it appears that if the 10 Kt mass limit has to be reached using "hit calorimeter" modules, the solution proposed is still the best. This conclusion is reinforced by the great choice of read-out systems[4,16] offered by such a solution[4], including serial magnetostrictive[11] and capacitive[12] read-outs. However, for a 10 Kt detector, formidable mechanical problems still need to be solved, especially if a horizontal disposition is chosen for the module chamber and counter planes. Also, the need to sensitize such massive detector modules by the application of a ~ 5 KVolt high voltage pulse, as well as the selection of the triggering signal, represent technical challenges which should not be underestimated.

PHYSICS AND ASTROPHYSICS WITH "GUD"

A 10 Kt detector of the type outlined in the previous section (see Ref. 4) for more details) would only occupy a fraction of the GSL volume, but it would still offer the possibility of attacking a number of fundamental problems of physics and astrophysics, as discussed already at the GUD Workshop. In what follows I shall only briefly mention some of the open possibilities.

NUCLEON INSTABILITY

The problem of the detection of individual nucleon decay modes and their separation from background induced reactions has been discussed at the GUD Workshop[17] by comparing three proton decay detectors of different space

---

(*) It is worthwhile to stress that the extruded plates of polypropylene currently used in large flash calorimeters are commercially available at very low cost in sheets 1.5 m wide, of arbitrary length. Each sheet contains some 350 tubular cells of $(4\ mm)^2$ cross-sectional area and any desired length and thus it provides quite naturally, in the form of <u>wireless</u> detectors, the counters needed for the "hit calorimeter" counter planes.

resolution and granularity: the Kolar Gold Field detector[18], the Nusex calorimeter[1,19] and a fine grained track detector of the GUD type but without time information, as originally proposed by Grant and Tallini[20]. The conclusion of this analysis is that the detector space resolution is at least as important as fine sampling in radiation length, and that adding the time-of-flight information, as proposed for GUD, could allow one to separate out decay modes at the 10-20% level.

I shall not discuss again the "usual" electromagnetic proton decay mode ($p \rightarrow e^+ \pi^\circ \rightarrow e^+ \gamma\gamma$) but merely recall[4] that GUD would not only allow one to "see" the topological structure of the event, but also to determine the energy of the final state with a resolution of ± 17%. By the method explained in Ref. 4) the GUD fiducial mass for this type of event can be made nearly equal to the total mass.

In the so-called supersymmetry ("Susy") theories[21] the nucleon decay modes $p \rightarrow e^+ \pi^\circ$, $n \rightarrow e^+ \pi^-$,... are strongly suppressed, whereas decays involving strange mesons are favoured. The decay mode $p \rightarrow \mu^+ K^\circ \rightarrow \mu^+ \pi^+ \pi^-$, which is expected to occur in both standard and supersymmetric theories, could appear in GUD as shown by the Monte Carlo simulation shown in Fig. 5. The track of the positron from the $\pi^+ \rightarrow \mu^+ \rightarrow e^+$ decay sequence (not shown in the figure) could also be observed in GUD if the flash chambers could be efficiently operated with a few μs sensitive time.

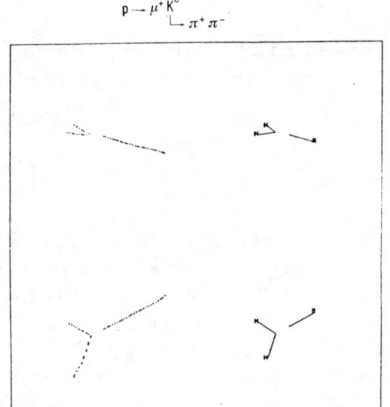

Fig. 5 - Simulation on GUD[17] of the proton decay mode $p \rightarrow \mu^+ K^\circ \rightarrow \mu^+ \pi^+ \pi^-$. The track of the positron from $\mu^+ \rightarrow e^+$ decay (not shown) can be observed too if the chambers are operated with a few μs sensitive time.

Fig. 6 shows sketches of some "Susy" decay modes as expected to appear in GUD under favourable conditions. In addition to the topological information, GUD would also provide measurements of the K^+ and μ^+ decay times and of the energy release. Of course the "energy signature" is not as good here as in the previous cases, due to the large amount of "invisible energy" carried off by ν's.

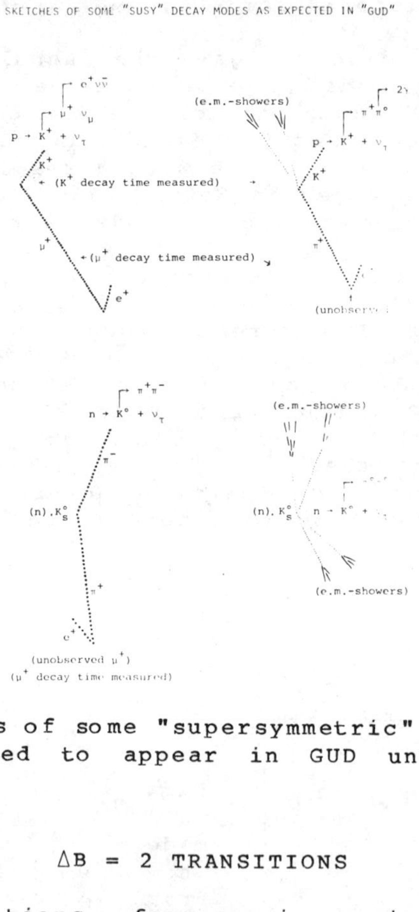

SKETCHES OF SOME "SUSY" DECAY MODES AS EXPECTED IN "GUD"

Fig. 6 - Sketches of some "supersymmetric" nucleon decay modes as expected to appear in GUD under favourable conditions.

$\Delta B = 2$ TRANSITIONS

Such transitions, forseen in various theoretical models[22], should be mediated by some massive boson in the "SU(5) desert" from ~100 GeV to ~10^{15} GeV. The events searched for (NN → pions) should be characterized by a clear signature due to the large amount of energy release. These transitions, however, are better investigated as a 1st order effect searching for neutron-antineutron oscillations in reactor experiments. As an example, a 10 Kt detector of the GUD type has an estimated sensitivity comparable to that expected for the $n \leftrightarrow \bar{n}$ oscillation experiment in preparation at the Pavia reactor[14].

NEUTRINO OSCILLATIONS

Neutrinos are generated in the Earth's atmosphere (as sketched in Fig. 7) by primary cosmic rays mostly through pion production and the subsequent decays $\pi \to \mu \nu_\mu$ and $\mu \to e \nu_e \nu_\mu$. Hence these atmospheric neutrinos are prevalently ν_μ's. Since the Earth is virtually transparent to all ν's of the energies involved (which are in the GeV region) a detector like GUD will be equally exposed to ν's coming from the nearby atmosphere and from the opposite side of the Earth's atmosphere. These latter ν's may have travelled over distances of the order of 10^4 km and might therefore have changed their flavour through the phenomenon of neutrino oscillations[23], which is possible for massive neutrinos.

GUD can then be applied, of course, to search for these neutrino oscillations. Its sensitivity, as reported at the GUD Workshop[24], should allow one to record a 3 σ effect after ~ 1 year of data taking for the quantity $\Delta^2 = |m_1^2 - m_2^2|$ in the range $10^{-3} - 10^{-4}$ $(eV)^2$, m_1 and m_2 being the masses of the neutrino base states. This represents a considerable improvement with respect to the present limit of ~ 1 $(eV)^2$ for Δ^2.

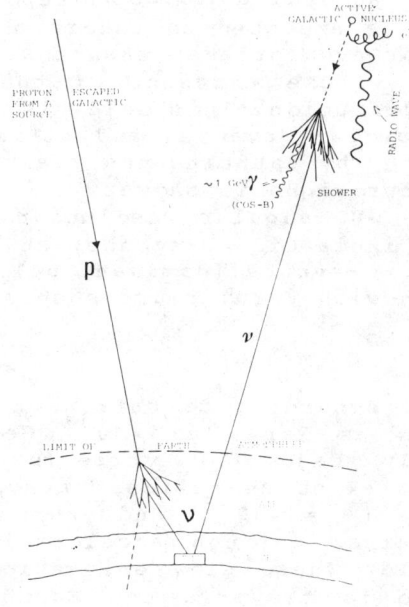

Fig. 7 - Illustrating generation of "atmospheric" and "extraterrestrial" neutrinos. GUD can also be oriented in the GSL so as to investigate neutrino-oscillations with neutrinos from CERN, e.g., as suggested in the last section of Reference 10.

SEARCH FOR EXTRA-TERRESTRIAL NEUTRINO SOURCES

Most luminous objects in the sky (pulsars in our galaxy, and extragalactic "active nuclei") radiate electromagnetic energy with a non-thermal spectrum extending from radio waves to gamma rays. The energy is liberated from "point-like" sources (on an astronomical scale), in the form of relativistic particles: e^-, which radiate the observed electromagnetic spectrum, and protons. Protons are accelerated up to several TeV and generate energetic ν's and γ's through $\pi^{\pm}$ and π° production and decay. The discovery of tens of γ-ray sources by satellite observations (COS-B, etc.) has confirmed this general picture.

Observation of ν's from such sources (see Fig. 7) would be of the greatest astrophysical interest, since these ν's would transport "unaltered information" from the interior of the source due to their very small cross section. A new field of neutrino astronomy in the energy region from about 0.1 GeV to ~100 GeV, could thus be investigated, complementing the eventual exploration made in the TeV region by DUMAND. (The 10^6t "Deep Underwater Muon And Neutrino Detector" under construction in Hawaii, detects Cerenkov light radiated by ν's from very high energy charged-current (c.c.) ν_μ-events.)

According to what was reported at the Rome Workshop[25] GUD is expected to record a rate of ν_μ-events comparable to or even greater than that expected for DUMAND, depending on the exponent of the assumed power-law ν-energy distribution. It should also detect medium and high energy c.c. ν_e-events, and allow one to measure the electron energy by counting the flashes produced by the associated electromagnetic shower.

GUD angular resolution is estimated to be[25] $4° \times 10°$ for inelastic ν-events, and of order $E_\nu/M_p c^2$ for elastic c.c. ν-events (dominant below $E_\nu \sim 1$ GeV). Hence, correlation with γ-ray sources should be possible.

MONOPOLES

According to current theoretical ideas monopoles of mass as large as $\sim 10^{16}$ GeV/c^2, generated at the very early stage of Universe formation ($< 10^{-35}$s), might nowadays exist as relics characterized by very small velocities ($\beta = \sim 10^{-3}$) and extremely high energies ($\simeq 10^{20}$ eV) acquired through acceleration in the galactic magnetic field. These slow energetic "heavy monopoles" should be able to traverse the Earth with no appreciable energy loss.

The arguments presented by Glashow at the Rome Workshop for "no monopoles at GUD"[26] need of course to be revised if the single event consistent with one Dirac unit of magnetic charge recently reported[27] is not considered as an exceptionally large time fluctuation.

The GSL in its present concept has an useful area of a few 10^3 m^2; vast enough to install large-area apparatus to search for such heavy monopoles under favourable background conditions. No specific proposal for such a search has yet been put forward.

GUD is not suitable for detecting very slow monopoles. However, in traversing matter, monopoles with beta equal to 10^{-3} or larger should undergo ionization losses[28] large enough to be detected by GUD RPC-planes as a sequence of electric pulses well separated in time, giving the monopole direction of motion. For the area covered by GUD in the new module disposition, and for a monopole flux corresponding to the new "astrophysical upper limit", GUD should detect some 10 events per year. If no event were found in a one-year exposure, GUD would yield an upper limit for the flux of such monopoles at least 10 times smaller than the present limit from the Bakstan Laboratory[29].

NEUTRINO BURSTS FROM STELLAR COLLAPSE

Gravitational collapse is expected to occur for stars of mass in excess of about 1.2 solar masses as the last phase of stellar evolution, when the nuclear fuel is exausted. The subsequent "neutronization process" ($e^-p \to n\nu_e$) then leads to a "neutron star", or to a "blackhole" if the star mass is large enough. The phenomenon is very complex[30]. It is presumably accompanied by emission of electromagnetic waves, bursts of neutrinos and other particles, and also of gravitational waves if the collapse does not occur under conditions of spherical symmetry (i.e. if $d^3 Q_{\alpha\beta}/dt^3 \neq 0$, where $Q_{\alpha\beta}$ is the mass quadrupole tensor).

In general theoretical models agree in predicting the emission of neutrino bursts peaked at the instants of the successive "bouncing points" characteristic of the implosion mechanism. Thus, neutrino bursts of ~ 1 ms duration, separated by time intervals of the order of 0.1 s, should accompany each stellar collapse. The total neutrino energy, $E_{tot}^{(\nu)} = 10^{53}-10^{54}$ erg, is nearly independent of the mass of the collapsing star.

The energy E_ν of the single ν's is expected to extend from ~ 10 MeV to ~ 60 MeV, with an average value $<E> = $ ~20 MeV. Under the experimental conditions of the GUD project a fraction of the ν's belonging to this spectrum could be recorded as electrons (from $\nu_e+Fe \to Co+e^-$) of energy greater than the GUD energy threshold. For a stellar collapse at the center of our galaxy (8.5 Kpc = 2.6 x 10^{22} cm) and for a 10 Kt GUD total mass there should be ~ 10^3 such electrons, and their space and time distributions could be recorded by the d.c. operated

RPC's. The correlation of such an event with one recorded elsewhere by a detector of gravitational waves would clearly be of the greatest interest[31].

However, the rate of stellar collapses within our galaxy is expected to be only ~0.5/year according to "optimistic estimates" based on pulsar data. On the other hand, with the known distribution of the surrounding galaxies, no substantial increase can be achieved for this rate, unless one can reach distances that include the Virgo Cluster (d = 1.9×10^7 pc = 5.9×10^{25} cm). However, the signal of such a neutrino burst could only be recorded, unfortunately, by detectors of mass in the megaton region.

REFERENCES

1) See E.Fiorini: Proceedings of Second Workshop on Grand Unification, Ann Arbor (1981), Birkhauser Press, p.55.
2) P.Picchi: Report at the present Workshop.
3) R.Barloutaud: see volume cited in Ref. 1, p.70.
4) "Physics and Astrophysics with a Multikiloton Modular Underground Track Detector": Workshop held in Roma, 29-31 October, 1981. Proceedings edited by G.Ciapetti, F.Massa and S.Stipcich (hereafter indicated for short "Proc. G.W."): see p.10 for a description of the GUD project.
5) A.Zichichi: Proc. G.W., p.141 (1981).
6) A.Zichichi: Report at the present Workshop.
7) G.Brosco: Thesis, University of Rome, 1972 (unpublished; with main results reported by present author in Sect. 8 of Rivista del Nuovo Cimento 3, 273 (1973)). It is worth pointing out that in this new type of calorimeter (in which the determination of the energy of the primary showering particle is not based on ionization measurements but on the recorded number of total counter hits), the response does not depend on the type but only on the transversal dimensions of the counters utilized for the counter planes. Flash tubes have the basic advantage of behaving as wireless counters of typically 0.5 cm linear transversal dimensions.
8) M.Conversi and L.Federici: NIM 151, 93 (1978).
9) R.Santonico and R.Cardarelli: NIM 187, 377 (1981).
10) L.Federici et al.: NIM 151, 103 (1978).
11) F.E.Taylor et al: Proc. IEEE Nucl. Sci. Symp. San Francisco (19-21 October 1977).
12) H.H.Chen: UCI-Neutrino n.20 UCI-10P 19-119 (June 1977).
13) R.Santonico: Proc. G.W. p.108 (1981).
14) G.Bressi et al.: Proc. of ICOBAN Workshop, Bombay (January 1982), reported by S.Ratti.

15) Yu. N.Pestov and G.V.Fodotovich: Preprint IYAF 77-78; SLAC Translation n.184 (1978).
16) H.Meyer: Prof. G.W. p.115 (1981).
17) A.Grant: Proc. G.W. p.122 (1981).
18) V.S.Narasimham: Proc. G.W. p.38 (1981).
19) E.Fiorini: Proc. G.W. p.46 (1981).
20) A.Grant and B.Tallini: DPhPE 80-10 (CERN Internal Report dated 4/2/1980).
21) See Proceedings of CERN Workshop on "Sypersymmetry versus Experiment", organized by D.N.Nanopoulos, A.Savoy-Navarro and Ch.Tao (Geneva, 21-23 April, 1982). TH.3311/EP.82/63-CERN.
22) See e.g.: J.C.Pati, A.Salam and J.Strathdee: Nuovo Cimento $\underline{26A}$, 77 (1975); R.N.Marshak and R.E.Mohapatra: Phys. Lett. $\underline{94B}$, 193 (1980); Phys. Rev. Lett. $\underline{44}$, 1316 (1980).
23) B.Pontecorvo: Sov. Phys. JEPT $\underline{6}$, 429 (1958); $\underline{7}$, 1972 (1958); $\underline{26}$, 984 (1968).
24) N.Dardo: Proc. G.W. p.131 (1981).
25) G.Auriemma: Proc. G.W. p.101 (1981).
26) S.Glashow: Proc. G.W. p.1 (1981).
27) B.Cabrera: Phys. Rev. Lett. $\underline{47}$, 1378 (1982).
28) V.A.Rubakov: JETP Lett. $\underline{33}$, 644 (1981; C.G.Callan: Princeton Preprint (1982). See also S.Dimopoulos, S.L.Glashow, E.M.Purcell and F.Wilczek: Nature $\underline{298}$, 825 (1982) for the possible generation of high-energy ν_e by monopoles in the Sun, at a level detectable by GUD.
29) See, e.g., E. N. Alexeyev, et al.: reported at the 8-th European Cosmic Ray Symposium (Rome, September 1982) by A. E. Chudakov.
30) Ya.B.Zeldovitch and O.Kh.Guseinov: Soviet Phys. JEPT Letters $\underline{1}$, 109 (1963); D.Z.Freedman, D.N.Schramm, and D.L.Tubbs: Ann. Rev. Nucl. Sci. $\underline{27}$, 167 (1977); K.Lande: Ann. Rev. Nucl. Sci. $\underline{29}$, 395 (1979).
31) See E.Amaldi and G.Pizzella: Atti del I Convegno Nazionale di Fisica Cosmica (Lecce, 11-12 February 1982), p.115.

THE GRAN SASSO PROJECT

A. Zichichi
INFN, Rome, Italy

ABSTRACT

This paper presents the final solution adopted for the "Gran Sasso" Underground Laboratory.

INTRODUCTION

At the Rome Meeting I presented the first two versions of the proposed Gran Sasso project.[1] The basic point of both versions was a "unique-large-volume" experimental hall. Further studies, and a more detailed analysis of all the problems connected with the excavation of such an immense volume inside a mountain (a volume never excavated before), have brought two conclusions: i) the money needed exceedes by far the original 20×10^9 Italian Lire ($\sim \$20 \times 10^6$ USA); ii) the time estimate (2 years from the date of approval) could not be realized, because of the many unknowns related to the large volume to be excavated.

An intense series of studies was undertaken in order to reach the final edition of the project. Seven alternatives have been considered, in addition to the first two already presented at the Rome Meeting.[1]

The present project is the alternative no. 7 of the second series. Let me illustrate its main features.

THE "GRAN SASSO" PROJECT: SOLUTION NO. 7

The Underground Lab

Let me remind you that "Gran Sasso d'Italia" is a mountain located near Rome ($\sim 10^2$ km). Figure 1 shows the schematic profile of the mountain. The position of the underground laboratory is shown in black. The horizontal lines indicate the highways already excavated. The structure of the mountain in terms of its main constituents - detritus, limestone, dolomite - is shown in Fig. 2, where again the INFN underground Laboratory is indicated. The impermeability, the porosity, and the rock density of the various strata are illustrated in Fig. 3.

Figure 4 indicates the stations A,B,C, where the measurements of the cosmic ray background were performed.

The Laboratory has on top of it 1.5 km of rock; i.e., 4×10^3 meters of water equilvalent.

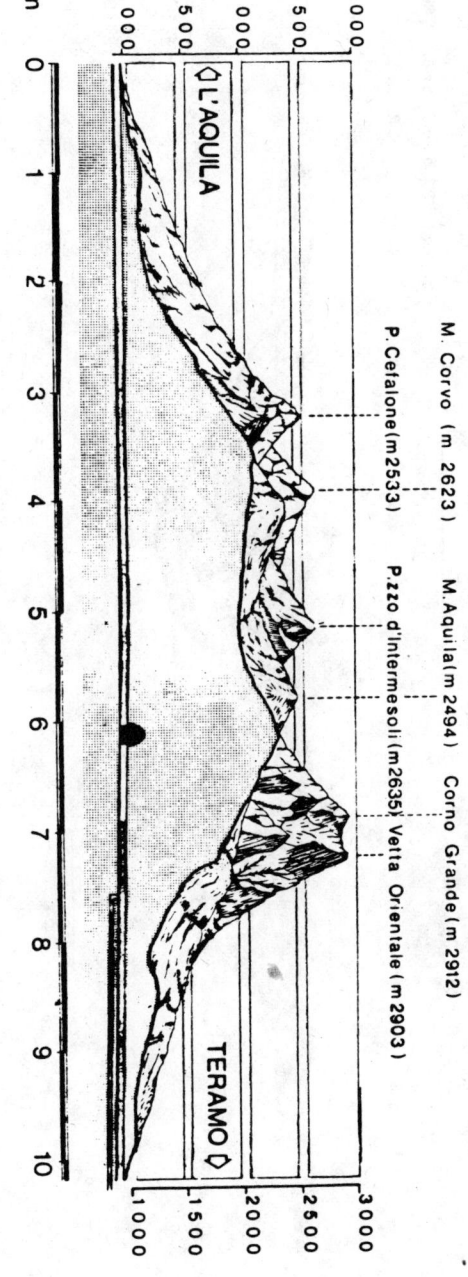

Figure 1

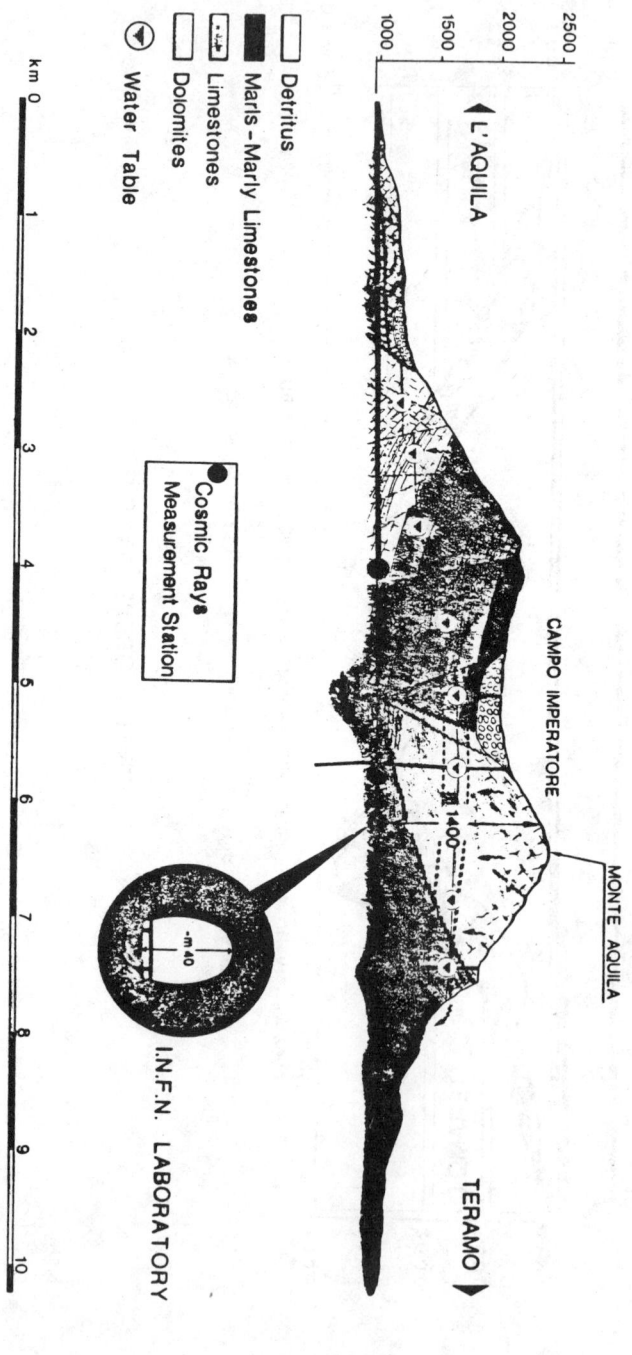

Figure 2

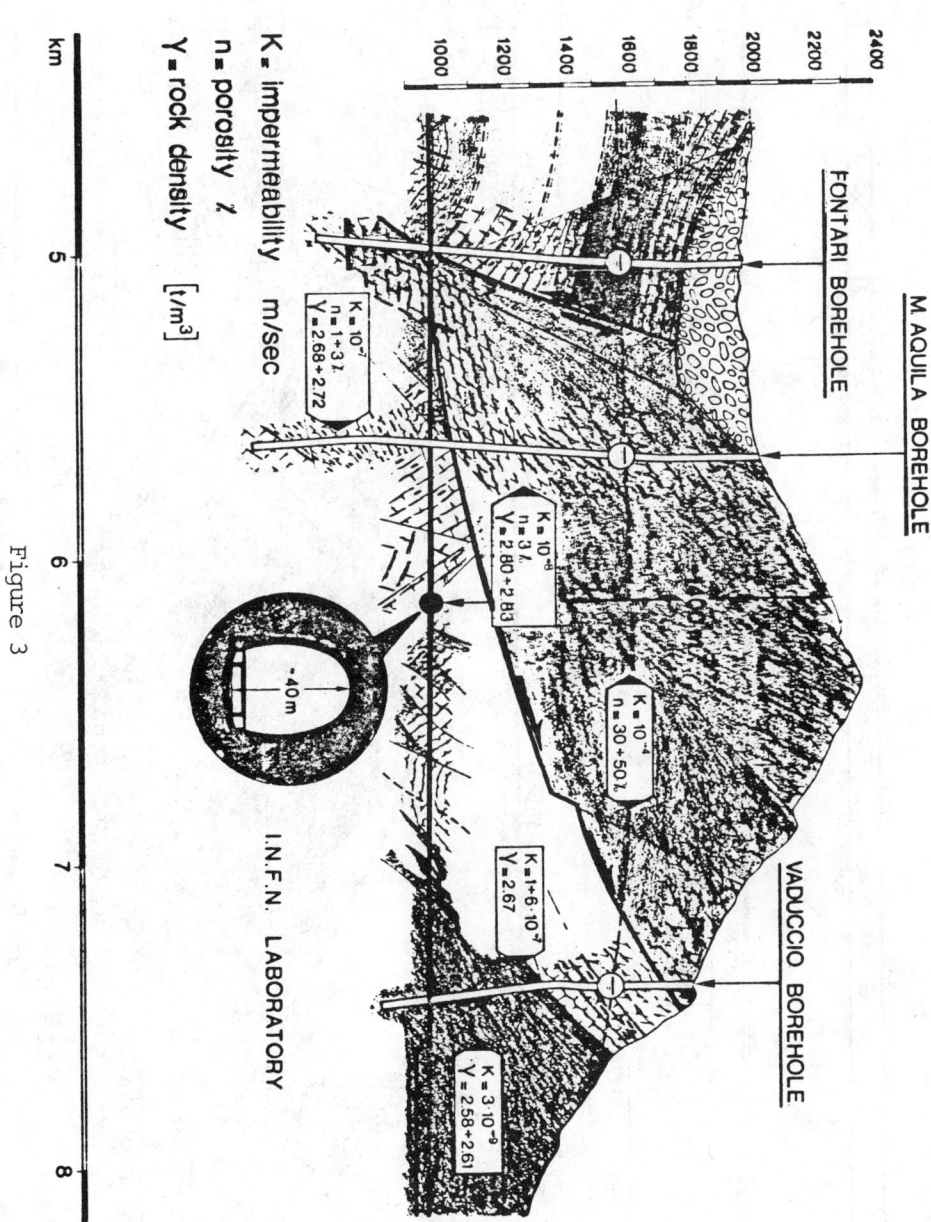

Figure 3

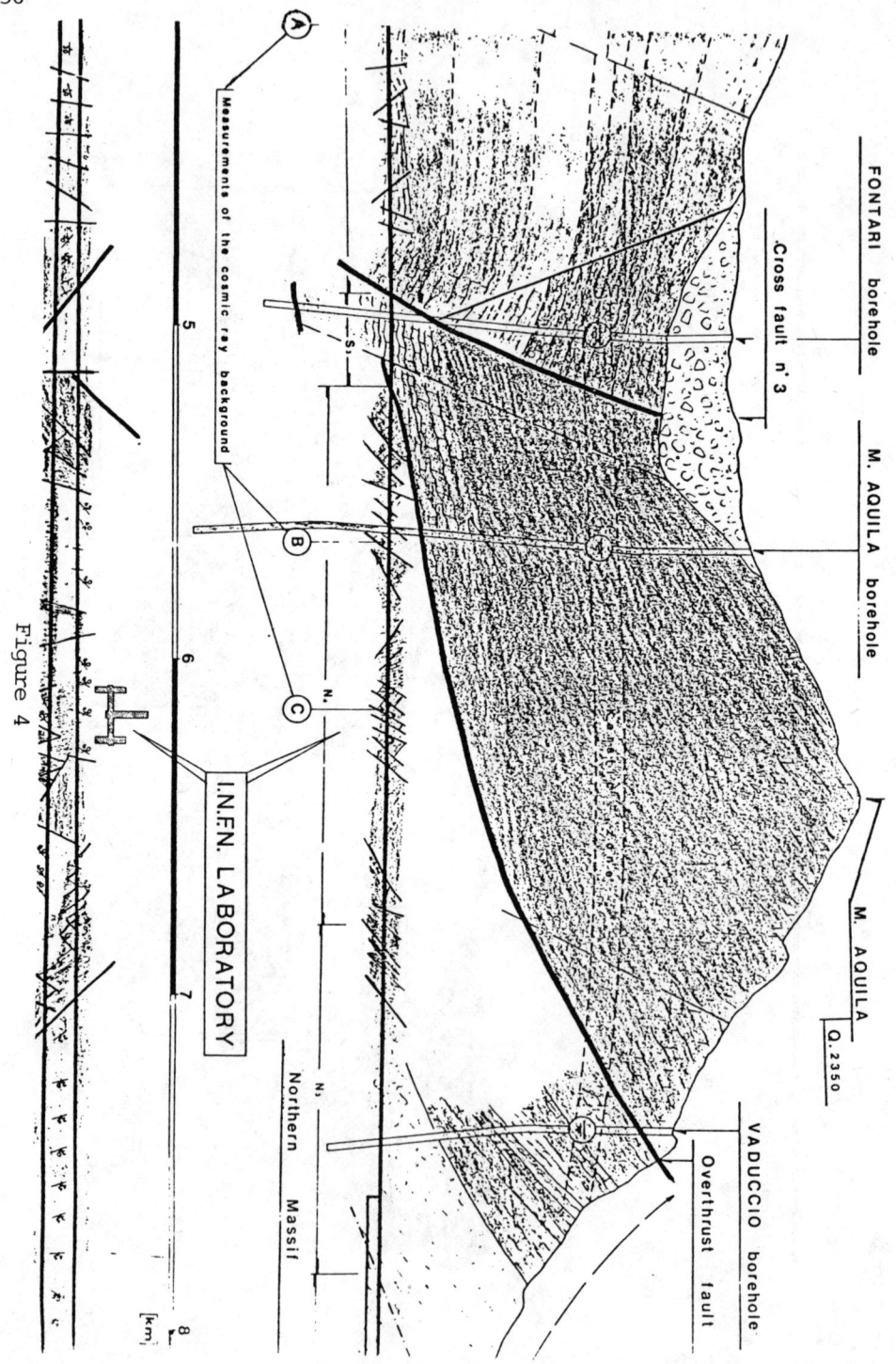

Figure 4

The plan of the Laboratory is shown in Fig. 5. The Laboratory consists of three main experimental halls.

The length of the central hall (B) is <u>127 meters</u>. The other two lateral halls (A and C) are each 92 meters long. The separation of these two halls (A and C) from the central one (B) is 60 meters. The positioning of these halls and their separation is the best compromise between our two main considerations:
i) a large surface covered by the three halls;
ii) a separation such as not to complicate the excavation or to make the realization of the project more expensive.

The three halls have their axis oriented towards Geneva.

The cross section of the halls is such that equipment $(9\times9)m^2$ can be installed, free of technical problems. Physicists can work around this equipment, displace it, change components, etc. Notice the location of the recesses in Fig. 5: four on each side in hall B; two in Halls A and B. Some details of the halls are shown in Fig. 6. Notice that the effective cross sectional area of the halls is 290 m^2.

The cross section of the "connection" between the three Halls (A,B,C) is shown in Fig. 7 (top left). The effective cross sectional area of the connection gallery is 106 m^2. This connection can also be used, if needed, as an experimental hall. An experimental set up of $(6\times6)m^2$ can be installed without problems. (See Fig. 6, top right, as an example of how the connection gallery could be transformed into an experimental hall.) The only change would be an increase in the presently planned cross section from 106 m^2 to 155 m^2. The option for this further enlargement will be decided during the execution of the connection gallery, which will be drilled after all other elements are finished.

There are two types of access to the Laboratory halls. One is for "TIR-like" lorries (Fig. 7, top right). This is essential for big equipment and for large quantities of material to be delivered. In this case the lorry must first traverse all the highway from L'Aquila to Teramo (bottom way in Fig. 5). Once out of the highway tunnel, there is a connection with the other highway, in the direction Teramo to L'Aquila. On this side of the underground highway, there is a direct access to the Laboratory halls C-B-A, and the way out and towards L'Aquila (Fig. 5).

For every day access - physicists and technicians - there is a standard car access. Here one doesn't need to traverse all the highway, L'Aquila-Teramo and then back, as for the lorry case. A connection gallery which goes above both highways allows one to have a direct access to the Laboratory halls, as shown in Fig. 5.

Finally, the three experimental halls are connected by by-passes as shown in Fig. 5. Their cross sectional areas are shown in Fig. 7, bottom-center. Each hall is also directly connected with the highway by emergency exits.

58

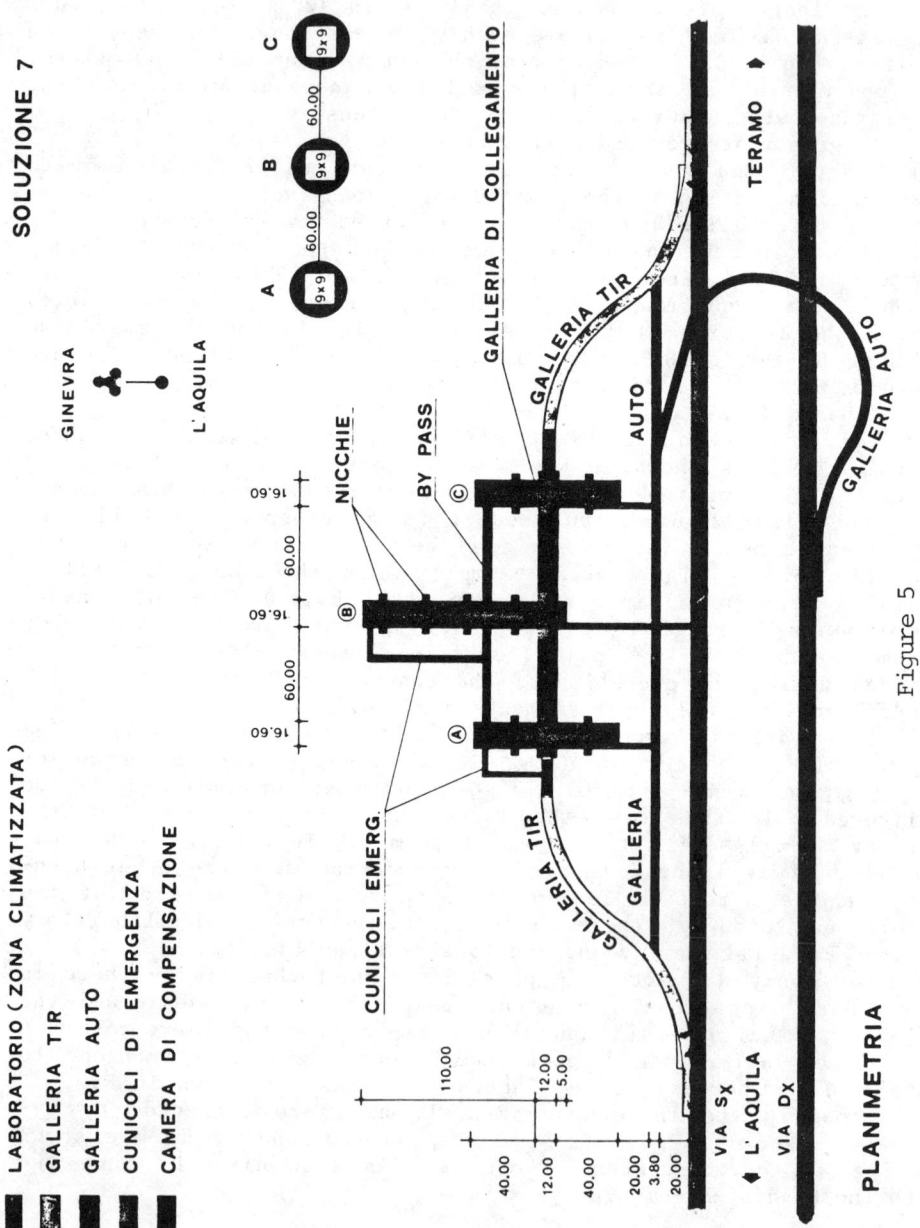

Figure 5

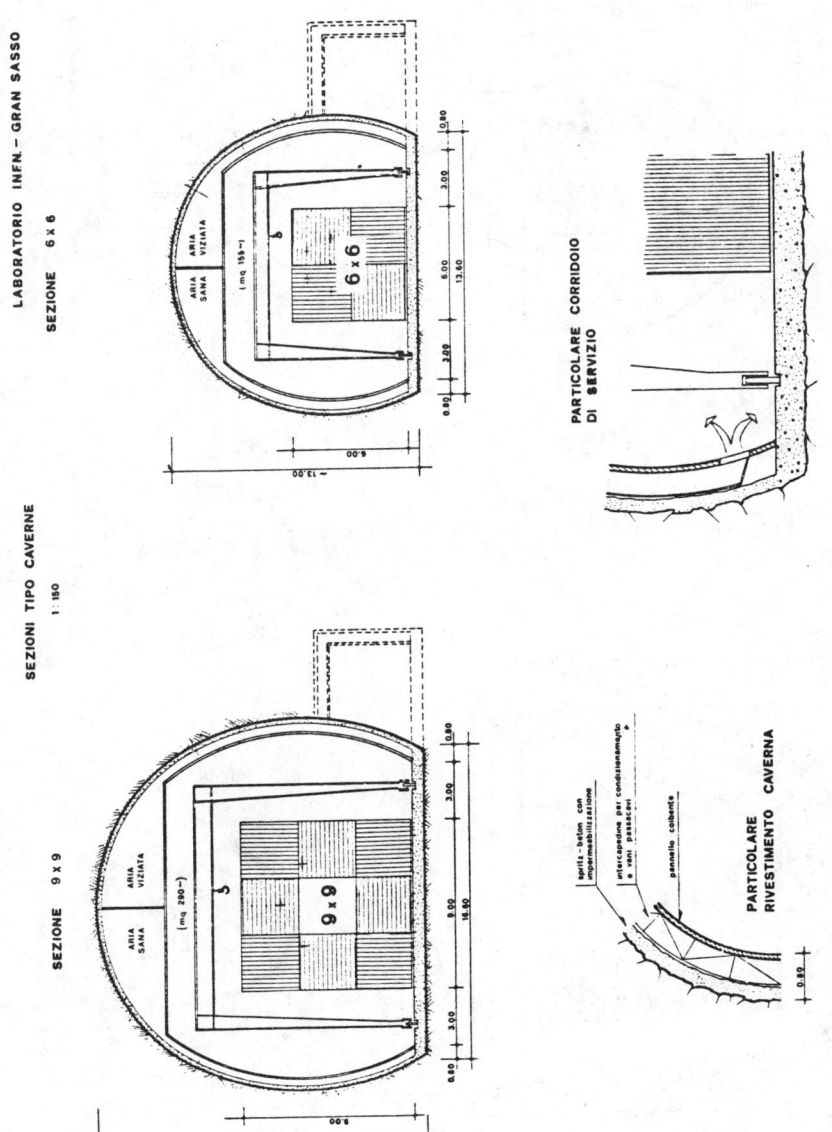

Figure 6

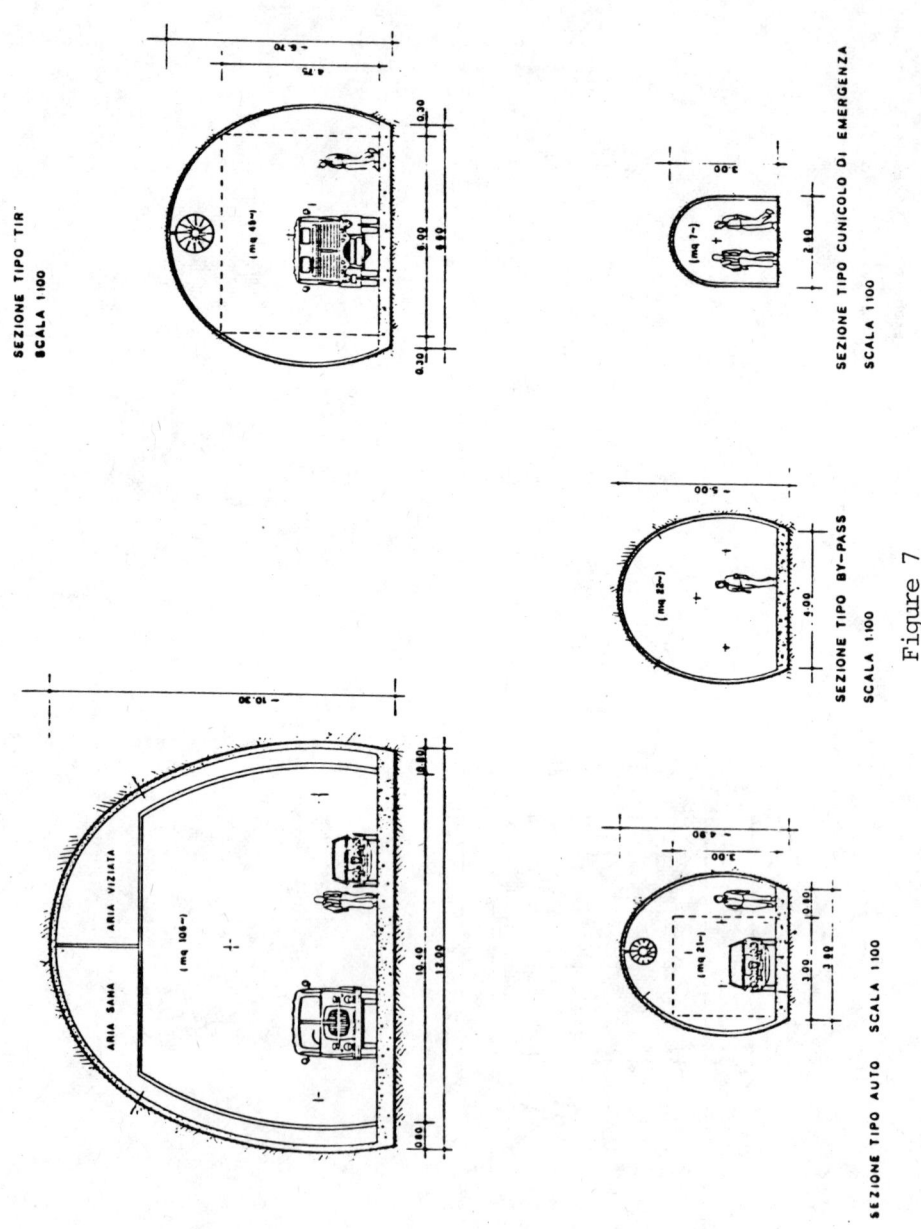

Figure 7

The view of the emergency exits is shown in Fig. 5. The cross sectional area is shown in Fig. 6, bottom-right. A general artistic view of the underground Lab is shown in Fig. 8.

The Top-Lab

On top of hall B, centered over the total area covered by the three Halls A-B-C, there will be a small laboratory of 10 m^2 surface, 3 m high, for TOF studies connected with the main detectors underground. The access to this Top-Lab will only be by helicopter.

The Outside Lab

On the L'Aquila side of the highway there will be an outside series of building for workshops, lecture halls, a library, guest rooms, and offices. The outside buildings have been designed for 100 people, physicsts and technicians.

THE RESEARCH PROGRAM

A vast range of topics are of interest for the "Gran Sasso" Laboratory. Let me briefly quote the main research fields about which I was thinking when I started to work on this new INFN project. These fields are closely connected with the remarkable development in the understanding of the fundamental laws of nature and of cosmic events.

Nobody would have doubted, just a few decades ago, that the baryon number conservation law is exact. However, now there are several reasons for believing that baryon number is no longer a sacred quantity. This is firstly because if we want - one day - to unify all the forces of nature, then all the basic constituents must necessarily be in the same multiplet. Once we put quarks, antiquarks, and leptons in the same multiplet, transformations among them are no longer forbidden. Furthermore, we now live in the age of "gauge" forces. We all think that the only conservation laws that are sacred are those coming from local "gauge" symmetries. As baryon number is not linked with any local "gauge" symmetry, its sacred nature vanishes.

Any model grand unified theory (GUT) predicts nuclear instability. The various models only differ in the total decay rate, and in the relevance of some decay channels with respect to others. For example, supersymmetry would suggest that the decay mode $p \to K + \nu$, is going to be dominant over $p \to \pi + \nu$.

Another exciting research topic is the study of cosmic collapses producing neutrino bursts. According to some astrophysical models, this phenomenon should not be so rare, and it would allow our eyes to be opened to a very interesting new horizon in astrophysical research.

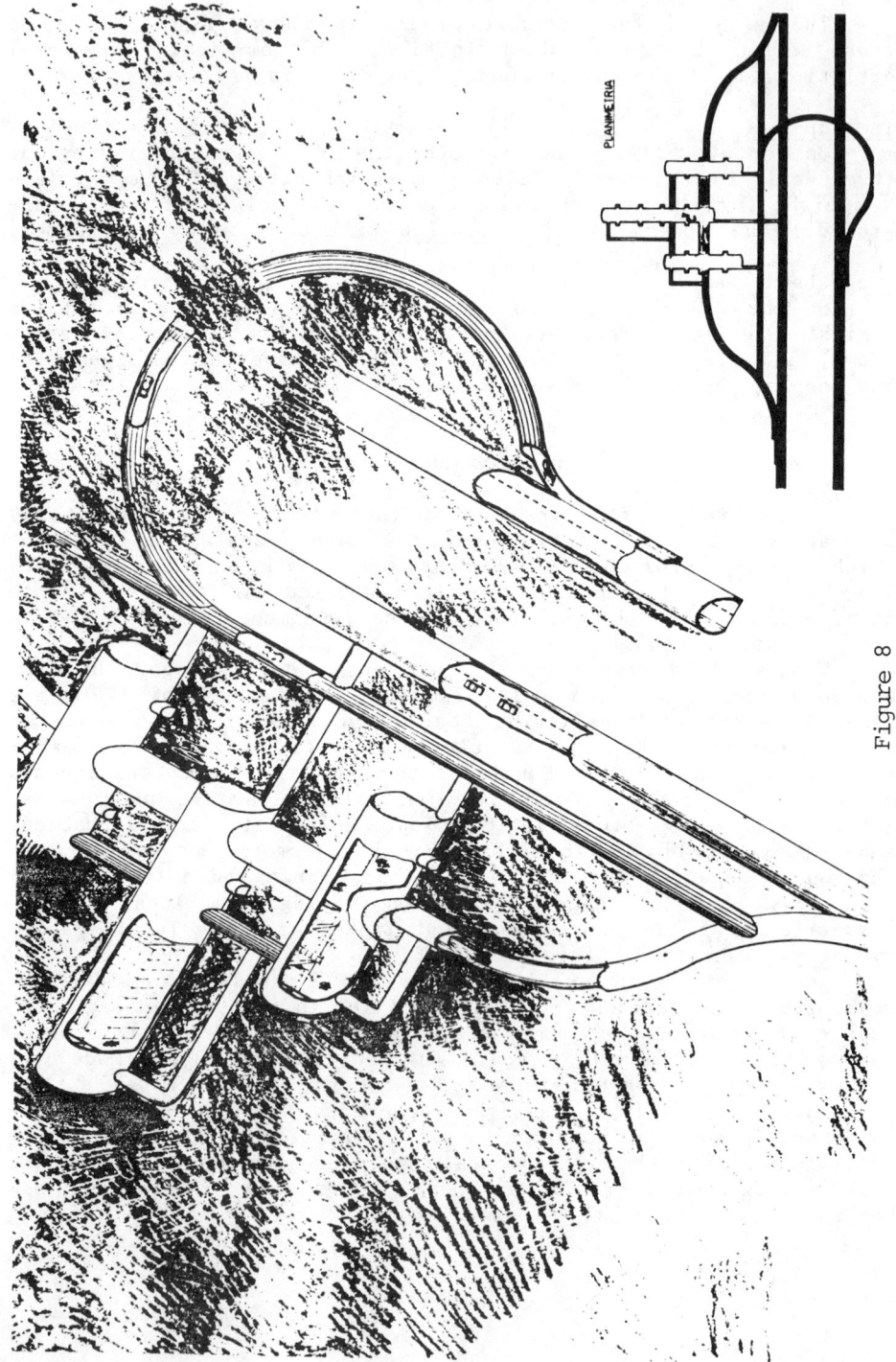

Figure 8

Another field is the search for "new" and particularly penetrating cosmic-ray particles, whose properties are suitable for detection using the facilities of the "Gran Sasso" Laboratory. This field of exploration could be called "new cosmic phenomenology." In this field are monopole searches.

The Laboratory's detectors could be oriented towards the CERN Super Proton Synchrotron (SPS) in order to study the hot problem of neutrino oscillations. The interest in this search is particularly relevant since the SPS is a source of all sorts of neutrinos (ν_e, ν_μ, ν_τ) and antineutrinos ($\bar{\nu}_e$, $\bar{\nu}_\mu$, $\bar{\nu}_\tau$). Furthermore, "coincidence" searches could be envisaged, just in case...

The study of biologically active "living" matter, in this very low cosmic-ray background, is open to a systematic series of investigations.

Finally, the problem of ground stability is a topic which has generated a lot of interest amongst physicists at the University of L'Aquila.

To summarize, the scientific aims of the "Gran Sasso" Laboratory are the study of:
1) nuclear stability;
2) neutrino astrophysics;
3) new cosmic phenomenology;
4) neutrino oscillations;
5) biologically active matter;
6) ground stability.

THE INTERNATIONAL CHARACTER OF THIS NEW INFN LABORATORY

1. The laboratory will be open to anyone who wants to exploit its unique properties.
2. The international scientific committee will be the consultive body for deciding the experiments to be financed. It will consist of 15 members, of which only 3 will be from Italy.
3. The INFN will consider how to solve the problem of offering staff positions to physicists from other countries who want to do experiments in this new laboratory.
4. The international structure of the lab has to be such that even a small group in a small university could contribute to the research activity of the lab.

So far, physicists in the following countries have expressed an interest in doing experiments in the "Gran Sasso" Laboratory: Canada, the People's Republic of China, the United Kingdom, USA, USSR, the Federal Republic of Germany.

In Italy, physicists from 15 universities have expressed an interest in working in the "Gran Sasso" Underground Laboratory. These are at L'Aquila, Frascati, Rome, Torino, Milano, Trieste, Padova, Bologna, Firenze, Pisa, Bari, Napoli, Cagliari, Palermo, Messina, Catania, and Parma.

Needless to say, INFN wants to encourage anyone, from the smallest to the largest scientific institutions, to work in the "Gran Sasso" Laboratory. We consider this lab a further contribution to promoting among physicists from different countries the world over, new friendships, mutual understandings, and collaborations.

DECISION TO BE TAKEN SOON

There are already several proposals being considered by groups of physicists who want to do experiments.

The most urgent decision to be taken is the choice of the 10^4 tons detector. This very large instrument must be "modular" in conception, in order to allow even a very small group of physicists at a small university to contribute; for example, by constructing and testing part of the equipment.

PRESENT STATUS

The first experimental area (hall B) is being excavated. It is already 3 times the Frejus volume. Hall B will be fully excavated by the end of the summer, at the latest.

New money is needed by the end of the year: $30 million. If this money comes in time, the Laboratory should be ready by the end of 1985.

REFERENCE

1. A. Zichichi, "The Gran Sasso Project," International Workshop on Physics and Astrophysics with a Multikiloton Modular Underground Track Detector, Rome 29-30 October, 1981, p. 141.

Chapter II

SOLAR NEUTRINOS

There are strange things done in the midnight sun
By the men who moil for gold; ...

Robert Service,
The Cremation of Sam McGee

THE ANALYSIS OF SOLAR MODELS -- NEUTRINOS AND OSCILLATIONS

Roger K. Ulrich[†], Edward J. Rhodes, Jr.[††]
Steven Tomczyk[†], Philip J. Dumont[†], Wendee M. Brunish[††]

Introduction

The theory of stellar structure and evolution is used to calculate the properties of a variety of objects from red giants and supernova precursors to white dwarfs and neutron stars. Accurate tests of the theory in the context of these applications are generally not available. The sun as the nearest star provides a unique example of a star which can be subjected to a variety of precise tests not possible with remote stars. We will concentrate on two of these tests -- solar neutrinos and solar oscillations -- which currently indicate that there is something seriously wrong with our standard solar model. Although we do not yet know what the source of the error is, it is quite possible that the correction of this error will require some modification of the results of other applications of stellar structure theory. It now seems unlikely that the difficulty with the solar neutrino experiment lies in the experiment itself. The combination of the difficulty with the solar neutrino flux and the difficulty with the solar oscillation frequencies suggests that the solar neutrino problem is a failure of stellar structure theory rather than a failure of weak interaction theory, although this latter possibility cannot yet be firmly ruled out.

In addition to the solar neutrinos and solar oscillations, we note that two other tests of the sun yield results which are not completely understood. First are the abundances of light elements which indicate that the convective envelope extends to 2.8×10^6 K, whereas the standard solar model[1] has this temperature equal to 1.9×10^6 K. Blake and Schramm[2] have considered the possibility of convective overshoot and found that this process cannot explain the discrepancy without an <u>ad hoc</u> assumption. Second is the shape of the solar surface. Apart from phenomena occurring in the visible layers of the solar surface, the apparent solar surface corresponds to an equipotential surface and provides information concerning the angular distribution of matter in the solar interior. The standard solar model presumes this distribution to be spherically symmetric. The measurements by Dicke and Goldenberg[3] have been interpreted by Dicke[4] to imply a rotating internal deviation from spherical symmetry. The measurements by Hill and Stebbins[5] are not in contradiction with Dicke's conclusion, although they are not supportive either. If confirmed, Dicke's work would indicate a rather nonstandard property of the solar interior.

The solar neutrino difficulty[6,7] is the best known example of a test of the standard solar model which has failed. For a period it was common for workers in each of the three areas related to the problem -- stellar interior theorists, particle physicists and

---
[†] University of California, Los Angeles, CA 90024
[††] University of Southern California, Los Angeles, CA 90089

experimental physicists -- to hope and occasionally believe that the
solution lay in the other fellow's camp. The experiment has survived
a very rigorous examination and seems an unlikely place to find the
cause of the discrepancy. Particle physics, particularly in the area
of neutrino oscillations, could still play a role in the solution,
but as pointed out by Bahcall et al.[8], oscillations among three
particles would be required and even then some discrepancy would
remain. Within the context of the standard solar model there are
uncertainties which do not relate to the assumptions defining the
standard model which nonetheless lead to uncertainties in the
calculated neutrino flux. Chief among these uncertainties are the
possible errors in the calculated opacities, in the measured nuclear
cross sections and in the abundances of the heavy elements. Bahcall
et al.[7,8] have examined the uncertainties and give a predicted
neutrino counting rate of $7.2 \pm 3.3 \times 10^{-36}$ captures/target atom/s
(this unit is called the solar neutrino unit or SNU). The
uncertainty is specifically interpreted as a 3σ limit with the
definition of 3σ meaning "if a result falls outside the 3σ range,
then a mistake has been made," when 3σ is applied to a theoretical
result such as the opacity. Since mid-1968, the theoretical
predictions have fallen within this range, although both the
prediction and its error have varied. Prior to 1968, the nuclear
reaction rates were too poorly known to permit an accurate
prediction. The discrepancy between these predictions and the
observations (which now yield 2.2 ± 0.3 SNU) has been relatively
unchanged since the announcement of the earliest experimental results
in 1968.[9]

We wish to emphasize in our discussion the new field of solar
seismology. This method of studying the solar interior involves the
measurement of the structure and frequencies of global solar
oscillations. The oscillations have been most readily detected
through the measurement of line-of-sight velocities[10], although there
have been some measurements of solar limb position variations. The
largest amplitude oscillations have periods near five minutes. So
far the oscillations have been studied in three ranges of spatial
structure -- nearly radial, highly nonradial and most recently
intermediate size. The eigenmodes are classified according to the
spherical harmonic $Y_\ell^m(\theta, \phi)$, which describes the angular distribution
of velocity   according to the distribution and number of nodes in
the radial component of the velocity eigenfunction. The eigenvalue
associated with this latter classification corresponds roughly to the
principal quantum number of atomic physics and we use the quantity n
to denote this number. The nearly radial modes correspond to degrees,
or ℓ values, between 0 and 4, the highly nonradial modes have $\ell \gtrsim 150$
and the intermediate modes have $4 < \ell < 150$. The highly nonradial
modes may not be truly global in the sense that ℓ may not be a good
quantum number as a result of perturbations produced by the solar
convection zone. The nearly radial oscillations have been detected
for individual values of n and ℓ so that they must be global in
character. These modes penetrate to the center of the sun and measure
an average of the sound velocity throughout the solar interior. In
fact, the spacing of the frequencies from one value of

n to the next depends on the inverse of an integral of dr/c where c is the sound speed. This integral tends to emphasize the solar surface regions, but the deep solar interior is also included. Because periods of oscillation can be measured with high accuracy, the constraint on the deep solar interior imposed by the nearly radial oscillation frequencies is useful in testing both standard and nonstandard solar models.

Observations

We will concentrate on the nearly radial and intermediate modes. In order to resolve individual eigenfrequencies, an observing span in excess of 24 hours is required. Such long duration observations can be obtained by connecting together sequences from separate days. This method was used by Claverie et al.[10] working from the Canary Islands to show first that the global oscillations are detectable and resolvable. Their observations were interrupted regularly by the diurnal cycle and the lack of continuity of observation complicates the analysis of the power spectrum. The diurnal limitations can be avoided by going to the geographic south pole during the Austral Summer. Informal reports indicate that a clear spell in the weather lasting up to about a week typically occurs at the beginning and end of each Austral Summer. During the second of these clear spells in 1980, Grec, Fossat and Pomerantz[10] obtained the power spectrum shown in Figure 1. These remarkable data are clearly worthy of very close analysis.

The sharpness of the power spectrum peaks and their close spacing make a detailed examination of the data difficult when they are presented in the format of Figure 1. For this reason, it has become common practice to use the nearly regular spacing of the frequencies to restructure Figure 1 into what has been termed an echelle format. The scale of the abscissa is stretched and then chopped into equal length intervals which are then displaced vertically from one another. Figure 2 from Grec, Fossat and Pomerantz[10] shows the positions of the eigenfrequencies displayed in an echelle format. The length of the strips has been chosen to be 136 μHz, the average spacing of the eigenfrequencies. Figure 2 compares the frequencies from observations by several groups and the overall agreement is good. The identification of the appropriate ℓ value has been made on the basis of comparison to theory. The pattern of the eigenfrequency spacing is a more reliable theoretical result than is the numerical value of either the spacing or zero point of the frequencies.

Comparison of Theory to Observations

The identification of the appropriate value of the principal eigenvalue n with each power spectrum peak is required before a comparison between theory and observation can be carried out. The highly nonradial modes have frequencies which are so widely spaced for differing values of n that the identification of n can be made easily. By tracing the locus of frequency versus ℓ from large ℓ to small, it is then possible to identify the appropriate value of n for the nearly radial modes. Until recently, the intermediate modes were

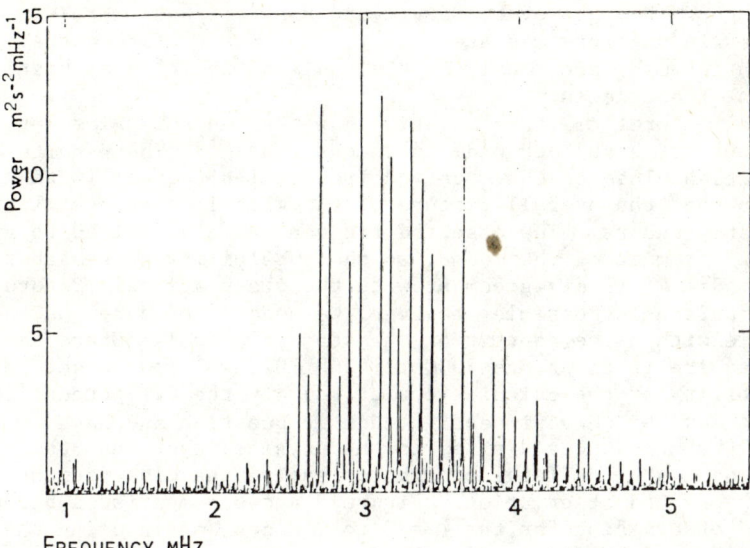

Figure 1. The power spectrum of solar oscillations by Grec, Fossat, and Pomerantz.

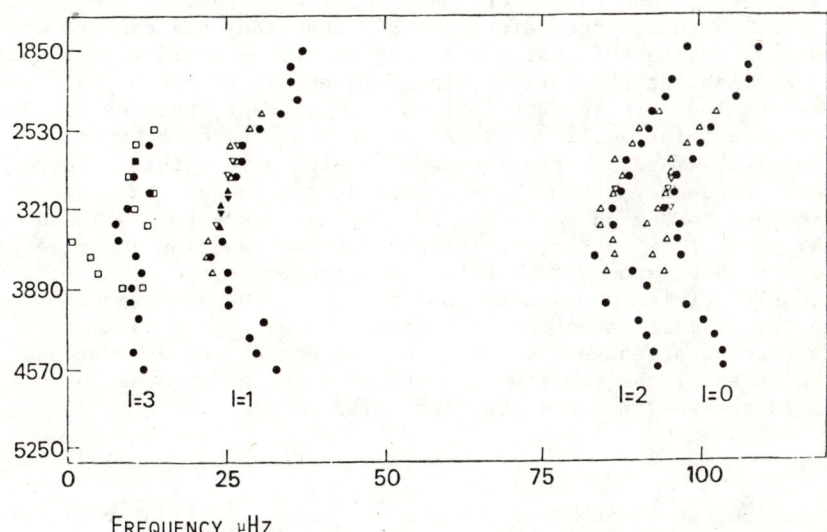

Figure 2. The positions of central frequencies for the peaks in the power spectrum of Figure 1 displayed in an echelle format. Figure 1 has been divided into 136 μHz intervals which are then displaced above each other. The different shaped points represent results from different groups.

not observed and this locus could not be used as a means of identifying n. Now, however, Duvall and Harvey[12] have obtained frequencies in the intermediate ℓ range and have confirmed the identification made previously on the basis of the closest fitting theoretical frequencies.

Several theoretical investigations of the theoretical solar eigenfrequencies have been made in recent years.[13] The results have generally been close to the observed frequencies (within 10 to 30 µHz) so that the overall picture of the global solar oscillations is reasonably secure. The observed frequencies are defined to within 1 to 2 µHz for most of the modes, so the possibility arises that the standard model is in disagreement with the observations. Before a valid comparison is possible, however, we need to be sure that theory is reliable within the context of the standard model. Since the frequencies are in the range 2000 to 4000 µHz, we need to determine the reliability of the calculations at roughly the 0.1 percent level. Ulrich and Rhodes[14] have investigated this question and have concluded that the uncertainties in the physics input into the standard model are not large enough to resolve the disagreement between theory and observation. Figure 3 shows comparisons between theory and observation for the ℓ = 0 to 3 modes, again using the echelle format. The discrepancy between theory and observation takes the form of both an error in the zero point of the frequency sequence and an error in the spacing. The theoretical zero point is too low and the theoretical spacing is too large. The outer boundary condition of the eigenvalue problem as well as some of the details of the structure of the solar convection zone just below the photosphere are all uncertain and can alter the comparison. However, the spatial coincidence of these uncertainties means that they could introduce at most one new parameter. Using this parameter, we could adjust the spacing between the theoretical eigenfrequencies to match the observed spacing. If we carry out that adjustment, then we could make the zero point smaller and the discrepancy becomes worse. Conversely, if we adjust the boundary condition to obtain a larger zero point frequency, then the spacing becomes larger and again we are no closer to a good fit. The range in the uncertainty in such physical parameters as the opacity and nuclear reaction cross section is also not large enough to permit a resolution of the discrepancy. Consequently, Ulrich and Rhodes concluded that the discrepancy between theoretical and observed frequencies of global solar oscillations constitutes a significant failure of the standard solar model which should be considered along with the solar neutrino problem in the search for a modified solar model.

Nonstandard Solar Models

In the course of the search for a solution to the solar neutrino problem, several nonstandard solar models have been proposed. Table 1 lists a number of the more popular of these models. The altered composition models (high Z and low Z) reflect the possibility that the solar nebula could have undergone element segregation in the condensation phase of evolution and this segregation may not have

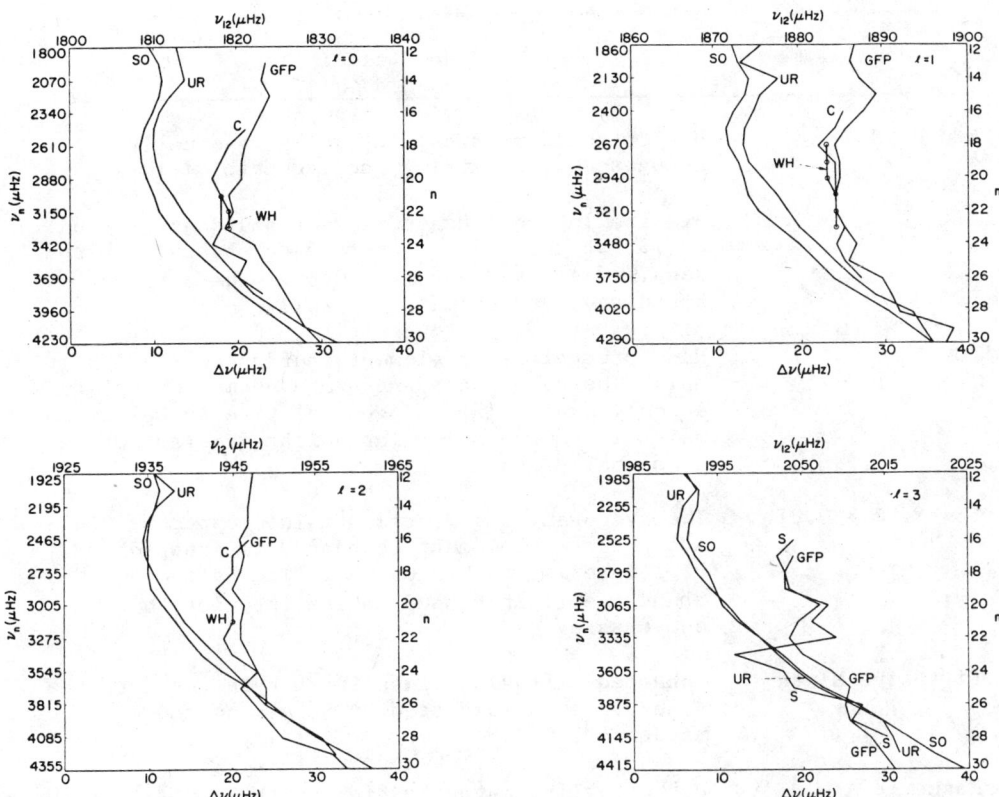

Figure 3. A comparison of frequencies from the standard models of Ulrich and Rhodes (UR) and Shibahashi and Osaki (SO) for ℓ=0 and $12 \leq n \leq 30$ with the observed frequencies of Claverie, et al. (C), Grec, Fossat, and Pomerantz (GFP), Woodard and Hudson (WH), and Scherrer, et al. (S). Successive slices of the power spectrum separated by 135 μHz are displaced vertically. The left axis shows the starting frequencies of the slices corresponding to even values of n. The right axis shows those values of n. The bottom axis gives the frequency offset from the starting frequency of each slice.

Table 1

Model Characteristics

Model Name	Model Description
Standard	Has standard physics, initially chemically homogeneous, no mixing, no magnetic fields.
High Z	Has its internal heavy element abundance Z equal to 1.19 times Z_{RA} (Z_{RA} refers to abundances from Ross and Aller[18] and is 0.018). The surface abundances are normal ($Z = Z_{RA}$).
Low Z	Has internal heavy element abundances $Z = 0.3\ Z_{RA}$. The outer abundances are normal except for a small excess in hydrogen relative to helium to maintain a mean molecular weight gradient which is normal.
Low X, Z = 0.01	The hydrogen mass fraction X is dropped by 0.04 and the heavy element abundance is dropped to 0.01 interior to $M = 0.7\ M_\odot$. The helium abundance is increased in the interior to compensate.
Diffusive Mixed	Enhanced diffusion mixes fresh hydrogen into the solar core as suggested by Schatzman and Maeder.[19]
Magnetic A	Has a Cowling magnetic field with $B_o = 3.16 \times 10^8$ gauss. The first zero in B is at $r = 0.7\ R_\odot$. The field is assumed to remain zero exterior to this point.
Magnetic B	Like Magnetic A but with the zero at $r = 0.5\ R_\odot$
Inner 0.05 $M_\odot$ mixed	The inner 0.05 of the sun's mass is artificially mixed on a time scale rapid enough to homogenize the ^{3}He as well as the hydrogen.
S_{34}=0.3 KeV-Barnes	The lower value of the $^3He + ^4He \to ^7Be + \gamma$ reaction suggested by the Munster group is adopted.
Enhanced opacity	The opacity is artificially multiplied by 1.2 throughout the model.
No scattering states	The second virial coefficient due to scattering states is set to zero.

been smoothed out by a fully convective phase. Neither supposition is supported by our present theories of star formation, but the early stages of stellar evolution are sufficiently complex that such segregation cannot be ruled out. Although as we will see below both the high and low Z models fail, we feel the idea of element segregation is attractive. Another version of modified element distribution was suggested by Schatzman and Maeder[15] and involves diffusive mixing. Their hypothesis is for a small scale turbulence field to enhance the diffusion coefficient enough that the gradient of hydrogen abundance is smoothed out. This model is attractive because of its low neutrino flux; however, there is no known mechanism which could enhance the diffusion coefficient enough to be consistent with the model. Another version of element segregation involves modifying the initial distribution of the hydrogen and helium throughout the sun. This could have occurred if, for example, very thick molecular hydrogen mantles were formed on the surfaces of grains in the coldest part of the protosolar cloud. Segregation between the grains and the residual gas is possible if the grains are charged and bound to a magnetic field while the gas is neutral. We might expect the solar core to be relatively depleted in both hydrogen and heavy elements according to this scenario. Finally, we have considered a truncated Cowling[16] magnetic field of high strength. The central field is large enough that the magnetic pressure $|B|^2/8\pi$ is a few percent of the gas pressure. This requires $B = 3 \times 10^8$ gauss at the solar center. Various authors[17] have argued that 1) a field larger than 10^7 gauss will produce an unacceptably large solar oblateness and 2) magnetic buoyancy will cause a field larger than 10^6 to 10^7 gauss to rise to the solar surface. We do not know of an argument which could counter the first point, although it is possible in principle that some combination of dipole and toroidal fields could avoid producing significant distortion on the solar surface. We believe that a large enough scale field may not be subject to the buoyancy considerations, at least for a long enough time that nuclear burning can enhance the helium abundance to the point where it can counterbalance the magnetic buoyancy. Several additional nonstandard models are also listed in Table 1, which differ only slightly from the standard model. These are included as tests of the effect of uncertainties in the basic physical input on various derived model properties.

Tables 2 and 3 give some of the important derived properties of the models listed in Table 1. The neutrino fluxes in the third and fourth columns are larger for the standard model than those given by Bahcall et al.[7] because the code used to calculate solar models for the purpose of studying eigenfrequencies uses a less accurate treatment of the abundances of the minor nuclear constituents than does the code used for the solar neutrino problem. Ratios between the fluxes given in Table 2 for different models should be reliable even though the absolute value is high. The frequencies in Table 3 differ slightly from those plotted in Figure 3 for the standard model. This is because Table 3 includes the results of a new radiative interaction theory. Also, the radius is corrected to the more precise value given below.

Table 2

Model Derived Properties

Model	T_c (10^6 K)	ϕ_8 (10^6 cm^{-2} s^{-1})	ϕ_7 (10^9 cm^{-2} s^{-4})	X (surface)	ℓ/H	T_{CEB} (10^6 K)	M_{CE} ($10\,M_\odot$)	r_{CEB} (R_o)
Standard	15.629	7.18	4.45	0.718	1.55	1.867	1.46	0.749
High Z	16.113	12.69	5.38	0.684	1.62	2.051	1.99	0.740
Low Z	14.466	1.68	2.23	0.830	1.62	1.802	1.20	0.738
Low X, Z = 0.01	15.367	4.15	3.53	0.787	1.98	2.217	3.00	0.699
Diffusive Mixed	14.740	2.60	2.50	0.705	1.39	1.642	0.87	0.774
Magnetic A	15.346	5.01	3.71	0.749	1.40	1.569	0.72	0.776
Magnetic B	15.547	6.71	4.33	0.723	1.59	1.913	1.63	0.741
Inner 0.05 $M_\odot$	15.652	7.59	4.41	0.717	1.54	1.852	1.45	0.749
S34 = 0.3 Kev-Barnes	15.589	4.71	2.82	0.721	1.56	1.892	1.50	0.745
Enhanced Opacity	16.059	11.99	5.52	0.689	1.49	1.881	1.54	0.748
No Scattering States	15.598	7.02	4.33	0.702	1.63	1.963	1.78	0.737

Table 3

Model Frequencies ($\nu_{n,\ell}$ in μHz)

Model		$\nu_{17,0}$	$\nu_{18,0}-\nu_{17,0}$	$\nu_{22,0}$	$\nu_{23,0}-\nu_{22,0}$	$\nu_{17,0}-\nu_{16,2}$
	Observed:	2496.5	134.5	3169.5	135.2	9.4
Standard		2481.86	134.12	3154.77	134.79	10.55
High Z		2490.39	134.31	3164.75	135.16	8.15
Low Z		2471.01	134.24	3142.49	134.90	11.30
Low X, Z = 0.01		2496.67	134.69	3170.17	135.57	11.22
Diffusive Mixed		2474.27	133.76	3147.84	134.71	13.44
Magnetic A		2499.22	135.85	3179.15	136.63	10.86
Magnetic B		2485.47	134.34	3159.77	134.96	10.39
Inner 0.05 $M_\odot$		2487.62	134.13	3159.55	134.34	17.54
S34 = 0.3 KeV - Barnes		2481.87	134.16	3154.68	134.81	10.00
Enhanced Opacity		2482.74	133.75	3155.23	134.49	9.63
No Scattering States		2484.60	134.22	3157.41	134.79	10.17

The behavior of the physical parameters at the base of the convective envelope for the different models is not intuitively obvious. This is because the envelope is adjusted to force the solar model radius to be 6.9625×10^{10} cm^{19} at an optical depth of 10^{-3} for normal incidence rays. Some of the quantities varied, such as the opacity, would cause the envelope depth to change if all other quantities were held constant. For example, the higher X in the envelope of the low X and low Z model would tend to increase the temperature and decrease the density at a fixed depth because of the lower mean molecular weight. However, the total volume available to the model is fixed so that the convective efficiency is increased in order to achieve a satisfactory model. It is the interplay between convective efficiency, average hydrogen abundance and the introduced model variations which causes the results to be difficult to deduce without the detailed calculations.

We see from Table 3 that the model with a discontinuity in hydrogen abundance as well as in heavy element abundance yields frequencies in good agreement with observation. Although the general pattern of abundance modification which this model represents is attractive on the basis of this result, we believe that it almost certainly is not unique. There are two parameters which are adjustable -- the initial X abundance gradient and the position within the sun where it occurs. Our incomplete results for other models not listed in Tables 2 and 3 indicate that these two parameters can be adjusted simultaneously so that the frequencies of the nearly radial modes are unchanged. We note, however, that this form of model is the only one we have yet found which appears capable of yielding frequencies in agreement with the observations.

Finally, we wish to emphasize the potential importance of the intermediate degree modes. For a fixed frequency, these have very similar eigenfunction structure in the outer parts of the sun, but as ℓ decreases their inner turning point penetrates progressively more deeply. Thus, the comparison of solar frequencies with the theoretical models at ℓ values between 0 and 100 can be used to probe solar structure in a selective fashion. The frequencies are dependent on the sound velocity as a function of depth. Figures 4 and 5 illustrate the similarity of the frequency changes as a function of ℓ to the sound velocity changes as a function of $r/R_\odot$. Clearly, there is great potential for using solar seismology to choose among the various nonstandard solar models.

This research was supported in part by N.S.F. Grant AST 80-19745 to U.C.L.A., by N.S.F. Grant ATM 80-09469 to U.S.C., and by NASA Grant NAGW-13 to U.S.C.

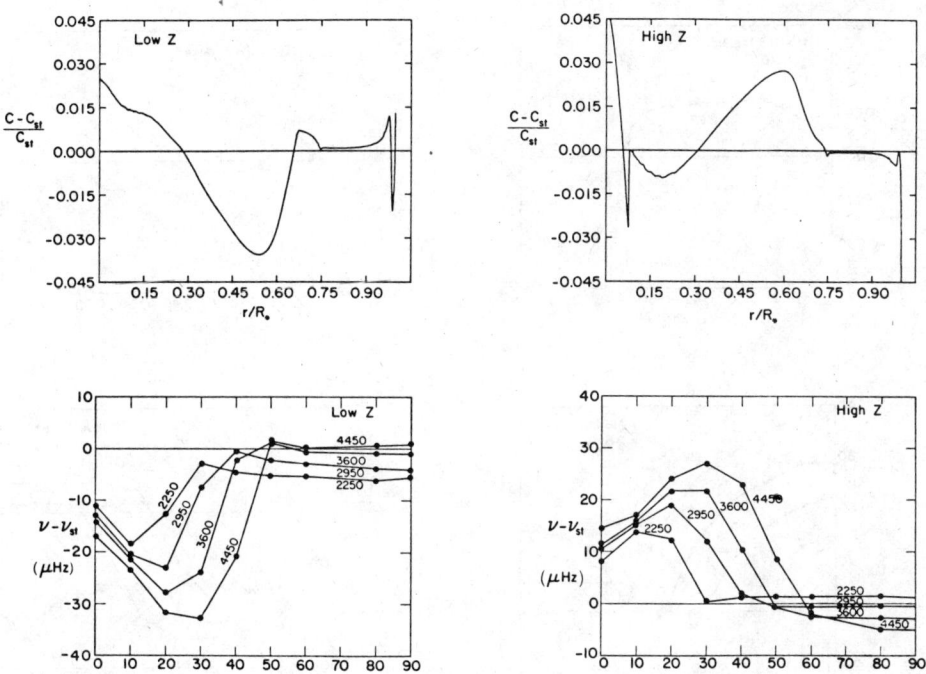

Figure 4. A comparison of the change in sound speed ($C - C_{st}$) to the change in eigenfrequency ($\nu - \nu_{st}$) induced by going from the standard model to two nonstandard models. Note the similarity in functional form for these two varibles. The nonstandard models in this figure are the low Z and high Z cases of Table 1.

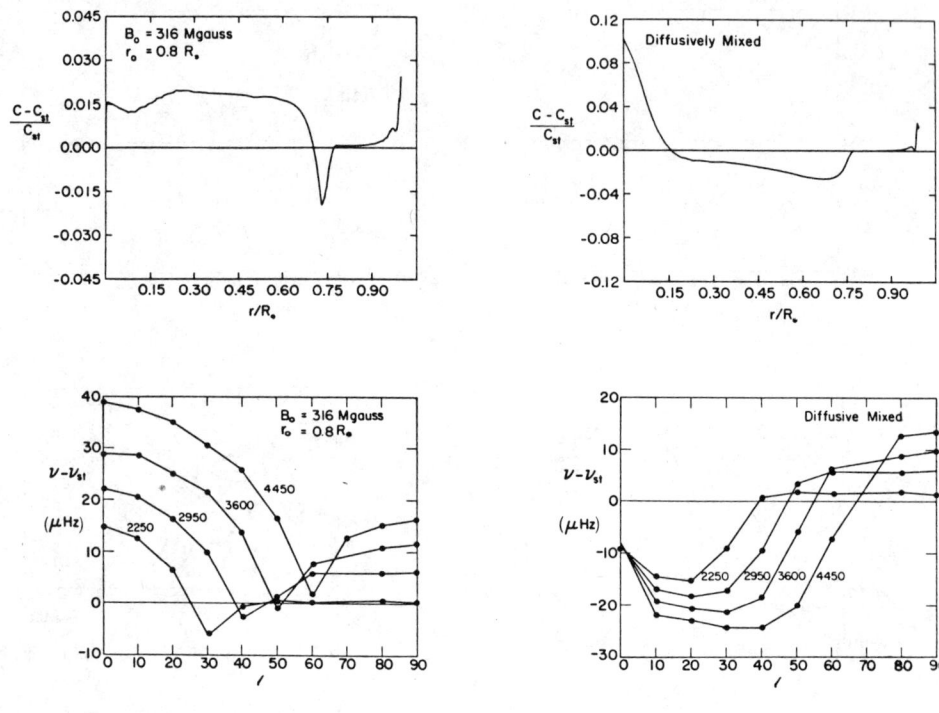

Figure 5. Like Figure 3 but for the magnetic A and diffusive mixed nonstandard cases.

References

1. R.K. Ulrich, Ap. J., 258, 404 (1982).
2. J.M. Strauss, J.B. Blake and D.N. Schramm, Ap. J., 204, 481 (1976).
3. R.H. Dicke and H.M. Goldenberg, Phys. Rev. Lett., 18, 313 (1967); Ap. J. Suppl., 27, 131 (1974).
4. R.H. Dicke, Proc. Natl. Acad. Sci. USA, 78, 1309 (1981).
5. H.A. Hill and R.T. Stebbins, Ap. J., 200, 471 (1975).
6. B.T. Cleveland, R. Davis, Jr., and J.K. Rowley, in "Weak Interactions as Probes of Unification," G.B. Collins, L.N. Chang and J.R. Frence, eds. (AIP Conf. Proceedings No. 72), p. 322 (1981); R. Davis, Jr., these proceedings.
7. J.N. Bahcall, W.T. Huebner, S.H. Lubow, P.D. Parker and R.K. Ulrich, Rev. Mod. Phys., 54, 767 (1982).
8. J.N. Bahcall, W.T. Huebner, S.H. Lubow, N.H. Magee, Jr., A.L. Merts, P.D. Parker, B. Rozsnyai, R.K. Ulrich and M. Argo, Phys. Rev. Lett., 45, 945.
9. J.N. Bahcall, N.A. Bahcall and G. Shaviv, Phys. Rev. Lett., 20, 1209 (1968); R. Davis, Jr., D.S. Harmer and K.C. Hoffman, Phys. Rev. Lett., 20, 1205 (1968).
10. R.B. Leighton, R.W. Noyes, and G.W. Simon, Ap. J., 135, 474 (1962); F.-L. Deubner, Astron. Astropdys., 44, 371 (1975); E.J. Rhodes, Jr., R.K. Ulrich and G.W. Simon, Ap. J., 218, 901 (1977); F.-L. Deubner, R.K. Ulrich and E.J. Rhodes, Jr., Astron. Astrophys., 72, 177 (1979); A. Claverie, G.R. Isaak, C.P. McLeod, H.B. van der Raay and T. Roca Cortes, Nature, 282, 591 (1979); 293, 443 (1981); Astron. Astrophys., 91, L9 (1980); G. Grec, E. Fossat and M. Pomerantz, Nature, 288, 541 (1980); Solar Physics, Dec., 1982; P. Scherrer, J. Wilcox, J. Christensen-Dalsgaard and D. Gough, Solar Physics, Dec., 1982.
11. R. Bos and H.A. Hill, Solar Physics, Dec., 1982.
12. T.L. Duvall Jr., and J.W. Harvey, submitted to Nature, Sept., 1982.
13. J. Christensen-Dalsgaard and D.O. Gough, Nature, 288 (1980); M.N.R.A.S., 198, 141 (1982); H. Shibahashi and Y. Osaki, Solar Physics, Dec., 1982; R. Scuflaire, M. Gabriel and A. Noels, Astron. Astrophys., 99, 39 (1981).
14. R.K. Ulrich and E.J. Rhodes, Jr., Ap. J., Feb., 1983.
15. E.J. Rhodes, Jr., and R.K. Ulrich, Proceedings of the Conference on Solar and Stellar Pulsations, June, 1982.
16. E. Schatzman and A. Maeder, Astron. Astrophys., 96, 1 (1981).
17. T.G. Cowling, M.N.R.A.S., 105, 166 (1945).
18. E.N. Parker, Astrophys. Space Sci., 31, 261 (1974).
19. J.E. Ross and L.H. Aller, Science, 191, 1223 (1976).
20. A. Wittmann, Astron. Astrophys., 61, 225 (1977).

THE CASE OF THE MISSING SOLAR NEUTRINOS[*]

William A. Fowler
Will Keith Kellogg Radiation Laboratory 106-38
California Institute of Technology, Pasadena, California 91125

ABSTRACT

Observations employing the Cl-Ar technique find only one-fourth the neutrino flux from the Sun at the earth expected on the basis of standard solar models and the rates of the nuclear reactions which fuse hydrogen into helium as measured in terrestrial laboratories. The current situation concerning nuclear reaction rates is discussed. It is concluded that the Ga-Ge technique which will detect the model and reaction rate independent flux of pp-neutrinos from the Sun should be put into full-scale operation as soon as possible. Other techniques which will probably require a National Underground Science Facility are discussed briefly.

INTRODUCTION

The Case of the Missing Solar Neutrinos is still unsolved.[1] It is appropriate to discuss the Case at this Workshop on Science Underground because it is eloquent testimony to the superb scientific detective work which can be — nay, must be — done deep underground, free from God- and man-made interference in the form of background effects at the surface of the earth.

The Case rivals the most celebrated detective cases of the 19th and 20th centuries from those involving Sherlock Holmes all the way to Perry Mason. The Chief Inspector on the Case, Raymond Davis, Jr., started it all in the 1950's when he developed a radiochemical technique for the detection of the neutrino, one of the jewels in the crown of elementary particle physics. In his detective work Inspector Davis used the nuclear reaction $^{37}Cl(\nu,e^-)^{37}Ar$ - 0.814 MeV which had been suggested independently by Pontecorvo and Alvarez.

He was one of the first to show that the antineutrino was not identical to the neutrino by discovering that reactor antineutrinos would not trigger the Chlorine-Argon transformation. For those of you who are science history buffs, Davis and John Bahcall have given an account of the development of the solar neutrino problem in Chapter 12 of that excellent (-Adv) volume Essays in Nuclear Astrophysics (1982).[2] As you all know, John Bahcall has been Superintendent of the Theoretical Detail almost from the beginning of the Case and you have just heard Roger Ulrich, his right-hand man, discuss this enthralling aspect of the problem.

I have been a detective on the case for one year short of twenty-five. In a way Al Cameron (1958)[3] and I (1958)[4] uncovered the real horror of the case when we independently deduced from the work of Holmgren and Johnson (1958, 1959)[5] on the copious rate of the

---

[*]Supported in part by the National Science Foundation [PHY79-23638].

$^3\text{He}(\alpha,\gamma)^7\text{Be}$ reaction that the subsequent $^7\text{Be}(p,\gamma)^8\text{B}$ reaction might well produce substantial quantities of ^8B in the Sun. It then followed that the relatively high energy neutrinos from the ^8B beta-decay should be readily detectable by the Davis-Pontecorvo-Alvarez scheme. I have used the word horror advisedly since in spite of herculean efforts in installing a 100,000 gallon tank of perchloro-ethylene ($C_2{}^{35}Cl_3{}^{37}Cl$) almost one mile deep in the Homestake Gold Mine in Lead, South Dakota and patiently milking it for the precious ^{37}Ar every several months since 1968, Davis (1982)[6] and his collaborators find only about one-fourth the neutrino flux from the Sun at the earth expected on the basis of standard solar models and the rates of the nuclear reactions which fuse hydrogen into helium as measured in terrestrial laboratories. Therein lies a fascinating aspect of the Case of the Missing Solar Neutrinos — along with the underground observations there are experiments at the accelerator and theoretical calculations at the computer on the earth's surface. Those who would build a National Underground Science Facility must not forget this three-pronged aspect of scientific detective work — observation, experiment, theory.

Let me say right at this point that many professionals and amateurs alike have come up with answers to the solar neutrino puzzle. I have suggested a few myself. Suffice it to say that all raise more problems than the one they attempt to solve. In spite of this, I think we can look forward to the future with great optimism. We know how to break the case. It will be done by the combined use of a number of additional detectors: ^2H which captures neutrinos with the production of two protons and an electron (all detectable in principle); ^7Li, ^{71}Ga, ^{79}Br, ^{81}Br, ^{97}Mo, and ^{98}Mo, all of which capture neutrinos with electron emission to form a product which then decays by electron capture; ^{115}In, which captures neutrinos with electron emission to an excited state in ^{115}Sn, which then γ-cascades with a delay of 3.3 µs to its ground state; and finally atomic electrons which scatter neutrinos and are then detected as they recoil.

The use of ^{71}Ga as a target to produce radioactive ^{71}Ge through $^{71}\text{Ga}(\nu,e^-)^{71}\text{Ge}$ - 0.236 MeV is essentially ready to go now.[6] This reaction will detect the model independent flux of neutrinos — 60 billion per cm^2 per sec at the earth — from the primary reaction in the Sun between two protons, $p + p \rightarrow d + e^+ + \nu + 0.420$ MeV, which occurs twice and thus produces two neutrinos as 26 MeV is produced in the overall hydrogen to helium conversion. From the Q-values it will be noted that almost one-half the pp-neutrinos are above the ^{71}Ga threshold. If these neutrinos are not detected we will know that the fusion of hydrogen into helium does not occur in the Sun through the pp-chain or, for that matter, through the CN-cycle which produces even more energetic neutrinos. The other alternative is that the electron-type neutrinos from hydrogen fusion transform into other particles, perhaps muon- or tauon-neutrinos, on their way to the earth. Today the search for neutrino transformations or oscillations is very active and the preliminary results very controversial but becoming more negative with time. Again this illustrates the close and complex connections between underground science and the rest of scientific endeavor.

The catch is that 50 tons of gallium are required and 50 tons of gallium cost $25 million dollars. In spite of the competition for the available dollars from other worthwhile scientific endeavors I strongly favor pushing ahead with the full-scale ^{71}Ga-observations. I pray the observations will be made in my lifetime. I don't want to go up above or down below and get the answer by private communication.

You will hear more about some of the other experiments today. I myself am partial to the In-Sn on-line experiment which will in principle be able to measure the energy spectrum of the pp-neutrinos above its threshold. I also like the electron-neutrino scattering because in principle it can measure directionality and establish the Sun as the source or not. The Ga-observations can probably be done at the Homestake Gold Mine in parallel with the Cl-observations before a National Facility comes into being, but I feel that the other experiments requiring underground sites will come along on a time scale of the few years needed to bring the National Facility into being.

THE NUCLEAR PHYSICS OF HYDROGEN-HELIUM FUSION IN THE SUN

The puzzling fraction of approximately one-fourth between observations and expectations arises in the application of our full range of physical and chemical knowledge to the Sun — atomic physics, gravity, hydrodynamics, nuclear physics, etc. etc. In the remainder of this report I will concentrate on the nuclear physics and the direct impact on this fraction of the accelerator experiments performed by my colleagues at the Will Keith Kellogg Radiation Laboratory at Caltech and at other laboratories throughout the world. Kellogg was built for Charles C. Lauritsen using $100,000 which Robert A. Millikan obtained from the corn flakes king (-Adv). For fifty years the laboratory has devoted a substantial fraction of its experimental activity to the study of the nuclear processes which take place in the big bang, Sun, other stars, novae and supernovae.

First we must look at the nuclear processes we think are taking place in energy generation in the Sun via hydrogen to helium fusion. We are convinced from the relevant cross-section measurements that the pp-chain dominates the fusion over the CNO-cycle ($< 5\%$). Furthermore, the CNO-cycle gives a model and reaction rate independent neutrino flux more than 15 times that observed. Thus the reactions of interest are those given in Table I. It will be noted that the pp-chain is really a triple chain with three modes for the final production of ^{4}He after ^{3}He is produced. Inspection of the table will indicate that the production of the neutrinos from electron-capture on ^{7}Be depends directly on the competition between the ^{3}He$(\alpha,\gamma)^7$Be-reaction and the ^{3}He$(^3$He,2p$)^4$He-reaction. The ^{7}Be-neutrinos are detectable in the ^{37}Cl-^{37}Ar observations while, it must be remembered, the pp-neutrinos are not. Further inspection of the table will indicate that the production of the neutrinos from the positron-decay of ^{8}B also depends on this competition but in addition on the competition between the capture of electrons by ^{7}Be via ^{7}Be$(e^-,\nu)^7$Li and the capture of protons via ^{7}Be$(p,\gamma)^8$B. The ^{8}B-neutrinos are rare but their large energies in a typical beta-decay spectrum up to ~ 14 MeV lead to large detection cross sections and are the main type ($\sim 80\%$) contributing to the expected flux detectable via ^{37}Cl-^{37}Ar.

Table I The pp tri-chain

^{1}H + ^{1}H $\rightarrow$ ^{2}D + e^+ + ν	^{7}Be + e^- $\rightarrow$ ^{7}Li + ν
^{1}H + e^- + ^{1}H $\rightarrow$ ^{2}D + ν (pep)	^{7}Li + ^{1}H $\rightarrow$ ^{4}He + ^{4}He
^{2}D + ^{1}H $\rightarrow$ ^{3}He + γ	OR
^{3}He + ^{3}He $\rightarrow$ ^{4}He + ^{1}H + ^{1}H	^{7}Be + ^{1}H $\rightarrow$ ^{8}B + γ
OR	^{8}B $\rightarrow$ ^{8}Be* + e^+ + ν
^{3}He + ^{4}He $\rightarrow$ ^{7}Be + γ	^{8}Be* $\rightarrow$ ^{4}He + ^{4}He

The point which must be stressed is that the absolute values of the cross sections and thus the reaction rates involved in these competitions cannot be calculated accurately with current theoretical nuclear models and must be measured in the laboratory. The ^{7}Be(e^-,ν)^{7}Li-rate in the Sun must be calculated from the measured terrestrial rate of the ^{7}Be-decay and even this calculation is somewhat dependent (not significantly according to consensus of the experts) on the central density and temperature determined from solar models.

The five independent experimental studies of the competing reaction, ^{7}Be(p,γ)^{8}B, are summarized in Bahcall et al. (1982).[1] These authors conclude that there is a 12% uncertainty (1σ) in the rate of this reaction so it is a candidate for further study and, in fact, the Argonne National Laboratory is well on the way to obtaining new results. For me it is another indication of the continuing vitality of classical nuclear physics and its applications to nuclear astrophysics. It must be borne in mind that the measurements of small low energy cross sections are extraordinarily difficult and new developments in accelerators and detectors lead to the possibility of exciting new measurements of increased sophistication and precision. In Kellogg we are now involved in the final acceptance tests of a new high resolution, high current 3 MV tandem electrostatic accelerator which will not only enormously extend our capabilities in nuclear astrophysics but permit a broad spectrum of other experiments including, for example, a highly sensitive search for free quarks.

The greatest excitement and controversy resides in the recent measurements of the cross section of the ^{3}He(α,γ)^{7}Be-reaction against a background extending back to 1958.[5] We note that the results of such measurements are traditionally expressed in terms of the so-called cross-section factor at zero energy, $S(0)$, obtained by dividing the measured cross sections by the explicit Gamow factor in the Coulomb-barrier penetration function to obtain a slowly varying $S(E)$

and then extrapolating to zero energy either by a parametrized fit to the data or theoretically. Of course, the Gamow factor is put back in before calculating the final reaction rate. It must be remembered that the action in the Sun occurs in the energy range of a few tens of keV-not far from zero energy. It is generally believed that $^3\text{He}(\alpha,\gamma)^7\text{Be}$ is a direct-capture reaction and all calculations and measurements of the energy dependence closely agree with the analytical results of Williams and Koonin (1981)[7] who found $S(E) = S(0) \exp(-aE)$ with $a = +0.575$ MeV^{-1} and a very small higher order term in E^2.

To make a long story short, we can summarize the current situation as discussed in detail in Bahcall et al. (1982)[1] and add the results of an experiment just completed at the Los Alamos National Laboratory (1982).[8] Parker and Kavanagh (1963)[9] found $S_{34}(0) = 0.47 \pm 0.05$ keV barns while Nagatani, Dwarakanath, and Ashery (1969)[10] found $S_{34}(0) = 0.57 \pm 0.06$ keV barns. These values are the results of a reanalysis of the old data by Parker in Bahcall et al. (1982)[1] which changed the original results by less than 10%. The excitement and controversy started when Rolfs (1979)[11] reported that preliminary results from Münster had indicated a nearly flat energy dependence for $S(E)$ and a value $S_{34}(0) \approx 0.3$ keV barns on normalization of these preliminary results to the measurements of Parker and Kavanagh (1963)[9] near 1.5 MeV. More recently the Münster group has reported in Kräwinkel et al. (1982)[12] that $S(E)$ does decrease with energy but that their absolute calibration of the gamma-ray yield extrapolates to $S_{34}(0) = 0.30 \pm 0.03$ keV barns. Stimulated by the preliminary Münster results, groups in Kellogg and Los Alamos National Laboratory returned to the accelerator. In Kellogg, Osborne et al. (1982)[13] found $S_{34}(0) = 0.51 \pm 0.03$ keV barns from the gamma-ray yield and $S_{34}(0) = 0.56 \pm 0.03$ keV barns from the ^{7}Be activity. Since the ^{7}Be-measurements were not as extensive as the gamma-ray measurements, Osborne et al. (1982)[13] gave a weighted average $S_{34}(0) = 0.52 \pm 0.03$ keV barns. At LANL, Robertson et al. (1982)[8] found $S_{34}(0) = 0.63 \pm 0.04$ keV barns from the ^{7}Be activity. In both of the activation experiments, which detect the cascade gamma-ray from the capture to the first excited state of ^{7}Li at 0.478 MeV, the branching ratio to this state was taken as $10.4 \pm 0.1\%$ from the review analysis by Ajzenberg-Selove (1979).[14] Because of the importance of this branching ratio in the solar neutrino problem and because of its application in many calibrations of neutron flux using $^7\text{Li}(p,n)^7\text{Be}$, where the gamma-ray from $^7\text{Be}(e^-,\nu)^7\text{Li}^*(\gamma)^7\text{Li}$ is counted to give a standardized neutron source, Kellogg and several other laboratories are remeasuring the branching ratios with the superior techniques and detectors which have become available since the early measurements; once again, nuclear astrophysics is active and exciting.

$^3\text{He}(^3\text{He},2p)^4\text{He}$ has been measured down to 30 keV in the laboratory, less than 50% above the effective operating solar energy. The accepted[1] cross-section factor is $S_{33}(0) = 4.7 \pm 0.3(1\sigma)$ MeV barns which is approximately $10^4 \times S_{34}(0)$. However, the $^3\text{He}(\alpha,\gamma)^7\text{Be}$ to $^3\text{He}(^3\text{He},2p)^4\text{He}$ reaction-rate ratio is approximately 0.14 in the Sun because ^{4}He is approximately 10^3 as abundant as ^{3}He. It is the ratio of $n_4 S_{34}(0)$ to $n_3 S_{33}(0)$, where the n's are number densities, which counts and n_4/n_3 is model dependent.

THE SOLAR-NEUTRINO FLUX

The detectable solar-neutrino flux is conventionally expressed in terms of the solar-neutrino unit, the SNU. The number of SNU's is just the total expected events triggered per second by the neutrinos in 10^{36} target nuclei. A flux of 10^6 neutrinos cm^{-2} sec^{-1} at the earth with a detection cross section of 10^{-42} cm^2 yields one SNU.

We conclude by summarizing in Table II, after listing the most recent ^{3}He$(\alpha,\gamma)^7$Be measurements, the expectations[1] and the observations[6] on the solar-neutrino flux as expressed in SNU. We note that none of the measurements on $S_{34} \equiv S_{34}(0)$ are consistent within their 1σ uncertainties but that all differ from 7σ to 14σ from the value necessary[1] to yield the SNU observations by Davis (1982).[6] The Case of the Missing Solar Neutrinos is still unsolved by nuclear detective work!

Table II Solar neutrino yields in SNU (10^{-36} sec^{-1} per ^{37}Cl target)
Theoretical SNU calculations: Bahcall et al. RMP 54, 767 (1982).
All 1σ errors

^{3}He$(\alpha,\gamma)^7$Be Rate	SNU
CALTECH (Osborne et al. 1982)[13] $S_{34} = 0.52 \pm 0.03$ keV-b (γ and ^{7}Be-decay)	7.6 ± 1.1*
MÜNSTER (Kräwinkel et al. 1982)[12] $S_{34} = 0.30 \pm 0.03$ keV-b (γ-only)	5.0 ± 0.7*
LANL and MSU (Robertson et al. 1982)[8] $S_{34} = 0.63 \pm 0.04$ keV-b (^{7}Be-decay only)	$8.8^\dagger \pm 1.3$*
OBSERVED (Davis 1982)[6] REQUIRES (1982)[1]: $S_{34} = 0.09$ keV-b 7σ from Münster / 14σ from Caltech / 14σ from LANL/MSU	1.8 ± 0.3
*USED $1\sigma(S_{34}) = \pm 0.05$ keV-b. DECREASE BY $\sim 10\%$ FOR ± 0.03 keV-b. †EXTRAPOLATED BY PRESENT AUTHOR.	

If the early measurements cited above are included there are five measured values for $S_{34}(0)$, namely, 0.30 ± 0.03, 0.47 ± 0.05, 0.52 ± 0.03, 0.57 ± 0.06, and 0.63 ± 0.04 keV-barns. If the first listed value is excluded the average value is 0.558 keV-barns with an internal error of 0.037 and an external error of 0.033. This yields 8.0 SNU. If the first listed value is included the average value is 0.533 keV-barns with an internal error of 0.035 and an external error of 0.050. This yields 7.8 SNU. The average values were computed by weighting each measurement inversely as the square of its fractional error. There is little comfort in any of it. It is imperative that a technique which will detect the model independent pp-neutrinos be put in operation as soon as possible. The Ga-Ge technique is ready to go. I strongly urge prompt full-scale implementation.

For the record the stellar reaction rate for $^3\text{He}(\alpha,\gamma)^7\text{Be}$ can be best expressed as

$$N_A \langle \sigma v \rangle_{34} = 1.038 \times 10^7 \times S_{34}(0) \times T_{9A}^{5/6} T_9^{-3/2}$$

$$\times \exp\left(-12.826\, T_{9A}^{-1/3}\right) \text{ cm}^3 \text{ mole}^{-1} \text{ sec}^{-1}$$

where

$$T_{9A} = T_9(1 + akT_9)^{-1} = T_9(1 + 0.050\, T_9)^{-1}$$

with $T_9 = T/10^9 K$, $a = 0.575$ MeV^{-1} and $k = 0.08617$ MeV/10^9K. The reader may make his own choice for $S_{34}(0)$ which must be in keV-barns. My choice is 0.558 ± 0.037 keV-barns which yields a numerical coefficient in $N_A \langle \sigma v \rangle_{34}$ above equal to 5.79×10^6. It is appropriate to end this piece by citing the reaction-rate of a key nuclear process in emanations from the Sun.

REFERENCES

1. J. N. Bahcall, W. F. Huebner, S. H. Lubow, P. D. Parker, and R. K. Ulrich, Rev. Mod. Phys. 54, 767 (1982).
2. J. N. Bahcall and R. Davis, Jr., Essays in Nuclear Astrophysics, eds. C. A. Barnes, D. D. Clayton, and D. N. Schramm (Cambridge University Press, New York, 1982), p. 243.
3. A. G. W. Cameron, Ann. Rev. Nucl. Sci. 8, 249 (1958).
4. W. A. Fowler, Astrophys. J. 127, 551 (1958).
5. H. P. Holmgren and R. Johnson, Bull. Am. Phys. Soc. II, 3, 26 (1958); Phys. Rev. 113, 1556 (1959).
6. R. Davis, Jr., Proceedings of the Workshop on Science Underground at LANL (1982).
7. R. D. Williams and S. E. Koonin, Phys. Rev. C23, 2773 (1981).
8. R. G. H. Robertson, P. Dyer, T. J. Bowles, R. E. Brown, N. Jarmie, C. J. Maggiore, and S. M. Austin (1982 preprint submitted to Phys. Rev. C).
9. P. D. Parker and R. W. Kavanagh, Phys. Rev. 131, 2578 (1963).
10. K. Nagatani, M. R. Dwarakanath, and D. Ashery, Nucl. Phys. A128, 325 (1969).

11. C. Rolfs, Proceedings of the Bethe 40th Anniversary Symposium at BNL (1979).
12. H. Kräwinkel et al., Z. Phys. A304, 307 (1982).
13. J. L. Osborne et al., Phys. Rev. Lett. 48, 1664 (1982) and private communication.
14. F. Ajzenberg-Selove, Nucl. Phys. A320, 1 (1979).
15. B. W. Filippone, A. J. Elwyn, C. N. Davids, and D. D. Koetke, submitted for publication (1983).

Note added in proof. The rate of the ^{7}Be(p,γ)^{8}B reaction has been recently remeasured by Filippone et al. (1983)[15] and their new value averaged with previous values to yield $S_{17}(0) = 0.0238 \pm 0.0023$ keV-b. With the new value for S_{34} described above this changes the expected ^{37}Cl counting rate from 7.6 ± 1.1 SNU in Table II to 6.9 ± 1.0 SNU.

THE GALLIUM SOLAR NEUTRINO DETECTOR

W. Hampel
Max-Planck-Institut für Kernphysik
Heidelberg, Fed. Rep. Germany

Gallium Solar Neutrino Collaboration[+]

ABSTRACT

This paper reports on the present status of the Gallium Solar Neutrino Experiment performed by an international collaboration. The scientific goals of the experiment and the general experimental approach are described first, then details of the chemical procedure, the counting technique and the influence of background effects are discussed, and finally a brief outline of the future plans is given.

INTRODUCTION

Solar neutrino experiments are among those difficult experiments which aim at the detection of very rare events, and like most of these (e.g. proton decay or double beta decay experiments) they have to be operated in a protected environment in order to shield the detector from all external radiation which could mask the solar neutrino signal. So far, the only experiment which actually has been set up at full scale to measure the solar neutrino flux is the Brookhaven Chlorine Experiment by Ray Davis and his collaborators[1]. This detector, located about 1 mile underground in the Homestake Gold Mine in South Dakota, has now been in operation for about 15 years. It is well known that the measured neutrino capture rate, 1.8 ± 0.3 SNU (1 SNU = 10^{-36} captures per second and target atom), is about a factor of 4 lower than the theoretical prediction from the so-called Standard Solar Model, 7.6 SNU[2,3]. This discrepancy (called the solar neutrino problem) has been discussed in detail at this workshop by W.A. Fowler[4] (see also Refs. [2,5-7]).

However, the overwhelming part of the expected signal in the Cl detector is due to the high energy ^{8}B neutrinos generated in a rare side branch of the proton-proton fusion chain. Their flux is strongly

---

[+] R. Davis Jr., B.T. Cleveland, G. Friedlander, S. Katcoff, L.P. Remsberg, J.K. Rowley, J. Weneser (Brookhaven National Laboratory);
T. Kirsten, W. Hampel, G. Heusser, M. Hübner, J. Kiko, E. Pernicka, R. Schlotz, R. Wunderlich (Max-Planck-Institut für Kernphysik, Heidelberg);
I. Dostrovsky (Weizmann Institute of Science, Rehovot);
J.N. Bahcall (Institute for Advanced Study, Princeton);
K. Lande, R.I. Steinberg (Univ. of Pennsylvania, Philadelphia)

dependent on details of the solar model. On the other hand, the flux of the bulk part of solar neutrinos, the low energy pp neutrinos, is almost independent of uncertainties in the solar model calculations as long as hydrogen burning is assumed to be the principal energy source of the sun. There is therefore general agreement that an experiment capable of observing these pp neutrinos could solve the problems raised by the outcome of the Cl experiment.

This has been the main motivation for our collaboration to start work on a radiochemical solar neutrino detector based on gallium about three years ago. In this article we shall report on the present status of this experiment.

BASIC PRINCIPLES OF THE Ga DETECTOR

The gallium detector is based upon the neutrino capture reaction $^{71}Ga(\nu_e,e^-)^{71}Ge$ (see Fig. 1). The energy threshold (236 keV) is well below the maximum energy of the pp neutrinos. ^{71}Ge decays back to ^{71}Ga by electron capture. Table I lists the neutrino fluxes for the different sources in the sun (according to the standard model)[2], the ^{71}Ga neutrino capture cross sections (averaged over the energy for the continuum sources)[3] and the resulting capture rates. The cross section values have been increased by 3% in order to allow for the recent redetermination of the ^{71}Ge halflife (11.43 days)[8]. The total capture rate of 106 SNU is dominated by about 70 SNU from the pp and pep neutrinos.

It should be noted that the cross section values of Table I and thus the given capture rates apply only for transitions to the ^{71}Ge ground state. The more energetic neutrino sources can in principle populate levels at 175, 500 and 708 keV excitation energy in ^{71}Ge. The matrix elements corresponding to these transitions are not known; however, a conservative estimate of their maximum possible contribution to the total capture rate yields only about 9 SNU[2,3]. Experimental information on the contribution of these three excited states will be obtained from measurements of the $^{71}Ga(p,n)$ forward scattering differential cross sections presently carried out by C.D. Goodman and co-workers[9] and from a calibration experiment with the gallium detector using an artificial neutrino source (see below).

The basic experimental procedure of the gallium solar neutrino detector is as follows. In order to obtain 1 capture per day from pp and pep neutrinos about 50 tons of gallium in the form of a $GaCl_3$-HCl solution have to be exposed to solar neutrinos.

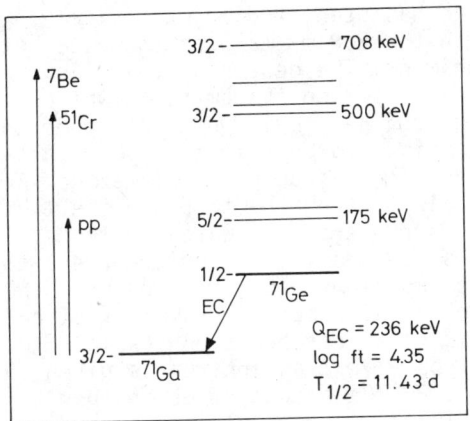

Fig. 1. ^{71}Ga-^{71}Ge energy levels.

Table I. Solar neutrino fluxes[2], cross sections[3] and capture rates for $^{71}Ga(\nu_e,e^-)^{71}Ge$.

Neutrino Sources and Energies [MeV]	Flux on Earth [cm^{-2}sec^{-1}]	Cross Section [cm^2]	Capture Rate [SNU]
p+p (0-0.42)	6.07x10^{10}	1.11x10^{-45}	67.2
p+e$^-$+p (1.44)	1.50x10^8	1.62x10^{-44}	2.4
^{7}Be decay (0.86)	4.3 x10^9	6.62x10^{-45}	28.5
^{8}B decay(0-14.06)	5.6 x10^6	$\sim$ 3 x10^{-43}	1.7
^{13}N decay(0-1.19)	5.0 x10^8	5.5 x10^{-45}	2.7
^{15}O decay(0-1.74)	4.0 x10^8	9.5 x10^{-45}	3.8
			106

In such a medium the neutrino induced ^{71}Ge atoms (as well as the inactive Ge carrier atoms which have been added in the beginning of the run) form the volatile compound germanium tetrachloride (GeCl$_4$) which at the end of an exposure (10 to 30 days) is simply swept out of the solution by bubbling He, Ar or air through it. The gas stream is passed through water where the GeCl$_4$ is absorbed. The germanium is then reduced to the gas germane (GeH$_4$) which is introduced into a small proportional counter. The number of ^{71}Ge atoms in the counter is finally determined by observing their radioactive decay.

A pilot experiment with 1.26 metric tons of gallium (4.65 tons of GaCl$_3$-HCl solution) has been successfully completed. It was performed in order to test the entire chemical procedure on a large scale, to develop a counting system for low-level counting of ^{71}Ge, and to investigate background effects in the Ga detector (production of ^{71}Ge by sources other than solar neutrinos). The results of these studies will be discussed in the following.

CHEMICAL PROCEDURE

The germanium extraction procedure has been studied extensively with the 2.8 m³ tank of the pilot experiment. The design of this tank and the extraction system is shown in Fig. 2. The liquid pump on top of the tank is used to stir the GaCl$_3$ solution. The gas pump circulates He, Ar or air through the tank and the two gas scrubbers. The water in the scrubbers is circulated countercurrent to the gas flow. GeCl$_4$ removed from the tank is absorbed in the water of one of the two scrubbers. An example for a typical extraction run is given in Fig. 3. The amount of Ge in the GaCl$_3$ solution is plotted versus the extraction time. More than 99% of the

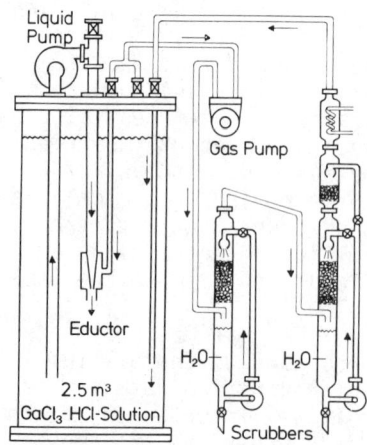

Fig. 2. Design of the pilot tank and the Ge extraction system.

Table II. Germanium extraction yields.

Ge carrier (mg)	Number of runs	Ge recovered (%)	
		Range	Average
> 2	11	94 - 101	97.5 ± 2.0
0.54	6	94 - 99	96.6 ± 2.1
0.11	4	98 - 101	99.0 ± 1.4

Ge introduced into the tank before the sweep (about 2 mg) were extracted from the solution and recovered in the scrubbers after a 28 hour run. Table II summarizes the results of 21 individual runs. It turns out that on the average the Ge extraction yields are 97% even when the amount of carrier is as small as 110 micrograms.

The second step in the chemical procedure is the conversion of the $GeCl_4$ to GeH_4. The $GeCl_4$ is first extracted from the scrubber solution into CCl_4, then back extracted into tritium-free water, and finally reduced to GeH_4 with potassium borohydride (KBH_4). The average efficiency for this conversion step is 97%. Thus it can be concluded that the entire extraction and conversion procedure can be performed with a ~95% overall yield.

COUNTING

The counting system must be able to detect ^{71}Ge decays at a rate of less than 1 decay per day with high efficiency. This can best be achieved in miniaturized proportional counters with ^{71}Ge being present as GeH_4 in the gas phase. ^{71}Ge decays by K(88%), L(10%) and M(2%) electron capture. Neglecting the M events, this results in an energy spectrum with an L peak at 1.2 keV and a K peak at 10.4 keV (see Fig. 4).

For the counter construction only ultrapure materials are used. The measurements are performed inside an anticoincidence shield (plastic scintillator plates) surrounded by 20 cm of lead and iron. The proportional counters (up to 8) are placed into the well of a well-type NaI pair spectrometer. The NaI serves as an additional anticoincidence in ^{71}Ge counting. On the other hand, a coincidence between a proportional counter pulse and a NaI event is applied for the identification of ^{69}Ge decays. ^{69}Ge (1.6 days) can serve as a monitor for background reactions in the Ga tank (see below).

For further background reduction we must be able to distinguish ^{71}Ge decays from background due to Compton electrons or electronic noise. This can be achieved by pulse shape analysis of the preamplifier output pulse.

Fig. 3. Example of Ge extraction from the pilot tank.

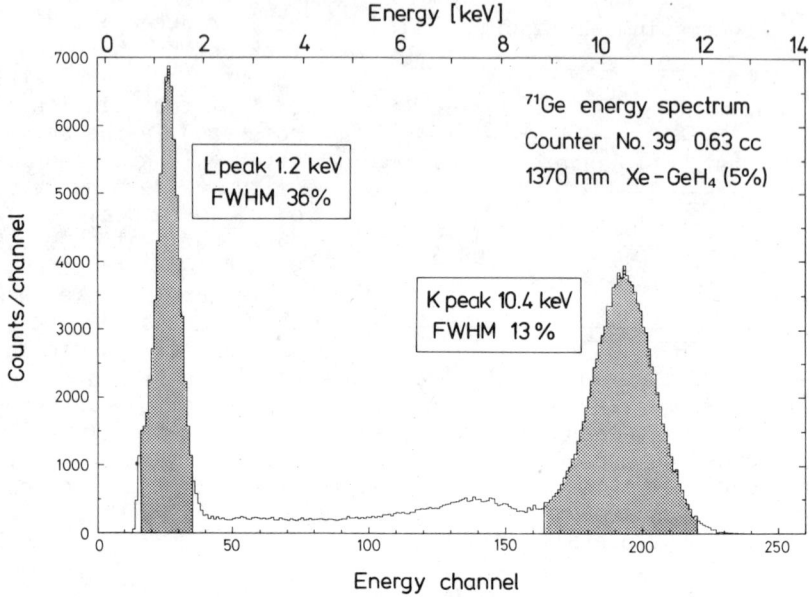

Fig. 4. ^{71}Ge energy spectrum in a 0.63 cc proportional counter.

Therefore the whole pulse shape of each proportional counter event is recorded by means of a transient digitizer. Fig. 5 shows a block diagram of the computer-controlled counting configuration. The data of all detectors (plastic scintillator, NaI, proportional counters) are transferred to the PDP 11/23 computer where an on-line evaluation of the data is performed. However, all data are finally stored on magnetic tape without any preselection.

In order to judge the background count rates reached so far, we have performed a series of Monte Carlo calculations simulating a full scale Ga experiment (50 single exposures within 2 years, 1 neutrino capture per day, chemical yield 95%). Fig. 6 is based on these calculations. The error of the final result is plotted as a function of the background count rate and the counting efficiency. We have carried out a long series of test and background measurements in order to find the best counting conditions (counter size, gas composition, gas pressure). The best results have been obtained with Xe-GeH$_4$ mixtures in 0.6 cc counters (see Fig. 4). The background

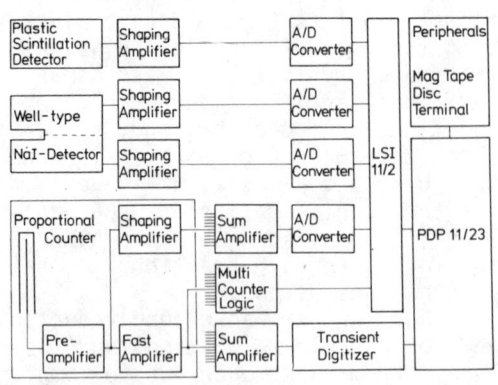

Fig. 5. Block diagram of the counting system.

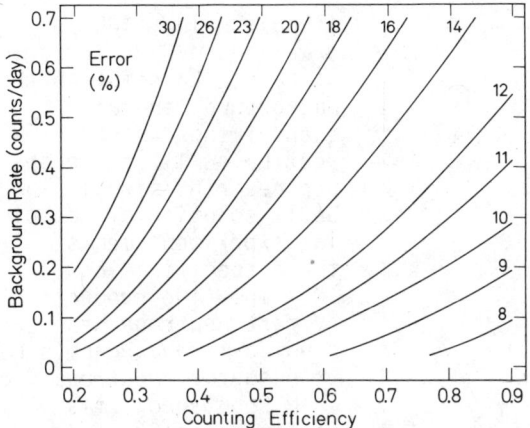

Fig. 6. Monte Carlo calculation: Error of the final result as a function of counting efficiency and background rate.

rate in the K peak region is 0.15 counts per day, while the counting efficiency is 42%. These values would allow a measurement of the solar neutrino signal to within 15% (Fig. 6). It should be noted, however, that we have used so far only the amplitude of the differentiated pulse (ADP) as background rejection criterion. There is hope that a pulse shape analysis by more sophisticated methods (e.g. Fourier analysis) will improve the results. In addition, work is in progress to reduce the L peak background (at present 0.5-1 counts per day) so that the L peak can be included in the analysis of the counting data.

BACKGROUND REACTIONS

Radiochemical solar neutrino detectors provide no direct way to distinguish the neutrino signal from the signal due to background processes. It is therefore essential to examine carefully all possible background effects which also produce ^{71}Ge from Ga. The most serious source of background is the ^{71}Ga(p,n) reaction, the protons being generated in the GaCl$_3$ solution as secondaries by cosmic ray muon interactions, (α,p) or (n,p) reactions.

The depth dependence of the muon background for the gallium detector (plotted in Fig. 7) has been derived from measurements and calculations on the same effect for the chlorine detector and from the measured cross section ratio $\sigma(^{71}$Ge from GaCl$_3)/\sigma(^{37}$Ar from C$_2$Cl$_4)$ for 225 GeV muons[10,11]. The data point at 4400 hg/cm^2 corresponds to the result obtained with the full scale Cl detector (the excess above the extrapolated curve is the signal attributed to solar neutrinos). It follows that the muon background in the Ga detector is less than 1% of the signal expected from pp and pep neutrinos if the experiment would be placed at a depth similar to that of the Cl detector.

At shallow depth the cosmic ray production of ^{71}Ge is dominated by the nuclear active component. The cosmic ray signal actually measured in the Ga pilot tank (shielding depth about 4 hg/cm^2) is in agreement with the sealevel production of ^{37}Ar from C$_2$Cl$_4$[11] (assumed mean attenuation length for the nuclear active component 1.5 hg/cm^2).

The yields for ^{71}Ge production from alpha particles have been measured as a function of the α energy. It follows for the content of α emitting impurities in the GaCl$_3$ solution that with upper limits of

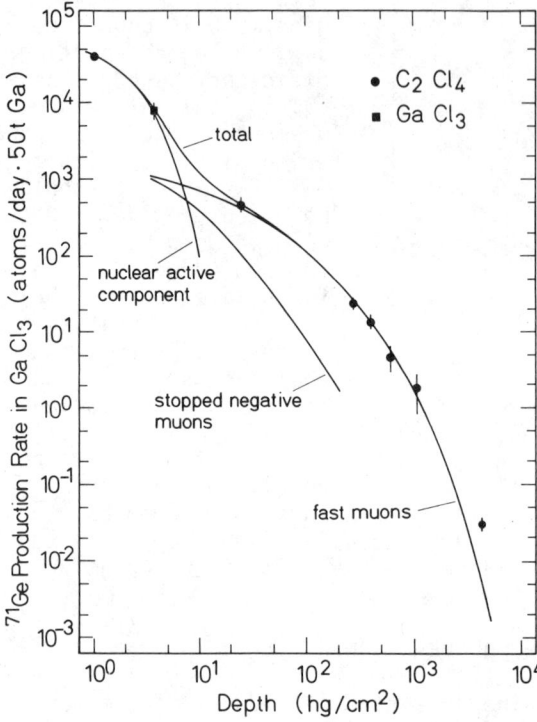

Fig. 7. Cosmic ray production of ^{71}Ge in GaCl$_3$ versus shielding depth.

1 ppb thorium, 100 ppb uranium and 0.2 picocurie ^{226}Ra per kg solution the background generated by each of these sources is kept below 1% of the pp+ pep neutrino signal. The GaCl$_3$ solution of the pilot experiment meets these specifications.

The ^{71}Ge production by fast neutrons has been studied with a Pu-Be neutron source. Neutrons from ^{238}U spontaneous fission in the Ga solution are unimportant at U levels below 100 ppb. Fast neutrons from the surrounding rock at the detector site produce $\sim$0.01 atoms/day in a 50 ton Ga detector if the U content of the rock is 10 ppb. However, the fast neutron flux from the rocks can simply be reduced to the required level by application of a water shield.

In any event, the Ga detector provides a built-in monitor for all (p,n) induced background since natural gallium has two isotopes, ^{69}Ga(60.1%) and ^{71}Ga (39.9%). Any source of such a background will therefore also produce ^{69}Ge by the ^{69}Ga(p,n) reaction at a rate which is higher than the corresponding ^{71}Ge rate for all processes discussed above. As mentioned earlier, our counting system is designed for the simultaneous measurement of both ^{71}Ge and ^{69}Ge.

Finally, it should be noted that muon and electron neutrinos produced by cosmic ray interactions in the earth's atmosphere could in principle contribute to the ^{71}Ge background rate. However, an estimate of this effect for the Ga detector results in a production rate corresponding to 0.03 SNU[12] which is negligible compared to the solar neutrino signal.

FUTURE PLANS

The outcome of the pilot experiment has demonstrated that a full scale gallium experiment can be performed, especially in view of the fact that it is planned on a modular basis. We therefore anticipate no serious problems from the upscaling. The time schedule for this full scale experiment depends essentially on the progress in gallium acquisition. In the meantime, we will perform two important further tests.

A tank containing 1.7 tons of gallium will be set up at a location with intermediate shielding depth in order to measure the muon background and to verify the curve in Fig. 7.

The second experiment is a test of the entire Ga detector by means of an artificial ^{51}Cr neutrino source. ^{51}Cr (27.8 days) emits monoenergetic neutrinos of 751 keV (90%) and 432 keV (10%). Such a test provides a measurement of the ^{71}Ga neutrino capture cross section from which information on the contribution of the 175 and 500 keV levels in ^{71}Ge to the total cross section can be obtained (see Fig. 1).

The source will be produced by neutron irradiation of chromium in the High Flux Isotope Reactor (HFIR) at Oak Ridge National Laboratory. The expected ^{71}Ge production rate in an experiment with 7 tanks containing a total of 12.8 tons of gallium is 4 atoms per day per Megacurie of ^{51}Cr.

At present we have 2 tons of gallium on hand. However, with a gallium amount considerably less than 12.8 tons the source experiment is sensible only if a source appreciably stronger than 1MCi could be provided. Test irradiations in the HFIR revealed that about 1MCi of ^{51}Cr can be produced in that reactor. An additional 0.4 MCi may be obtainable from exposures in the Oak Ridge Research Reactor. It therefore is necessary to acquire more gallium before the source experiment can get started.

REFERENCES

1. R. Davis Jr., Underground Science: A Personal View, Workshop on Science Underground, Los Alamos National Laboratory, September 27-October 1, 1982.
   J.N. Bahcall and R. Davis Jr., Science 191, 264 (1976).
2. J.N. Bahcall, W.F. Huebner, S.H. Lubow, P.D. Parker and R.K. Ulrich, Rev.Mod.Phys. 54, 767 (1982).
3. J.N. Bahcall, Rev.Mod.Phys. 50, 881(1978).
4. W.A. Fowler, The Case of the Missing Solar Neutrinos, Workshop on Science Underground, Los Alamos National Laboratory, September 27-October 1, 1982.
5. W. Hampel, Proc.Int.Conf. Neutrino Physics and Astrophysics 1980, Erice (Italy), E.Fiorini, Editor, Plenum Publishing Corporation, New York, p. 61 (1981).
6. V. Barger, K. Whisnant, D. Cline and R.J.N. Phillips, Phys.Rev. Lett. 93B, 194 (1980).
7. J.N. Bahcall, Proc.Int.Conf. Neutrino Physics and Astrophysics 1981, Maui, Hawaii (USA), R.J. Cence, E. Ma, A. Roberts, Editors, Univ. of Hawaii, Honolulu, Vol. II, p. 253 (1982).
8. W. Hampel and L.P. Remsberg, to be published.
9. C.D. Goodman, C.A. Goulding, M.B. Greenfield, J. Rapaport, D.E. Bainum, C.C. Foster, W.G. Love and F. Petrovich, Phys.Rev. Lett. 44, 1755 (1980).
10. A.W. Wolfendale, E.C.M. Young and R. Davis Jr., Nature Phys.Sci. 238, 130 (1972).
11. R. Davis Jr., private communication.
12. M.A. Rudzskij and Z.F. Seidov, Sov.J.Nucl.Phys. 30 (4),553 (1979).

A RADIOCHEMICAL SOLAR NEUTRINO EXPERIMENT
USING $^{81}Br(\nu,e^-)^{81}Kr$

G. S. Hurst, C. H. Chen, S. D. Kramer, and M. G. Payne
Chemical Physics Section, Health and Safety Research Division
Oak Ridge National Laboratory, Oak Ridge, TN. 37830

R. D. Willis**
Scripps Institution of Oceanography
University of Caifornia at San Diego
La Jolla, CA 92093

ABSTRACT

Both geochemical and radiochemical experiments based on the interaction $^{81}Br(\nu,e^-)^{81}Kr$ to detect ^7Be solar neutrinos have been suggested as a logical extension of the ^{37}Cl experiment of Davis et al. The ^{81}Br experiment, however, requires the development of a direct counter for the slowly decaying ^{81}Kr. Progress toward such a detector based on Resonance Ionization Spectroscopy (RIS) is discussed.

INTRODUCTION

The Davis radiochemical experiment[1] using the interaction $^{37}Cl(\nu,e^-)^{37}Ar$ proved that the flux of ^8B neutrons from the sun was much lower than predicted from stellar models of the solar interior. A great amount of careful work[2] has gone into elucidation of this "solar neutrino problem." A very interesting and important aspect of this effort is the probing of areas of weak interaction physics which are more general than the solar processes themselves; they include questions of neutrino particle classification (Dirac vs Majorana paticles), neutrino mass, neutrino oscillations, and lepton conservation in weak interactions.

Various additional solar neutrino experiments, neutrino terrestrial source experiments (to look for possible neutrino oscillations), and ββ-decay studies have been proposed to answer questions raised by the Davis chlorine experiment in the large radiochemical tank at the Homestake mine. New proposals involving the Homestake facility include the $^{71}Ga(\nu,e^-)^{71}Ge$ detector which measures neutrinos from the important solar pp process, the $^{81}Br(\nu,e^-)^{81}Kr$ detector which responds mainly to the ^7Be solar source, and the $^7Li(\nu,e^-)^7Be$ detector that responds to many of the important

---

* Research sponsored by the Office of Health and
Environmental Research, U.S. Department of Energy under
contract W-7405-eng-26 with Union Carbide Corporation.

** On assignment to Oak Ridge National Laboratory, Oak
Ridge, Tennessee

neutrino sources but not the pp source. A geochemical experiment was suggested by Cowan and Haxton,[3] using the interaction ^{99}Mo(ν,e$^-$)^{99}Te to measure the ^{8}B flux over the past few million years.

The gallium experiment[4] is a very attractive one since it measures the basic pp process. Also, very convincing arguments are made for a comparison of the present ^{8}B neutrino flux (measured by the chlorine experiment) with the aged ^{8}B neutrino flux (measured by the molybdenum experiment). In this paper a bromine experiment to measure the ^{7}Be flux by counting ^{81}Kr atoms is discussed. As Davis points out, a bromine experiment is certainly a logical sequel to the chlorine radiochemical measurements and would closely follow the procedures developed for the Homestake facility except that, unlike ^{37}Ar, ^{81}Kr cannot be counted by radioactive decay methods. However, recent Oak Ridge National Laboratory (ORNL) work on resonance ionization spectroscopy (RIS) strengthens our belief that the ^{81}Kr atoms could be counted by using laser techniques. Serious thought should now be given to the bromine experiment because of its clear physical significance and because it is a relatively inexpensive experiment. While some work remains to be done in completing the ^{81}Kr atom counter, we believe that most of the basic concerns have been addressed. The purpose of this paper is to summarize the progress on ^{81}Kr counting and to emphasize any gaps which need further consideration.

Before returning to these questions we first discuss some other types of uncertainties involved in the bromine experiment – namely the cross section for the ^{81}Br(ν,e$^-$)^{81}Kr interaction, backgrounds, and expected signal levels. Figure 1 shows some energy levels and the important transitions involved in neutrino capture in ^{81}Br. Neutrinos from the ^{7}Be source in the sun have an energy of 0.87 MeV, sufficient to activate the 7/2$^+$, 1/2$^-$, and 5/2$^-$ states of ^{81}Kr. However, the 7/2$^+$ has a low σ, log(ft) = 11.2, compared to the 1/2$^-$ state for which log(ft) = 4.88 as deduced from a measurement[5] of the inverse process of electron capture from the 1/2$^-$ metastable state to ^{81}Br(3/2$^-$). Questions arise[6,7] concerning the contribution from ^{81}Br(3/2$^-$) $\rightarrow$ ^{81}Kr(5/2$^-$) in the solar neutrino capture process. Based on a (^{3}He,T) experiment[8] by a Princeton group, Liu and Gabbard[9] have estimated that the solar neutrino capture rate into the 5/2$^-$ state is about half of the capture rate into the 1/2$^-$ state. Furthermore, (p,n) experiments in progress[10] to measure Gamow-Teller (GT) transition strengths should improve the cross section estimate.

Rowley et al.[11] show the requirements needed in order that background rates (in a geochemical experiment) due to muon, neutron, and alpha-particle production of ^{81}Kr atoms be less than 0.1 of those produced by solar neutrinos. Clearly, these conditions are far more easily met in a radiochemical experiment located underground and in a controlled chemical environment.

According to estimates of Raymond Davis,[12] one expects that approximately 500 atoms of ^{81}Kr would be produced in one year in the Homestake tank containing 862 tons of ethylene bromide ($C_2H_2Br_2$), and further that the tank could contain about 10^8 atoms of krypton (no ^{81}Kr, but 10^7 atoms of the neighboring ^{82}Kr) due to atmospheric contamination. We need to accurately count 500 atoms of ^{81}Kr even when the sample contains 10^7 atoms of ^{82}Kr. Decay counting will not work since the half-life of ^{81}Kr is 2×10^5 years. Direct counting is possible because laser schemes are just now available for resonance ionization of krypton.

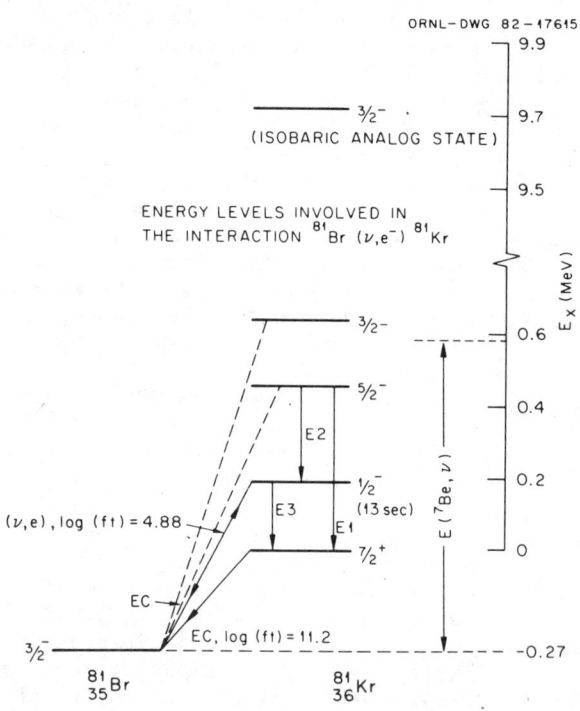

Fig. 1. Energy levels in ^{81}Br and ^{81}Kr relevant to the cross section for the reaction ^{81}Br(ν, e^-)^{81}Kr.

COUNTING ATOMS WITH RESONANCE IONIZATION SPECTROSCOPY

For a number of years[13] the ORNL Photophysics Group has been developing the RIS laser technique for selective and efficient ionization of atoms. Basically, a laser is pulsed through a region of space containing free atoms of various types. After the laser pulse, atoms of the selected type are ionized with efficiency approaching unity, while the atoms not selected are left in their ground state. In a <u>Reviews of Modern Physics</u> article[14] we showed that RIS can be applied to every known atom except helium and neon. Saturated and selective ionization of atoms is a prerequisite to actually counting atoms (direct counting, in contrast to radioactive decay counting); thus, we explored[15] "Atom Counting" in general terms.

For applications in weak interaction physics (such as neutrino capture and $\beta\beta$ decay), noble gases as product atoms form a special class, since the mild chemistry of these atoms allows them to be separated from huge (many tons) targets, as clearly shown in the Davis radiochemical method. Therefore, our attention has been directed toward counting individual atoms of argon, krypton, and xenon with isotopic selectivity. Our method combines RIS with mass spectrometers to make it possible to select out each atom of ^{81}Kr even when ^{82}Kr is more abundant by many orders of magnitude. We call this device Maxwell's demon because of the obvious parallels in the logic involved.

GENERAL DESCRIPTION OF MAXWELL'S DEMON

The concept for counting noble gas atoms with isotopic selectivity is shown in Fig. 2, and utilizes a laser for producing ions from Z-selected atoms and a quadrupole mass filter for A (atomic mass) selection. After atoms have been Z and A selected, the ions

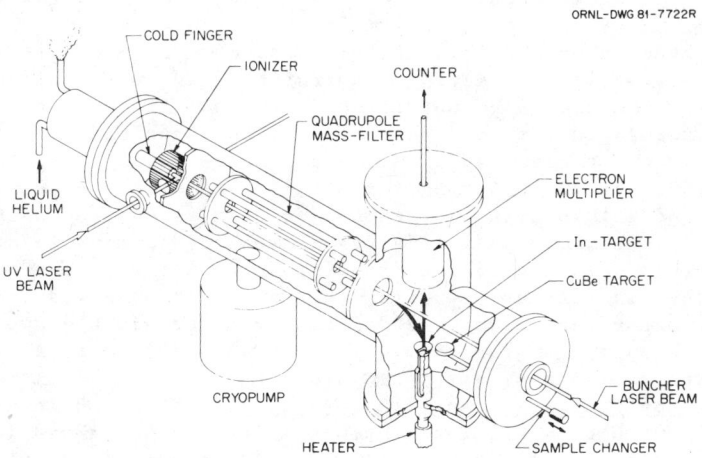

Fig. 2. Concept for counting noble gas atoms with isotopic selectivity.

are accelerated to a few keV onto a target where implantation occurs. One hopes to have a target in which all ions will be implanted and which will emit a pulse of several electrons each time. The ideal target will also have a low melting point and a low vapor pressure, so that all of the atoms can be recovered for repeated operations. An electron multiplier is used with sufficient gain and signal-to-noise ratio to count each implanted ion from the electrons emitted by the target. Finally, an "atom buncher" is provided so that atoms have a much enhanced probability of finding themselves in the RIS laser beam at the time of the laser pulse. Such an instrument is intended for "static" operation in the sense of the Reynolds' static mass spectrometer for sensitive noble gas detection.[16] In the simplest case, the RIS laser is fired until all atoms of selected Z and A have been implanted and counted. The sorting involved here is just the process involved in an intelligent "Maxwell Demon." In practice, a sample containing 10^8 krypton atoms will have to be cycled through the machine more than one time. Suppose that the abundance sensitivity of the quadrupole is only 10^4; then, the first time through, the 10^7 atoms of ^{82}Kr would give 10^3 counts. If the gaseous part of the sample is now pumped out and the target is then melted, it will release the 500 ^{81}Kr atoms along with the 10^3 atoms of ^{82}Kr into the static system. Obviously, on the next cycle of this process the ^{82}Kr contribution to the second count will be negligible.

STATUS OF THE MAXWELL DEMON

The <u>atom buncher</u> consists of a cold finger held at low temperature by means of liquid helium in contact with a copper conductor that connects onto a thin stainless steel disc which terminates in a heat sink. In this way, a temperature gradient is established such that a 2-mm diameter spot is held at about 15°K. Atoms of krypton condense on the cold spot and can be vaporized by flashing a dye laser onto the surface. All atoms leave quickly and travel about 1 mm before encountering the slightly time-delayed RIS laser pulse. After a thin skin of the buncher disc is heated to about 80°K, it cools back to 15°K in a few microseconds; thus, atoms not ionized return with a characteristic recurrence time. The entire atom buncher was tested[17] by using ^{82}Kr and a thin-walled G-M tube. All atoms were released when 120 mJ/cm^2 from the visible dye laser hit the cold spot, as predicted. The recurrence time in a 0.9-liter system was 15 sec, also predicted.

<u>Resonance ionization spectroscopy of krypton</u> has been achieved[18] at ORNL by using a scheme similar to that shown in Fig. 3. The most difficult step was to generate vacuum ultraviolet (VUV) radiation at 1164.9 Å to excite the $5s[1/2]_{J=1}$ state in krypton, a resonance state with good oscillator strength (0.19) for one-photon absorption from the ground state. This radiation was generated by four-wave mixing in xenon gas. The sum process $2\omega_1 + \omega_2$, with ω_1 corresponding to 2525 Å and ω_2 corresponding to 15,073 Å,

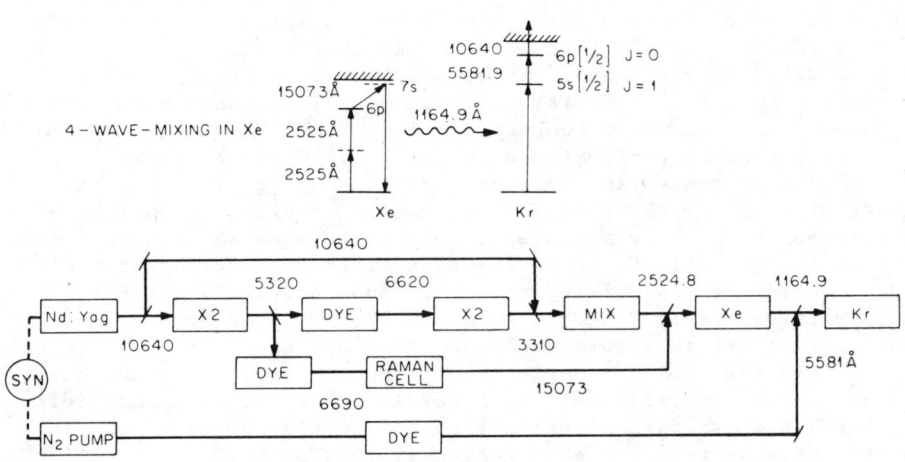

Fig. 3. Laser schemes suitable for RIS of krypton atoms. In the actual demonstration thus far, 2525-Å photons were used to ionize the 5s state. However, p states have a much larger photoionization cross section, thus the more elaborate scheme shown will be used for ^{81}Kr counting.

produced the 1164.9-Å radiation near a 7s level in xenon. With a xenon pressure of 90 Torr and with an argon pressure of 900 Torr added to the xenon cell (to produce phase matching in the constructive interference process for plane-polarized light), and with 210 μJ at 2525 Å and 0.5 mJ from the dye-pumped Raman cell (15,073 Å), we found that 50 nJ was produced at 1164.9 Å. This amount of radiation (at a line width FWHM = 0.2 cm^{-1}) would saturate the resonance excitation step[19] of the RIS process only over an area of about 5×10^{-3} cm^2 or an effective volume of only about 10^{-3} cm^3 for a 2-mm acceptance aperture of a quadrupole mass spectrometer. With a spectrometer volume of three liters, this represents a probability of 3×10^{-7} that an atom moving at random will be ionized in one laser pulse. With the buncher as described above, this probability is increased to 0.1. This is ideal for effective digital counting of each ^{81}Kr atom.

The **mass spectrometer** used is an Extranuclear Corporation quadrupole mass spectrometer equipped with Extranuclear's electron ionizer. Initially, some isotopic enrichment experiments[20] were successfully done using normal abundance krypton isotopes and the electron ionizer. Later, much difficulty was encountered with enrichment. A separate experiment using a simple ion source to ionize and implant ^{85}Kr revealed that the enrichment problem was due to sample "memory" in the ion source. Further, it was found that memory was due to implantation of hot neutral atoms which stayed in the collector walls so long that atoms from the filament could cover them. This mechanism is used to advantage in ion pumps. We found how to decrease the source pumping (hence memory) to a satisfactory level. Our concern had been memory elsewhere in the machine which would have compromised the counting of ^{81}Kr atoms. Fortunately, the solar neutrino experiment does not require "electron enrichment" as an initial step to the "laser enrichment"; hence, even a small residual source memory is of no concern.[21]

We can now return to a discussion of the target, an important element since it is the door for the Maxwell sorting demon. From the standpoint of the gamma coefficient (electrons emitted per ion striking the target), Cu-Be is the standard and is well characterized.[22] However, its melting point is inconvenient (1,000°C). Thus, we investigated[23] indium which melts at 157°C; however, we found that it has an accommodation coefficient or sticking fraction (SF) of less than 0.5. A better choice may be aluminum which we find has a gamma coefficient for 5 keV Kr$^+$ equal to 3.3. According to published works, its sticking fraction is near unity,[24] and it has unit release fraction when melted at 660°C.

Finally, some comments on the single ion detector are needed. With a gamma coefficient of 3, the probability of missing a count should be only about 5%. However, it is convenient to have the target at -5 kV and the electron multiplier near ground potential. Thus, the electrons actually striking the front stage of the electron multiplier are so energetic that the delta coefficient (electrons emitted per incident electron) is only about 0.5; therefore, an unacceptable number of ions would not be counted. To overcome this we have developed an "electron preamplifier," containing two plates of MgO, which gives six electrons at low energy for each high-energy electron. The minimum value of the product $\gamma\delta$ should be more than three; hence, less than 5% of the ions would be lost.

In conclusion, we believe that progress made toward counting individual noble gas atoms with isotopic sensitivity justifies our continued optimism for applications to weak interaction physics, particularly solar neutrino experiments, neutrino oscillations, and $\beta\beta$ decay (^{82}Se to ^{82}Kr, ^{128}Te to ^{128}Xe, and ^{130}Te to ^{130}Xe).

We wish to thank our colleagues (especially Ray Davis and Bruce Cleveland) at Brookhaven National Laboratory for their support of our detector development and for their interest in having us as collaborators on these very exciting solar neutrino experiments.

REFERENCES

1. R. Davis, Jr., D. S. Harmer, and K. C. Hoffman, Phys. Rev. Lett. **20**, 1205 (1968).

2. J. N. Bahcall, W. F. Huebner, S. H. Lubow, P. D. Parker, and R. G. Ulrich, Revs. Mod. Phys. **54**, 767 (1982).

3. G. A. Cowan and W. C. Haxton, Sci. __216__, 51 (1982).

4. J. N. Bahcall, B. T. Cleveland, R. Davis, Jr., I. Dostrovsky, J. C. Evans, Jr., W. Frati, G. Friedlander, K. Lande, J. K. Rowley, R. W. Stoenner, and J. Weneser, Phys. Rev. Lett. __40__, 1351 (1978).

5. C. L. Bennett, M. M. Lowry, R. A. Naumann, F. Loeser, and W. H. Moore, Phys. Rev. C __22__, 2245 (1980).

6. J. N. Bahcall, Phys. Rev. C __24__, 2216 (1981).

7. W. C. Haxton, Nucl. Phys. A __367__, 517 (1981).

8. R. T. Kouzes, M. M. Lowry, and C. L. Bennett, Phys. Rev. C __24__, 1775 (1981); R. T. Kouzes, M. M. Lowry, and C. L. Bennett, Phys. Rev. C __25__, 1076 (1982).

9. K. F. Liu and F. Gabbard, to be published.

10. The (p,n) differential cross at 0° is used by the Indiana University cyclotron group to estimate (GT) strengths [for $^{37}Cl(p,n)^{37}Ar$, see J. Rapaport, T. Taddeucci, P. Welch, C. Gaarde, J. Larsen, C. Goodman, C. C. Foster, C. A. Goulding, D. Horen, E. Sugarbaker, and T. Masterson, Phys. Rev. Lett. __47__, 1518 (1981)]. Similar studies are planned at Indiana University and at the University of Kentucky for the reaction $^{81}Br(p,n)^{81}Kr$.

11. J. K. Rowley, B. T. Cleveland, R. Davis, Jr., W. Hempel, and T. Kirsten, in Proceedings of Conference on Ancient Sun (1980), pp. 45-62

12. Raymond Davis, Jr., private communication.

13. G. S. Hurst, M. G. Payne, M. H. Nayfeh, J. P. Judish, and E. B. Wagner, Phys. Rev. Lett. __35__, 82 (1975).

14. G. S. Hurst, M. G. Payne, S. D. Kramer, and J. P. Young, Rev. Mod. Phys. __51__, 767 (1979).

15. G. S. Hurst, M. G. Payne, S. D. Kramer, and C. H. Chen, Phys. Today __33__(9), 24 (1980).

16. J. H. Reynolds, Rev. Sci. Instrum. __27__, 928 (1956).

17. B. E. Lehmann, G. S. Hurst, J. W. Dabbs, R. C. Phillips, and M. G. Payne, to be published.

18. S. D. Kramer, C. H. Chen, M. G. Payne, and G. S. Hurst, to be published.

19. After excitation of the resonance state in krypton, a variety of laser schemes can be used to complete the RIS process.

20. R. D. Willis, C. H. Chen, S. L. Allman, and G. S. Hurst, to be published.

21. Electron enrichment on a continuous basis is a way to overcome the space charge limitation (about 10^7 ions in one laser pulse) and is required for other applications such as groundwater dating with ^{81}Kr and ocean water circulation using ^{39}Ar.

22. I. Mitteilung and R. Baumhakel, Z. Phys. <u>199</u>, 41 (1967).

23. R. D. Willis, S. L. Allman, G. D. Alton, and G. S. Hurst, to be published.

24. D. Lal, W. F. Libby, G. Wetherill, J. Leventhal, and G. D. Alton, J. Appl. Phys. <u>40</u>, 3257 (1969).

A PROPOSED GEOLOGICAL SOLAR NEUTRINO MEASUREMENT

George A. Cowan and Wick C. Haxton
Los Alamos National Laboratory, Los Alamos, NM 87544

ABSTRACT

It may be possible to measure the boron-8 solar neutrino flux, averaged over the past several million years, from the concentration of technetium -98 in molybdenum-rich ore. This geochemical experiment could provide the first test of nonstandard solar models that suggest a relation between the chlorine-37 solar neutrino puzzle and the most recent glacial epoch. The necessary conditions for achieving a meaningful measurement are identified and discussed.

INTRODUCTION

The standard solar model[1] postulates a complex set of thermonuclear reactions which evolve and change with time. The rate at which energy is generated is determined by conditions in the deep solar interior. Although there is considerable reason to believe that the model is generally correct, there is, unfortunately, almost no way to probe the solar interior to check its details.

A general solar model should comprehend processes leading to solar variability. Evidence for variability exists in the cycle of sunspot activity which appears to affect the terrestrial climate. This cycle is presently explained in terms of the dynamics of convection and rotation and does not call into question the details of energy production. However, long term changes have occurred in the terrestrial climate and, if these are due to solar variability, they imply possible changes in the rate of energy production which are not presently explained by the standard model.

Dramatic changes in climate have occurred several times in terrestrial history when warm periods of hundreds of million years duration have been interrupted by major glacial epochs which last a few million years. There is, as yet, no consensus that glacial epochs are related to energy changes in the sun. Such a relationship is hypothesized because the standard solar model can be modified to produce a cyclic change in solar temperature with a pattern corresponding to that of the glacial epochs. For example, the "solar spoon," proposed by Dilke and Gough[2] ten years ago, postulates that the ratio $^3He/^1H$ varies sharply with radius when equilibrium burning is achieved. This large concentration gradient induces large amplitude, non-radial oscillations with a characteristic period of about 1 hour. Such oscillations produce a mixing of the core with colder material at larger radii. The approach to equilibrium prior to the onset of instability requires about 200 million years. The mixing is relatively sudden and the consequent core expansion produces a cooling trend with a 5% drop in solar luminosity for a period of about 3 million years.

The last glacial epoch, the Pleistocene, began only three million years ago. If it were due to a solar instability, the core should presently be cooler than predicted by the standard model. Thus, this hypothesis can be used to help explain the solar neutrino anomaly, i.e., the fact that the high energy neutrino component measured by the ^{37}Cl-^{37}Ar reaction is nearly a factor of four lower than the flux predicted by the standard model. The hypothesis can be tested by measuring a product which is proportional to the high energy neutrino flux integrated over several million years.

The neutrino measurement being developed at Los Alamos involves the separation and analysis of ^{98}Tc, with a half-life of 4.2 million years, made by neutrino absorption in ^{98}Mo contained in a deeply buried ore deposit. An additional, potentially useful product is ^{97}Tc with a half-life of 2.6 million years. A number of requirements must be met in order to perform this measurement with useful accuracy and at acceptable cost. These include:

1. Accurate neutrino capture cross sections must be available from theory and experiment.

2. A molybdenum ore deposit must exist in a working mine which is deep enough to reduce cosmic-ray-induced backgrounds to acceptable levels.

3. Backgrounds induced by naturally radioactive elements must be minor and measurable.

4. Separative and analytical chemistry procedures must be developed to measure extremely small quantities of ^{97}Tc, ^{98}Tc, and ^{99}Tc.

5. Geochemical immobility of Tc in the chosen ore body must be demonstrated.

These requirements are discussed in more detail in the following sections of this paper.

NEUTRINO CAPTURE CROSS SECTIONS

The threshold for the ^{98}Mo$(\nu,e)^{98}$Tc reaction is 1.68 MeV but, because transitions from the 0^+ ground state of ^{98}Mo to the $(6)^+$ ground state and $(4,5)^+$ first excited state in ^{98}Tc are strongly hindered, the effective threshold is $\gtrsim 1.74$ MeV and only high energy ^{8}B neutrinos contribute significantly.

The model-independent Fermi strength contribution to the formation of ^{98}Tc is negligibly weak. With sufficient accuracy for our present calculations, we estimate the model-dependent GT strength distribution and calculate a production rate of 5.0 SNU (1SNU=10^{-36} captures per target atom per second) based on the standard solar model neutrino flux of Bahcall et al.[3] The GT transition can be mapped experimentally with a typical accuracy of 5% based on measurements of forward angle (p,n) reactions.[4] Such measurements are planned for Mo and should be available relatively soon.

From similar calculations for the productions of ^{97}Tc, we obtain a total production rate of 5.9 "equivalent" SNU, the result including a contribution from deexcitation by neutron emission in excited ^{98}Tc.

THE ORE DEPOSIT

Although a number of unexploited Mo ore deposits may be known which are buried at a sufficient depth, the amount of ore needed per sample is so great that we are limited by practical considerations to the use of working mines. We know of only one such active mine. It is located at the Henderson ore body at Red Mountain in Clear Creek County, Colorado. The ore averages approximately 0.5% molybdenite (MoS_2) and comes from a depth in excess of 1000 m. The ore body extends to a depth > 1500 m below the surface[5]. At the time of ore body formation (∼25 million years ago) the depth of burial was 1500-1800 m. The present minimum depth of overburden is due to glacial scouring which occurred about 10,000 years ago.

We calculate that 1410 m of overburden at a density of 2.9 reduces the cosmic ray background to 10% of the solar neutrino signal.

NATURAL RADIOACTIVITY BACKGROUNDS

The most common Mo mineral at Henderson is molybdenite embedded in altered granitic rock. A representative sample contains 11.8 ppm of uranium. The associated radioactivity is the source of a number of potentially bothersome reactions which produce ^{97}Tc and ^{98}Tc. Direct (α,p) reactions are suppressed to negligible levels by high thresholds and Coulomb barriers. However, (α,p) reactions on S followed by (p,n) reactions on ^{97}Mo and ^{98}Mo may be serious. If uranium is found only in the veinlets containing molybdenite crystals, the ratio of neutrino-induced to alpha-induced Tc is given by R<d>/3.75U where R is the neutrino capture rate in SNU, U the uranium content in ppm, and < d > the mean thickness of molybdenite in micrometers. For ^{98}Tc, if U is 11.8 and <d> is 100, the ratio is 11.3. The result is more favorable if the U in concentrated in macroscopic grains of host minerals which degrade the energy of the alpha particle.

Fission neutrons produce (n,p) reactions on S followed by (p,n) reactions on Mo. The signal-to-background ratio is 510R/U or about 216.

Neutron capture on ^{96}Ru can make ^{97}Tc to the extent that 0.11 ppm of Ru in the ore concentrate will generate a 10% background. The (α,γ) reaction on ^{93}Nb generates a similar background level of ^{97}Tc at a Nb content of a few ppm. We conclude that background levels of ^{97}Tc may prove to be unacceptable.

SEPARATIVE CHEMISTRY AND ANALYSIS FOR Tc

Our calculations indicate that the untreated ore will contain 5×10^4 atoms of ^{99}Tc per gram, the secular equilibrium level for mass 99 produced by the spontaneous fission of ^{238}U at a concentration of 12 ppm. Neutrino-produced ^{97}Tc and ^{98}Tc concentrations should be a factor of 10^7 lower. When the molybdenite is concentrated from 0.5% to 90%, the ^{238}U concentration should fall to 1.2 ppm if all the U is associated with minerals outside the molybdenite. A representative sample of concentrate actually contained 1.8 ppm. Thus, the concentrate can be expected to contain $\sim 7 \times 10^3$ atoms of ^{99}Tc per gram and $\sim 10^4$ less ^{98}Tc, the most troublesome mass number due to the ubiquity of Mo contamination. The concentrate from 2600 tons of ore divided by the recovery efficiency will be required to produce such a sample. This represents considerably less than one day's production at the mine.

In the industrial process, the concentrate is roasted to produce molybdenum oxide at a controlled temperature of $\sim 700^\circ$C. An effluent gas stream, consisting largely of SO_2 and excess air, is scrubbed with water to remove particles and volatile impurities before conversion of the SO_2 to sulfuric acid. The scrub water, know as the acid wash, should contain most of the environmentally objectionable element selenium together with chemically similar rhenium and technetium. The selenium is precipitated as the sulfide for environmental reasons. This precipitate will contain the Tc. We are developing separative techniques to recover traces of Tc from kilograms of selenium sulfide. The necessary chemistry is known, but the efficiency must be determined.

THE GEOCHEMICAL MOBILITY OF Tc

Based on measurements of loss rates of decay products of uranium, particularly radiogenic lead, in a variety of uranium ores, we are hopeful that Tc loss rates will not exceed 10^{-9} per year in the reducing sulfide-rich environment of the Henderson ore. We intend to measure the degree to which ^{99}Tc is in secular equilibrium with uranium. However, final reassurance on this question will rest on the reproducibility of results obtained from more than one location in the ore body.

1. J. N. Bahcall et al., Phys. Rev. Lett. 40, 1351 (1978)
2. F. W. W. Dilke and D. O. Gough, Nature, 240, 262 (1972)
3. J. N. Bahcall et al., Phys. Rev. Lett. 45, 945 (1980)
4. G. A. Goulding and C. C. Foster, private communication; C. D. Goodman et al., Phys. Rev. Lett. 44, 1755 (1980)
5. D. E. Ranta et al., in "Studies in Field Geology," R. C. Epis and R. J. Weimer, Eds. (Colo. School of. Mines, Golden, 1976) p.477
6. G. A. Cowan and W. C. Haxton, Science, 216, 51-54 (1982)
7. G. A. Cowan and W. C. Haxton, Los Alamos Science, 3, No.2, 46-56 (1982)

Chapter III

PROTON DECAY

Why did these three learned gentlemen, Weyl, Stueckelberg, and Wigner, feel so sure that baryons are conserved? Well, you might say that it's very simple; they felt it in their bones. Had their bones been irradiated by the decays of nucleons, they would have noticed effects considerably exceeding "permissible radiological limits" if the nucleon lifetime were $<10^{16}$ years . . .

M. Goldhaber, in *Unification of Elementary Forces and Gauge Theories*, ed. by D. B. Cline and F. E. Mills (Harwood, London, 1978), p. 535.

THEORETICAL PREDICTIONS FOR BARYON NUMBER VIOLATION[1]

Paul Langacker
University of Pennsylvania, Philadelphia, Pa. 19104

ABSTRACT

In this talk I describe the theoretical predictions for proton decay and other baryon number violating processes, emphasizing that there are many models and theories involving baryon number violation and that it is an experimental problem to distinguish between them. I first review the theoretical predictions for the unification mass M_X and for the weak angle $\sin^2\theta_W$. It will be seen that the class of models involving an $SU_3 \times SU_2 \times U_1$ invariant desert between M_W and M_X are strongly favored. I then turn to baryon number violation. The proton lifetime and branching ratio predictions for the SU_5 and other 3-2-1 desert models are reviewed, with emphasis on distinguishing between models and on the implications of the small value of the QCD parameter $\Lambda_{\overline{MS}}$ that seems to be favored by the data. I then discuss the consequences of low energy supersymmetry for proton decay, nuclear effects, and models with low mass scales. Finally, I briefly mention several other topics, including the possibility of observing baryon number violation at accelerators, baryon number violation induced by magnetic monopoles, the violation of Lorentz invariance in proton decay, and antiproton cosmic rays.

A. THE UNIFICATION MASS AND $\sin^2\theta_W$

Grand Unified Theories (GUTs) are gauge field theories in which the strong, weak, and electromagnetic interactions (described by the standard model gauge group $SU_3^c \times SU_2 \times U_1$, with gauge couplings g_s, g, and g', respectively) are embedded in a simple group G with gauge coupling g_G. GUTs generally predict the existence of new gauge bosons X which can mediate baryon number violating processes such as proton decay through diagrams such as shown in Figure 1.

The expected proton lifetime τ_p is of order

$$\tau_p \simeq \frac{1}{\alpha_G^2} \frac{M_X^4}{m_p^5}, \qquad (1)$$

where $\alpha_G \equiv g_G^2/4\pi$ and M_X is the unification mass (mass of the X boson). The existing limits ($\tau_p > 10^{30}$ yr) therefore require $M_X > 10^{14}$ GeV, which is twelve orders of magnitude larger than the predicted masses of the W and Z bosons which are believed to mediate the weak interactions! Fortunately, however, M_X can be independently predicted from the ratios of the observed low energy coupling constants ($\alpha_s = g_s^2/4\pi$, $\sin^2\theta_W = g'^2/(g^2 + g'^2)$, and $\alpha = e^2/4\pi = g^2 \sin^2\theta_W/4\pi$). For a large class of theories M_X indeed turns out to be $\gtrsim 10^{14}$ GeV.

The coupling constant predictions of grand unified theories depend on the details of the spontaneous symmetry breaking (SSB) pattern.

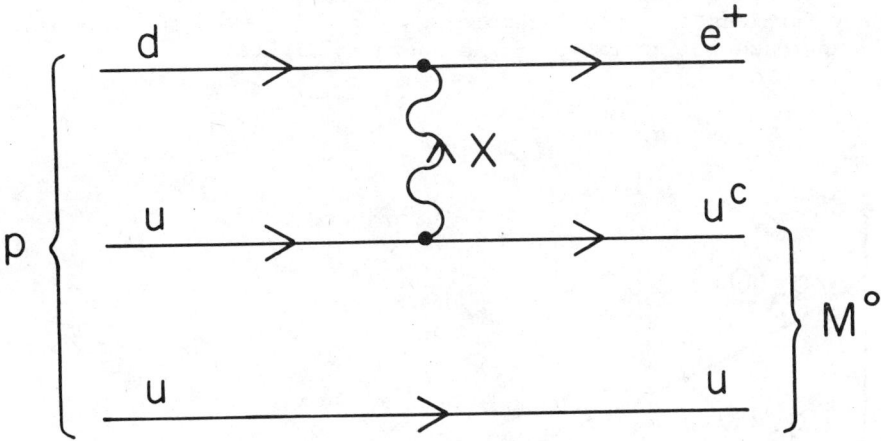

Fig. 1. A typical diagram for proton decay. M^0 represents any combination of neutral mesons (π^0, ρ^0, η, ω, $\pi^+\pi^-$, etc.). The upper (lower) vertices are referred to as lepto-quark (diquark) vertices.

In practice, clean predictions only emerge for theories involving just two mass scales, M_X and M_W, with a desert (a region with no gauge, Higgs, or fermion thresholds) in between. That is, suppose the masses of the gauge bosons in G are all M_X, M_W, or zero. One then writes

$$G \xrightarrow[M_X]{} G_1 \xrightarrow[M_W]{} SU_3^c \times U_1^{EM} \quad , \qquad (2)$$

where G_1 is the effective gauge group obtained by omitting the superheavy (mass M_X) gauge bosons from the theory and where it is assumed that all of the bosons in G_1 except the massless color gluons (in SU_3^c) and the photon have mass $\simeq M_W \simeq M_Z$. In this case, the low energy running coupling constants g_s, g and g' all approach[2,3,1] g_G up to computable normalization factors as $Q^2 \to M_X^2$. One can therefore use the observed values of either α/α_s or $\sin^2\theta_W$ to predict M_X. Alternately, M_X can be eliminated, yielding a relation between α/α_s and $\sin^2\theta_W$. This relation is just a consistency condition that the three couplings all meet at the same point. In order to evaluate this relation (i.e. predict $\sin^2\theta_W$) one must know the intermediate energy group G_1 and the G_1 assignments of all of the light (mass $\leq M_W$) fermions and Higgs bosons (in order to determine the renormalization group equations (RGE) for the running couplings). One must also know the G_1 assignments of an entire family (light and heavy) of fermions (in order to evaluate the normalizations). An explicit knowledge of G is not required however.

For the case[2,3,1] that $G_1 = G_s = SU_3^c \times SU_2 \times U_1$ with F families of ordinary fermions, for example, $g_3 = g_s$, $g_2 = g$, and $g_1 = \sqrt{5/3}\, g'$ all approach g_G for large Q^2, as shown in Figure 2.

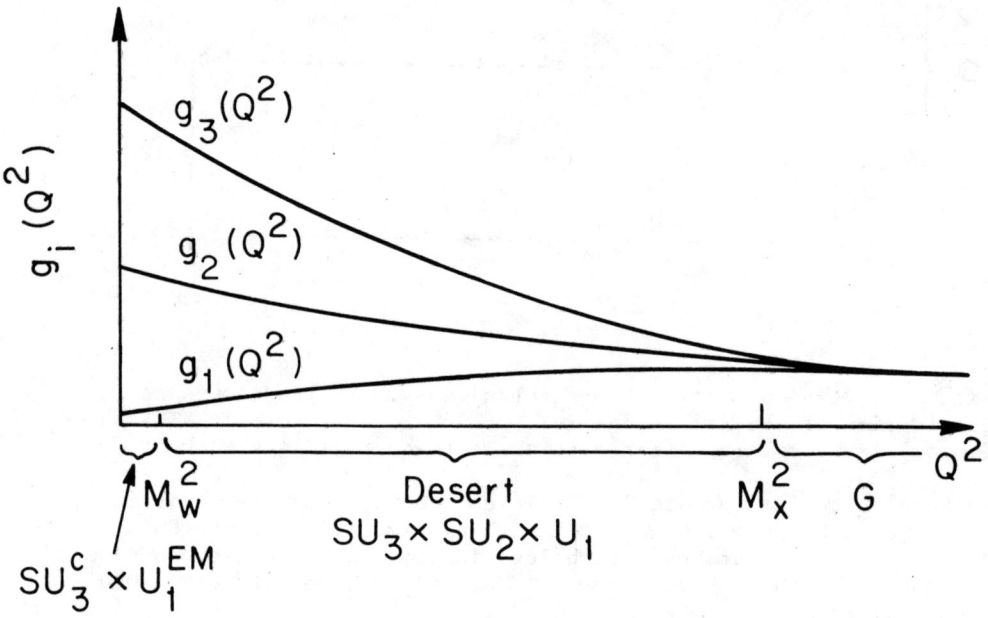

Fig. 2. The behavior of the running coupling in 3-2-1 desert models. The apparent symmetry in each region is indicated.

Detailed calculations of M_X and $\sin^2\theta_W$ including higher order and threshold effects have been completed by several groups.[4-6,1] The results (for F ordinary fermion families and n_H light Higgs doublets) can be summarized as

$$M_X(\text{GeV}) = 2.4 \times 10^{14} \times (1.5)^{\pm 1} \times \left[\frac{\Lambda_{\overline{MS}}}{0.16\text{ GeV}}\right]$$

$$\times \left[\frac{1}{1.5}\right]^{n_H - 1} \times (1.2)^{F-3} \quad (3)$$

$$\sin^2\hat{\theta}_W(M_W) = 0.2138 \pm 0.0025$$

$$+ 0.006 \ln\left[\frac{0.16 \text{ GeV}}{\Lambda_{\overline{MS}}}\right] + 0.004 (n_H-1) - 0.0004(F-3) \quad (4)$$

The stated errors in (3) and (4) are due to uncertainties in higher order terms, thresholds, heavy Higgs bosons[1], and the top quark mass ($15 < m_t < 50$ GeV is assumed). $\Lambda_{\overline{MS}}$ is the QCD parameter used to characterize the Q^2 dependence of the strong coupling α_s (at one loop level $\alpha_s(Q^2) = 12\pi/25 \ln (Q^2/\Lambda_{\overline{MS}}^2)$). A number of recent analyses[7,1] of deep inelastic scattering, e^+e^- jet production, the upsilon width, and charmonium hyperfine splitting suggest a low value for $\Lambda_{\overline{MS}}$ (although the uncertainties are large). I will use the average

$$\Lambda_{\overline{MS}} = \left\{ 160 \begin{array}{c} +100 \\ -80 \end{array} \right\} \text{ MeV} \quad (5)$$

given by Buras[7]. For this range, the log in (4) is $0 \begin{array}{c} -0.0029 \\ +0.0042 \end{array}$.

The prediction (4) is in remarkable agreement with the experimental value

$$\sin^2\theta_W(M_W)\bigg|_{\exp} = 0.215 \pm 0.012 \quad , \quad (6)$$

(which corresponds to $M_W = (83.0 \pm 2.4)$ GeV and $M_Z = (93.8 \pm 2.0)$ GeV) obtained from the neutral current data[8,6,1]. This gives strong encouragement that grand unified theories (in particular those with $SU_3 \times SU_2 \times U_1$ invariant deserts) may be relevant to the real world.

B. BARYON NUMBER VIOLATION[1]

<u>The SU_5 Model</u>

In a single fermion family version of the Georgi-Glashow[2] SU_5 model, the left-handed fermions are assigned to the reducible representation $5^* + 10$:

$$\begin{pmatrix} \nu^e \overset{X,Y}{\longleftrightarrow} \\ d^c \\ e^- \end{pmatrix}_L \quad \begin{pmatrix} & \overset{X,Y}{\longleftrightarrow} u \\ e^+ & u^c \\ & d \end{pmatrix}_L \updownarrow W^{\pm} \quad (7)$$

The notation in (7) is that fields arranged in columns transform as SU_2 doublets (i.e. they are transformed into each other by the emission or absorption of W^+ or W^- bosons), while fields in adjacent columns are transformed into each other by the emission or absorption of the superheavy bosons X and Y ($M_X \simeq M_Y \simeq 2.4 \times 10^{14}$ GeV). Vertices involving the X and Y bosons, which have electric charges 4/3 and 1/3, respectively, can transform quarks into antiquarks or antileptons. The X and Y can therefore mediate proton (or bound neutron) decay, through diagrams such as those shown in Figure 1. The allowed decay modes are $p \to e^+ M^0$ or $\nu^c M^+$ and $n \to \nu^c M^0$, $e^+ M^-$, where M^0 ($M^\pm$) represents a neutral (charged) meson state such as π^0, ρ^0, η, ω, $\pi^+\pi^-$ ($\pi^\pm, \rho^\pm, \pi^\pm\pi^0$).

The Proton Lifetime

The naive estimate (1) for τ_p yields $\tau_p \simeq 2 \times 10^{29}$ yr for $\Lambda_{\overline{MS}} \simeq 0.16$ GeV ($M_X \simeq 2.4 \times 10^{14}$ GeV) and $\alpha_5 = g_5^2/4\pi \simeq 0.022$. As we will now see, more detailed calculations yield similar results.

These more detailed analyses encounter the three classes of diagrams shown in Figure 3.

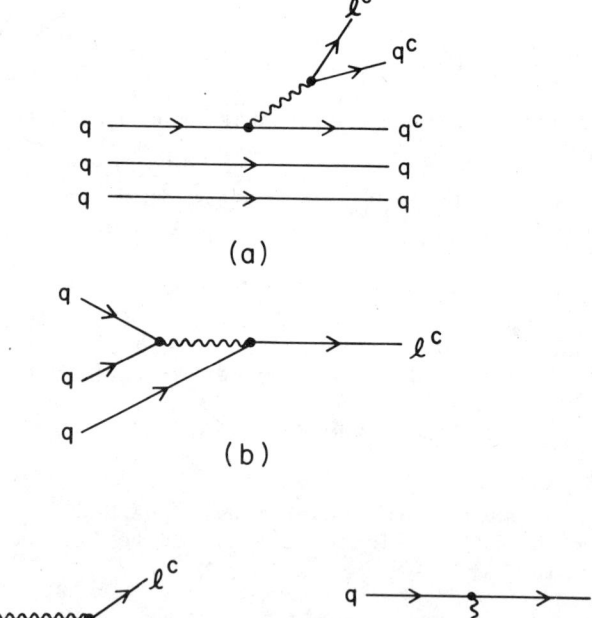

Fig. 3. (a) Quark decay. (b) Three-Quark fusion. (c) Two-quark fusion.

A number of authors[1] have estimated the proton and bound neutron lifetimes assuming that the two-quark fusion diagrams of Figure 3c dominate. (The quark decay diagram of 3a is suppressed by phase space and the three-quark fusion diagram of 3b, in which the initial quarks emit gluons or mesons, will be discussed below.) These calculations indicate that τ_p varies approximately as M_X^4 and yield

$$\tau_p(\text{yr}) \simeq (0.8-13) \times 10^{29} \left[\frac{M_X}{2.4 \times 10^{14} \text{ GeV}}\right]^4 \qquad (8)$$

The spread of values in (8) is mainly due to uncertainties in computing the hadronic matrix elements (e.g. some calculations use SU_6 with a phenomenological wave function $\psi(0)$ for two quarks to overlap; others use bag model wave functions) and must be considered a genuine theoretical uncertainty. There are also smaller differences due to the treatments of the second family, mixings, and phase space (e.g. inclusive or exclusive sums on hadronic states, effective quark masses, etc.) The results of the various calculations have been normalized to common anomalous dimensions and $\psi(0)$. I will use the range of values in (8) as an estimate of the theoretical uncertainties (due to the hadronic physics) in $\tau_p(M_X)$.

Combining (8) with (3) one has

$$\tau_p(\text{yr}) \simeq 3.2 \times 10^{29 \pm 1.3} \left[\frac{\Lambda_{\overline{MS}}}{0.16 \text{ GeV}}\right]^4 \qquad (9)$$

where the error is due to $\tau_p(M_X)$ and $M_X(\Lambda_{\overline{MS}})$. For $\Lambda_{\overline{MS}}$ in (5) one has finally

$$\tau_p(\text{yr}) = 3.2 \times 10^{29 - 2.5}^{+2.1} , \qquad (10)$$

where the error includes the uncertainty in $\Lambda_{\overline{MS}}$. Furthermore[1], the bound neutron lifetime is predicted to be $0.8 < \tau_n/\tau_p < 1.5$.

We see that the small values of $\Lambda_{\overline{MS}}$ that are now becoming accepted lead to the prediction of a very "short" proton lifetime. For example, $\Lambda_{\overline{MS}} = 200$ MeV implies τ_p in the range $(3.8 \times 10^{28} - 1.5 \times 10^{31})$ yr., while $\Lambda_{\overline{MS}} = 100$ MeV yields $(2.4 \times 10^{27} - 9.6 \times 10^{29})$ yr. Clearly it should be easy to verify or disprove the prediction (10) of the minimal SU_5 model with small $\Lambda_{\overline{MS}}$ in forthcoming experiments, which will be sensitive to lifetimes of $10^{31} - 10^{32}$ yr.

The two-quark fusion diagrams for most other 3-2-1 desert models are expected to yield lifetimes similar to (8) - (10).

The results summarized in (8) - (10) are for the two-quark fusion diagrams of 3c only. However, several authors[9] have recently examined the three quark diagrams of Figure 3b using either soft-pion techniques or by saturating the 3-quark (antilepton) channel with a nucleon pole. These calculations actually yield a proton lifetime shorter than (8) - (10), with τ_p in the range $(0.5 - 5) \times 10^{29}$ yr for $M_X = 2.4 \times 10^{14}$ GeV. If the lower estimates in this class are correct then the 3-2-1 desert models are in serious trouble. However, there may be important corrections to these calculations because the pion is in fact very energetic and the antilepton mass is far from the nucleon pole.

There are several theoretical loopholes that could allow longer nucleon lifetimes. These include (a) modifying the light and/or heavy Higgs and fermion representations of the standard model (which affect M_X). These could probably not change τ_p by more than $\simeq 10^{\pm 2}$ without also destroying the successful prediction of $\sin^2\theta_W$ (an exception is supersymmetric models). (b) τ_p could be increased by large mixing effects (e.g. by associating the light quarks and heavy leptons in multiplets - this does not occur in the simplest models, however[1]). Even if the proton coupled predominantly to heavy channels, it is unlikely that τ_p would be increased by much more than $\sin^{-2}\theta_c \simeq 20$, however, unless the proton is made absolutely stable by imposing an extra symmetry (see below). (c) Models with three or more mass scales can give arbitrarily long proton lifetimes. In such models the value of $\sin^2\theta_W$ is an input (to determine the extra mass scale), so the success of (4) would be an accident.

Branching Ratios in SU_5

A number of estimates have been made of the proton and neutron branching ratios.[1] These differ in the hadronic matrix elements, mixings, and phase space treatments used, and in the method of projecting the spin of the recoiling q^c onto meson wave functions. Typical ranges for the branching ratio (%) predictions are

$$p \to e^+\pi^o, \quad e^+\rho^o, \quad e^+\eta, \quad e^+\omega,$$
$$36-38, \quad 2-11, \quad 0-7, \quad 21-26$$
$$\bar{\nu}\pi^+, \quad \bar{\nu}\rho^+, \quad \mu^+K^o, \quad \bar{\nu}K^+$$
$$14-15, \quad 1-4, \quad 5-18, \quad 0-1$$
$$n \to \bar{\nu}\pi^o, \quad \bar{\nu}\rho^o, \quad \bar{\nu}\eta, \quad \bar{\nu}\omega,$$
$$7-8, \quad 1-2, \quad 0-2, \quad 5,$$
$$e^+\pi^-, \quad e^+\rho^-, \quad \bar{\nu}K^o \quad\quad (11)$$
$$68-79, \quad 6-19, \quad 1-3$$

(There may be additional suppression of the pionic modes due to recoil effects).

Other Models

The baryon number violating interactions in a general model can be described[10] by an effective Lagrangian

$$L = CM^{4-d} O, \quad\quad (12)$$

where O is an effective operator of dimension d, constructed from the light fields of the theory, M is the (superheavy) unification scale and C is a dimensionless constant that is typically of order e^{n-2} for n external fields in O. This is analogous to the effective four-fermion operator $L_W = G_F JJ^\dagger/\sqrt{2}$ of the weak interactions, for which d = 6, n = 4, and $G_F/\sqrt{2} = g^2/8 M_W^2$. For M >> M_W, it suffices to consider $SU_3 \times SU_2 \times U_1$ invariant operators in (12). $SU_2 \times U_1$ breaking effects (suppressed by powers of M_W/M) can be incorporated by including the Higgs field (which can be replaced by its vacuum expectation value) in O.

3-2-1 Desert Theories

The lowest dimension baryon-number violating, color singlet, Lorentz scalar operators are four-fermion operators with d = 6. Only the d = 6 operators are relevant for 3-2-1 desert theories because of the extremely large value of M > 10^{14} GeV. (d>6 interactions are suppressed by $(M_W/M)^{d-6}$). Weinberg and Wilczek and Zee have shown that, up to family indices, there are only six such d = 6 operators, and that (suppressing Lorentz, family, and G_s indices) all are of the form $qqq\ell$. These operators can be generated by gauge or Higgs exchange diagrams as in Figure 1.

It is remarkable that these six operators all imply the following selection rules, which must be respected in all 3-2-1 desert theories:

(a) $\Delta B = \Delta L$. For example $p \to e^+\pi^0$ is allowed, but $p \to e^-\pi^+\pi^+$ is forbidden;

(b) $\Delta S/\Delta B = -1$ or 0. Thus, $p \to \nu^c K^+$ but $n \not\to e^+ K^-$;

(c) the $\Delta S = 0$ operators transform as strong isospin doublets, implying relations such as

$$\Gamma(p \to e^+\pi^0) = \frac{1}{2} \Gamma(n \to e^+\pi^-) \tag{13}$$

For operators involving (u, d, ℓ^-, ν_ℓ) only, where ℓ = e or μ, there are only four operators, O_i, i=1, ... 4. Hence, the most general effective Lagrangian for decays involving (u, d, e^-, ν_e) only in 3-2-1 desert theories is of the form[10]

$$L = \sum_{i=1}^{4} C_i O_i + H.C., \tag{14}$$

where of course the C_i depend on the model.

Gauge Boson Mediated Processes

If the underlying mechanism is gauge boson exchange, then it can be shown that $C_3 = C_4 = 0$, so that

$$L = C_1 O_1 + C_2 O_2 + H.C., \tag{15}$$

Hence, only the ratio $r \equiv C_2/C_1$ can distinguish between models in this class for (u, d, e^-, ν_e) decays. (The hadronic uncertainties in τ_p are too large to effectively discriminate between models).

Furthermore, the relative branching ratios for $e^+\pi^0/e^+\rho^0/e^+\eta/e^+\omega$ and $\nu^c\pi^+/\nu^c\rho^+$ are independent of r. That is, measurements of the specific hadronic final state in $\Delta S = 0$ decays will not distinguish between gauge-mediated models. On the other hand, the ratio of positrons to anti-neutrinos depends on r and is useful for distinguishing between models. For example,

$$\frac{\Gamma(O^{16} \to e^+ X)}{\Gamma(O^{16} \to \nu^c X)} = 1 + r^2. \qquad (16)$$

Other predictions, including the polarization of prompt muons and the μ/e and K/π ratios, are detailed in the literature.[1]

Higgs Mediated Decays in 3-2-1 Desert Theories[1]

Proton decay can also be mediated by the exchange of Higgs bosons, which can generate all four operators in (14). The relative branching ratios $e^+\pi^0/e^+\rho^0/e^+\eta/e^+\omega$ and $\nu^c\pi^+/\nu^c\rho^+$ depend on the ratio $\lambda \equiv (c_4^2 + c_3^2)/(c_1^2 + c_2^2)$ (the ρ and η modes are enhanced for $\lambda > 0$), so in principle it is possible to distinguish between gauge and Higgs mediated theories by careful measurements of these branching ratios. In practice this is difficult because of the hadronic uncertainties and also because λ can assume any value ≥ 0 in scalar mediated theories.

A much more dramatic and obvious signal of Higgs mediated decays would be an enhancement of the branching ratios for muons and strange particles. This would occur in most models because of the tendency for Higgs particles to couple preferentially to heavier fermions. In fact, Golowich[11] has shown that in the minimal SU_5 model with a single Higgs 5 the branching ratio for $p \to \mu^+ K^0$ would be $\gtrsim 80\%$, with $P(\mu^+) \simeq -1$ (the other important modes are $\nu_\mu^c K^+$ and $\mu^+\omega$). This would be a dramatic signal indeed! In this model, the Higgs mediated decay is expected to be observable for $M_H \simeq 10^{10}$-10^{11} GeV.

Supersymmetry at Low Energy

There has been much recent interest[1] in combining grand unified theories with supersymmetry that is unbroken down to low energies (susy GUTs). In such theories there are new fermion partners (in the Gev-Tev range) of the standard gauge and Higgs bosons and new spin-0 partners of the ordinary fermions. These new particles affect proton decay in two ways:

(a) they modify the renormalization group equations (RGE) so as to affect the predictions of M_X and $\sin^2\theta_W$. The results[12] as a function of $\Lambda_{\overline{MS}}$ and the number of light Higgs doublets n_H ($n_H \geq 2$ are needed in susy GUTs in order to generate fermion masses) are

n_H	$\Lambda_{\overline{MS}}$(GeV)	M_X^{susy} (GeV)	τ_p (yr)	$\sin^2\theta_W$
2	.1	4.8×10^{15}	2×10^{34}	.239
2	.2	1.1×10^{16}	4×10^{35}	.235
4	.1	2.6×10^{14}	1×10^{29}	.260
4	.2	5.5×10^{14}	3×10^{30}	.258

Table 1. Predictions (Marciano and Senjanovic[12]) for M_X^{susy}, τ_p, and $\sin^2\theta_W$ in susy GUTs at the two loop level. The τ_p estimate is for the ordinary (X,Y exchange) diagrams only.

For $n_H = 2$ the lifetime due to X, Y exchange is probably too long to observe. For $n_H = 4$ the τ_p and M_X^{susy} predictions are similar to the non-supersymmetric theory, but the prediction for $sin^2\theta_W$ is probably too large.

(b) The new particles also lead, at the one loop level, to new d=5 diagrams for the proton to decay into ordinary particles.[13] These are potentially very dangerous because the amplitude is proportional to $1/M_X^{susy}$, rather than $1/M_X^2$. However the diagrams are higher order and are suppressed by small Yukawa couplings and anomalous dimension factors[14], so that the lifetime from the new diagrams may be compatible with present experimental limits. In the simplest SU_5 susy model, the dominant decay modes are $p \to \nu_\ell^c K^+$ and $n \to \nu_\ell^c K^0$, where $\ell = \tau$ or μ.

Proton decay (into $\mu^+ K^0$, $\nu^c K^+$, etc.) may also be mediated by "light" ($M \gtrsim 10^{10}$ GeV) colored Higgs particles in susy GUTs.

Nuclear Effects

All foreseeable experiments involve nuclei rather than free nucleons. Nucleon decays within a nucleus are complicated by Fermi motion and also by the possibility that produced pions may be absorbed or elastically or charge-exchange scattered before getting out of the nucleus.[15]

There may also be desirable effects. In the process $p \to e^+$ + (virtual π^0), for example, the virtual π^0 can be absorbed by another nucleon producing an N, N* or Δ. The nucleon lifetime in nuclei may be shortened by anywhere from 5% to 50%, depending on short range correlations. Also, the three-quark fusion diagram of Figure 3b (with gluons radiated from the quarks absorbed by other nucleons) may be important in nuclei (although the uncertainties are large).

All three of these mechanisms (interactions of a real meson, of a virtual meson, and 3-quark fusion) would generate recoiling nuclear fragments and possibly associated pions. The last two would lead to a continuous momentum spectrum (from 0 to m_p) for the emitted lepton, with values larger than $m_p/2$ favored.

Finally, $\Delta B = 2$ processes associated with $d > 6$ operators may be observable in nuclei.

Models with Low Mass Scales

Operators with $d > 6$ could be important in models with mass scales $M \ll 10^{14}$ GeV. These operators may be distinguished experimentally from the $d = 6$ operators relevant to the 3-2-1 desert theories by their very different selection rules (e.g. the $d = 7$ operators conserve $B + L$ rather than $B - L$). A number of low dimension operators are listed in Table 2, along with typical processes that they initiate, their selection rules, and the mass scales for which they would lead to nuclear or nucleon lifetimes in the observable $10^{30} - 10^{33}$ yr range. (Unlike the 3-2-1 desert theories, there is no independent reason to expect the mass scales to actually be in this range. Nor, in general, is there a prediction of $sin^2\theta_W$.) Many of the models which lead to these operators have Higgs mediated interactions.

The $d = 9$ six quark operator leads to the interesting processes of neutron-antineutron oscillations[16], and to dinucleon annihilation

d	O	Process	ΔB, ΔL, ΔS	M (GeV)
6	$qqq\ell$	$p \to e^+\pi^0$	$\Delta B = \Delta L = -1$ $\Delta S = 0, 1$	$4 \times 10^{14} - 2 \times 10^{15}$
7	$qqq\ell^c\phi$	$n \to e^-K^+$	$\Delta B = -\Delta L = -1$ $\Delta S = 1$	$2 \times 10^{10} - 10^{11}$
7	$qqq\ell^c D$	$n \to e^-\pi^+$	$\Delta B = -\Delta L = -1$	$4 \times 10^9 - 2 \times 10^{10}$
10	$qqq\ell^c\ell^c\ell^c\phi$	$n \to \nu\nu e^-\pi^+$	$\Delta B = -1/3$ $\Delta L = -1$	$(3 - 7) \times 10^4$
11	$qqq\ell\ell\ell\phi^2$	$p \to e^+\nu^c\nu^c$	$\Delta B = +1/3$ $\Delta L = -1$	$(2 - 4) \times 10^4$
9	$qqqqqq$	$n \leftrightarrow \bar{n}$	$\Delta B = -2$	$4 \times 10^5 - 10^6$
		$pn \to \pi^+\pi^0$	$\Delta L = 0$	
		$nn \to 2\pi^0$		
12	$qqqqqq\ell\ell$	$pp \to e^+e^+$	$\Delta B = \Delta L = -2$	>500
		$H \leftrightarrow \bar{H}$		

Table 2. Baryon number violating operators of dimension d. q, ℓ, ϕ, and D represent quarks, leptons, Higgs fields, and derivatives, respectively.

in nuclei. The operator can be rewritten as an effective dineutron operator $\delta m \, n^T n$, where $\delta m \sim e^4 \, m_p^6/M^5$ is the neutron-antineutron transition mass. $n\bar{n}$ annihilations lead to a nuclear lifetime $\Gamma_{nuc} \sim a \times (\delta m)^2/m_p$. For $\tau_{nuc} > 10^{30}$ yr and $a \sim 10^{-2}$ (for wave-function overlap effects) one requires $\delta m > 10^{-20}$ eV, leading to a free neutron-anti neutron oscillation time $\tau_{n\bar{n}} = 1/\delta m > 10^5$ s. However, a number of recent more detailed estimates[1] of Γ_{nuc} find more stringent lower limits on $\tau_{n\bar{n}}$, ranging from 3×10^6 s to 5×10^7 s.

C. OTHER TOPICS

There have been speculations[17] that Poincaré invariance could be violated on the very short distance scales ($M_X^{-1} \sim 10^{-28}$ cm) relevant to proton decay. Proton decays could conceivably violate energy, momentum, or angular momentum conservation (e.g. $p \to e^+\nu$!). Unfortunately, it would probably be impossible to identify such final states as being due to proton decay.

Grand unified theories predict the existence of superheavy ($M_M \sim M_X/g^2 \sim 10^{16}$ GeV) magnetic monopoles. Such GUT monopoles could induce baryon and lepton number violations in matter through such processes as $Mp \to Me^+$ or $Me^+\pi^0$. One would naively expect the cross sections to be negligibly small ($O(M_X^{-2}) \sim 10^{-56}$ cm^2)), but several authors[18,19] have recently argued that the cross sections will be enormously enhanced by fermion cloud effects, perhaps to strong interaction size[18] ($\sigma \sim 10^{-26} cm^2$). In this case a monopole passing through a proton decay detector could induce a burst of energy every few meters (or cm)! Also, limits on x-ray fluxes from neutron stars[20,21] would then severely limit the possible universal flux of monopoles (at a level thirteen orders of magnitude more stringent than the flux suggested by the Stanford event[22]).

In models with composite fermions[23], quarks and leptons are generally made up out of the same constituents. Baryon number violating processes can then generally occur via the interchange of constituents, but it is difficult to make any general statements concerning the lifetime or selection rules.

There have been a number of models[1] in which the proton is made absolutely stable by imposing a globally conserved quantum number. This quantum number corresponds to baryon number for the ordinary fermions but not for a new class of exotic quarks and leptons. Such theories therefore allow a baryon asymmetry for the universe and predict the possibility of observing baryon number violation at accelerator energies[1,24] through such reactions as $e^+e^- \to Q\bar{Q} \to B\ell\ell\bar{E}^0\pi$'s, where Q is a new heavy (m ~ 50-100 GeV) quark and E^0 is a quasi-stable massive (m ~ 1-10 GeV) neutrino.

In the integer-charge-quark versions of the Pati-Salam models[1] (e.g. SU_4^4) there is a $M_S \sim 10^4 - 10^6$ GeV leptoquark boson that can lead to quark decays such as $q \to \pi\nu$ with $\tau_q \sim 10^{-6} - 10^{-9}$ s by mixing with the W. Proton decays into $3\nu \pi^+$, $3\nu\pi^+\pi^+\pi^-$, $\nu\nu_e^- \pi^+\pi^+$. etc., with $\tau_p \sim 10^{29} - 10^{34}$ yr can then occur via the simultaneous decay of all three quarks.

Cosmic Ray Antiprotons

There have been several recent observations[25] of cosmic ray antiprotons with a flux ($\bar{p}/p \gtrsim 10^{-4}$) an order of magnitude larger than expected from secondary production in the interaction of cosmic ray primaries with interstellar gas. Especially surprising is the report of low energy (100's of MeV) antiprotons, because essentially no secondary $\bar{p}$'s are expected in this region.

Possible explanations of this large flux of low energy $\bar{p}$'s, if it is confirmed by subsequent measurements are:[25]
(a) interesting new astrophysics, such as a modification of the standard picture of cosmic ray propagation, new deceleration mechanisms, or some sort of explosion in the galactic center.
(b) The existence of antimatter galaxies, although this view encounters severe cosmological difficulties.
(c) n-$\bar{n}$ oscillations (i.e., secondary neutrons oscillate into $\bar{n}$'s which then β decay). However, the observed $\bar{p}$ flux would require an oscillation time $\tau_{n\bar{n}} \simeq 10^4$ s, which is apparently ruled out by direct measurement[26] ($\tau_{n\bar{n}} > 10^5$ s) and nuclear stability ($\tau_{n\bar{n}} > 10^5 - 10^7$ s).
(d) The evaporation of primordial black holes.
(e) The decay of other exotic relics of the big bang, such as the heavy (m~1-10 GeV) quasi-stable neutrinos E^0 ($\to \bar{p}\ e^+\nu$) that occur in some models with a stable proton.

Summary

There are many possibilities for baryon number violation, the different models being characterized by different mass scales and selection rules. One large class of theories, those (except for susy GUTs) involving an $SU_3 \times SU_2 \times U_1$ invariant desert, predict $M_X \simeq 2.4 \times 10^{14}$ GeV and $\tau_p \simeq 3 \times 10^{29\pm2}$ yr for the small value of $\Lambda_{\overline{MS}}$ that is currently favored. The correct prediction of $\sin^2\theta_W \simeq .214$ lends strong support to this class of theories.

Other possible manifestations of baryon number violation include $n\bar{n}$ oscillations, cosmic ray antiprotons, baryon number violation at accelerators, and monopole induced reactions.

It is clearly desirable to search for these (and other) exciting possibilities as vigorously as possible.

REFERENCES

1. A more detailed discussion and a complete set of references may be found in my earlier reviews: P. Langacker, Proceedings of the 1982 Summer Workshop on Proton Decay Experiments (Argonne, 1982) ed. D. S. Ayres, and Int. Conf. on Baryon Nonconservation (Tata Institute, Bombay, 1982), Pennsylvania preprint UPR-0186T; Proc. of 1981 Intl. Symposium on Lepton and Photon Interactions, ed. W. Pfeil, (Bonn) p. 823; Proc. of Second Workshop on Grand Unification, eds. J. P. Leveille, L. R. Sulak, and D. G. Unger (Birkhäuser, Boston, 1981) p. 131; Physics Reports 72, 185 (1981).
2. H. Georgi and S. L. Glashow, Phys. Rev. Lett. 32, 438 (1974).
3. H. Georgi, H. R. Quinn, and S. Weinberg, Phys. Rev. Lett. 33, 451 (1974).
4. T. D. Goldman and D. A. Ross, Nucl. Phys. B171, 273 (1980); Phys. Lett. 84B, 208 (1974).
5. W. J. Marciano, Phys. Rev. D20, 279 (1979).
6. W. J. Marciano and A. Sirlin, Second Workshop on Grand Unification, p. 151.
7. A. J. Buras, 1981 Int. Sym. on Lepton and Photon Interactions, p. 636.
8. J. E. Kim, et al., Rev. Mod. Phys. 53, 211 (1981); W. J. Marciano and A. Sirlin, Phys. Rev. Lett. 46, 163 (1981); Nucl. Phys. B189, 442 (1981); Rockefeller preprint COO-2232B-206.
9. V. S. Berezinsky, B. L. Ioffe and Ya. I. Kogan, Phys. Lett. 105B, 33 (1981); Y. Tomozawa, Phys. Rev. Lett. 46, 463 (1981); 49, 507 (E) (1982), and Michigan UMHE 81-57; K. V. L. Sarma and V. Singh, Phys. Lett. 107B, 191 (1981); J. F. Donoghue and E. Golowich, U. Mass. UMHEP-166; A. W. Thomas and B. H. J. McKellar, TH-3376-CERN; N. Isgur and M. B. Wise, HUTP-82/A029.
10. S. Weinberg, Phys. Rev. Lett. 43, 1566 (1979), Phys. Rev. D22, 1694 (1980); F. Wilczek and A. Zee, Phys. Rev. Lett. 43, 1571 (1979), Phys. Lett. 88B, 311 (1979); A. Weldon and A. Zee, Nucl. Phys. B173, 269 (1980).
11. E. Golowich, Phys. Rev. D24, 2899 (1981).
12. S. Dimopoulos and H. Georgi, Nucl. Phys. B193, 150 (1981); L. E. Ibáñez and G. G. Ross, Phys. Lett. 105B, 439 (1981); M. B. Einhorn and D. R. T. Jones, Nucl. Phys. B196, 475 (1982); W. Marciano and

G. Senjanovic, Phys. Rev. D25, 3092 (1982).
13. S. Weinberg, Phys. Rev. D26, 287 (1982); N. Sakai and T. Yanagida Nucl. Phys. B197, 533 (1982).
14. J. Ellis, D. V. Nanopoulos, and S. Rudaz, Nucl. Phys. B202, 43 (1982); S. Dimopoulos, S. Raby, and F. Wilczek, Phys. Lett. 112B, 113 (1982).
15. See (1) and A. Nishimura and K. Takahashi, KEK 82-6.
16. See (1) and R. N. Mohapatra, CCNY-HEP-82/7; A. Raychaudhuri and P. Roy, Tata TIFR/TH 82-16; S.Rao & R.Shrock, Phys. Lett.116B,238.
17. J. Ellis, M. K. Gaillard, D. V. Nanopoulos, and S. Rudaz, Nucl. Phys. B176, 61 (1980); A. Zee, Phys. Rev. D25, 1864 (1982).
18. V. Rubakov, JETP Lett. 33, 644 (1981) and Nucl. Phys., to be published; C. G. Callan, Jr., Phys. Rev. D25, 2141 (1982), and to be published.
19. F. Wilczek, Phys. Rev. Lett. 48, 1146 (1982).
20. E. W. Kolb, S. A. Colgate, and J. A. Harvey, Princeton preprint.
21. Cosmological limits are considered by J. Ellis, D. V. Nanopoulos and K. Olive, Phys. Lett. 116B, 127 (1982).
22. B. Cabrera, Phys. Rev. Lett. 48, 1378 (1982).
23. For a review, see M. E. Peskin, 1981 Int. Symposium on Lepton and Photon Interactions, p. 880.
24. P. Langacker and D. Sahdev, to be published; P. Langacker and Y. Oh, to be published.
25. See (1); F. K. Gaisser and B. G. Mauger, Astr. Journal 252, L57 (1982); R. J. Protheroe, Astr. Journal 251, 387 (1982); A. Buffington and S. M. Schindler, Astr. Journal 247, L105 (1981).
26. M. Baldo-Ceolin, Padova preprint.

NUSEX EXPERIMENT AND THE FUTURE OF THE MONT BLANC LABORATORY

(CERN[1]-Frascati[2]-Milan[3]-Turin[4] Collaboration)

G. Battistoni[2], E. Bellotti[3], G. Bologna[4], P. Campana[2],
C. Castagnoli[4], V. Chiarella[2], D. C. Cundy[1],
B. D'Ettore Piazzoli[4], E. Fiorini[3], E. Iarocci[2],
G. Mannocchi[4], G. P. Murtas[2], P. Negri[3], G. Nicoletti[2],
P. Picchi[4], M. Price[1], A. Pullia[3], S. Ragazzi[3], M. Rollier[3],
O. Saavedra[4], L. Trasatti[2] and L. Zanotti[2]

1 - CERN, Geneva, Switzerland.
2 - Laboratori Nazionali dell'INFN, Frascati, Italy.
3 - INFN-Sezione di Milano, and Dipartimento di fisica dell' Università, Milan, Italy.
4 - Istituto di cosmogeofisica del CNR and Istituto di Fisica Generale dell'Università, Turin, Italy.

(Presented by P. Picchi)

INTRODUCTION

For 20 years, garages in the Mont Blanc road tunnel have been used as cosmic-ray underground laboratories. In 1973 [1] in one of these laboratories we carried out a proton decay experiment using one ton of liquid scintillator surrounded by a 4π anticoincidence shield. Of course we found no proton decays and gave a lifetime limit of $> 10^{29}$ years. However, at the same time we found no background events, and it was clear that the Mont Blanc laboratory was ideal for this type of experiment.

The laboratory is located deep underground; the minimum depth is ~ 5000 m.w.e. Figure 1 shows the azimuthal thickness profiles of the Mont Blanc as seen from the laboratory. There is only one problem: the typical size of the garages is $(8.6 \times 8 \times 6.5)$ m^3, which limits the mass of the detector.

Actual proton decay detectors fall into two classes: totally active water Čerenkov detectors, and tracking calorimeters.

In October 1979, when we had decided to make a new proton decay experiment, our choice (Nusex, fig. 2, a digital calorimeter) was made for several reasons: small laboratory size; our experience in digital tracking calorimeter techniques; and the desire to get quick and good results.

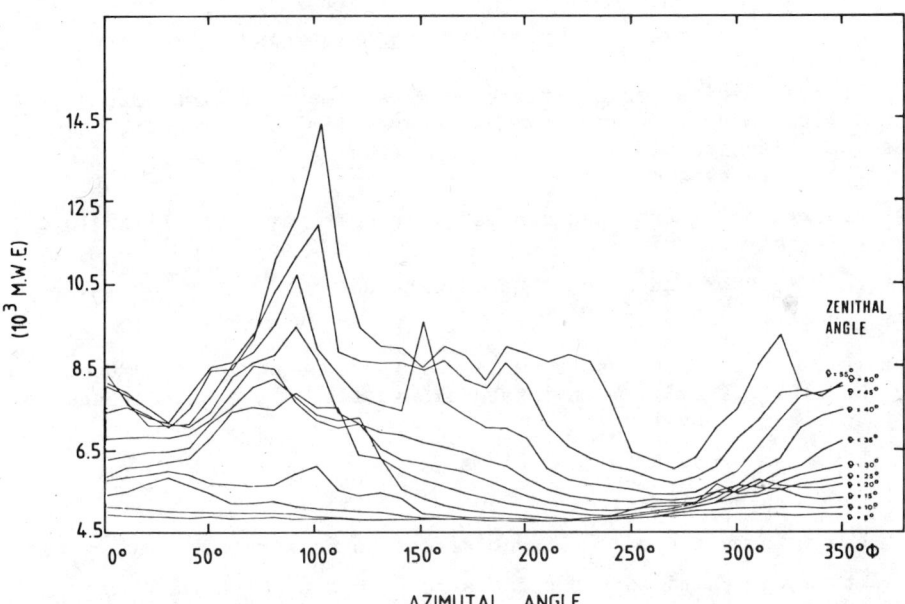

Fig. 1 Azimuthal thickness (m.w.e.) profiles of the Mont Blanc as seen from our laboratory.

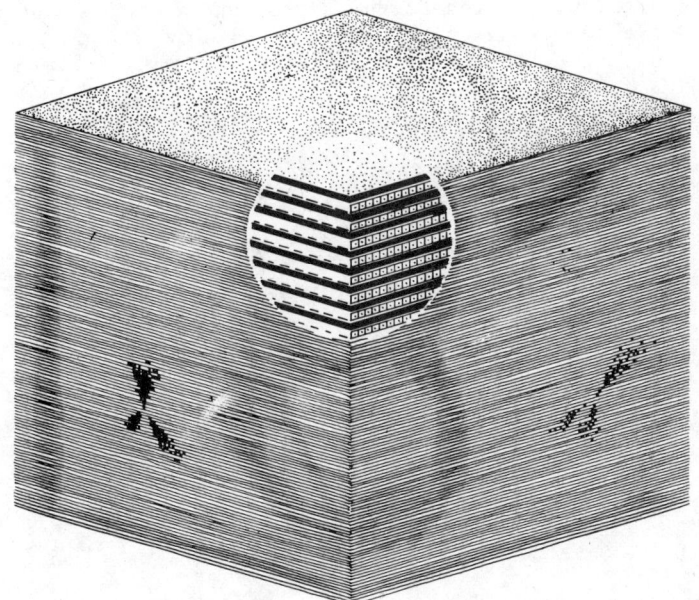

Fig. 2 Artist's view of the Nusex detector with three-prong nucleon decay.

FEATURES OF DIGITAL CALORIMETERS

Present digital calorimeters (DCs) are made of iron plates interleaved with saturated mode devices such as limited streamer, Geiger tubes, or flash chambers.

The advantages of DCs are:

i) tracks coming from nucleon decay products are visualized (pictures of events);

ii) the detector can be calibrated with beams;

iii) modularity;

iv) it is possible to use them as multipurpose detectors [muon bundles (fig. 3), muon interactions (fig. 4), exotic events, cosmic-ray physics],

but they also present some problems:

a) they can be expensive;

b) the information on the track direction is sometimes ambiguous;

c) it is very hard to use them as detectors of neutrinos from stellar collapse;

d) high-Z material introduces nuclear effects.

However, DCs are complementary to the Čerenkov devices. If in DCs it is sometimes difficult to locate the vertex and to prove the back-to-back nature of a two-prong event, their sensitivity to multibody decay modes is superior to that of Čerenkov detectors.

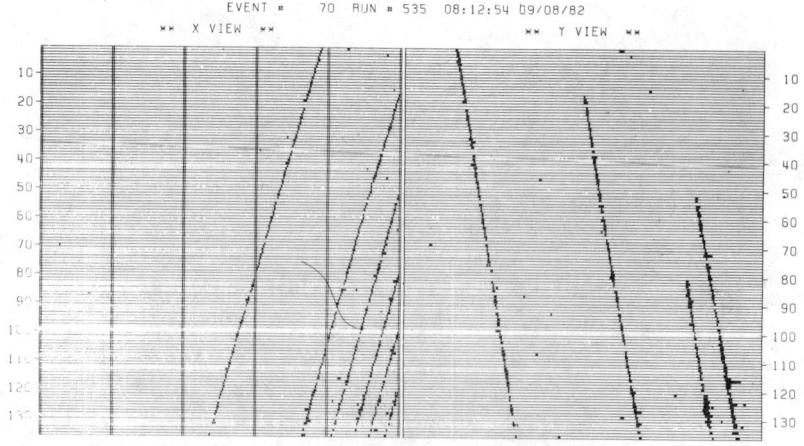

Fig. 3 The two orthogonal views of a six-muon bundle. In the y view we can see that the track 6 is coplanar with track 4, and track 5 is coplanar with track 3.

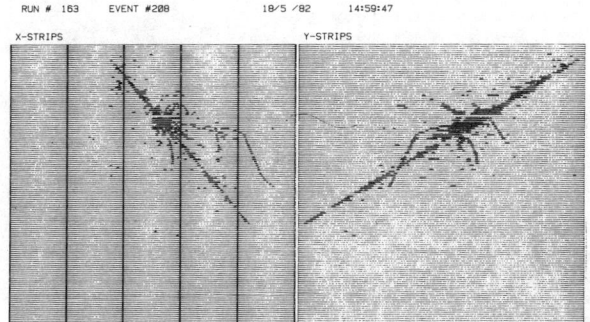

Fig. 4 The two orthogonal views of an inelastic hadronic muon interaction. The muon energy should be in the TeV region, and the energy lost in the apparatus is ~ 100 GeV.

THE DETECTOR

The detector consists of a cube (each side = 3.5 m) made of 134 horizontal iron plates, 1 cm thick, interleaved with planes of plastic streamer tubes (total number of tubes: 42,880). Each layer (iron + device) is 2.7 cm thick, with mean density $\rho = 3.5$ g/cm^3.

The basic element of our tubes (fig. 5) is a comb profile (3.5 m long) of extruded[2] PVC, which consists of eight open cells (cross-section 9×9 mm^2) coated with graphite (resistivity $\geq 5 \times 10^4$ Ω/square). Three sides of the tube cathode are made by the profile, the fourth by a PVC cover which is also coated. The wire thickness is 100 µm, and it is kept centred by PVC spacers every half-metre. A PVC container houses two profiles, and on the exterior there are only the connectors for the gas and the HV supply. The limited streamer is a self-triggered and saturated mode localized in a few millimetres, favoured by the use of thick sense wires and characterized by a large signal (~ 1 mA peak current, 40-50 ns wide) and exhibiting noiseless operation in a wide HV range. The large streamer signals and the plastic tubes with resistive cathode allow external pick-up electrodes, such as strips, to be used. In fact the readout is not through the wires but through the x and y strips which pick-up the pulses induced on the cathode. This bidimensional readout (fig. 6) makes it possible to record the detailed spatial pattern.

The gas mixture used under the Mont Blanc for streamer operation is: one volume of argon, two volumes of CO_2, and one volume of N-penthane. The performance is equivalent to that of the standard argon + isobutane mixture (1:3).

The electronic chain to process the strip signal (triangular shape, pulse height ≥ 3 mV/50 Ω and width ~ 40 ns) is very simple: an amplifier, a comparator with variable threshold, a one-shot, and a shift register bit.

The basic building block of our readout is the LeCroy model 4200 streamer tube card. This card, designed for easy mounting directly

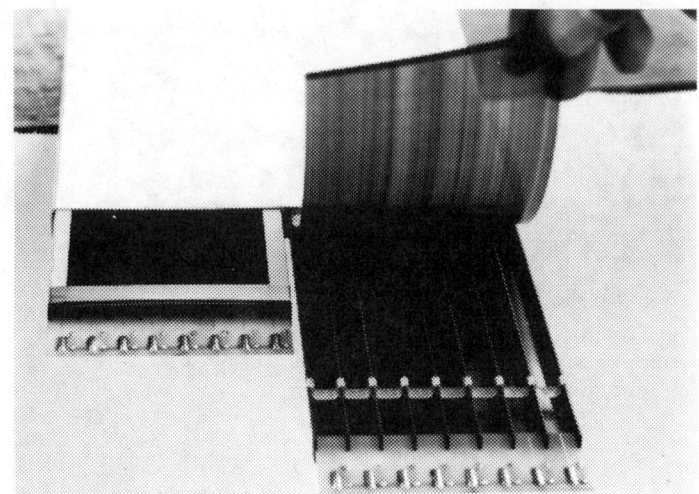

Fig. 5 Plastic streamer tube module.

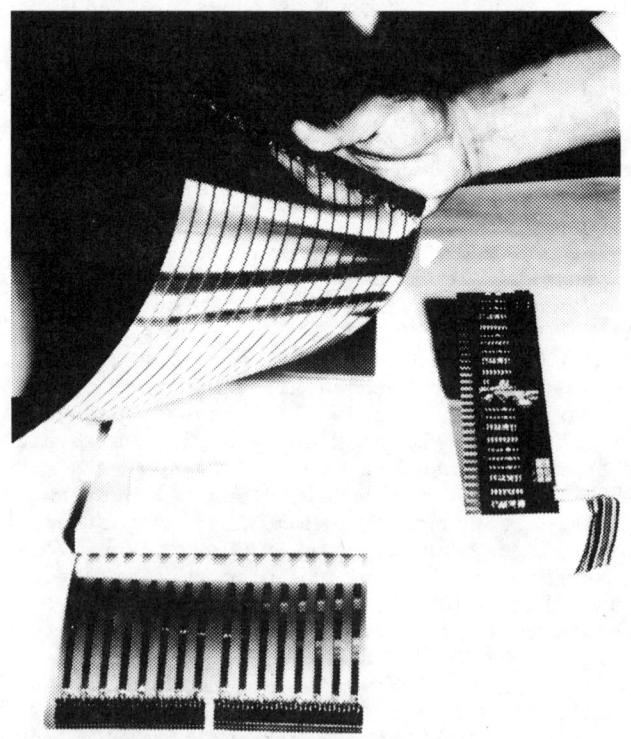

Fig. 6 Plastic streamer tube module with x and y pick-up strips and LeCroy Model 4200 streamer tube card.

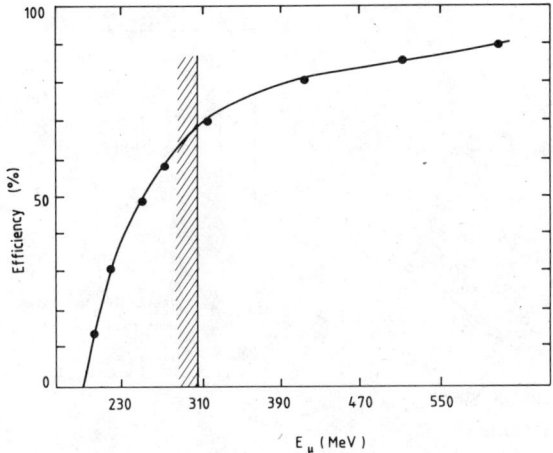

Fig. 7 Efficiency averaged throughout the isotropic angular distribution to detect a muon as a function of the total energy.

on the detector, contains all the necessary circuits to detect, amplify, discriminate, store, and read out 32 strip signals. The electronics cards on the edge of the x and y strips of the same plane are interconnected by a bus. The bus cables provide the readout control signals and the data path to the processor and to the trigger. This trigger is made by the OR's [$\Sigma(x+y)$ strips of the same plane] of the planes.

The loosest trigger that we take is the "AND" between four contiguous planes. The other comparable combinations (three contiguous planes -- AND -- two contiguous planes anywhere, etc.) are tighter. The minimum penetration of our trigger is 5 cm iron and the rate is $\sim$ 10 events per hour. In fig. 7 we show the efficiency for detecting a muon as a function of total energy E_μ. At E_μ = 300 MeV the efficiency (averaged throughout an isotropic angular distribution of the muons) is 80-90%.

BACKGROUND AND CALIBRATION

If the laboratory is located deep underground and a peripheral veto region is defined in the detector, the only significant background is the neutrino background. (In Nusex, the total flux per year of neutrons with energy E_n > 0.7 GeV, and uncorrelated with muons traversing the detector, is $\simeq$ 2.5). The best way to eliminate the neutrino background is by the identification of the event vertex.

In fig. 8 the patterns in which it is possible to identify particles in DCs are shown. The energy for π and μ is given from the visible range, for electrons and photons, from the number of hits. These patterns can sometimes be ambiguous and therefore the vertex identification could be difficult. Additional information is coming

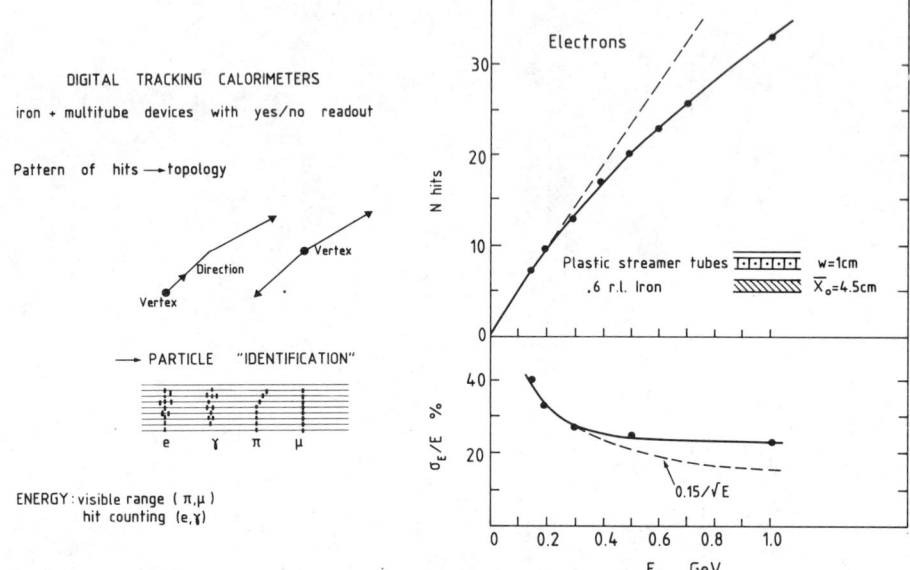

Fig. 8 Patterns from which it is possible to identify particles in a digital calorimeter

Fig. 9 Linearity and energy resolution as a function of electron energy

from stopping muons. The signal from the delayed positron provides at the same time the direction and the charge of the particle (μ, π, K) (98% of μ^- are absorbed in iron). The efficiency in our apparatus to detect the decay positron is $\simeq 46\%$.

It was important to calibrate the set-up with beams in order to measure its sensitivity. Two years ago we prepared a test module $[V = (3.5 \times 1 \times 1)$ m^3, M = 15 tons, 32 layers$]$ having the same geometry as the final detector, and we made different tests using π, e, and ν_μ beams. The small detector was exposed to the beam at both 0° and 45°.

In fig. 9 the calibration with electrons is shown; it exhibits the typical features of a DC: a linear behaviour at low energy, then a smooth deviation from linear response (10% non-linearity at $\sim$ 500 MeV). The energy resolution is about 20% $\sqrt{E}$, saturating at a constant value.

The response to 500 MeV π^- is the following: 85% pion identification, $\sim$ 10% π-μ ambiguity, and 5% π-e confusion. The most important test is with the neutrino beam (fig. 10). The energy spectrum (fig. 11) of our beam is close to the spectrum of cosmic-ray antineutrinos; however, a relevant difference is the negligible admixture of electron-neutrinos, which is about $\frac{1}{3}$ of the total atmospheric neutrino flux.

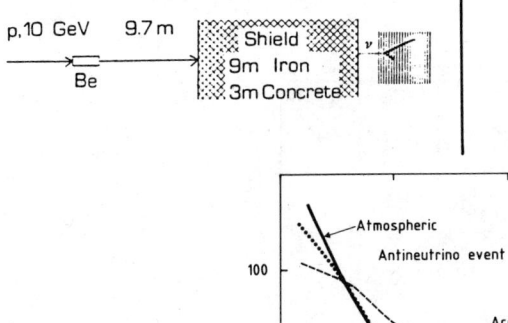

Fig. 10   Neutrino beam.

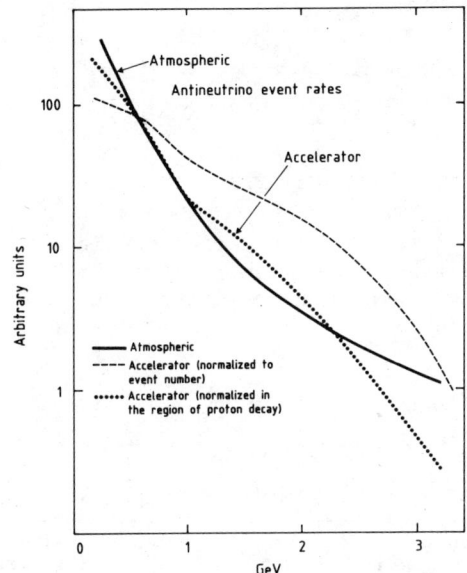

Fig. 11 Comparison of the atmospheric antineutrino spectrum with that of the beam obtained from 10 GeV protons on a bare target.

Our test set-up[3] has been exposed to neutrino beams produced by 1.8×10^{18} and 1.5×10^{18} protons on a beryllium target for the 0° and 45° incidence, respectively. Candidates for neutrino interactions have been selected on the basis of the following criteria:

a) At least three contiguous planes fire in the time gate.

b) No hits are recorded in the first two planes.

c) The vertex is contained between planes 3-25 and must be seen in both the x and the y views. This requirement gives a fiducial mass of two tons.

d) The visible energy E_{vis} is more than 300 MeV.

The total number of events in the various topologies are shown in table I.

It is now interesting to apply these results in order to estimate the background. The simplest definition of a nucleon decay candidate would be topological, but owing to Fermi momentum and π rescattering

Table I Experimental event topologies

	0°	45°
One prong	99	110
Two prongs	49	49
Three prongs	13	12
Four prongs	3	4
One prong + shower	11	8
Neutral current candidates	17	13
Total	192	196

inside the iron nucleus the correct definition is more difficult. To evaluate this difficulty we have made[4] a Monte Carlo study of some decay modes in which π's are born inside the nucleus. The results are summarized in table II.

Table II Pion reinteractions inside the nucleus

Decay type	π history (in %)		Energy and momentum conservation (%)	
$e^+\pi^0$	Absorbed	39.4	A)	43.9
	Elastic scattered	11.8	B)	41.8
	Charge exchanged	14.4	C)	39.9
	Not interacting	34.4	D)	37.9
$e^+\rho^0$	2π's absorbed	23.6	A)	16
	1π absorbed + 1π scattered	29		
	1π absorbed + 1π free	16.6	B)	11.4
	2π's scattered	9.7	C)	8.7
	1π scattered + 1π free	15.1	D)	7.0
	no π's interacting	6.0		
$e^+\omega$	3π's absorbed	5.1	A)	51.3
	2π's absorbed	16.4	B)	45.9
	1π absorbed	21.7	C)	42.5
	no π absorbed	56.8	D)	39.9
K^* ↳ $K + \pi$	π absorbed	41.3	A)	54.2
	elastic scattered	21.3	B)	42.2
	charged exchanged	10.1	C)	26.5
	not interacting	27.4	D)	15.4
where	A) → E_{tot} > 700 MeV, ρ < 450 MeV/c			
	B) → E_{tot} > 750 MeV, ρ < 400 MeV/c			
	C) → E_{tot} > 800 MeV, ρ < 350 MeV/c			
	D) → E_{tot} > 850 MeV, ρ < 300 MeV/c			

We define topologically a nucleon decay candidate as:

a) a vertex with two tracks subtending an angle $\geq 120°$;

b) a vertex with three tracks or more in which at least one track is backward in respect to the vector sum of the others.

Only nine of our neutrino events can simulate nucleon decays of type (a). No neutrino events with three or more tracks can simulate decays of type (b).

This background of 3% drops to 0.6% if the condition $E_{tot} >$ > 700 MeV, p < 450 MeV, in energy and momentum is respected. Thus in Nusex we have only 0.1 neutrino background event per year which can simulate nucleon decay. But considering the reinteractions inside the nucleus (table II), the maximum lifetime τ_{max} which can be measured in our set-up should be $\sim 10^{32}$ years.

RESULTS

The detector has been operated for the equivalent of 40 tons per year. We have observed 2870 muons crossing the apparatus, 23 µ's stopping in it, and 36 µ bundles.

In fig. 12 we show the depth-intensity curve in standard rock as obtained from our muons, and it agrees well with other measurements. Four fully contained events have been observed and are shown in figs. 13 to 16. The most natural interpretation of the first three of these is as interactions of atmospheric neutrinos; the fourth event is very difficult to explain as being neutrino induced. Before discussing them we would like to make two remarks:

1) our apparatus is able to distinguish between clear neutrino events and proton decay candidates;

2) the neutrino interaction rate observed is in agreement with theoretical predictions.

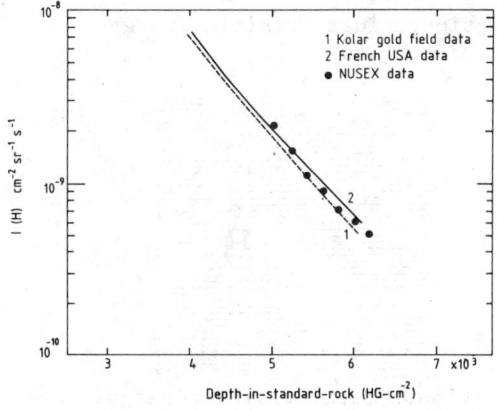

Fig. 12 Comparison of the intensity-depth curve in standard rock obtained by Nusex, with those of the Kolar Gold Field and the French-USA Collaboration

The two events shown in figs. 13 and 14 are elastic neutrino interactions. The first one has total energy 360 ± 15 MeV, zenithal angle $\theta_z = 35°$ or $145°$ (probably the neutrino is coming from below). The second one has total energy 330 ± 15 MeV and $\theta_z = 45°$ or $135°$.

The third event (fig. 15) is a clear example of inelastic upgoing neutrino interaction. The event consists of two charged tracks and an electromagnetic shower. Two interpretations are possible:

a) $\nu_e + N \to e + (\pi/p) + (\pi/p)$

b) $\nu_\mu + N \to \mu + \pi^0 + (\pi/p)$.

The total energy is 1.5 ± 0.4 GeV, the total momentum is 1.2 ± 0.3 GeV/c, and the zenithal angle is 166°.

In fig. 16 we show the two magnified views of the fourth event. If this event really has three prongs (vertex at A) it is very difficult to interpret it as a neutrino event; it could be seen as a true proton decay event. (In our neutrino test we saw no three-prong balanced event.) If this event has two prongs we have an inelastic neutrino interaction with the vertex at D producing two almost parallel charged particles (DB = μ, DAC = π, scattering at A). In this case the visible energy is 1.2 ± 0.3 GeV and the total momentum 1.2 ± 0.3 GeV/c, which is kinematically consistent with a neutrino interaction, such as $\nu + N \to \mu^- + \pi^+$ plus a low-energy nucleon.

Fig. 13 Elastic neutrino interaction. Total visible energy 360 ± 15 MeV, zenithal angle $\theta_z = 35°$ or $145°$.

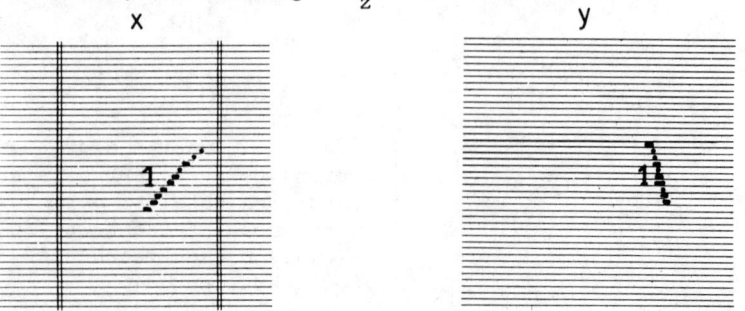

Fig. 14 Elastic neutrino interaction. Total visible energy 330 ± 15 MeV, zenithal angle $\theta_z = 45°$ or $135°$.

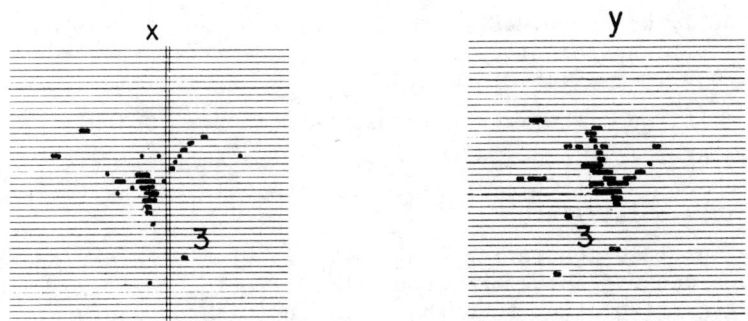

Fig. 15 Up-going inelastic neutrino interaction. Total visible energy: 1.5 ± 0.4 GeV; total momentum: 1.2 ± 3 GeV/c; zenithal angle θ_z: 166°.

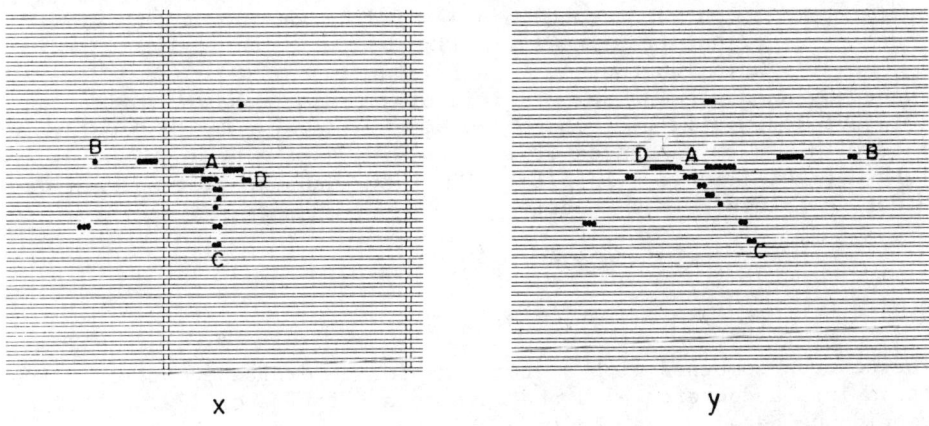

Fig. 16 The two orthogonal magnified views of the proton decay candidate.

This event (three prongs) can be interpreted as a nucleon decay in various modes. We prefer to choose $p \rightarrow K^0 + \mu^+$. The $m^*_{\pi\pi}$ (tracks AC and AD) is 0.55 ± 0.06 GeV; the muon and kaon momenta are respectively 0.38 ± 0.15 and 0.34 ± 0.1 GeV/c. The total momentum imbalance is 0.4 ± 0.2, in good agreement with proton decay if Fermi momentum is taken into account. Decay modes containing a ρ are also possible, but (table II) would be suppressed by π reinteractions in the nucleus.

CONCLUSIONS

The important conclusion is that we are waiting for other candidates.
It is clear that the real problem in a DC is the identification of the vertex.

Actually we are aided in this problem by the detection of the delayed positron from stopping μ^+. In order to improve our second-generation DC, we will implement the following modifications.

1) Finer grain: the iron thickness should be ~ 0.5 cm.

2) Time-of-flight counters: in fig. 17 we show the time resolution which it is possible to get, with the help of the tubes[5], by a typical scintillator.

3) The multiple scattering can sometimes give information about the track end and therefore about the direction. We need a better spatial resolution than that achieved with a tube hit. If we consider the streamer tube plane as a drift multiwire chamber, then using only a few multihit TDC channels per plane it is possible to get, on the x view, a resolution better than 1 mm.

In the Mont Blanc we have two laboratories: Nusex is located in the first one; in the second we are building another set-up (fig. 18) to detect neutrinos from stellar collapse.

We will have 100 tons of liquid scintillator (72 tanks) surrounded by an anticoincidence shield of streamer tubes covering 4π. To reduce radioactivity from the rock, this laboratory is completely lined with a 3 cm thickness of iron. Owing to the hydrogenous nature of the target (liquid scintillator) we have unbound protons, and we detect neutrinos by the reaction:

$$\bar{\nu}_e + p \to n + e^+$$
$$\hookrightarrow n + p \to d + \gamma .$$

(The neutron combines with a proton giving an excited deuteron which decays into a deuteron plus photon with a lifetime of 170 µs.) The coincidence between the positron and the photon strongly reduces the background from radioactivity and any type of noise.

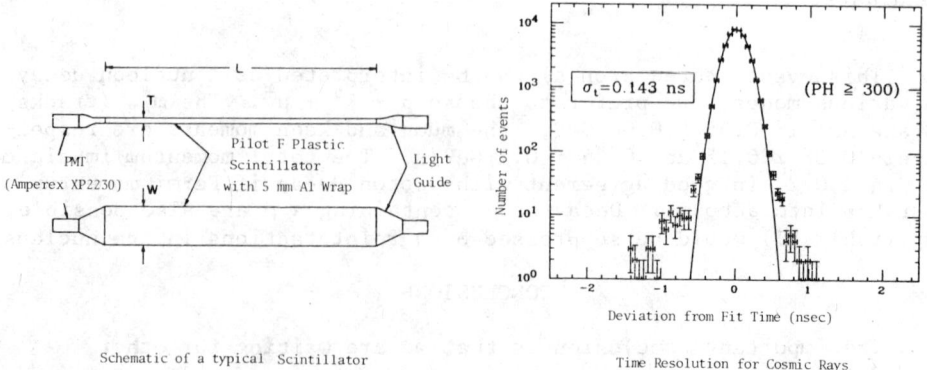

Fig. 17 Time resolution obtained with a typical scintillator.

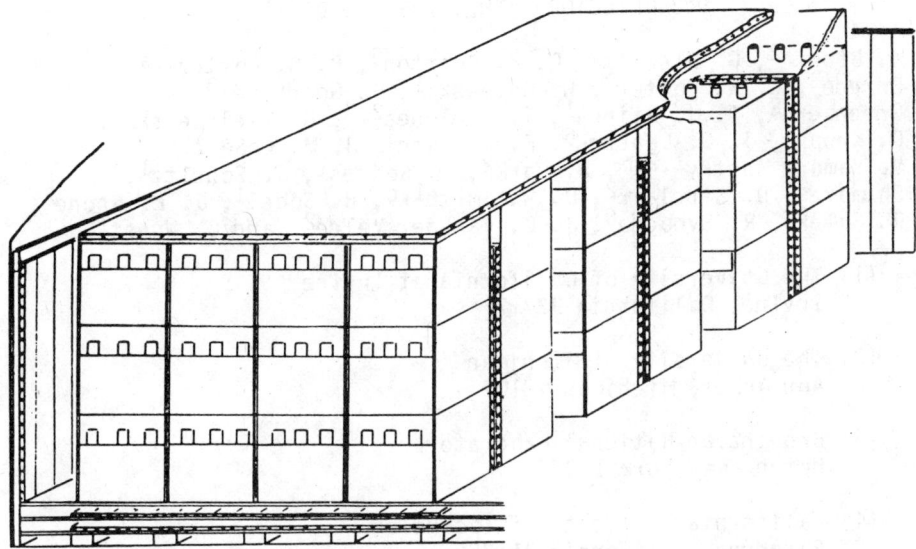

Fig. 18 Liquid scintillator detector with streamer tube anticoincidence.

Before the summer of 1983 the apparatus will be ready, and we want to synchronize in real time (error < 500 µs) the two detectors under the Mont Blanc and the gravitational-wave experiment at CERN. A world-wide network in real time of all experiments which are looking for rare events would be very useful.

It is possible that in the near future one will observe stellar collapse, and determine the neutrino mass and some time correlations in muon bundles with these detectors.

REFERENCES

1. L. Bergamasco and P. Picchi, Lett. Nuovo Cimento 11, 636 (1974).
2. G. Battistoni et al., Resistive cathode transparency, Frascati Report LNF-82/6 (1982), and Nucl. Intrum. Methods (in press).
3. G. Battistoni et al., The Mont Blanc experiment, Proc. Icoben Bombay Suppl. (1982) Pramana (in press). A more detailed report on the results of the neutrino run and the analysis of the background from the atmospheric neutrinos is in preparation.
4. S. Ragazzi, private communication.
5. S. Marini and F. Ronga, private communication.

IMB DETECTOR - THE FIRST 30 DAYS[*]

R. M. Bionta[2], G. Blewitt[4], C. B. Bratton[5], B. G. Cortez[2,a],
S. Errede[2], G. W. Foster[2], W. Gajewski[1], M. Goldhaber[3],
J. Greenberg[2], T. J. Haines[1], T. W. Jones[2,7], D. Kielczewski[1],
W. R. Kropp[1], J. G. Learned[6], E. Lehmann[4], J. M. LoSecco[4],
P. V. Ramana Murthy[1,2,b], H. Park[2], F. Reines[1], J. Schultz[1],
E. Shumard[2], D. Sinclair[2], D. W. Smith[1,c], H. Sobel[1], J. L. Stone[2],
L. R. Sulak[2], R. Svoboda[6], J. C. van der Velde[2], and C. Wuest[1].

(1) The University of California at Irvine
    Irvine, California 92717

(2) The University of Michigan
    Ann Arbor, Michigan 48109

(3) Brookhaven National Laboratory
    Upton, New York 11973

(4) California Institute of Technology
    Pasadena, California 91125

(5) Cleveland State University
    Cleveland, Ohio 44115

(6) The University of Hawaii
    Honolulu, Hawaii 96822

(7) University College
    London, U.K.

(a) Also at Harvard University

(b) Permanent address: Tata Institute of Fundamental
    Research, Bombay, India

(c) Permanent address: University of California, Riverside,
    California

* Talk presented by D. Sinclair

ABSTRACT

A large water Cherenkov detector, located 2000 feet below ground, has recently been turned on. The primary purpose of the device is to measure nucleon stability to limits 100 times better than previous measurements. The properties of the detector are described along with its operating characteristics.

INTRODUCTION

In this talk I shall describe the properties of the IMB detector and summarize the first 30 days of operation.

Figure 1 shows a schematic diagram of the detector. It consists of a rectangular volume of water bounded by six planes of photomultiplier tubes, 2048 in all. The tubes have hemispherical photocathodes, 5" in diameter. The spacing is about 1 meter. The volume of water contained within the six planes is 6800 m^3, while the volume held by the cavity is somewhat larger, ~ 8000 m^3.

The cavity was hewn from the rock and salt of the Fairport Mine of the Morton-Thiokol Corporation. It is lined with a double layer of black, high density polyethylene sheet, each 2.5 mm in thickness. A pump-out between the two layers removes the small amount of water which leaks out of the inner liner, about 10 gal. per hour. The cavity is at a depth of 1940 feet (1600 meters water equivalent).

SUMMARY OF DETECTOR PROPERTIES

The detector operates by sensing the Cherenkov light emitted by charged particles with velocities in excess of $\beta = .75$. Each photomultiplier (PM) tube senses the intensity and time of arrival of the Cherenkov wave front. This information is recorded and may be used to reconstruct the position and direction of the tracks of charged particles. The gamma rays from $\pi°$ decay produce electromagnetic showers of $e^{\pm}$ in the water and are thus detected. The detector is therefore well-suited for the detection of proton decay to $e^+\pi°$.

The properties of the detector are summarized in the following table:

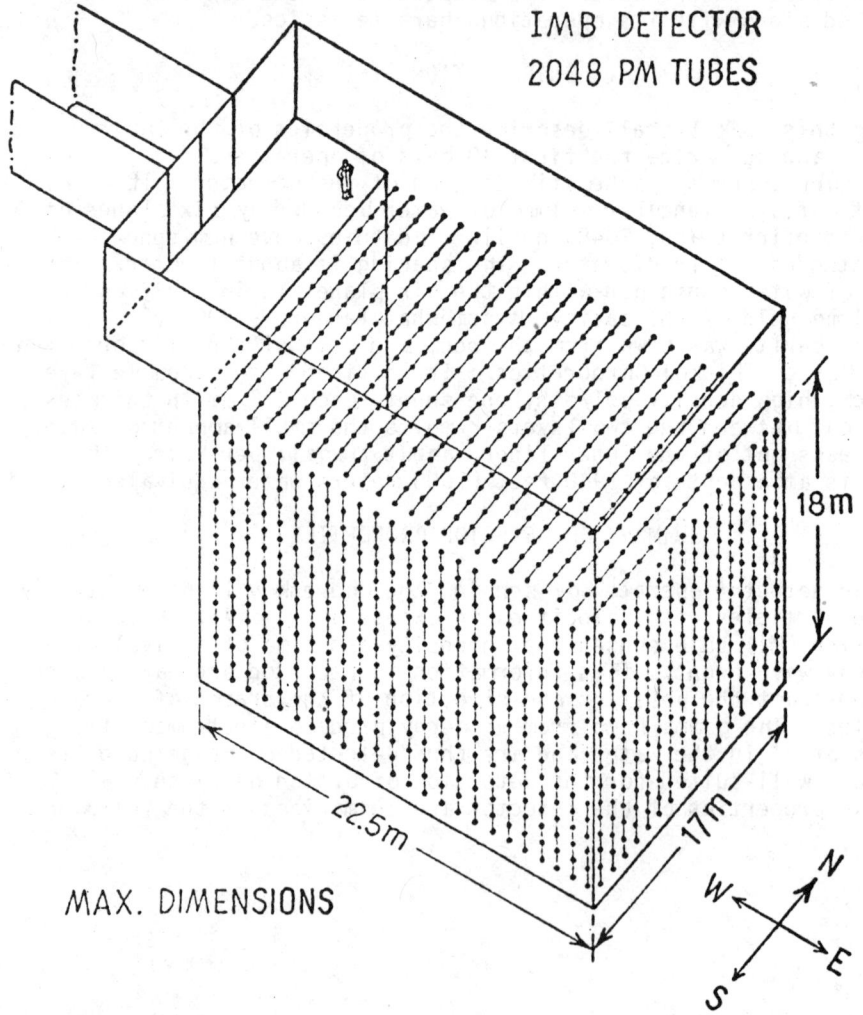

Figure 1. A schematic view of the IMB detector.

TABLE 1. PROPERTIES OF THE IMB DETECTOR

SIZE:

    Between PM-tube planes: 4×10^{33} nucleons
    Estimated fiducial volume: 2×10^{33} nucleons

ENERGY RESOLUTION:

    500 MeV shower: $\sigma = 11\%$
    500 MeV $\pi\pm$, $\mu\pm$: $\sigma = 15\%$
    1 GeV shower: $\sigma = 8\%$

VERTEX LOCALIZATION:

    Two tracks, wide opening angle: $\sigma \sim .5$ m
    Single track: $\sigma \sim 2$ m

ANGULAR RESOLUTION:

    Showers ($e\pm$, $\pi°$): $\sigma = 10°\text{-}20°$
    Charged tracks: $\sigma = 5°$

TRIGGERING:

    Noise triggers: $\lesssim 1\%$
    Cosmic rays: 2.7 ev./sec.
    Energy threshold: ~ 30 MeV

OPERATION OF THE DETECTOR

The detector was filled with water in July '82 and the PM tubes were installed immediately thereafter. Serious data taking had started by the end of August. At the time of this talk (Sept. 28, '82) the detector has taken 30 days of data suitable for the study of nucleon decay. This corresponds to a duty cycle of $\sim 70\%$. All but 1% of the PM tubes are working. The quality of the data is excellent, corresponding closely with our Monte Carlo simulations.

A. TRIGGERING

The discriminator threshold for each PM tube is set at a level corresponding to $\sim .5$ of the pulse height for a single photoelectron. The trigger logic divides the detector into 32 patches, each patch consisting of 64 PM tubes in an 8 x 8 square. A patch trigger consists of a discriminator pulse from 3 PM tubes in the same patch within a 50 nsec coincidence. A detector trigger consists of either 2 patch triggers in 150 nsec coincidence or 10 PM tubes anywhere in the detector in 50 nsec coincidence.

The trigger rate is due almost entirely to cosmic rays and is normally 2.7 triggers/sec. Noise triggers are usually less than 1%.

The minimum energy necessary to trigger the detector is 30 MeV if the particle is an electron.

B. RECORDED DATA

For each PM tube the following is recorded:

1. The time of the pulse. There are two time scales:

    a. T_1 scale; 0-512 nsec in 1 nsec intervals.
    b. T_2 scale; 0-7.5 nsec in 15 nsec intervals.

2. The charge (Q) collected from the photocathode. The Q scale ranges from 1 to 512 and this corresponds approximately to the range 1 to 50 photoelectrons though it is very non-linear.

C. CALIBRATION

The T_1, T_2 and Q scales are calibrated by means of a pulsed light source located at the center of the detector. This light source consists of a diffusing ball connected via an optical fiber to a nitrogen laser. This laser produces pulses of u.v. light (λ = 330 nm) whose time variance is less than 1 nsec. The intensity of the light may be varied by the insertion of neutral filters. With this system we obtain for each PM tube:

1. Absolute time calibrations of the T1 and T2 scales for different light intensities.

2. Relative Q calibrations in the range 1 to 50 pe.

A single overall constant is needed to provide the absolute Q calibration. This is obtained by computing the total Q expected from vertical muons traversing the center of the detector and comparing this with the data for such muons.

THE HARVARD-PURDUE-WISCONSIN BARYON DECAY
EXPERIMENT AT PARK CITY, UTAH

David B. Cline[*]
CERN, Geneva, Switzerland

ABSTRACT

A large tank of water, instrumented with photomultipliers and surrounded by an active shield using proportional wire tubes is being constructed in an underground laboratory at Park City, Utah. Baryon decay can be separated from background using the Cherenkov light pattern to a lifetime of approximately 4×10^{32} years in three years of running time. The status of the experiment is reported here.

PROTON STABILITY

The subject of proton stability has been of interest since at least 1929 Recently there has been renewed interest in this experimental measurement for three reasons:
1. <u>Baryon Number</u> is not coupled to a massless gauge field and hence need not be conserved.
2. <u>The Universe</u> appears to be <u>asymmetric</u> in baryon number and $N_B/N_{photons} = 10^{-9\pm1}$.
3. <u>Unified Theories</u> - Theories (GUTS) of Strong/Weak and Electromagnetic Interactions imply nonconservation of baryon number.

We will not attempt to reference the latest theoretical calculations since by now the list is very large and well known to experts in that field. However it is essential to make some assumptions about the decay products of protons or bound neutrons if they are to be detected experimentally. We assume that the energy of the proton or neutron is visible in either e^-, μ^- or pions or kaons. Decays with one or more neutrinos could be directed in some cases if the proton lifetime were short enough but would be extremely difficult to use as prime evidence for a finite proton lifetime.

Beyond the assumption of "detectable" charged or neutral particles in the final state the other essential assumption used in the detector design is the nature of the background events, including neutrino interactions in the detector and cosmic ray induced hadronic interactions. For the neutrino background extensive Monte Carlo calculations have been carried out using actual neutrino events observed in bubble chambers. These calculations are used as an input to the detector design. The case

---

[*]Permanent address: University of Wisconsin, Madison, Wisconsin.

of hadronic background is more complex and will be discussed further in a later section.

There are three important components of the detector: the volume array of 5" photomultipliers; a 4π box of gas proportional counters and a large "active shield" surrounding the centered detector.[1]

The detector can also be used to observe low energy cosmic ray neutrino interactions. These neutrinos traveling terrestrial distances across the earth in principle contain information about neutrino mixing through neutrino oscillations. The range of δm^2 that can be reached is about 10^{-3} eV^2.

The signature of a proton decay occurring in the sensitive volume is:

1. The total energy released in the event is approximately 938 MeV.
2. The decay products have no net momentum.
3. The event originates inside the fiducial volume.

Recognition of this signature in water is possible because the proton decay products will produce Cherenkov radiation. There is considerable experience with the use of water calorimeters to measure total energy as required by criterion 1. The experimental data are in good agreement with Monte Carlo electromagnetic shower calculations and indicate the general validity of the calculations which we have made. These results show that we can obtain total energy resolution of order 20 percent ($\Delta E/E$ at 1 GeV) by collecting 5-10 percent of the available Cherenkov light (i.e. the light which would be collected by a photocathode a very small distance from the light source). The directionality of the Cherenkov light permits the use of criteria 2 and 3 to distinguish proton decay events from background. Reconstruction of the Cherenkov light cones will permit a determination of the event vertex to better than 0.3 m in our design and provide a measurement of the net momentum.

The most important advantage of the volume lattice detector is that it collects light more efficiently than the surface detector when both are filled with water with a finite attenuation length. This larger efficiency means that for a given number of photodetector devices and a given fiducial volume, the lattice detector will make a better measurement of the total light and thus the total energy of an event. The realistic ratio of efficiencies between the two detectors is approximately a factor of five, which implies more than a factor of two better resolution for the volume lattice detector. This enhanced resolution significantly improves the ability of the detector to discriminate between proton decay and background events.

The validity of these claims can be proven analytically in an exercise which is useful for understanding how the volume detector works.[1,2] Consider a volume detector of volume V with N phototubes of cross-sectional area A. The volume of the unit cell v is then given by the relation $v = V/N$. Assume that the detector is filled with water which has an attenuation length L. A spherical shell of radius r and thickness dr, centered on a light source has a volume $dV = 4\pi r^2 dr$. The number of phototubes in volume dV is dV/v and the total surface area they present to the light source is $dS(p) = (dV/v)A$. The efficiency for collecting randomly generated photons is then

$$E = \int_V [dS(p)/S(s)]e^{-r/L} = NAL/V$$

This approximate scaling rule has been verified by detailed Monte Carlo calculations.

The Cherenkov radiation is generated by Monte Carlo techniques photon by photon. The number of photons per cm of path length for particles with a velocity of c is calculated by the equation

$$N = \int_{\lambda_1}^{\lambda_2} dN = 2\pi a L [1 - \frac{1}{\beta^2 n^2}] (\frac{1}{\lambda_1} - \frac{1}{\lambda_2})$$

= 500/cm between 250 and 710 nm.

Many detector geometries were simulated.[1] Parameters entering into the simulation include: lattice spacing of detectors, actual and effective radius of detectors, wavelength dependence of efficiency of detectors, size and shape of lattice, reflectivity of the detector walls (with or without mirrors), and position of reflecting walls. The number of photoelectrons collected in all the photomultipliers should be a measure of the total energy of the showers created by the e+ and π^0 from proton decay. Good resolution in this number will sharpen any peak corresponding to true proton decay events over most backgrounds. It has also been verified that with a mirrored detector, the light collected in a given event is independent of the position of the event in the detector and the volume of the detector for a wide range of those parameters.

We now turn to more specific calculation. The properties of the various decay modes of the proton are given in Table 1: these serve as the signature for the detection of proton decay. Using these decay modes we have carried out Monte Carlo simulation. The expected number of collected photoelectrons per decay mode is given in Table 2.

In order to collect adequate light and directional information from the $e^+p/e^+\omega$ decay modes the detector must be "close packed." Fortunately, the same configuration now allows the detection of the $\mu \to e$ sequence, which provides an additional signature for these decay (i.e., $\rho^0 \to \pi^+\pi^-$; $\pi \to \mu \to e$; $\omega^0 \to \pi^+\pi^-\pi^0$; $\pi \to \mu \to e$). We expect the neutrino backgrounds to be reduced because neutrino event with multiple pion production provide the major background and the rates are expected to be very low.

TABLE 1

Decay Type	Lepton	Meson	
$p \to e^+ \pi^°$	$\langle E_{e^+}\rangle > $ ~ 450 MeV	$\langle E_{\pi^°}\rangle$ ~ 480 MeV $\langle E_{\gamma 1,2}\rangle$ ~ 240 MeV $\langle \theta_{\gamma 1 \gamma 2}\rangle$ ~ 45°	Back to back Event. Energy balance between 1 forward and 2 backward in showers
$p \to e^+ \eta^°$	$\langle E_{e^+}\rangle$ ~ 310 MeV	$\langle E_{\eta^°}\rangle$ ~ 630 MeV $\langle E_{\gamma 1,2}\rangle$ ~ 320 MeV $\langle \theta_{\gamma 1 \gamma 2}\rangle$ ~ 125°	"Triangular event". 3 em showers of ~ same energy
$p \to e^+ \rho^°$	$\langle E_{e^+}\rangle$ ~ 150 MeV	$\langle E_{\rho^°}\rangle$ ~ 790 MeV $\langle E_{\pi^\pm}\rangle$ ~ 400 MeV $\langle \theta_{\pi^+\pi^-}\rangle$ ~ 150°	"Triangular event" 3 low energy. Few Cherenkov light tracks e^+ from $\pi^+ \to \mu^+ \to e^+$ decay chain.
$p \to e^+ \omega^°$	$\langle E_{e^+}\rangle$ ~ 145 MeV	$\langle E_{\omega^°}\rangle$ ~ 800 MeV $\langle E_{\pi^\pm}\rangle$ ~ 270 MeV Isotropic 3 body decay	Isotropic event. 5 low energy. Few Cherenkov light tracks. e^+ from $\pi^+ \to \mu^+ \to e^+$ decay chain.
$p \to \mu^+ \pi^°$	$\langle E_{\mu^+}\rangle$ ~ 465 MeV	$\langle E_{\pi^°}\rangle$ ~ 475 MeV $\langle E_{\gamma 1,2}\rangle$ ~ 235 MeV $\langle \theta_{\gamma 1,2}\rangle$ ~ 45°	Back to back event. No light balance between low C light back and e.m. show. e^+ from μ^+ decay.

TABLE 2

Decay Mode	Signature	Number of Photoelectrons (wavelength shifter added)	Major Backgrounds
$p \to e + \pi^\circ$	$E_e, E_\pi \sim 460$ MeV $\theta_{e\pi} < 6^\circ$ $E_{vis} \sim 940$ MeV	1300±150	$\nu_e N \to e\pi X$ +nuclear scattering
$p \to e + \eta$ $\gamma\gamma/3\pi^\circ$	$E_e \sim 300$ MeV $E_{vis} \sim 940$ MeV	1300±150	$\nu_e N \to en(\pi^\circ)X$
$p \to e + \omega$ $(\mu \to e)*$	$E_e \sim 150$ MeV $E_{vis} \sim 450$ MeV	675±100 (50±20)*	$\nu_e N \to en(\pi)X$ n>1
$p \to e + \rho^\circ *$	$E_e \sim 150$ MeV $E_{vis} \sim 380$ MeV	530±75	
$p \to \mu + \pi^\circ *$	$E_\mu, E_\pi \sim 460$ MeV $\theta_{\pi\mu} < 10^\circ$ $E_{vis} \sim 650$ MeV	900±150	$\nu_\mu N \to \mu\pi X$ +nuclear scattering
$p \to \mu + K_1^\circ *$	$E_\mu \sim 300$ MeV $E_{vis} \sim 325$ MeV $E_e \sim 460$ MeV	450±65	$\nu_\mu N \to \mu\pi\pi X$ $\mu K_1^\circ X$
$n \to e + \pi^-$	$\theta_{e\pi} < 6^\circ$ $E_{vis} \sim 800$ MeV	1100±140	$\nu_e N \to e\pi X$
$n \to e + \rho^-$	$E_e \sim 150$ MeV $E_{vis} \sim 470$ MeV	650±90	$\nu_e N \to e\pi\pi X$
$n \to \mu + \pi^- *$	$E_\mu, E_\pi \sim 460$ MeV $E_{vis} \sim 615$ MeV	850±125	$\nu_\mu N \to \mu\pi X$

*Events with $\mu^+ \to e^+$ decays yielding $\sim$ 50 photoelectrons within a few microseconds.

We can define the following properties of a proton decay event:
1. Asymmetry = <asym> = $(N^+ - N^-)/(N^+ + N^-)$, where $N^+(N^-)$ are the number of photon electrons collected forward (backward) of the assigned vertex.
2. Sphericity of the event <sph>, defined in the same way as for events in e^+e^- experiments.

The <asym> and <sph> values for the various decay modes are listed in Table 3 (for the nonwavelength shifter case). For comparison this average value of these quantities for neutrino interactions is
<asym> = 0.7
<sph> = 0.1

TABLE 3

Nonwave Length Shifted Case

Decay Mode	<ASYM>	<SPH>	<D>*cm
$p \to e^+ \pi^°$	.16	.52	.29
$p \to e^+ \eta^°$	.17	.75	.35
$p \to e^+ \rho^°$	.24	.67	.35
$p \to e^+ \delta$	.32	.78	.37

*RMS error on vertex location

These values are considerably different from the baryon decay values and provide a technique for discrimination. In addition the pulse height spectrum provides an additional discrimination.

The technique for identifying (p/n) decays, put simply, is
1. Reconstruct events.
2. Calculate <asym>, <sph> and pulse height.
3. For p/n decay <asym> <0.4, <sph> > 0.5.
4. The e^+ spectrum has unique structure associated with the $\pi^°e^+/\omega e^+/\rho e^+$ decays, i.e., for nonwavelength shifted water the number of photoelectrons associated with the e^+ will be ~ 330 ($\pi^°e^+$) or ~ 110 (ωe^+, ρe^+) or, for the wavelength shifted case, ~ 700 and ~ 250, respectively.
5. The $\pi^+(\mu^+)$ tracks from $\rho^°$, $\omega^°$ ($\mu^+\pi^°$) decay have a unique range.

The spacing of the phototubes in the detector is approximately one meter, the time taken for light to pass from one tube to another is ~ 3 ns whereas the time taken to cross the full detector is approximately 50 ns. Thus it is possible to separate direct light and light reflected from the surface of the detector -by mirrors - since the water is very transparent it is thus possible to increase the light collection by using a mirror system. Figure 1 shows the time spectrum of the reflected light. It is also necessary to provide time digitizing with 3-6 ns resolution and multihit capacity. In addition the pulse height from each tube will be recorded by a LeCroy 2282A ADC.

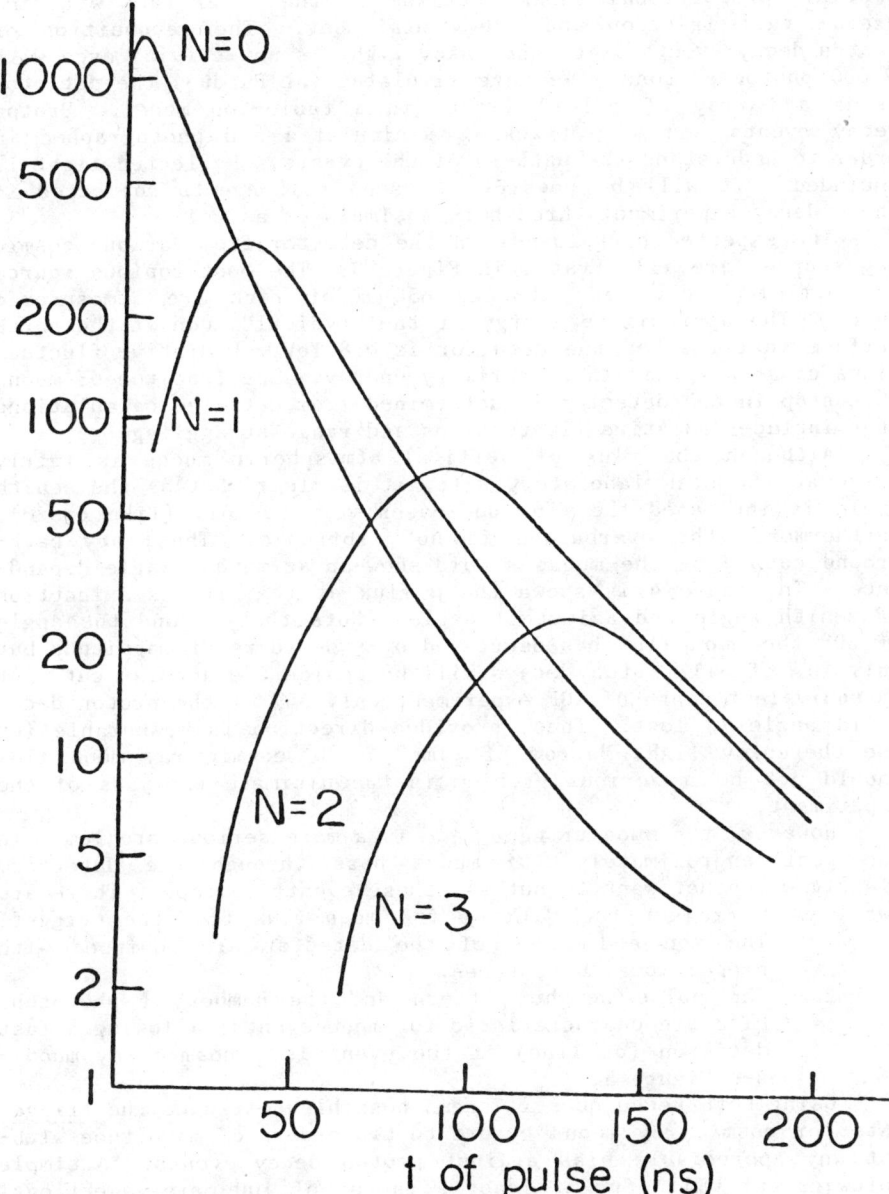

Figure 1. Time structure of the Cherenkov photons after reflection from the mirrors on the detector walls.

Monte Carlo calculation of the detection efficiency and energy containment for proton decay have been carried out - the results are shown in Figure 2. For 90% of all proton decay events 70% of the energy is contained in the detector. Including detectors (proportional tubes) outside of the water tank will increase this fraction and the containment. The recognition of proton decay events using Cherenkov light is not trivial even with $\sim$ 600 photoelectrons. We have simulated (at Purdue) the detector using an array of pulsed lights in a table-top model. Proton decay events and muon tracks are simulated and photographed in order to understand the pattern of the events. Reflected light is included. It will be possible to send real events collected in the ρ decay experiment through this simulator as well.

The expected backgrounds in the detector from various cosmic ray sources are illustrated in Figure 3. The most copious source of particles that pass through 650 m of rock are atmospheric muons. The approximate energy of the "typical" muon at the earth surface that reaches the detector is 0.8 TeV. Radiative fluctuations cause a spread in the primary energy. The fraciton of muons that stop in the detector is determined from detailed calculations that include radiative fluctuations and range straggling.

Although the flux of vertical atmospheric muons is fairly large at the Utah laboratory site, it is clear that as the zenith angle is increased the flux decreases very rapidly (like $\cos^3\theta$). Furthermore, the overburden is not isotropic. Thus, any background caused by the muons should show an azimuthal angle dependence. In Figure 4 is shown the μ flux at the site as a function of zenith angle and azimuthal angle. Note that beyond the angle of 60° the muon flux has decreased by two orders of magnitude but only 25% of all proton decays will be inside the angular cut. At an equivalent depth of KGF experiment only 35% of the proton decay solid angle is lost. Thus, provided direction is measurable (by the Cherenkov light "arrow of time"), the cosmic ray muon flux should not be a serious problem in the ultimate analysis of the experiment.

However, the muon trigger rate is a more serious problem. In one year approximately 10^7 muons pass through the detector. Clearly we do not want to put all these events on tape. There are two ways to reject the bulk of the muon flux from the trigger.
1. The top and sides of the detector are equipped with proportional wire tubes.
2. The pulse height pattern and the number of phototube hits are characteristic for muon events, allowing a fast decision (on line) if the event is a cosmic ray muon - see Figure 5.

Using both techniques it seems possible to reduce the <u>trigger</u> rate for cosmic ray muons by one to two orders of magnitude without any appreciable bias against proton decay events. A simple software cut will give at least a factor of 100 more rejections. Detailed timing and pattern information will reduce this background to a negotiable level.

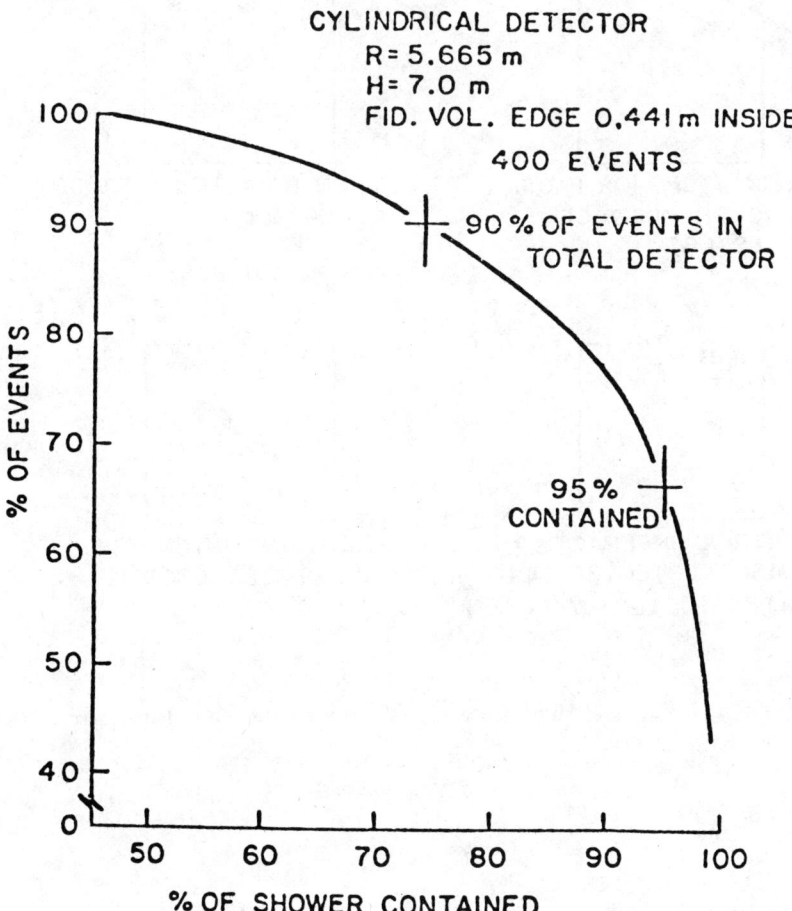

Figure 2. Shower containment efficiency for proton decays in the detector.

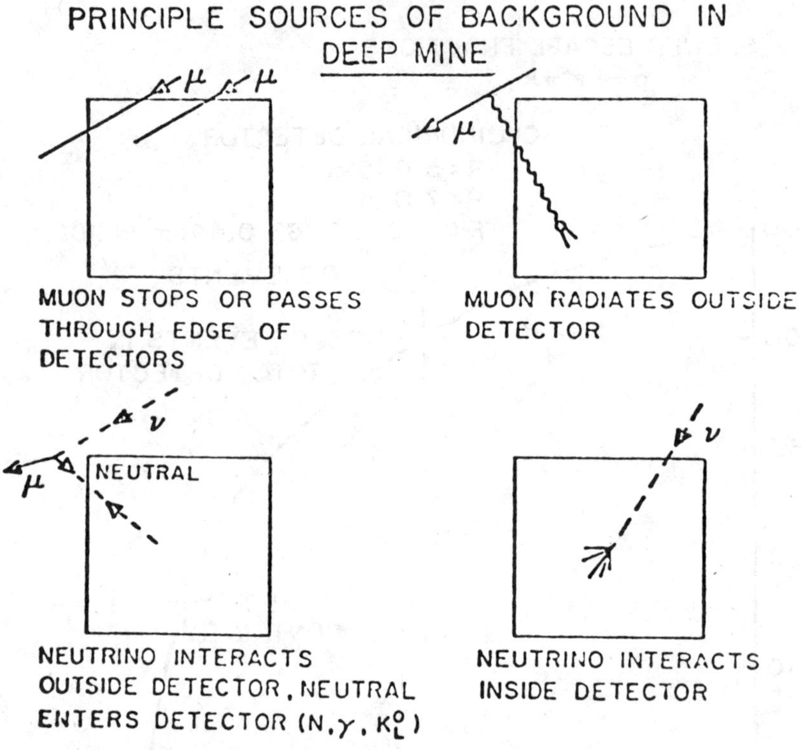
PRINCIPLE SOURCES OF BACKGROUND IN DEEP MINE
μ
μ
μ
MUON STOPS OR PASSES THROUGH EDGE OF DETECTORS
MUON RADIATES OUTSIDE DETECTOR
ν
μ
NEUTRAL
NEUTRINO INTERACTS OUTSIDE DETECTOR, NEUTRAL ENTERS DETECTOR (N, γ, K_L^0)
ν
NEUTRINO INTERACTS INSIDE DETECTOR

Figure 3. Various backgrounds to proton decay detector.

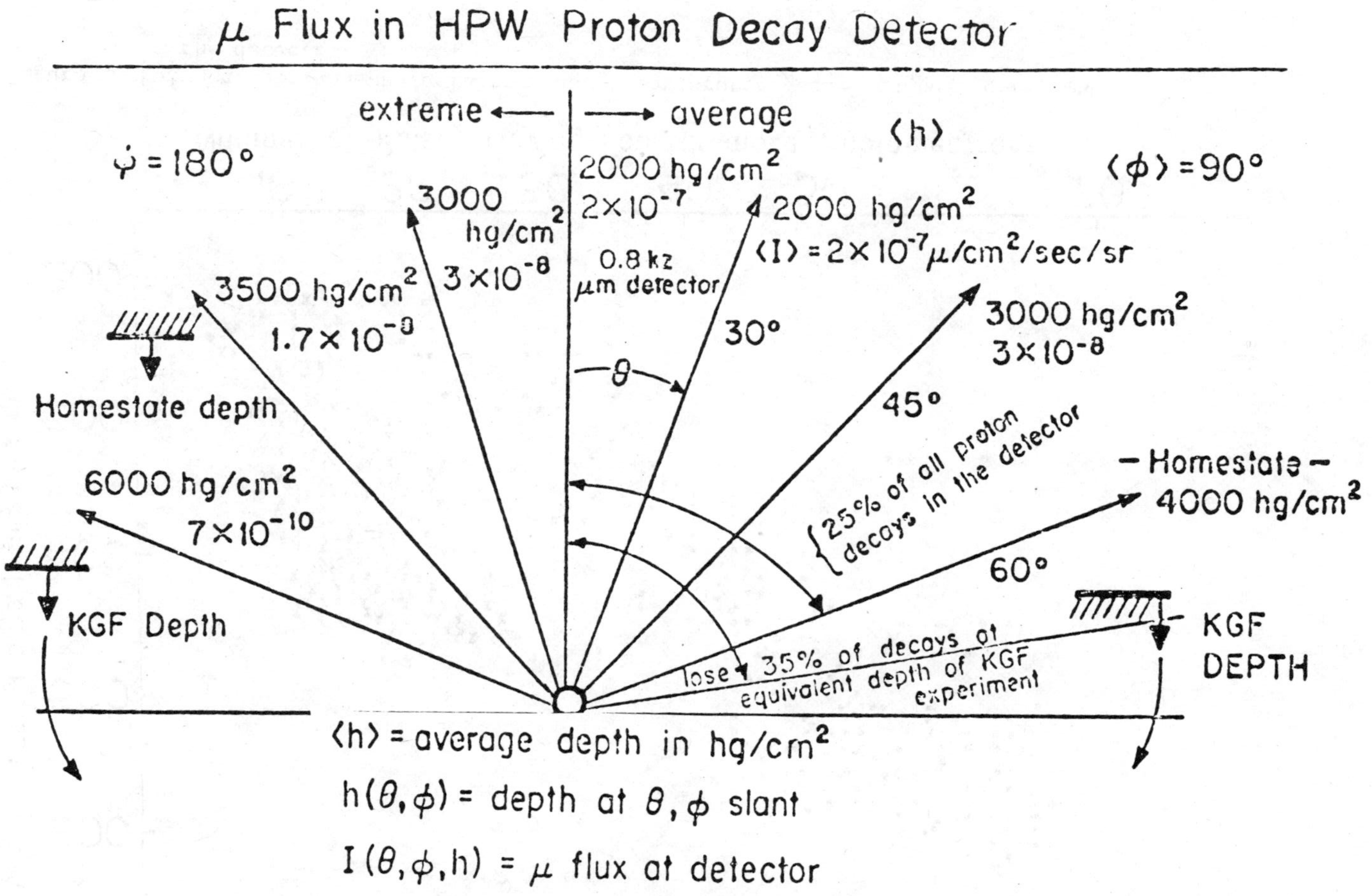

Figure 4. Muon flux as a function of angle at the Park City Location

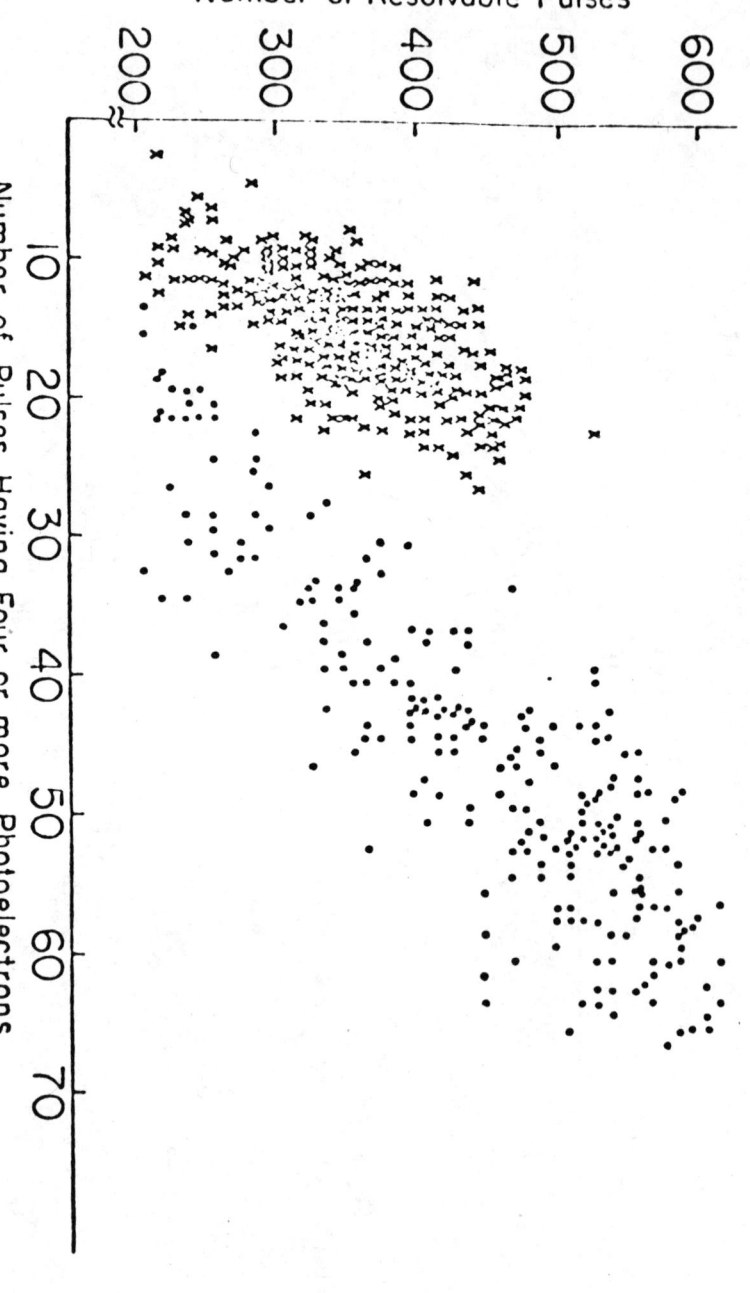

Figure 5. Pulse number and photoelectron count for proton decay and cosmic ray muons in the detector.

Atmospheric neutrinos comprise the other known source of background events in the detector. Unlike the accelerator produced neutrino beams, the cosmic ray flux contains an appreciable $\nu_e + \bar{\nu}_e$ component. Neutrino interactions in the detector in the 0.5-2 GeV energy region could produce final states that simulate proton decay; thus a detailed study of the background is necessary <u>before</u> the feasibility of a sensitive proton decay experiment can be ascertained. Fortunately the neutrino spectrum for the ZGS experiments using the 12' bubble chamber is very similar to the expected cosmic ray spectrum and we have used existing data from the ZGS experiments in our rate and background estimates. There are no detailed published estimates of the hadronic particles produced in the walls by either neutrino or muon interactions. These rates may amount to hundreds per year and are thus a potentially serious background for proton decay experiments. It is essential to protect the detector from these backgrounds by using an active shield in the same way that our earlier neutrino experiments at Fermilab and BNL used active shields.

All accelerator neutrino experiments have observed hadrons in the detector from the shield. Indeed, it was this background that inhibited the discovery of neutral currents for 12 years until large heavy liquid bubble chambers at CERN and counter experiments at Fermilab utilized active shields or techniques to suppress these incident hadrons. We expect these hadrons to be a major source of background for proton decay experiments as well.

The nuclear absorption length in water is 70 cm; we assume an active water shield around the inner detector with $\sim$ 5.5 hadron absorption lengths. This leads to an ideal suppression factor of 4×10^{-3}, which should be adequate to reject the hadronic background. The result is a background of less than 0.2 events/year for proton decay. In order to estimate an experimental limit on proton decay we use:

1. "N"$_{baryons} \cong 3 \times 10^{32}$

2. Expected background events $\lesssim$ 0.2 fake proton decays/year
3. Assume operation for three years and no events recorded

$$\tau_p > \frac{3 \times 3 \times 10^{32}}{2.3} \text{ years}$$

$\sim 4 \times 10^{32}$ years to 90% confidence.

The University of Utah group excavated a mine site in the Silver King Mine in 1964 using a horizontal tunnel. The effective cost of fabricating the mine today is guessed to be about $1,000,00. This site was operated for ten years as a neutrino laboratory and is fully roof bolted and fabricated (concrete floor and crane) as a laboratory. Figure 6 shows the location of the laboratory site, which is $\sim$ 5 km from the entrance to the tunnel.

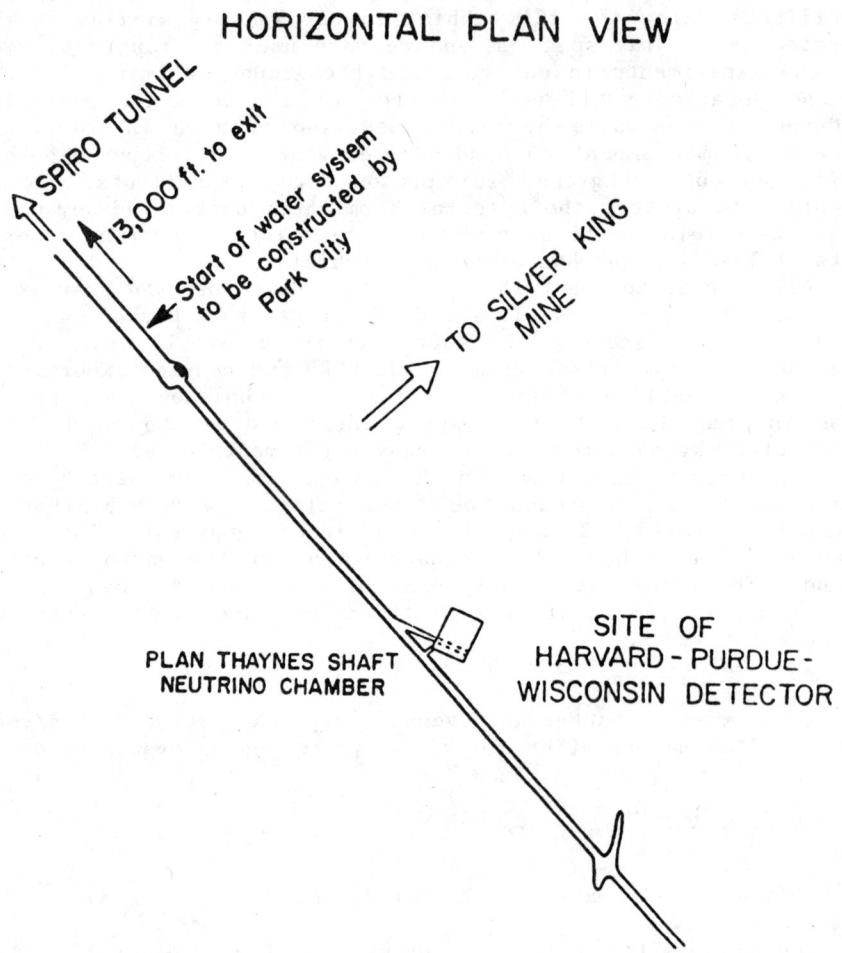

Figure 6. Tunnel to HPW Proton Decay Experiment.

A wooden tank has been installed in the mine which serves as a water container. The detector design has been optimized and influenced by several peices of information and requirements:
1. The available stope, which is already a laboratory (safety, crane, power, ventilation, water, floor, etc.).
2. Monte Carlo studies of the signals and backgrounds; in particular requirements of pulse height (p.m. spacing) time resolutions and water quality. It is especially important to optimize the signal/background. This implies attempting to be as sensitive as possible to all proton decay modes, as well as having high background rejection. It also implies as many cross checks and calibrations as feasible in the detector.
3. Availability of equipment; costs; time scales.

A schematic diagram of the detector is shown in Figure 7. The various elements of the detector and their functions are described in Table 4.

Figure 7 shows detailed drawings of the detector as it is now being constructed and installed in the Utah site.[3] Figure 7a shows the tank and the counting house. Figure 7b shows the phototube mounting and the proportional wire tube system on top, bottom, and the side. Figure 7c shows the addition of an active shield to be carried out during the summer of 1982. The mirrors are placed just outside the water container bag on the top, bottom, and sides. There is a data link between the LSI 11/23 computer inside the laboratory and a PDP 11/45 in the counting house outside the tunnel ($\sim$ 15,000 ft away) which is used to record data.

There appears to be two possibilities concerning proton decay.
1. The decay will be discovered with a lifetime of $< 10^{33}$ years using the present generation of detectors,
2. The lifetime limit will be pushed beyond 10^{32} years.

In either case there will very likely be the need for a very large detector (50,000 - 100,000 tons) to push beyond 10^{33} or to accumulate very large numbers of events. (Note that water detectors can be used to detect the various decay modes including those with μ's in the final state. See Tables 1, 2.)

Detectors of mass 50,000-100,000 tons must use water for economical reasons. To reduce background the detector should be placed as deep as the limit of strength of rock will allow. Recently the possibility of installing a proton decay detector in a deep lake or ocean using an active shield of water (mass $\sim 10^6$ tons) has been described.[4]

ACKNOWLEDGMENTS

I would like to thank the members of the HPW proton decay group for discussions. The members are: J. Blandino, U. Camerini, N. Duke, E. C. Fowler, W. F. Fry, J. A. Gaidos, W. A. Huffman, G. Kullerud, R. J. Loveless, A. M. Lutz, J. Matthews, R. McHenry, T. Meyer, R. Morse, T. I. Orosz, R. Palfrey, D. D. Reeder,

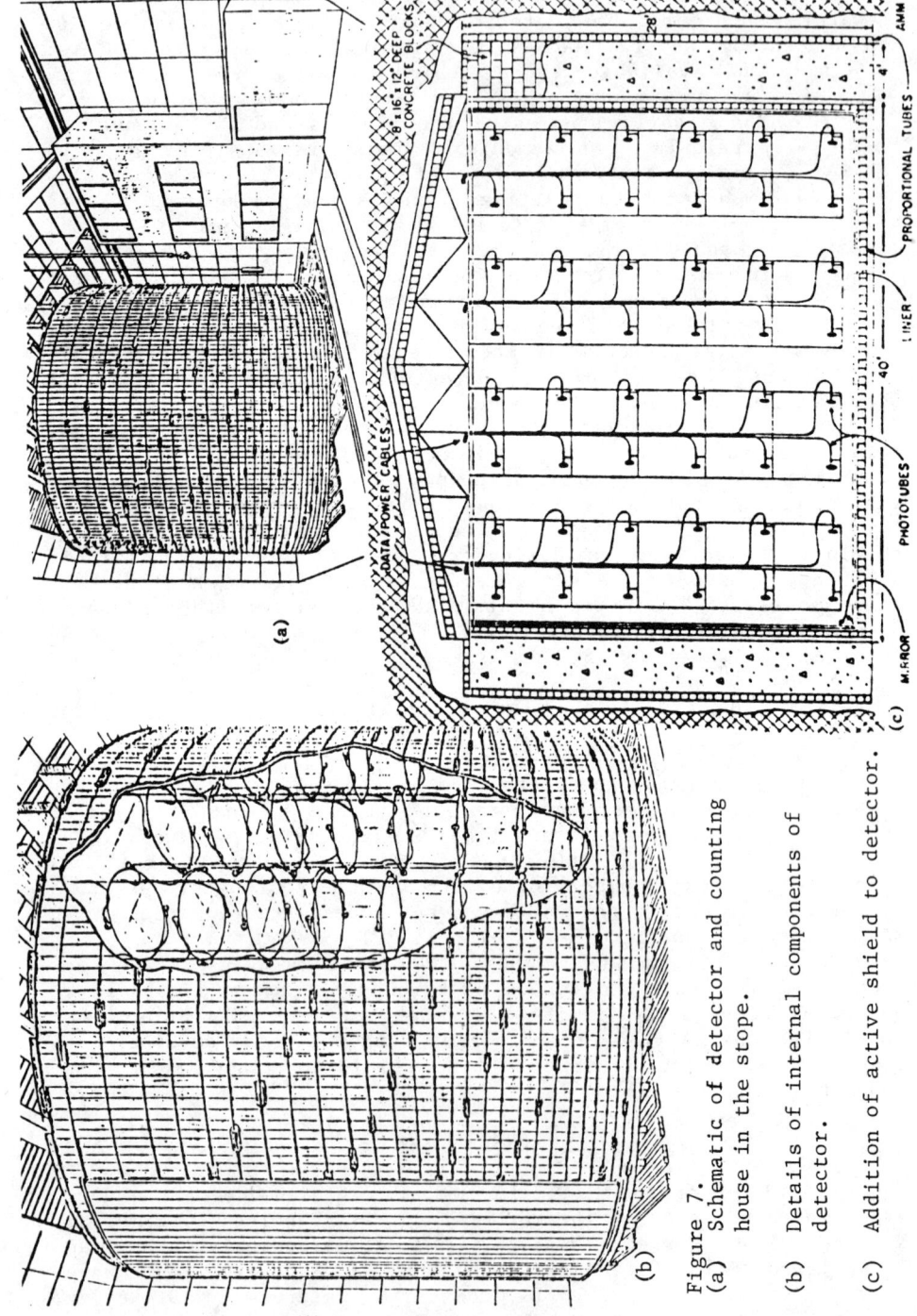

Figure 7.
(a) Schematic of detector and counting house in the stope.
(b) Details of internal components of detector.
(c) Addition of active shield to detector.

TABLE 4

ELEMENT	FUNCTION
1. Photomultiplier	Primary sig. & backgd. det. (EMI 9870)
2. Signal electronics	Timing, hits & pulse height of p.m. signals
3. Trigger logic and monitoring electronics	Event trigger for signal and background (tube hit multiplicity in 50 nsec gate) using p.m. and proportional tube information
4. Recording Electronics	Recording on magtype 2., 3., above, generating diagnostics controlling H.V. and thresholds
5. Counter proportional layers	a) active shield signal b) calibration signal (muon trajectories) c) event containment signal d) Al proportional tubes 6"x2"x(30 or 20') e) two layers on walls, roof, floors (crossed) f) 1280 tubes; 32,000' of Al readout, gas, HV
6. Water containment and active shield	a) Defines water volume and surrounding area for counters b) active shield absorber c) structural support
7. Structural elements	a) Roof proportional tube support b) p.m. tube support c) cable support d) mirror support
8. Water bag	a) water quality b) water containment
9. Water supply	a) supply water b) purify water
10. Mirrors	Light collection

C. Rubbia, A. H. Szentgyorgyi, R. B. Willmann, C. L. Wilson, D. R. Winn, J. Wist. In addition we wish to thank the Physics Department of the University of Utah for their help.

REFERENCES

1. J. Blandino et al., "A Decay Mode Independent Search for Baryon Decay," Harvard-Purdue-Wisconsin. Proposed to the Department of Energy (1979), unpublished.

2. C. Rubbia, private communication (1979).

3. J. Gaidos, "The HPW Proton Decay Experiment." Published in the Proceedings of the 1981 Neutrino Conference, Maui, Hawaii.

4. D. Cline, "Very Massive Water Detectors (10^5-10^6 Tons) for Proton Decay and Low Energy Cosmic Ray Neutrino Experiments." Talk given at the GUD meeting, Rome, October 1981, and University of Wisconsin Preprint UW DUMAND #4 (1981).

THE SOUDAN NUCLEON DECAY EXPERIMENTS

L. E. Price*
Argonne National Laboratory, Argonne, IL 60439

ABSTRACT

Two nucleon decay experiments using the Soudan mine are discussed. Soudan 1 is currently taking data. It contains 31 tons, and is based on proportional tubes. It has set an upper limit of 1.0×10^{30} years on the lifetime of the nucleon. Soudan 2 is a proposal based on long drift chambers with an initial mass of 1000 tons, to be expanded to 5000 tons.

INTRODUCTION

The most experimentally-testable consequence of grand unified models of the weak, electromagnetic and strong interactions is that protons and bound neutrons are unstable[1]. Although proton lifetime predictions vary with the specific model, many estimates are in the range of 10^{32} years. A number of experiments have previously tested the assumption of proton stability. The most sensitive of these were two deep underground experiments, one which could detect only stopping muons[2] and one which had more general sensitivity but coarse spatial resolution[3]. The stopping-muon experiment could measure the total nucleon decay rate only to the extent that theoretical models and Monte Carlo calculations could determine the fraction of decays which yielded stopping muons. The experiment in Ref. 3 has reported three proton decay candidate events which are wholly contained within the detector.

THE SOUDAN 1 DETECTOR

For the past year, we have operated the Soudan 1 nucleon decay detector, a 31.5 metric-ton tracking calorimeter, which is instrumented with 3,456 gas-filled proportional tubes. The detector is located 595 m underground in northeastern Minnesota. The overlying rock provides an attenuation of the cosmic rays equivalent to that of 1800 m of water. The detector is rectangular, 2.9 m x 2.9 m by 1.9 m high. As shown schematically in Fig. 1, it consists of 48 layers of 72 proportional tubes each. The tube axes are turned by 90° in alternate layers, in order to provide two views of each event. The 2.9 m-long proportional tubes are each 2.8 cm in diameter with an 0.8 mm-thick steel wall. The tube axes are spaced by 4 cm in the horizontal direction and adjacent tubes are staggered up and down by 0.45 cm from the layer center. The vertical distance between layers is 4.1 cm. The probability for a particle to traverse a layer without passing through the active volume of a tube varies from 0 to 40 percent, depending on angle. Most of the detector mass is provided by a matrix of heavy concrete in which the tubes are embedded. This substance was made from a mixture of purified iron ore (taconite), Portland cement and water. The total mass of protons in the detector is 15.1 tons; the neutron mass is 16.4 tons. The average density of the detector is 1.85 g/cm^3; the average radiation length is 9.3 cm. Additional discrimination against events which originate externally is provided by a scintillation counter shield which covers the top and four sides of the detector.

During normal operation, the outputs of all proportional tubes and shield scintillation counters are continually recorded in a buffer memory. A trigger is defined as a coincident signal above threshold in any one of the 72 proportional tubes in any 3 out of 4 adjacent layers. For each trigger with more than 5 proportional tube hits, the output of each proportional tube and each shield counter is permanently recorded for 32 time frames beginning about 370 nsec before the trigger and ending about 8 μsec after the trigger. This detailed time-structure data is used to determine the dE/dx ionization in the proportional tubes from the time over threshold, to detect stopped muon decays, to search for nonrelativistic particles and to diagnose malfunctioning channels. The rate for recorded events is about 0.1 Hz. The detector is monitored and controlled remotely, via a dial-up telephone line.

The data sample reported here contains about 10^6 recorded triggers and was collected between October 1981 and August 1982. The actual time that the detector was receptive to triggers during that period was 0.382 years, as measured by a 10 kHz, crystal-controlled clock. About 460,000 of these events were identified as single muon tracks; most of the other triggers were a result of radioactivity in combination with random proportional tube noise. The criteria adopted in the search for nucleon decay candidates were optimized to eliminate the muons which entered the detector. An initial selection of decay candidates was made by choosing those events in which one end of the event in each view is located at least 20 cm from a face of the detector. The direction of the event (as determined by a straight-line fit) was also required not to point at a shield scintillation counter with a coincident signal. Additional constraints on this first selection were the requirement of at least two proportional tube hits in each orthogonal view and of 8 to 40 total proportional tube hits in the entire event (corresponding to a total energy of 0.5 to 2.5 GeV).

These computer-applied cuts yielded a sample of 850 events, which were then scanned by physicists using a video terminal. Only one event, which is shown in Fig. 2, passed all of the selection criteria of both the computer and the physicist scans.

The efficiency of the Soudan 1 detector for identifying nucleon decays has been determined by a multi-step procedure. First, a sample section of the detector was exposed to tagged π, μ and e beams at the Argonne National Laboratory Rapid Cycling Synchrotron. Beam momenta from 150 to 400 MeV/c were used. The observed tracks were used to validate a Monte-Carlo model of particle transport in the Soudan 1 calorimeter. Fermi momentum in the parent nucleus and nuclear scattering effects in other nuclei (but not the parent) were combined with the transport model to simulate nucleon decays in the Soudan 1 detector. The decay point was randomly selected throughout the mass of the detector. This Monte Carlo model was then used to generate 300 events in each of several decay modes. Candidate events were required to meet the trigger criteria of the real detector and then analyzed using the same selection criteria which were applied to actual data. The percentage of Monte-Carlo events retained as decay candidates is typically 35 percent.

The Soudan 1 event shown in Fig. 2 meets all these selection criteria for a nucleon decay event. The total apparent energy, based on the number of proportional tube hits as calibrated in the detector test beam run, is 650±200 MeV. From Monte Carlo studies, the apparent energy expected for a decay event ranges from 820 to 940 MeV, depending on decay mode, with an uncertainty of 290 MeV. The event appears to have two prongs, which make an angle of about 135° with respect to each other. Ionization levels in the hit tubes together with absence of hits in layers 6, 8 and 10, suggest that the part of the event detected in layers 12 through 4 consists of an electromagnetic shower moving downward from layer 12. The event appears similar to Monte-Carlo simulations of two-pronged nucleon decays in several modes.

Three alternative hypotheses for this event are that it results from a neutrino interaction, that it is a product of some muon-induced interaction in the rock surrounding the detector or that it is a nucleon decay or other new phenomenon. Based upon the calculated atmospheric neutrino flux, the mass of the Soudan 1 detector, the live time and our analysis procedure, we expect to observe 0.5 contained, neutrino-induced events. However, less than one-third of the neutrino-induced events would appear similar to the event in Fig. 2, which seems to have both an electromagnetic shower and a substantial second prong. Based upon an extrapolation from the events rejected in the manual scanning process, we estimate a background of about 0.1 muon-induced events, similar to the one shown in Fig. 2. Our conclusion is that a cosmic-ray source for the event shown is somewhat unlikely but is still of sufficient probability that it cannot be ruled out.

If the event shown in Fig. 2 is not a nucleon decay, our exposure and detection efficiency can be used to set a lower bound for the lifetime of nucleons. For decay modes which do not produce final state neutrinos, the 90 percent confidence level bound obtained is 1.0×10^{30} years and applies to nucleons bound in the 60 percent iron, 30 percent oxygen and 10 percent other nuclei present in the Soudan 1 detector. The sensitivity to modes which include one neutrino is about one-fifth of this limit. The limits on lifetime divided by branching ratio (also 90 percent confidence level) are respectively 1.0, 0.9, and 1.1×10^{30} years for $e^+\pi^0$, $e^+\rho$, and μ^+K_s modes. These results are independent of any theoretical assumptions about branching ratios.

THE SOUDAN 2 DETECTOR

The Soudan 2 detector has been proposed[4] to the U.S. Department of Energy and to the U.K. Science and Engineering Research Council with an initial mass of 1000 tons, to be expanded later to 5000 tons.

The fine-grained tracking and ionization measurements will be made with long drift chambers, where the sensitive volumes are separated by a few millimeters of steel, which comprises most of the mass of the detector.

Two different drifting schemes are being investigated. One uses planar drift chambers mounted on 5mm thick sheets of steel. The other uses cylindrical drift spaces in a close-packed array embedded in steel, with the gas spaces 3 mm apart at the closest approach. In both schemes, the average density will be 2 gm/cm^3 and the maximum drift length will be 50 cm.

Since the planar drift chamber system is at present further developed[5], I will describe it in more detail. A schematic cross-section of chambers in the detector is shown in Fig. 3. The drift field is established by conducting lines on a dielectric substrate (implemented for tests with printed circuit board) which are attached to an external resistor chain. Ionization electrons drift along electric field lines in a 1 cm gap to the single anode wire at one edge of the chamber. The anode wire is surrounded on three sides by cathode, part of which is divided into pads with 2 cm spacing, in order to measure the coordinate along the anode wires. The detector is built up with two of these 50 cm wide chambers on each sheet of steel. Cathode pads from 250 of these chambers are bussed together to make strips. The chambers and steel sheets are 5 m high. A time history of the signal on each anode wire and cathode strip is read out, yielding a complete image of the event.

The use of imaging long drift chambers for particle physics has been pioneered by ISIS and TPC. What is new in our chamber is that all the drifting takes place close to the walls of the chamber. Drifting electrons can be lost because of imperfect electric fields, caused by charges on the dielectric surfaces between drift electrodes or by the grounded steel plates on both sides of the chamber (see Fig. 4a and 4b). They can also be lost by diffusion of the electrons into the walls. We have dealt with the stray field problem by covering the drift electrodes and intervening dielectric with resistive ink (10^{10} ohms/square). The resulting electric field is shown in Fig. 4c. The problem of diffusion has been improved by using a focussing electric field instead of a uniform one, as shown in Fig. 4d.

Tests of the chamber have been made both with a constant field and with a focussing field. The gas used was 90% Ar, 10% CO_2 with a total drift voltage of 10 kV. The results of attenuation measurements are shown for the two cases in Fig. 5. The attenuations, after a 50 cm drift, of 28% for the constant field and 12% for the exponential field are consistent with all losses being due to diffusion. Position resolution has been determined to be better that 2.5 mm (the size of the source spot) and two-particle separation after the maximum drift is 1 cm.

The full detector at the 1000-ton stage uses 10^4 of these drift chambers, arrayed as shown in Fig. 6. Chambers on steel plates are grouped into 50-ton modules, which are suspended on trolleys from structural steel. Each module is 1 m wide, 5 m deep, and 5 m high and can be rolled away from the neighboring modules to allow access to the chambers and readout wiring between modules. An active shield is provided by a double layer of proportional tubes on all sides of the detector except the bottom.

SENSITIVITY AND BACKGROUND REJECTION

Substantial Monte Carlo calculations have been made, simulating nucleon decay events as well as backgrounds from neutrino interactions and other sources. An example of a nucleon decay event is shown in Fig. 7. Typical nucleon decay modes (e.g. $e^+\pi^0$) show 60 drift chamber measurements. Energy resolution for modes with showering particles is $\sigma/E = 0.21$.

An important result from our Monte Carlo calculations is that the Soudan 2 detector has the ability, from ionization measurements, to determine the direction of most tracks. For muons from nucleon decay, we find that the direction is correctly determined in 90% of the cases. For pions, the possibility of hard scattering interferes with the use of ionization measurements, but if the approximately 25% of the tracks that show large angle scatters are removed, direction can be determined correctly for 85% of the remaining tracks. Even for the K^+ from the supersymmetric $K^+\nu$ decay mode, the direction can be correctly determined for 63% of the tracks. Overall sensitivity to common nucleon decay modes is, e.g., 30% for $e^+\pi^0$, 25% for $e^+\rho^0$ and 10% for μ^+K^0. The result is that for SU(5) branching ratios and a lifetime of 10^{32} years, Soudan 2 could expect to identify 1-2 nucleon decay events per year, after all cuts have been made to eliminate background. Background calculations show less than 0.5 event/year from neutrino interactions and less than 0.1 event/year induced by cosmic ray muons that will pass our cuts for nucleon decay.

SUMMARY

Soudan 2 will continue the Nucleon Decay program in the Soudan mine with a detector of mass initially 1000 tons, later 5000 tons. Fine-grained tracking and ionization measurements will give it the ability to observe all tracks from a decay and to determine the particle type and direction for most tracks. The sensitivity of the 1000-ton detector to nucleon decay, as determined by Monte Carlo simulations of nucleon decay and background processes, is 10^{32} years.

* The collaboration consists of J. Bartelt, H. Courant, K. Heller, M. Marshak, E. Peterson, K. Ruddick, M. Shupe (University of Minnesota); D. Ayres, K. Coover, J. Dawson, T. Fields, N. Hill, D. Jankowski, E. May, L. Price (Argonne National Laboratory); W. Allison, C. Brooks, J. Cobb, D. Perkins, B. Saitta (Oxford University); I. Corbett, S. Fischer, P. Litchfield, S. Yarker (Rutherford-Appelton Laboratory).

Soudan 1 research supported by U.S. D.O.E.

REFERENCES

1. J.C. Pati and A. Salam, Phys. Rev. Lett. **31**, 661 (1973); Phys. Rev. **D8**, 1240 (1973); Phys. Rev. **D10**, 275 (1974); H. Georgi and S.L. Glashow, Phys. Rev. Lett. **32**, 438 (1974).
2. J. Learned, F. Reines and A. Soni, Phys. Rev. Lett. **43**, 907 (1979); M.L. Cherry, et al., Phys. Rev. Lett. **47**, 1507 (1981).
3. M. R. Krishnaswamy, et al., Phys. Lett. **115B**, 349 (1982).
4. J. Bartelt, et al., "Soudan 2: A 1000 Ton Tracking Calorimeter for Nucleon Decay," University of Minnesota report COO-1764-410, unpublished.
5. L. E. Price, et al, Nucl. Instr. Meth. **119**, 499(1982)

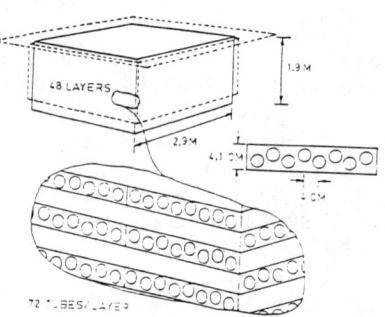

Fig. 1. Schematic diagram of Soudan 1 detector.

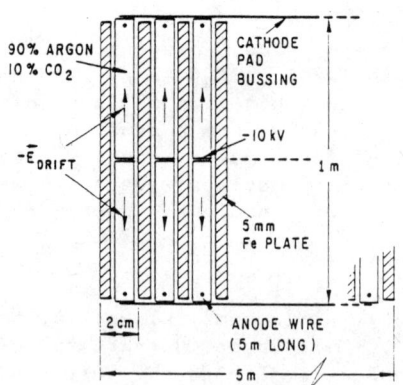

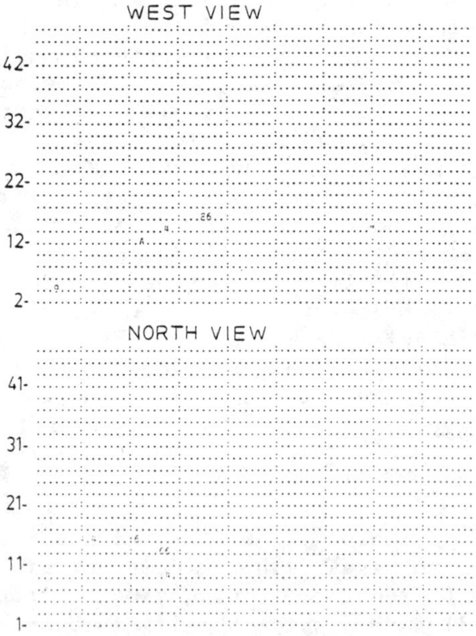

Fig 2. Soudan 1 event that survived nucleon decay cuts.

Fig. 3. Drift chambers mounted on steel plates for Soudan 2.

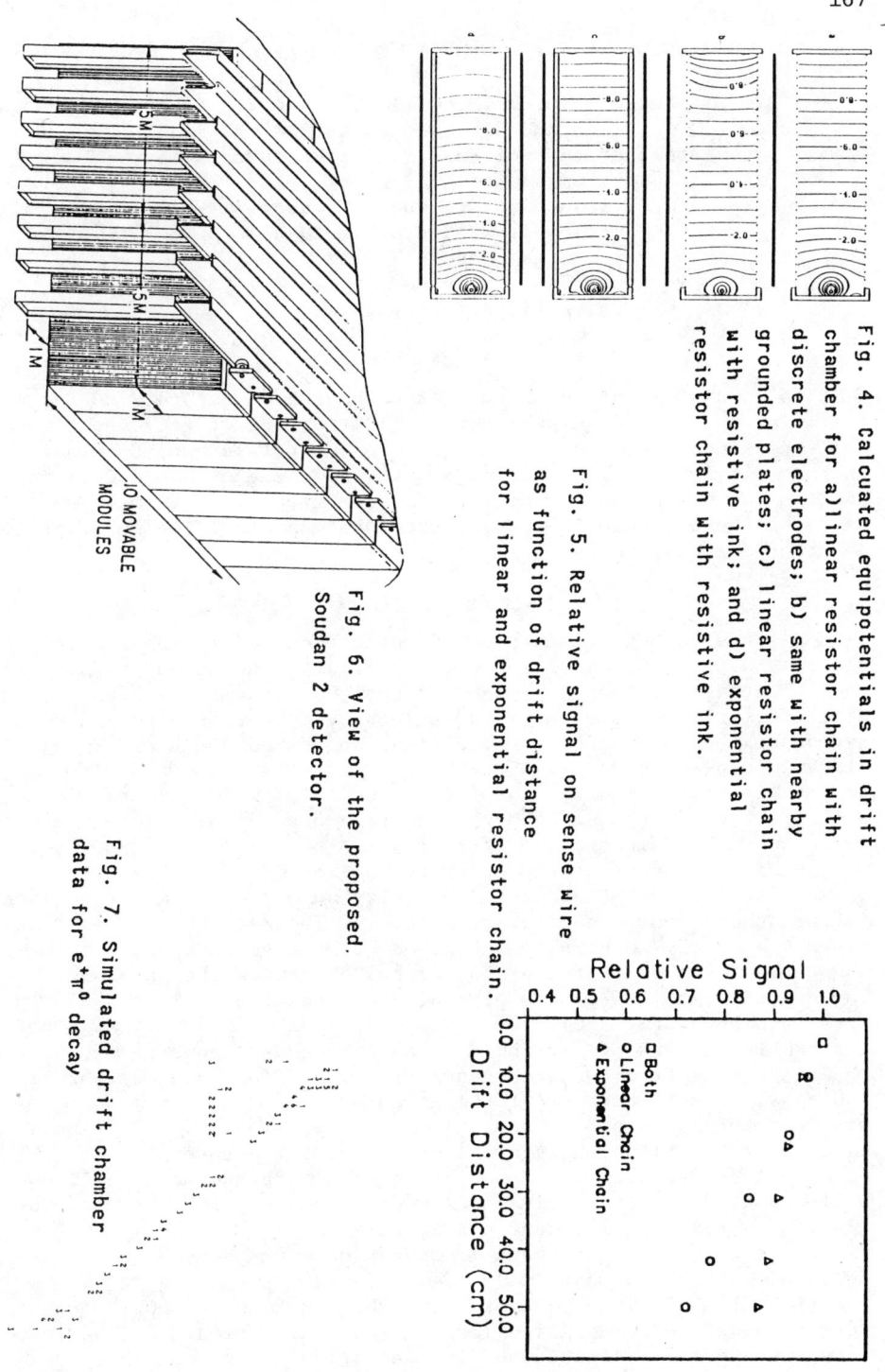

Fig. 4. Calcuated equipotentials in drift chamber for a) linear resistor chain with discrete electrodes; b) same with nearby grounded plates; c) linear resistor chain with resistive ink; and d) exponential resistor chain with resistive ink.

Fig. 5. Relative signal on sense wire as function of drift distance for linear and exponential resistor chain.

Fig. 6. View of the proposed Soudan 2 detector.

Fig. 7. Simulated drift chamber data for $e^+\pi^0$ decay

PROTON DECAY EXPERIMENT IN THE KOLAR GOLD FIELDS

M.R.Krishnaswamy, M.G.K.Menon, N.K.Mondal, V.S.Narasimham,
and B.V.Sreekantan
Tata Institute of Fundamental Research, Bombay

Y.Hayashi, N.Ito and S.Kawakami
Osaka City University, Osaka
and
S.Miyake
University of Tokyo, Tokyo

ABSTRACT

In total 6 events of special type have been recorded so far in the Kolar Gold Fields experiment. Out of them, three events are fully confined to the detector volume. It is shown that their characteristics are in conformity with the decay of bound nucleons and the background due to neutrino interactions is extremely small. Based on these events, it is suggested that a mean lifetime of protons is about 8×10^{30} years.

INTRODUCTION

A series of experiments has been carried out in Kolar Gold Mine since 1960. In the recent experiment, with the detector which has about 12 tons in fiducial volume, we noticed 2 examples which have characteristics consistent with nucleon decay, during a period of operation of about 5 years. They have suggested the possibility of finding nucleon decay with such an experiment. The estimated proton lifetime is of the order of 3×10^{30} years and the signal will be very much above the background of neutrino events. Based on such indication, we started a proton decay experiment with a medium size detector of the order of 100 tons.

A brief description of the detector for our experiment is given in the other report in this workshop. The detector of floor area 6 m x 4 m and 3.7 m high, is composed of 34 layers of proportional counters with iron plates of 1.2 cm thick between the layers. The counters, made of square pipe of cross section of 10 cm x 10 cm with thickness of 2.3 mm, have lengths of either 4 m or 6 m. The alternative layers are arranged in an orthogonal geometry so one can observe events from two orthogonal views. The total weight is 140 tons. The trigger of the detector is a coincidence of 5 layers with the threshold of $\sim 1/3$ of minimum ionization track. There is also an additional trigger of more than two counters in 2 in 3 consecutive layers. In every trigger, we record the position of a hit counter and the ionisation in each counter, in units of minimum ionising vertically incident particle.

This experiment has been in operation since November 1980 at the depth of 2300 m, equivalent to 7600 m.w.e. During the live-time of 550 days up to September 1982, about 1000 events have been recorded and classified into various categories. The intensity and angular distribution of penetrating tracks are in good agreement with expectations for atmospheric muons and neutrino-induced muons at this great depth. From the data itself, the layer detection efficiency is estimated as 97 % and the pulse height distribution of single tracks nearly vertical has a half width of ± 20%.

RESULTS

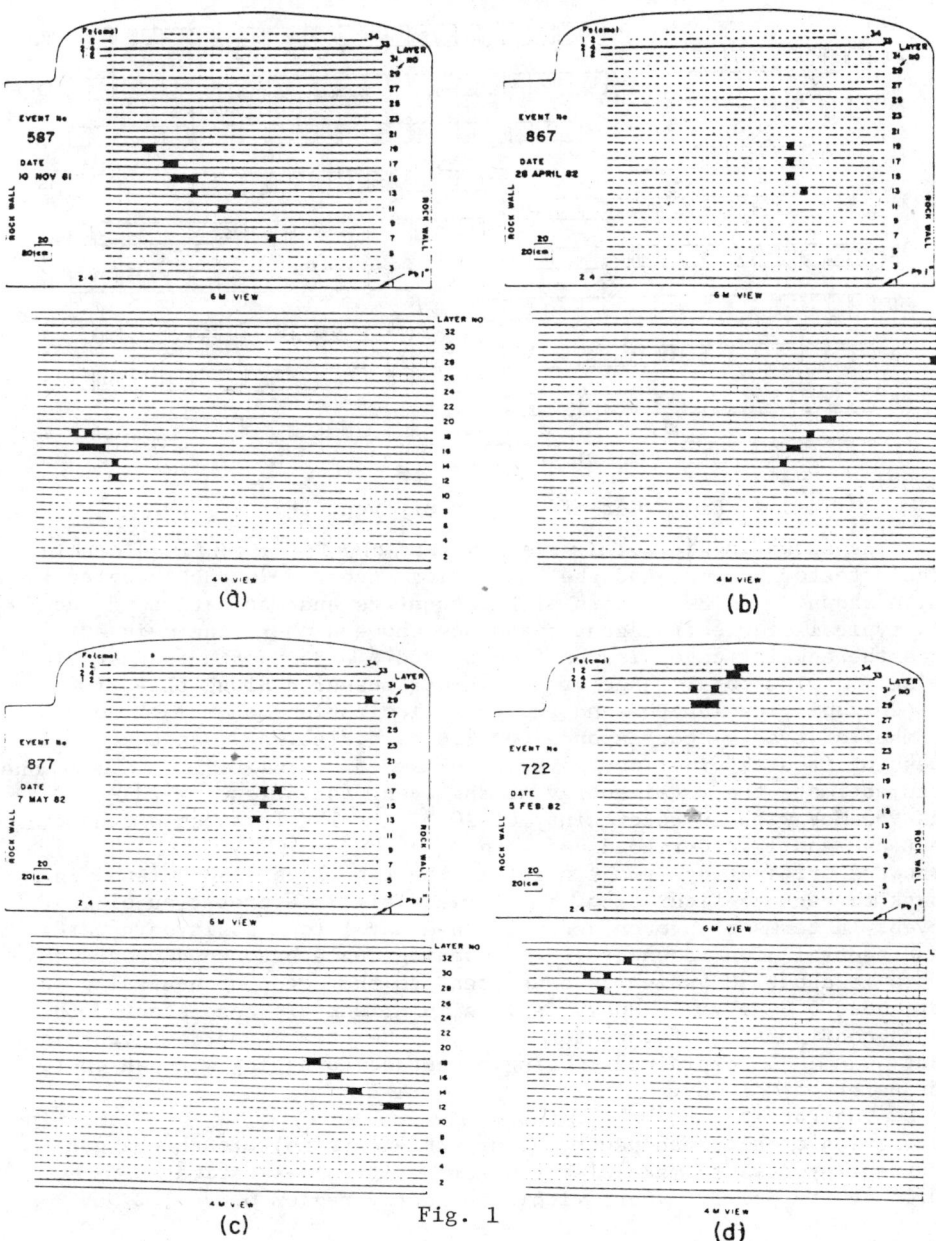

Fig. 1

Fig. 1 (a),(b) and (c) are confined events and (d) is one of the non-confined events. The data are analysed in terms of the pattern of the hit counters as well as the ionisation depositted in them. The details of individual events are given below.

Event No. 587

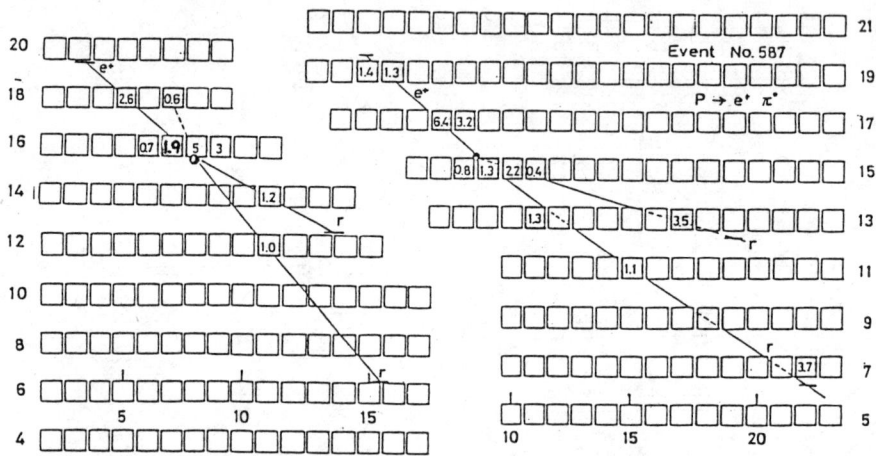

Fig.2 Cut-off figure of Event 587

As shown in Fig.2, this event has tracks fully confined within the detector volume and the ionisation recorded in each counter is also shown. The pattern of hit counters and ionisation in them is typical of electromagnetic cascade showers where the main axis has the angular co-ordinates θ ∼ 58° and φ ∼ 35° North to East clockwise. The total range of the shower measured along the axis is ∼ 20 radiation lengths and the total ionization (summation of the number of tracks) corresponds to 42.6 particles. Since the thickness of iron absorber per layer is close to one radiation length, one can estimate the total energy of the event (42.6/secθ x 23secθ MeV) as 980 MeV with an uncertainty of 20 %. The features of the event are easily understood as separate showers originating from a point in the 15th layer and emitted in a back to back configuration. A plausible interpretation of the event in terms of proton decay is a decay into a positron(upward) and neutral pion (downward). Opening angle between downward showers is about 30° which is consistent to the decay of neutral pion of about 500 MeV. Monte Carlo simulations are consistent with such configuration for proton decay, on the other hand, the event is not easily explained by single cascade downgoing with an energy transfer of ∼1 GeV.

The rate for the elastic interaction of electron neutrino (or anti-neutrino) through charged current process to make electron (or positron) is estimated to be 0.15 events per 1.5 years within an energy region 0.8 - 1.2 GeV.

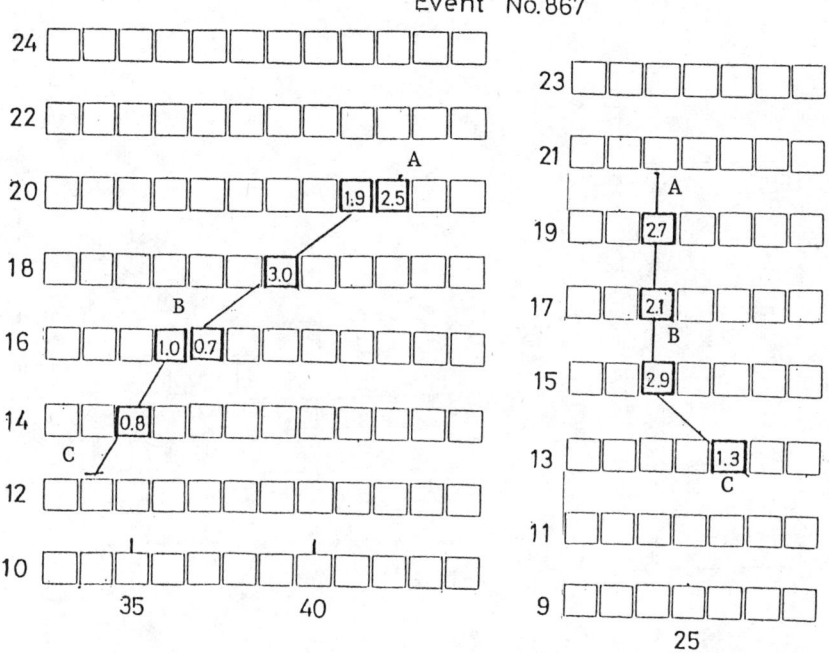

Fig. 3 (a)

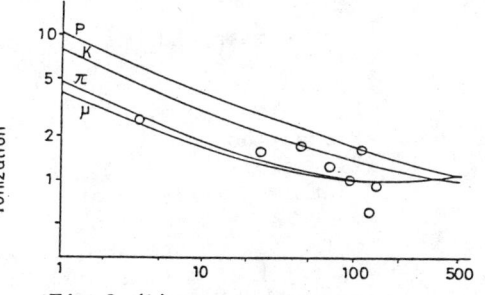

Fig.3 (b) Range g/cm³

The event shown in Fig. 3, besides being a fully confined one, has a kink at the point B with the angle of deflection of about 45°. An ionization along the path with range measured from point A, is shown in Fig. 3 (b). These features suggest the creation of a particle at the point C which slowed down to point A with a scatter along the path at the point B.

It is highly unlikely that this particle is a muon in view of the large scattering angle. It is possible to explain this event as a charged kaon of momentum ~450 MeV/c, traversing the path CB and decaying in flight at B into a muon and a neutrino. However, the decay in flight will make this less probable in comparison to the pion hypothesis discussed below. Assuming a nuclear elastic scatter at B, purely based on ionisation along the total path, we estimate the total energy of the pion as 435 MeV. The energy is in good agreement with that to be expected if the event is interpreted as a proton decay of the type $P \rightarrow \bar{\nu} + \pi^+$.

The background rate from ν-interactions is estimated as ~0.1/year

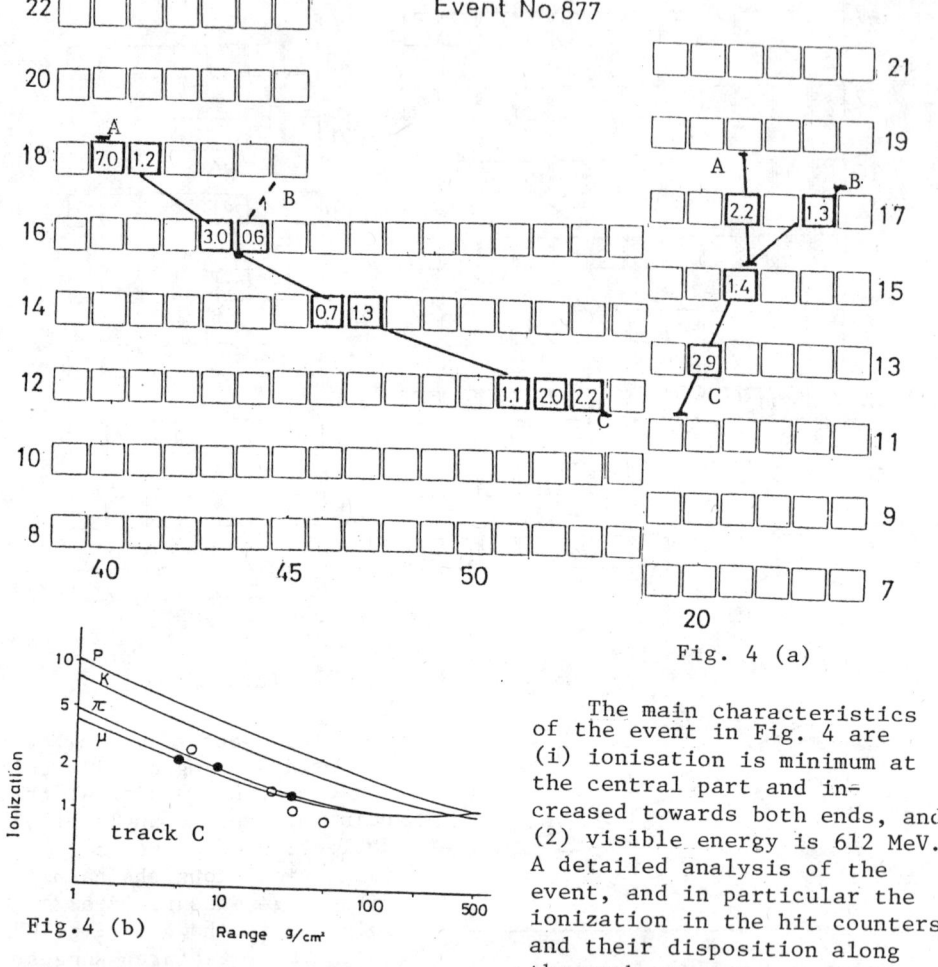

Fig. 4 (a)

Fig. 4 (b)

The main characteristics of the event in Fig. 4 are (i) ionisation is minimum at the central part and increased towards both ends, and (2) visible energy is 612 MeV. A detailed analysis of the event, and in particular the ionization in the hit counters and their disposition along the path suggests a reasonable fit to the decay mode of $P \to \mu^+ + K^\circ_s$ and $K^\circ_s \to \pi^+ + \pi^-$. Total energy including masses of these three particles comes to about 1 GeV. The momentum balance, 350 MeV/c for each, and the energies of π^+, π^- (320, 220 MeV) and their opening angle are consistent with decay of slow moving K°_s of momentum about 350 MeV/c.

The event is also possible to understand as a decay into positron and charged pion, if the upper part of the event is a cascade shower caused by a positron. A non-showering particle going downward is a pion or muon as shown in Fig. 4(b); This is a common identification for both decay possibilities. The former decay mode has a very low branching ratio in a majority of calculations based on standard GUTs, while the latter normally has a larger branching ratio.

The background due to ν-interaction is low for this event because of the observed back-to-back configuration. (~0.01 events/year)

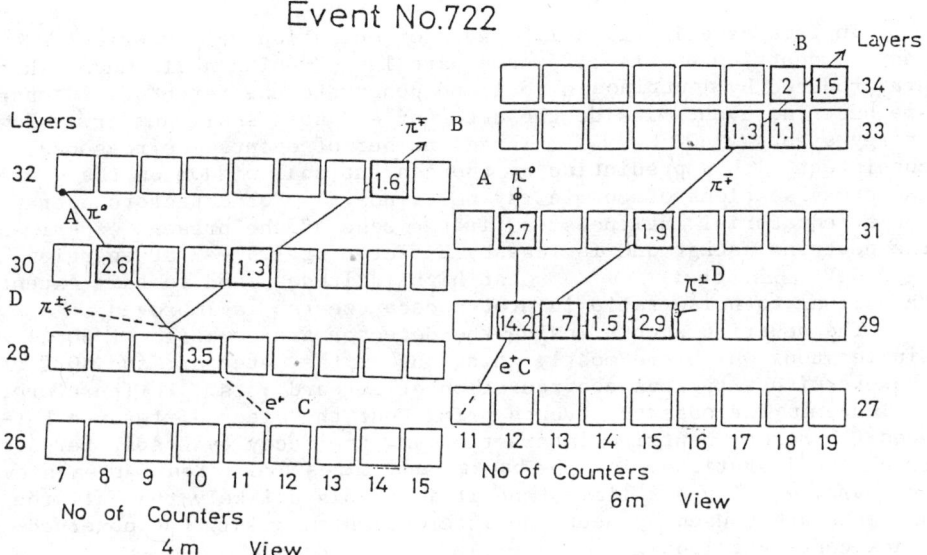

Fig. 5 (a)

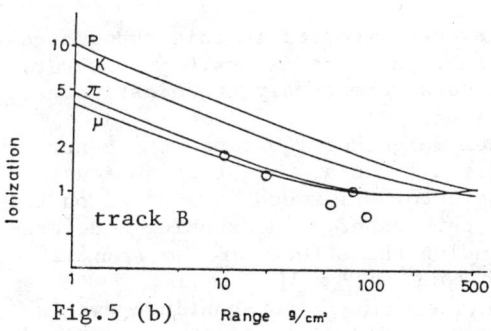

Fig.5 (b)

In the event of Fig. 5, we notice 4 tracks, of which track B has traversed 170 g/cm² of iron before emerging from the detector at the top; the other tracks A, C and D have stopped after traversing 52, 15 and about 10 g/cm², respectively. The striking features are; (i) the total visible energy of 700 MeV, based on integral number of tracks in the counters, and (ii) large opening angles of the particles from the vertex.

The relation between ionization and residual range of track B, is shown in Fig. 5(b). It belongs to the pion-muon group, moving upward and it would have stopped if the detector had 10g/cm² more absorber. This event can be an example of $P \rightarrow e^+ \omega^°$, if we assume track C to be an e^+ and the other tracks to be pions ($\pi^+ \pi^- \pi^°$). The energy division among the 3 pions as well as the approximate momentum balance makes this a viable hypothesis, particularly in view of the low background estimates.

As a background, an interaction by an upward moving neutrino could cause such an event with track B identified as a muon. However, at least 2 more pions need to be produced along with the muon. For a low energy ν-interaction, 2π production is suppressed by a factor of 10 compared to the single pion-inelastic process.

NEUTRINO BACKGROUND

In this experiment, in 1.5 years of operation, we have recorded many neutrino events; 25 cases are large zenith angle muons which are produced by neutrinos in rock and penetrate the detector, 10 cases are neutrino events inside the detector and some small number are low energy events (<300 MeV). These number of events are reasonably consistent with a prediction by the conventional wisdom on the fluxes and cross sections of cosmic ray neutrinos. For detectors operating at equatorial latitudes, as is the case of the present experiment, the neutrino background is less, by a factor of 1.5 - 2 at energies <2 GeV, than that at high latitudes, and to this extent the signal to noise ratio in proton decay search is improved.

The neutrino events inside the detector are composed of 60 % single muons which are mostly going out of the detector, and 40 % of showers with a typical configuration of forward jets. Therefore, a glance at the observed events brings out the clear difference between normal neutrino interactions and the decay events in terms of track configurations. The background rates predicted for each event are upper limits and it is highly unlikely that all the 4 events are caused by neutrino interactions mimicking the observed track configurations.

CONCLUSION

We consider that the events reported in this report and in our earlier paper are very probably nucleon decays. However, the small sample of events in our data permit only crude estimates on the lifetime on nucleons at present.

While the detector has a total weight of 140 tons, the central volume for confined events has only a fiducial weight of 60 tons, (cutting away 5 layers of the detector all round). Based on the three confined events reported in this paper, and assuming a detection efficiency of the events including the effects of the iron nucleus as 0.5, the lifetime of bound nucleons is 9×10^{30} years.

Considering all the 6 events (including non-confined events reported earlier), and with 100 tons of fiducial weight for efficient recognition of the event vertex, we estimate lifetime as 8×10^{30} y.

REFERENCES

M.R.Krishnaswamy et al, Phys. Lett. 106 B 339 (1981)
    ibid, Proc. Second Workshop on Grand Unification, Birkhauser Publications 11 (1981)
    ibid, Proc. of International Colloquium on Baryon Non-conservation PRAMANA Suppl. 1982 in press
    ibid, Phys. Lett. 115 B 349 (1982)

THE NUCLEON DECAY EXPERIMENT IN THE FREJUS TUNNEL

J. Ernwein
DPhPE, CEN-Saclay, 91191 Gif sur Yvette, France

ABSTRACT

This paper describes the 1.5 kiloton fine grained flash chamber detector which will be installed in an underground laboratory in the Fréjus tunnel near Modane, France. It will be dedicated to the search for nucleon decay. It will also be possible to search for monopoles and study multimuons.

INTRODUCTION

Grand unified theories predict nucleon decay with a lifetime of 10^{30-32} years [1]. Large Water Cerenkov detectors should soon produce data [2,3]. Modest size ($\sim$ 100 tons) fine grain tracking detectors have recorded events which can be interpreted as nucleon decays [4,5]. The Mont Blanc experiment [3] has demonstrated that a fine grained tracking detector can indeed detect events which are very clean and contained. The Fréjus detector which is under construction will be ten times more massive (1000 tons of fiducial mass) and its granularity will be three times better. It is designed to meet the challenge of recognizing nucleon decay events among neutrino interactions and providing better sensitivity to higher lifetimes.

In addition to this primary goal, this underground detector will record large numbers of multimuon events. Their abundance is relevant to the composition of the primary flux of cosmic rays entering the atmosphere. A change in composition from light to heavy nuclei around 10^{15} ev is suspected and should be investigated [6].

Atmospheric neutrinos will interact in the detector and constitute a major background for nucleon decay events. It will be possible to measure the neutrino flux and verify expected geographical anisotropies.

Because of the large area of the detector ($\sim$ 120 m^2) it will be possible to make a sensitive search for slow moving magnetic monopoles provided that they ionize in Argon.

1. THE LABORATORY

The detector will be located in a gallery specially excavated near the mid point of the 13 km long highway tunnel linking Modane, France to the region of Torino, Italy.

The average thickness of the rock above the laboratory is 1680 m and corresponds to 4500 meters of water equivalent. The atmospheric muon flux in the laboratory is $\sim 10^6$ times lower than at sea level.

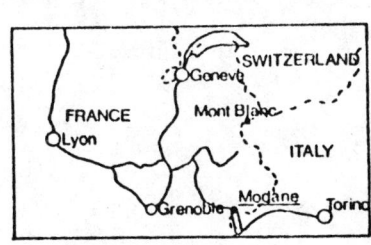

Fig. 1. Geography

It has been measured with large scintillators placed horizontally in the gallery : the measured muon rate is 6.2 ± 1.2 muons/day.m^2.

This rate corresponds to about one muon crossing the 1.5 kiloton detector every minute. Most of them will satisfy the trigger conditions and they will provide a convenient way to continuously check the performance of the detector without appreciable trigger efficiency loss due to the deadtime of the flash chambers (a few seconds).

2. DESCRIPTION OF THE DETECTOR

The detector is a very fine grained calorimeter designed to visualize tracks coming from nucleon decay products, measure their range, and identify them if possible. The dimensions of the detector are $6 \times 6 \times 20$ m^3. The structure of the detector is shown in Fig. 2. A module consists of four biplanes of flash chamber elements made of polypropylene with the cell direction alternately horizontal and vertical. Two consecutive flash chamber planes are separated by 3 mm of iron. The trigger planes which are encountered every 8 flash chamber planes are built with elements of eight extruded aluminium tubes. Table I summarizes the main characteristics of the detector.

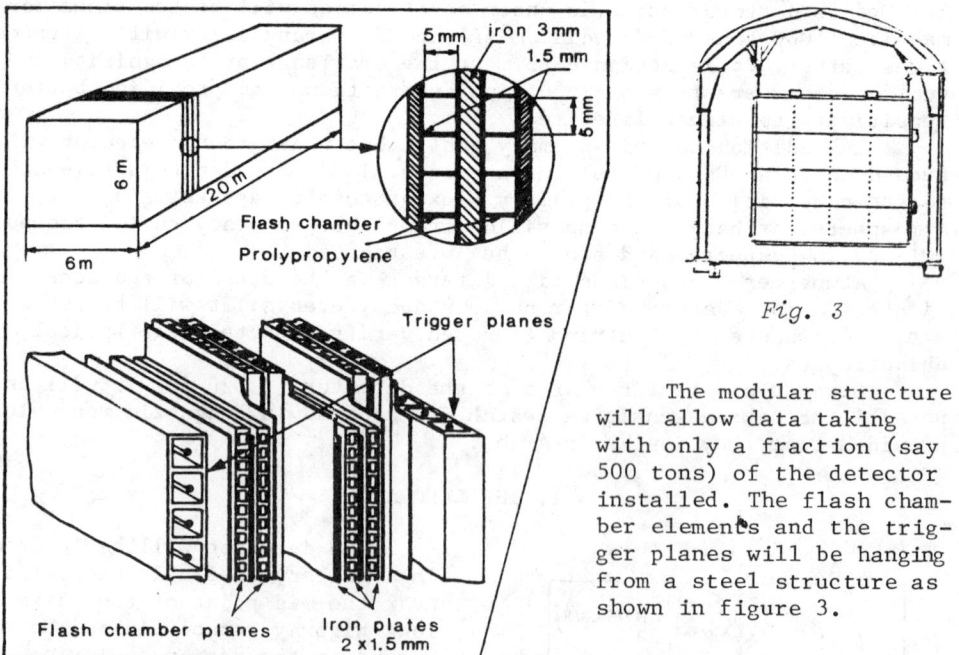

Fig. 3

The modular structure will allow data taking with only a fraction (say 500 tons) of the detector installed. The flash chamber elements and the trigger planes will be hanging from a steel structure as shown in figure 3.

Fig. 2. Structure of the detector.

Table I - Main features of the Fréjus detector

Total mass	1.5 kiloton
Fiducial mass	1 kiloton
Dimensions	6×6×20 m^3
Average density	2 g/cm^3
Flash chambers	
Total number of planes	1480
Internal size of cell	0.5×0.5 cm^2
Number of cells	$\sim$ 1.5 10^6
Sampling	3 mm of iron (.17 rad. length)
Geiger Tube Planes	
Number of planes	185
Internal Size of cells	1.5×1.5 cm^2
Number of cells	6.5 10^4
Longitudinal sampling	2.4 cm of iron (1.35 rad. length)
Distance between two trigger planes	$\sim$ 11 cm

Flash chambers

An element of flash chamber biplanes of 6×1.5 m^2 is made of two polypropylene planes sandwiched between two sheets of iron, 1.5 mm thick. Four elements are assembled to form a 6×6 m^2 flash chamber biplane. A mixture of Helium-Neon gas flows through the chambers and each plane in an element has its own gas distribution. A triggered high voltage pulse is applied to each element through a spark gap which discharges a distributed delay line placed between the two central iron plates. The high voltage pulse which produces an electrical field of $\sim$ 8 kV/cm has a rise time of $\sim$ 95 ns and a width of $\sim$ 800 ns. If the passage of a particle has produced ionization in a cell, a plasma forms, propagates over the full length (6 m) of the cell, and is detected by capacitive coupling to metal strips placed on top of each cell at the end of the chamber. Flashover from neighbouring cells is successfully eliminated by the use of black plastic foam at the two ends of the polypropylene plane. The signals induced by the plasma are directly fed into CMOS registers for read out and digitization of the position of the hit cells.

Extensive tests have been made using full size flash chamber elements and give results similar to those obtained in existing large flash chamber calorimeters [7,8].

Geiger tubes

The trigger planes consist of 44 elements of eight (1.5×1.5) cm^2 extruded aluminium tubes. They are filled with a mixture of Argon (97 %), Ethylal (3 %) and Freon (0.3 %) flowing at a rate of 0.01 volume/hour. With a 100 μm diameter wire, a plateau of 300 volts extends around 1700 volts. The signal which is obtained on the wire across a decoupling capacitor has a rise time of $\sim$ 30 ns and an amplitude of 10-30 mV. The position of the individual tube which was hit will be digitized and the firing time will be recorded with a resolution of $\sim$ 200 ns.

3. EXPECTED PROPERTIES OF THE DETECTOR

In order to evaluate the properties of this fine grained detector a test calorimeter has been exposed to a beam of pions and electrons.

The apparatus was made of planes of streamer tubes (2×2) cm^2 in cross section separated by 4 mm iron plates (0.22 radiation length). The experimental results obtained with 200 to 500 MeV/c pions and electrons hitting this detector at various angles were extrapolated to the actual granularity of the flash chambers (0.5×0.5) cm^2 and sampling (3 mm of iron). The following results were obtained.

3.1. Particle identification

Electron-pion separation which is essential to detect nucleon decay, at least in the mode $e^+\pi$, is based on the difference between showering and non showering particles. The average number of hit flash chamber cells per plane is quite different for pions and electrons and leaves less than 10 % ambiguous cases. The sign of the muon is deduced from the observation of the decay $\mu^+ \to e^+\bar{\nu}\nu$ (the μ^- is captured by the iron nucleus). The 200 ns resolution time of the trigger planes allows for the detection of delayed coincidences due to decay positrons with a $\sim$ 50 % efficiency. (The positron does not always reach the trigger plane because it is absorbed).

3.2. Energy resolution

With thin iron plates and in the energy range under consideration, the number of hit cells is proportional to the energy of the electrons. About 80 cells are hit by an electromagnetic shower of 500 MeV/c perpendicular to the plates. Averaging over all directions, this corresponds to an energy resolution $\Delta E/E \simeq 12 \%/\sqrt{E}$ (GeV).

For charged pions the energy is estimated from the range which is corrected in case a scattering is visible. The energy resolution for pions varies from 12 to 20 % for pion momenta between 200 and 300 MeV/c and for various incident directions. The energy of K^+ and μ^+ can be measured with the same technique.

3.3. Pattern recognition

The fine grain and fine iron sampling provide excellent pattern recognition capabilities. As an illustration, Fig. 4 represents four simulated nucleon decays as seen in the detector.

3.4. Sensitivity

The sensitivity of the detector to a given nucleon life time depends on the branching ratios of the modes which are seen, the global efficiency to detect these modes, the time during which the detector is active, and the fiducial mass.

For nucleon decays where all the products are seen as in $N \to e^+n\pi$ ($n \geq 1$), an event will have an extension in space of $\sim$ 70 cm. If the fiducial mass is defined as the region extending up to 50 cm from the sides of the detector, almost all those events will be contained. With this non fiducial region all around the detector, it is possible to make sure that all charged tracks originate from inside the detector, and that they are not due to interactions of neutrals coming from neutrino or muon interactions in the rock. Obviously, neutrino interactions which take place in the detector, and with

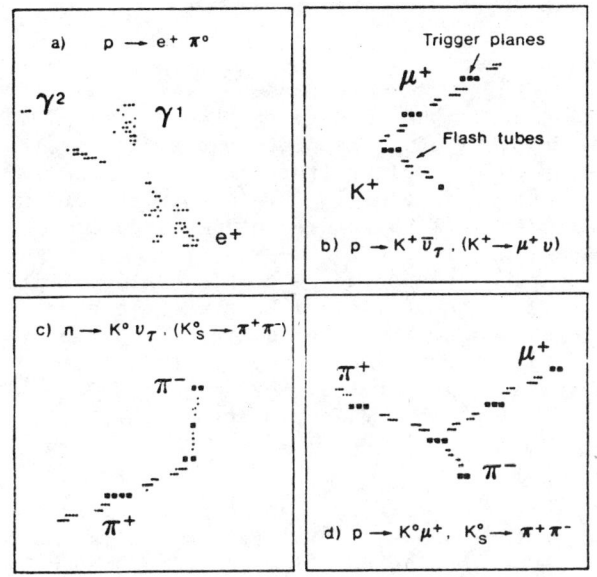

Fig. 4. Simulated nucleon decays as seen in the detector.

visible energy in the 1 GeV region, cannot be discriminated against. This neutrino background is one known limitation to reaching nucleon life times bigger than a few 10^{32} years.

The number of expected $e^+\pi$ and $e^+n\pi$ events which are well identified and neutrino background events are listed in Table II.

Table II - Number of nucleon decays which should be observed during one year in 1 kiloton fiducial mass of the Fréjus detector if $\tau = 10^{31}$ years.

Total number of decays	With branching fractions to $e^+n\pi$ (n⩾1)	Unabsorbed pion	After acceptance	ν background
60 { ~30 p	23	12.5	5 } 12	< 1
~30 n	27	15.5	7	

In these estimates the following has been taken into account :
The branching fractions as calculated in GUTS, the probability of non reinteraction of the decay pions in the iron nucleus (≃ 60 %) and the acceptance for a fully identified $e^+n\pi$ (n⩾1) event which is the product of three comparable fractions : trigger acceptance (0.8), good identification (∼ 0.8), and, in order to reduce the neutrino background, total energy and colinearity cut (∼ 0.8).

For nucleon decays such as $p \rightarrow K^+\nu_T$, or $n \rightarrow K^°\nu_T$ predicted in schemes using supersymmetry, the neutrino carries away much of the energy so that the kaons are left with ∼ 330 MeV/c and the range of their decay products (μ^+, π^+, $\pi^°$) is reduced. This results in a loss of trigger efficiency (two trigger planes are separated by 20 g/cm^2 of iron). However the signature of these events is very clear because of the good pattern recognition capability of the detector (see Fig. 4). In particular the decay $p \rightarrow K^°\mu^+$, ($K_s^° \rightarrow \pi^+\pi^-$ or $\pi^°\pi^°$) would produce a beautiful signature in the detector. The rejection of the neutrino background is expected to be roughly the same so that comparable sensitivities may be reached.

With the 1500 ton detector in operation a lifetime measurement of 10^{31} years can be made in one year of operation, while a lower limit of 10^{32} years may be achieved in a few years.

4. TRIGGERS AND SPECIAL MEASUREMENTS

Nucleon decay

The trigger will require that at least 4 tubes be hit within 5 consecutive planes, each plane contributing no more than 3 hits. This defines a geometrical volume typical of nucleon decay events. It also ensures that the contributions to the trigger rate from local radio-activity and random coincidences will be negligible within the 200 ns resolution time. The trigger rate will therefore be dominated by cosmic ray muons (1 per minute in the 1500 ton detector).

It is possible in principle to measure the time of delayed hits from $\mu^+ \rightarrow e^+$ decay ($\tau = 2.2$ μs) in the flash chambers which can be made sensitive for a few microseconds. The time of arrival of the plasma pulse can be measured with ~ 200 ns accuracy. Tests are in progress to demonstrate the feasibility of this measurement which would add a redundancy to the delayed hit measurements in the trigger planes.

Slow monopoles

The trigger planes will be sensitive to ionization equivalent to $0.1 \times$ "minimum". The trigger logic consists in "following" the monopole through the detector with successive coincidences between 5 consecutive planes. The resolution time of the coincidence corresponds to a particular range of β of the monopole (for instance $\Delta t = 1$ μs would correspond to $\beta = 5.10^{-3}$). The trigger rate due to randoms is < 0.25/minute.

If the monopole was heavily ionizing one could measure the ionization by looking at the average efficiency along the track in the flash chambers since the efficiency depends on the primary ionization. Given the area of the detector (~ 120 m^2) a flux limit of 2.10^{-15} cm^{-2} s^{-1} sr^{-1} could be reached in one year for $10^{-4} < \beta < 4.10^{-2}$ and ionization > 0.1 "minimum".

5. STATUS OF THE EXPERIMENT

The final approval and funding of the experiment occured in March, 1982. The excavation of the laboratory was completed in July, 1982 (Fig. 5) and will be ready for use in March, 1983. Extensive tests on full scale elements of flash chambers and Geiger tubes are nearing completion. The industrial organization to mass produce detector elements is being set up. The mounting of the detector elements in the tunnel is scheduled to begin in spring of 1983 and it is hoped that 500 tons will be instrumented by the end of 1983. One more year will be necessary to make the entire 1500 ton detector operational.

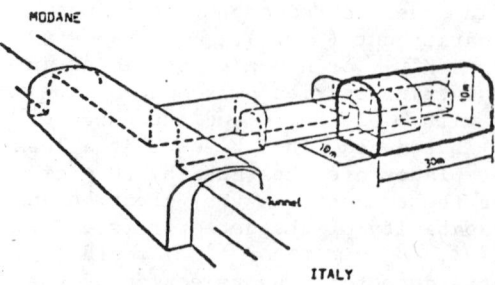

Fig. 5. Sketch of the underground laboratory.

We plan to build a test calorimeter of reduced size but with the same structure and expose it to beams of electrons and pions in order to check the resolutions. An exposure to a neutrino beam will also be made and will enable us to study the topology and kinematic properties of neutrino interactions which constitute the most serious background for nucleon decay.

CONCLUSION

Several detectors, specifically aimed at the measurement of nucleon decays are being installed in various underground laboratories around the world. They have fiducial masses one order of magnitude bigger than previous detectors. The Fréjus fine grained calorimeter made of 3 mm iron plates, plastic flash chambers, and Geiger tubes provides 1000 tons of fiducial mass ($\sim 6.10^{32}$ nucleons) and excellent pattern recognition capabilities.

In the next few years it should be possible to push the lower limit of the nucleon lifetime to a few 10^{32} years or definitively establish that the nucleon is unstable and measure its branching ratios.

Without adding much instrumentation it will be possible to use this detector to search for ionizing slow moving massive monopoles and to study multimuon events.

REFERENCES

1. For an extensive theoretical review, see P. Langacker, Physics Report 72, 185 (1981).
   See also P. Langacker, these proceedings
2. D. Sinclair, these proceedings.
3. D. Cline, these proceedings.
4. S. Miyake, these proceedings.
5. P. Picchi, these proceedings.
6. T. Gaisser, these proceedings.
7. F.E. Taylor et al., IEEE Trans. Nucl. Sci., NS 27, 30 (1980).
   D. Bogert et al., IEEE Trans. Nucl. Sci., NS 29, 363 (1982).
8. R.C. Allen, G.A. Brooks and H.H. Chen, IEEE Trans. Nucl. Sci., NS 28, 487 (1981).

A PROTON DECAY AND SOLAR NEUTRINO EXPERIMENT WITH
A LIQUID ARGON TIME PROJECTION CHAMBER*

Herbert H. Chen, Peter J. Doe and Hans-Jurg Mahler
University of California, Irvine, CA 92717

SUMMARY

Recent progress in development of the liquid argon Time Projection Chamber is reviewed. Application of this technique to a search for proton decay and ^8B solar neutrinos with directional sensitivity is considered. The steps necessary for a large scale application of this technique deep underground are described.

INTRODUCTION

The idea of the Time Projection Chamber (TPC) is familiar to most experimental particle physicists. Many gas TPC's exist and major physics problems are being addressed using such detectors. The liquid TPC is conceptually similiar, replacing the gas medium by a liquid. The liquid TPC has promise where a fine grained, totally live, high density source (target)/detector is required. The liquid argon TPC is particularly attractive where large detector mass is essential and where expected event rates are low, e.g. proton decay and neutrino detectors.

DEVELOPMENT STATUS

The operation of a liquid TPC requires the capability to drift ionization electrons in the medium over several tens of centimeters. Thus, electronegative impurities, e.g. oxygen, have to be reduced to the level of a few ppb. This has been achieved for several liquids (argon,[1] xenon,[2] methane[3]). In the case of argon, well known purification techniques were used.

The liquid argon TPC operates in the ionization mode and requires the use of low noise preamplifiers. These are available commercially at relatively low cost. Thus, two and three dimensional tracks have been observed using a 50 liter test detector.[4,5] The readout planes were either strips or a "woven" structure with 2.5 mm spacing etched on PC boards. Figure 1 shows examples of two dimensional tracks from cosmic rays.[4] The negative signals are from a pulser and serve as time reference and energy calibration (3 fC or about 0.6 MeV). Such tracks vividly demonstrate the capabilities of a fine grained, totally live detector.

---

*Presented at the Workshop on Science Underground,
 Los Alamos, New Mexico, Sepember 27 - October 1, 1982.

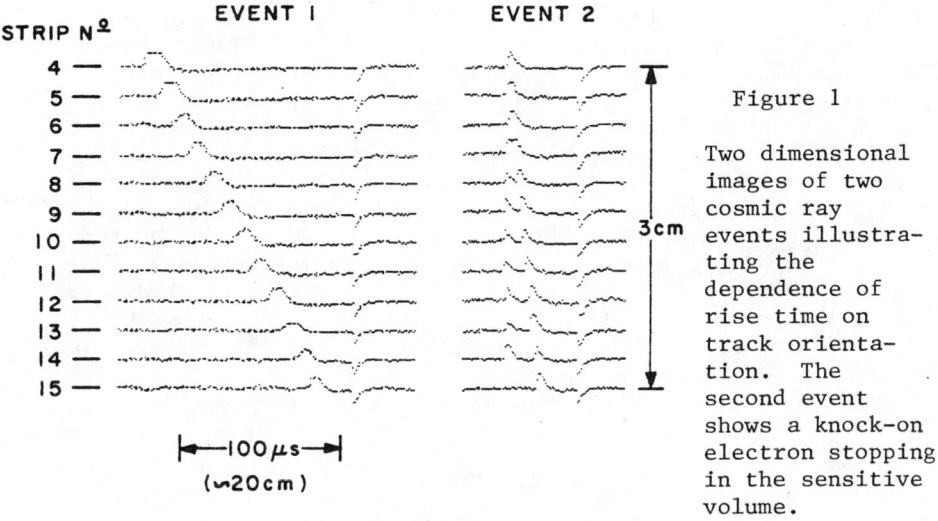

Figure 1

Two dimensional images of two cosmic ray events illustrating the dependence of rise time on track orientation. The second event shows a knock-on electron stopping in the sensitive volume.

DETECTOR DESCRIPTION

For an experiment sensitive to proton decay lifetimes in the range of 10^{32}-10^{33} years, the required detector must contain at least the same order of nucleons, i.e. the detector mass is of order 10^3-10^4 tons. For an experiment with directional sensitivity to ^{8}B solar neutrinos, the required detector mass is of order 10^3 tons if an event rate of 1/day is wanted from ν_e,e^- elastic scattering. This assumes a flux consistent with the ^{37}Cl radiochemical experiment of R. Davis et al.[6] The proximity of detector sizes and the high anticipated cost for any direct counting experiment sensitive to either process encourages the combination of both into a single detector.[7] This takes advantage of the unique sensitivity of liquid TPC's down to a few MeV.

It is useful to visualize the parameters of a detector at this scale. For simplicity, a single right circular cylinder of 20 m diameter and 20 m length is assumed. If the central $(10\ m)^3$ region were for solar neutrino detection and the entire volume for proton decay, one would have an appropriately large proton decay detector and an extra-ordinarily well shielded solar neutrino experiment. The density of liquid argon is 1.4 gm/cm^3, thus the mass of liquid argon in this cylinder is about 9,000 tons. If octagonal electrodes are used, the sensitive mass of liquid argon would be about 8,000 tons.

The number of electronics channels is determined by the required spatial resolution, and by the degree of multiplexing. For proton decay, 2 cm sampling should be adequate. A minimum ionizing particle deposits 4 MeV in 2 cm and creates $\sim$ 170,000 ionization electrons (27 fC) which results in a signal of $\sim$ 120,000 ionization electrons (20 fC) after recombination in an electric field of $\sim$ 1 kV/cm. Assuming a 1 m drift space and taking $(20\ m)^3$

as the readout volume, one sees that 40,000 channels are required. For solar neutrinos, 4 mm sampling may be adequate. This gives a signal of about 4 fC. Taking $(10 \text{ m})^3$ as the readout volume, one sees that 50,000 channels are required. A combined proton decay/solar neutrino detector would require 80,000 channels. Multiplexing by a factor of 4 appears reasonable and reduces the number of channels of electronics to levels already existing in present accelerator experiments.

Further investigations are essential to establish the detector system outlined above. Consider the physics issues: for proton decay, particle identification via dE/dx needs further work since ionization recombination is not yet a known function of dE/dx; for solar neutrinos, a better understanding of potential backgrounds, both cosmic ray and radioactive, is critical. If the detector is situated at sufficient depth, we believe that internal contaminants, e.g. in electrodes, remain as the most serious problem.

DESIGN STUDY

Before implementing a multi-thousand ton liquid argon TPC for proton decay and solar neutrinos, it is necessary to carry out a design study which addresses the problems associated with the construction, operation, cost and safety of such a deep underground detector. Several large engineering companies, e.g. Bechtel, Fluor, etc., have the expertise and the resources not only to carry out this study, but also to design, engineer and construct the required facilities. Fluor Engineers and Constructors, Inc. has shown substantial interest in such a design study, and has submitted[8] a cost competitive[9] proposal to UCI for this effort. Its relevant experience is abstracted from the proposal by taking titles of sub-sections, and these are shown in Table 1.

Table 1.   Section 2.0   RELATED EXPERIENCE
from Fluor Engineers, Inc. Design Study Proposal to UC Irvine

2.1	Cryogenic Experience
2.2	Processing Experience
2.3	Low Temperature Units in Ammonia Plants
2.4	Vessel Engineeringing-Cryogenic Storage Experience
2.5	Mining
2.6	Tunneling and Underground Excavation

PRESENT PLANS

Since the application of large liquid argon TPC's in such experiments as proton decay and neutrino interactions is extremely promising and we have been encouraged to speed the completion of this detector development effort, we have found some limited resources for and have started on the next scale test detector, i.e. a seven thousand liter system. Design and construction of the purification system which has a capacity of about 2 tons/day is well under way. This purification technique can be easily expanded to larger scales. We have also visited a manufacturer of commercial LN_2 dewars and have received quotations and delivery time for a standard 1920 gallon (7.27 m^3) dewar suitably modified for our needs. This system together will allow a study of argon purification and purity maintainance at the 10 ton level.

The readout electrodes for this detector would be constructed from "woven" printed circuit boards as was used in the demonstration of three dimensional tracking in the 50 liter system. The instrumented detector as envisaged here would have a sensitive volume of 1 to 2 m^3 and would require 200 to 400 channels of analog/digital electronics.

We are considering the possibility of moving this detector system to the neutrino beam at BNL. This operation will require a substantial effort (we would welcome contributions by others). Such a test will be a useful demonstration of the capabilities of this technique for experiments on proton decay and neutrino physics.

ACKNOWLEDGEMENTS

This research is supported in part by the U.S. Department of Energy under contract No. DE AT03-76ER71019.

REFERENCES

1. P.J. Doe, H.J. Mahler and H.H. Chen, IEEE Trans. Nucl. Sci. NS-29, 354 (1982).
2. K. Masuda, A. Hitachi, Y. Hoshi, T. Doke, A. Nakamoto, E. Shibamura and T. Takahashi, Nucl. Instr. and Meth. 174, 439 (1980).
3. A.S. Barabash, A.A. Golubev, O.V. Kazachenko, V.M. Lobashev and B.M. Ovchinnikov, Nucl. Instr. and Meth. 186, 525 (1981).
4. P.J. Doe, H.J. Mahler and H.H. Chen, Nucl. Instr. and Meth. 199, 639 (1982).
5. H.J. Mahler, P.J. Doe and H.H. Chen, IEEE Trans. Nucl. Sci. NS-30, to be published.
6. R. Davis Jr., Proc. Informal Conf. on the Status and Future of Solar Neutrino Res., BNL-50879 1, 1 (1978). G. Friedlander, editor; also, J.N. Bahcall, NEUTRINO '81 2, 253 (1981). R. J. Cence, E. Ma and A. Roberts, editors.
7. H.H. Chen, Proc. of Summer Workshop on Proton Decay Experiments, Argonne, Il. June, (1982). To be published.
8. R.J. Dugal, Advanced Technology Division, Fluor Engineers and Constructors, Inc., 2801 Kelvin Avenue, Irvine, CA.
9. J. Rogers and W. Quinn (LANL), Private communication.

MAGNETIC MONOPOLES, NUCLEON DECAY AND DUMAND*

P. C. Bosetti

III. Physikalisches Institut der TH Aachen, Aachen, West-Germany

ABSTRACT

A search for magnetic monopoles catalyzing baryon decay has been performed. The detector – a water filled tank equipped with photomultiplier tubes – was most sensitive for monopoles with velocities less than 0.001c and interaction lengths less than 50 cm. No monopole has been found leading to an upper limit of $F(M) \leq 1.6 * 10^{-2}$ m^{-2} sr^{-1} d^{-1}.

Magnetic monopoles were first hypothesized by Dirac (1) in 1931 and this work has led to many – so far unsuccessful – searches for their detection. With the advent of unified theories of electromagnetic and weak interactions and in particular of the Grand Unified Theories (GUTs) of strong, electromagnetic and weak interactions these monopoles have regained much interest in these days. In 1974 Polyakov and t'Hooft (2) showed that in a variety of these unified theories there exist solutions to the field equations representing particles with magnetic charge. In GUTs, these solutions always appear when there exists a simple gauge group, which is spontaneously broken down to $SU(3) \times U(1)$ like e.g. in $SU(5)$. The mass of these monopoles in Grand Unified Theories is given by $m(M) = 1/\alpha \, m(X)$, where $m(X)$ is the grand unification mass (10^{14} GeV), which is the mass of those lepto-quarks that lead to nucleon decay. These Grand Unified Monopoles (GUMs) are assumed to be produced very shortly after the Big Bang and nowadays presumably have velocities β of the order of 10^{-5} to 10^{-3}. (If their velocities were larger than 10^{-3}, they would escape our galaxy).

It is obvious that monopoles of this kind could not have been detected by early experiments designed to detect highly ionizing particles produced at accelerators or in cosmic rays. In recent months, new results have been obtained on the upper limit on the flux which apply also to slow magnetic monopoles. Cabrera (3) finds $F(M) < 0.5$ m^{-2} sr^{-1} d^{-1} independent of the velocity. Sokolowsky and Sulak (4) determine $F(M) < 0.1$ m^{-2} d^{-1} sr^{-1} for $10^{-2} < \beta < 10^{-4}$. Groom, Loh, and Ritson (5) find $F(M) < 0.02$ m^{-2} d^{-1} sr^{-1} for $3 * 10^{-2} < \beta < 0.75 * 10^{-4}$.

*The members of the Collaboration are
P. Gorham, F. Harris, J. Learned, M. McMurdo, D. O'Connor and S. Thompson
University of Hawaii, DUMAND-Center, Honolulu, Hawaii
K. Mitsui
University of Tokyo

In 1981 new theoretical developments due to Rubakov (6) and further discussed and elaborated by Callan (7), Wilczek (8), and Ellis, Nanopoulos and Olive (9) show that GUM's should have, in addition to their well known properties, baryon violating interactions leading to reactions of the type

$$\text{Monopole} + \text{Baryon} \longrightarrow \text{Monopole} + e^+ + X$$

with a cross section which could be comparable in strength with typical hadronic cross sections. The monopoles are expected to catalyze baryon decays when passing through matter, remaining themselves essentially undisturbed by the "interaction". The "interaction-length" is usually parametrized as $IL = 4300 \beta/\sigma$ (cm) for unit density, where σ is an unknown parameter and estimated by various authors to be between 1 and 10^{-5}.

The "Rubakov - effect" provides a new possibility to search for these monopoles. So far no search has been performed for GUM's with this signature (though nucleon decay detectors presently in operation or beeing built are - in principle - useful devices to search for monopoles with such characteristics).

We have built a detector to search for baryon decay catalyzing monopoles and give first results of the experiment here.

The detector consists of a cylindric water tank of 3.2 m diameter and 2.4 m height. It is equipped with 3 photomultiplier tubes (PMT) of 8" diameter placed on the bottom. (During the second half of the run period two more PMTs were installed). These PMT's can detect through-going muons as well as monopoles catalyzing baryon decay, via the Cerenkov light produced. The signals received by the PMTs are fed into the fast electronics. Each signal received by all PMT's in coincidence sets a clock and the time differences to any further signal is recorded over 8 consecutive time intervals, each of 277 nanosecond duration. All events were accepted as monopole candidates with at least four consecutive signals in the PMT's within these eigth time intervals. From muons we expect $7 * 10^{-4}$ accidentals per day, a negligible background.

The detector size and the time interval (the detector is sensitive for consecutive hits) determine the maximum interaction length and thus the minimum value of σ that can be detected.

The detection effeciency for different values of σ and β has been calculated using a Monte Carlo program. For $\sigma > 0.5$ it is larger than 90% for β between 10^{-5} and 10^{-3} and drops to 80% for $\beta = 3 * 10^{-2}$, and 50% for $\beta = 10^{-2}$.

The apparatus has run for 50 hours so far. We found no monopole candidate and thus we can set an upper limit (90% cl) for the flux of magnetic monopoles, catalyzing baryon decay, for $\beta < 10^{-3}$ and $\sigma > 0.5$ of

$$F(M) < 1.6 * 10^{-2} \, m^{-2} \, d^{-1} \, sr^{-1}.$$

This limit is calculated making the assumption that the monopole flux is isotropic.

This result is about one order of magnitude better for $\beta < 10^{-3}$ than any presently available limit and of the same order of magnitude as the one obtained in reference 4 for the intermediate region of velocities. It is, however, still some orders of magnitude higher than what is expected from the limit on the lifetime of the proton deduced in ref. 9. So far, no result from nucleon decay experiments has been obtained on the monopole flux. This could be due to the fact, that for such high cross sections, as the one proposed by Rubakov and slow monopoles, the identification of monopoles might cause difficulties for detectors designed mainly for the observation of single nucleon decays. Furthermore, our detector is the only one operating at sea level designed for the detection of monopoles catalyzing baryon decay, filling the gap between experiments at sea-level searching for currents induced in superconducting coils by monopoles and the nucleon decay detectors placed deep underground.

We will continue with this experiment and hope to improve the upper limit on the GUM flux by one or two orders of magnitude in the near future. It is further planned to operate the detector under conditions that allow to detect GUMs with lower cross sections.

We should mention that the DUMAND collaboration plans to deploy a string of 10 PMT's in the deep ocean within one year as a test experiment for the DUMAND detector and this will serve as a much larger detector for GUMs. The whole DUMAND array will be able to improve the present limit for these monopoles to about $10^{-8} \, m^{-2} \, d^{-1} \, sr^{-1}$ within one years time of operation.

ACKNOWLEDGEMENT

We like to thank our colleagues of the DUMAND group for their support and useful discussions. Crucial parts of the electronic set-up were kindly loaned to us by the Physics Department of the University of California at Irvine.

REFERENCES

1. P.A.M. Dirac, Pro. Roy. Soc. London, Ser A133, 60 (1931)

2. G. t'Hooft, Nucl. Phys. B79(1974)276, A. Polykov, JETP Lett.20(1074)194

3. B. Cabrera, Phys. Rev. Lett. 48(1982)1378

4. K. Sokolowsky and L. Sulak, University of Michigan Preprint

5. D. Groom et al. Talk presented at the International Conference on Particle Physics, Paris, 1982

6. V.A. Rubakov, JETP Lett. 33(1981)644

7. C.G. Callan, Princeton University Preprint (1982)

8. F. Wilczek, Phys. Rev. Lett. 48(1982)1146

9. J. Ellis, D. Nanopoulos, and K. Olive, CERN-TH 3323(1982)

Chapter IV

COSMIC RAYS

They then gave them the jewel-spear of Heaven.
Hereupon the two Gods stood on the floating bridge
of Heaven, and plunging down the spear, sought for land.

From Nihongi (Chronicles of Japan), in R. Van Over, *Sun Songs, Creation Myths from Around the World* (Mentor, New York, 1980), Pt. 10.

THE FLY'S EYE

R. Cady, G. L. Cassiday, J. Elbert*, E. Loh,
Y. Mizumoto, P. Sokolsky, D. Steck, M. Ye
University of Utah, Salt Lake City, UT 84112

ABSTRACT

We describe a high energy physics observatory, the Fly's Eye, designed to measure extensive air showers (EAS) in the energy range 10^{17}-10^{21} eV via atmospheric fluorescence. Preliminary results are presented for the following measurements: (1) the high energy cosmic ray spectrum at 10^{17}-10^{19} eV, (2) the total proton cross section σ_{pp}, (3) limits on the extra-galactic neutrino flux at 10^{20} eV.

INTRODUCTION

The "Fly's Eye" (see Fig. 1) is a high energy physics/astrophysics observatory designed to detect ultrahigh energy cosmic rays (UHCR; $E \gtrsim 10^{17}$ eV) via air fluorescence. It consists of two stations separated by 3.3km. The first station (Fly's Eye I) consists of 67 62-inch mirrors, 880 associated photomultipliers and Winston light collecting funnels arranged in clusters of 12 or 14 tubes

Fig. 1. View of Fly's Eye Mirror Units. Cosmic rays generate air showers viewed by an array of 67 mirrors and 880 PMT's.

mounted in the focal plane of each mirror. The second station, Fly's Eye II, is smaller, consisting of only eight mirrors and associated PMT clusters. Fly's Eye I is designed to image the entire night sky (2π steradians) and thus to detect the passage through the astmosphere of EAS generated by an incoming UHCR. Even though the atmosphere is a poor scintillator ($\gtrsim 0.1\%$ efficiency; see Fig. 2 and 3), the overwhelming amount of energy being liberated by the large number of charged particles in an EAS ($n \sim 10^7$-10^{12}) makes it possible to optically detect cosmic rays with energies exceeding 10^{20}eV out to distances of the order of 20km or so. Experimental measurements to be carried out with this detector include the following:

(1) $\sigma_{p\text{-air}}$
(2) Secondary multiplicity growth.
(3) Search for rare but potentially exciting events, i.e., "Centauros".
(4) Limits on the spectrum of the extragalactic neutrino flux.
(5) Composition of cosmic ray primaries.
(6) Cosmic ray spectrum.
(7) Cosmic ray anisotropies.
(8) Search for UHCR spectrum cut-off.[1]

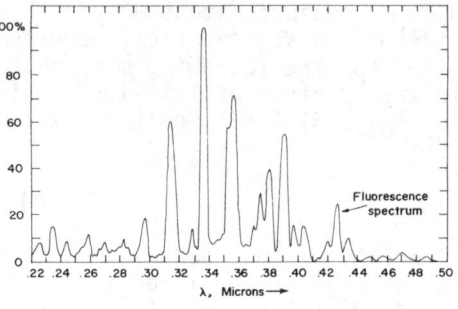

Fig. 2. Atmospheric fluorescence

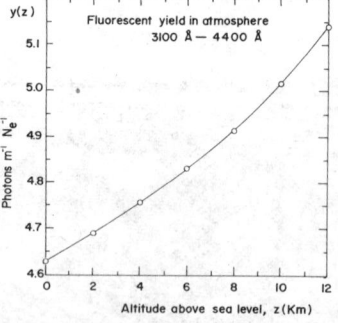

Fig. 3. Photon yield/m/electron vs. atmospheric altitude.

The goals listed above constitute a unique blend of high energy particle physics and astrophysics. No previous experimental program initiated to investigate the behavior of UHCR has completely succeeded in disentangling the effects of particle physics (i.e., cross section & multiplicity) from those of astrophysics (composition). The Fly's Eye detector represents an attempt to solve this problem in the UHCR regime. Relevent detector parameters are listed in Table I.

The low UHCR intensity constitutes a formidable problem for the experimentalist. Shown in Fig. 4 is a luminosity vs. energy diagram for a large number of existing or proposed accelerators. Also shown for comparison is the region of luminosity-energy accessible to the Fly's Eye. It's clear from such a plot that the low luminosity limits such a detector to the study of processes with cross sections at the millibarn level. However, the Fly's Eye detector alone occupies the energy regime $10^4 \text{GeV} \leq S^{1/2} \leq 10^6 \text{GeV}$ and this situation is likely to persist for a long time to come. Fortunately, the "beam" is free and one can counteract, somewhat, the low Fly's Eye luminosity by spending a long time taking data in order to obtain reasonably accurate cross section/ multiplicity measurements. Long observation time is necessary anyway in order to carry out the specified UHCR astrophysics experiments.

TABLE 1
Fly's Eye Parameters

		FE I	FE II
1.	# of Mirrors	67	8
2.	Diameter	1.6 meters	
3.	focal length	1.5 meters	
4.	Obscuration	~ 13%	
5.	Aberration	~ 20mrad	
6.	# PMT & (Winston Cones)	880	112
7.	PMT efficiency (EMI 9861B)	20%	
8.	Mirror/Winston Cone Reflectivity	85%	
9.	Overall light gathering ε	65%	
10.	Angular field of view/PMT	5°	
11.	#Electronics Analog Channels	3520	448
12.	Charge dynamic range	~ 10^5 (linear)	
13.	Time resolution	± 25nsec	
14.	Angular resolution	± 1°	

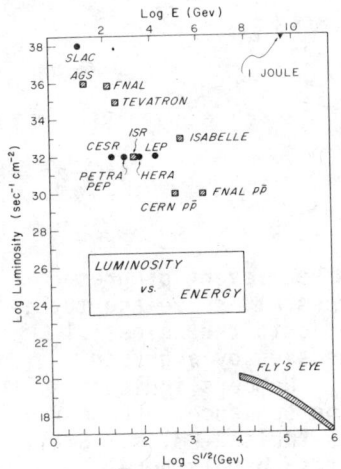

Fig. 4 Luminosity vs Energy

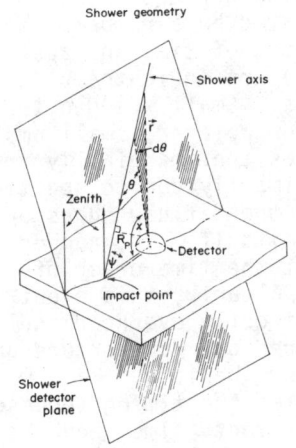

Fig. 5. Fly's Eye Shower Geometry

EVENT RECONSTRUCTION & SHOWER SIZE ANALYSIS.

Shown in Fig. 5 is a schematic of the geometry of an EAS as seen by the Fly's Eye. The location of the EAS track in space can be obtained by measuring four parameters: two parameters determine the plane in which the EAS lies while two additional parameters (R_p - the impact parameter and ψ - the ground impact angle) determine the orientation of the track in that plane. The plane can be obtained purely from the geometry of hit PMT's (see Fig. 6) while the other two parameters can be obtained from accurate timing given the kinematics of a light source propagating thru the sky at the speed of light. Consecutive PMT pulses arrive at the detector according to times given by the following expression
$ct = ct_0 + R_p \tan[(\chi_0 - \chi)/2]$ (χ_0 is the angle of shower observation at t_0, i.e., it represents the direction of shower approach; $\chi_0 + \psi = \pi$).

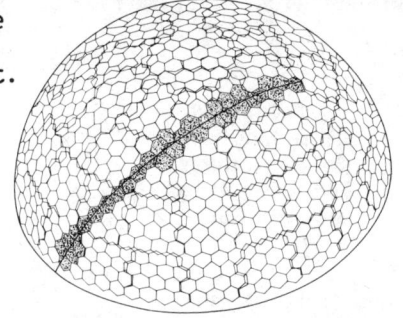

Fig. 6. Projection of Fly's Eye aperture onto "celestial" sphere. EAS track projects as a great circle.

Shown in Fig. 7 and 8 are the results of track reconstruction for a single event. In Fig. 7, the shower track has been projected onto the "celestial sphere", i.e., it represents the picture an observer would see looking up at a line source of light projected onto the night sky. The dashed line curve represents the horizon. Each hit PMT is indicated by a number. Noise PMT's, (out of time and spatial sequence) are indicated by large X's. Small x's denote barely noncoplanar, small amplitude tubes that marginally triggered primarily due to scattered light or the diffuse edges of the EAS, itself. The numbers represent the time order of firing. Clearly, this event passed directly overhead and disappeared out of aperture on the western horizon. Fig. 8 illustrates the timing sequence for this event (times have been converted to kilometers). The impact parameter for the event

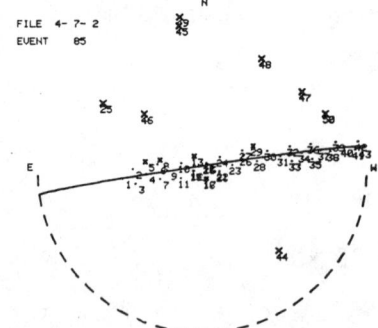

Fig. 7. Real event progressing across Fly's Eye aperture, projected onto the celestial sphere as seen by a ground observer. Numbers indicate PMT firing sequence. Large x indicates "noise" PMT's. Small x indicates, barely noncoplanar small amplitude PMT's.

was 1.52 ± .02 km while the zenith and azimuthal angles were 38.8 ± 1.3° and 353.6 ± 0.1° respectively.

Given the geometry of the event one can now use the recorded pulse integrals to convert the light yield received at the detector into shower size N_e as a function of distance along the shower's trajectory. Furthermore, given the atmospheric scintillation efficiency, one can then calculate the number of charged particles in the shower which generated that light. The photoelectron yield obtained by each hit PMT is:

$$N_{pe} = N_e Y \frac{\varepsilon A}{4\pi R^2} \exp(-R/\lambda) \Delta \ell$$

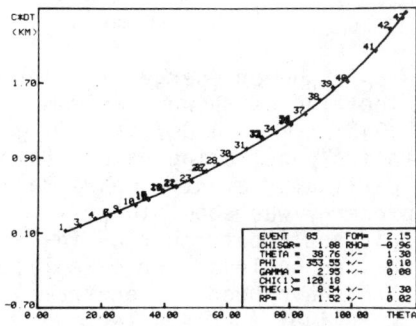

Fig. 8. Timing curve ct vs θ, shower emission angle for event shown in Fig. 7. Best fit values shown in insert.

where
N_e = shower size at observed location along trajectory
Y = fluorescent light yield (~ 5 photons/m/electron) (see Figure 3)
ε = combined light collection efficiency and photoelectron conversion efficiency (~0.17 ± 20%)
A = effective light gathering area (1.7m²)
λ = attenuation length of 3600Å photons in air(~18km)
R = distance of EAS to detector
$\Delta \ell$ = differential shower path length in field of view $\Delta\theta$ (Figure 5) since
$\Delta \ell = \Delta(R_p/\tan\theta) = R_p \Delta\theta/\sin^2\theta$ we have:

$$N_e = \frac{4\pi}{Y \Delta\theta} \frac{1}{\varepsilon A} R_p N_{pe} \exp(R/\lambda)$$

Shower size measurements are weakly dependent on angle except for angles less than 20-30° where Cherenkov light begins to dominate scintillation light. For angles larger than 30° shower sizes have been determined to ~±20%. Ultimately, we believe we can obtain ~±5% accuracy with improved calibration. Shown in Fig. 9 is the result of applying the above analysis to event 85. Shower sizes as a function of observation angle have been converted to size vs atmospheric penetration "slant" depth in gm cm⁻² along the shower's trajectory. Overlapping angular intervals are binned and averaged in order to obtain this curve. The solid line is the result of fitting the data with the Gaisser-Hillas parameterization of shower development given by the expression:[2]

$$N(E_0, x) = N_0 \frac{E_0}{\varepsilon} \left(\frac{x-x_0}{x_m-x_0}\right)^p \exp((x_m-x)/\lambda)$$

Where E_0 = shower energy, x_0 = location of first interaction
x_m = location of shower maximum
N_0 = .045, ε = .074 GeV, λ = 70 g/cm^2, $p = (x_m-x_0)/\lambda$
Due to its penetrating nature ($x_0 \sim 358$g cm^{-2}; $x_m \simeq 794$g cm^{-2}) this particular event appears to have been generated by a <u>proton</u> whose energy was about 10^{18}eV!
We would anticipate that an incoming iron nucleus, for example, would not have been so penetrating. By judiciously selecting such events we can insure a proton-enriched sample and then by plotting the distribution of event maxima we can estimate the proton-air interaction length and hence the pp inelastic cross section. This procedure is carried out in the last section.

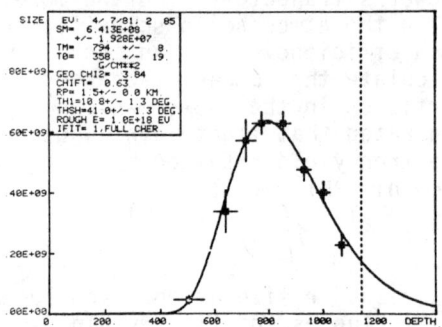

Fig. 9. Result shower size vs. atmospheric slant depth, in g/cm^2. Dotted line indicates the Earth's surface. Shower energy about 10^{18}eV.

CHECKS ON ANALYSIS

In order to insure that trajectories have been properly measured (depth perception with a single eye is difficult) and that recorded pulse integrals accurately reflect light yields, we have built and calibrated a high intensity pulsed xenon flasher permanently installed at Fly's Eye II and periodically fired over and above Fly's Eye I. This high intensity light pulse propagates up and out of the atmosphere and the scattered light it generates along the way (Rayleigh and Mie scattering) is picked up by the detectors at

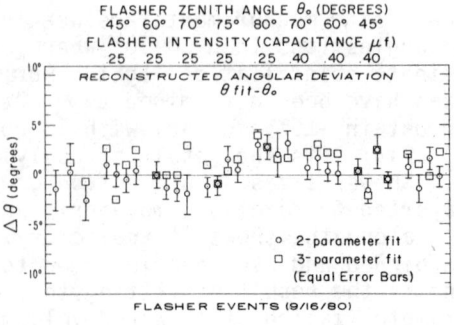

Figure 10. Angular difference between best fit zenith angle and known zenith angle for 32 "flasher generated" events. Observed deviations indicate angular accuracies $= \sim \pm 2°$.

Figure 10

Fly's Eye I. Thus, the event sequence strongly resembles an inverse EAS. By analyzing the received signals in the same way as for real events we can calculate both trajectories and light yields and in this case compare them to known values in order to assess the accuracy of track reconstruction and size analysis. Fig. 10 represents a summary of analysis of 32 "flasher" events.

We have also purchased and installed a pulsed nitrogen laser in order to generate much more accurate trajectories than could be generated with the xenon flasher. The laser generates a much shorter (100 psec) and better collimated (5m rad) light pulse than that of the xenon flasher. Indeed, shown in Fig. 11 is a timing curve for a trajectory generated by the nitrogen laser. The zenith angle setting for the laser was 70°. The fitted value was 69.89° ± 1.3°! A systematic analysis of 300 such laser events yields a zenith angle error of ± 1.5° averaged over zenith and azimuthal angles.

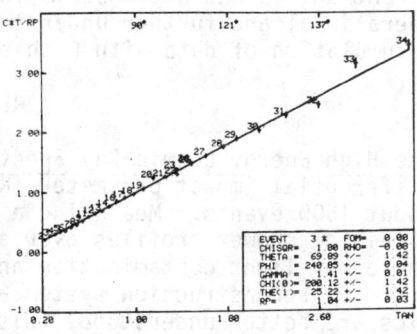

Fig. 11. Timing Curve ct/R_p vs $TAN(\theta/2)$ for laser generated event.

We show in Fig. 12, the result of shower "size" analysis applied to a single flasher event. No corrections were applied to the data. Each data point represents the conversion of the light received by a single PMT to the number of photons present in the propagating flasher beam. Conversion is based solely on track geometry and estimates of the Rayleigh and Mie scattered light received at the PMT. There were 2×10^{14} photons in the beam. Amplitude accuracy is about ± 20% as expected. This result gives us confidence not only in our overall calibration but also in knowledge of how

Fig. 12. Estimated number of photons in xenon flasher light pulse whose scattered light was recorded by Fly's Eye PMT's. Pulse heights recorded over emission angles θ ranging from 20° -160° yield correct number of photons to ± 20%.

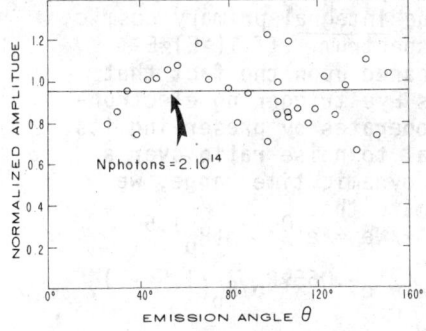

Figure 12

the atmosphere attenuates and scatters light!

Approximately 15% of events seen by Fly's Eye I will also be seen by Fly's Eye II. For these events, geometrical reconstruction is particularly simple, since the direction of the shower must lie along the intersection of the two planes defined by the two eyes. The additional timing information effectively allows us to do a 2-constraint fit to the trajectory. The subsample of events visible from both eyes will allow us to more fully understand both geometrical reconstruction systematics and our understanding of Cherenkov light contamination and propagation thru the atmosphere since the same part of an EAS will be viewed from two very different angles and distances. The second eye has recently become operational and further understanding of systematics awaits the accumulation of data with both eyes.

RESULTS

The High Energy Cosmic Ray Spectrum. Shown in Fig.13 is the differential impact parameter (R_p) distribution for a sample of about 1500 events. Measuring a shower's energy depends upon obtaining shower profiles over a rather long baseline. Until Cherenkov light contamination and residual reconstruction systematics are better understood, this procedure can be carried out only for a limited data sample at present. On the other hand, R_p can be precisely determined for a much larger data sample and since a shower's energy or, equivalently, its size is proportional to R_p, the R_p distribution should relate to the primary cosmic ray energy distribution. Quite simply we have $dN \propto I(>E) 2\pi R_p dR_p$ where $I(>E)$ is the integral primary cosmic ray spectrum. If $I(>E) \propto E^{-\gamma}$ and based upon the fact that Fly's Eye triggering electronics operates by preserving its signal to noise ratio over a wide dynamic time range, we estimate that
$$E \propto Ne \propto e^{.065 R_p}(R_p^{1.5})$$
Hence,
$$\frac{dN}{dR_p} \propto e^{-.065\gamma R_p}/R_p^{(1.5\gamma - 1)}.$$

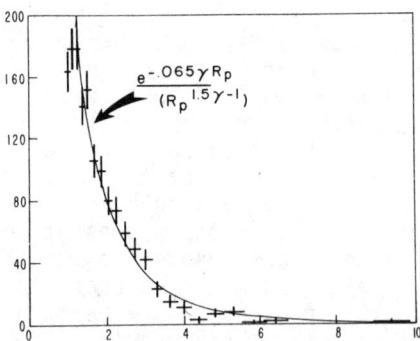

Fig. 13. Differential impact parameter distribution for 1500 events. Best fit to observed falloff indicates integral energy spectral slope $\gamma \sim 2.1 \pm 0.3$.

In addition to this rather rough estimate of the spectrum, we are in the process of developing a Monte Carlo program designed to completely simulate the response of the Fly's Eye to UHCR. We show in Figure 14 the resultant impact parameter distribution obtained from the Monte Carlo. The program was run by generating events at random whose energies were in excess of $2 \cdot 10^{17}$ eV and stopping the run when the total number of event triggers was identical to that in the real data sample. We find the best fit to the distribution shown in Fig. 13 yields a value of $\gamma = 2.0 \pm .1$ which is in agreement with the results of Watson[4] for shower energies less than 10^{19} eV. Our data sample spans the energy range of $2 \cdot 10^{17}$ eV – $5 \cdot 10^{19}$ eV. Also note that the smallest Monte Carlo event in the sample (E=$2 \cdot 10^{17}$ eV) occurred at 1km while the largest event $4 \cdot 10^{19}$ eV occurred at ~10km precisely as for the real data. Thus, we believe that our overall normalization is well determined! Watson[4] reports a spectral flattening for cosmic rays with energies >10^{19} eV. Such a flattening would show up as an enhancement in our R_p distribution at impact parameters $R_p \gtrsim$ 4-6km. We see only the tiniest hint--statistically insignificant--of such a flattening. However, the data reported here was obtained with the Fly's Eye "electronically cut-off" at $R_p \gtrsim$ 5-8 km (the cut-off is geometry dependent). This cut-off was instituted in order to optimize the nearby lower energy event rate. Currently, we are "electronically tuned" to greater distances with the obvious goal of examining the quoted spectral flattening at $E \gtrsim 10^{19}$ eV. Certainly, our preliminary spectral measurements for $E \gtrsim 10^{19}$ eV are consistent with those obtained by other workers.

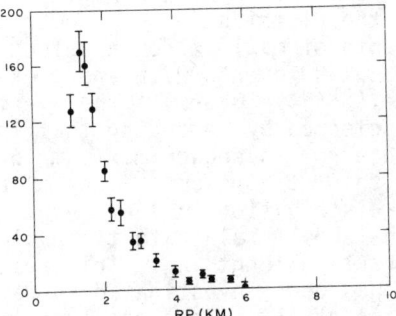

Fig. 14. Monte Carlo generated impact parameter distribution. Normalized to total event triggers. Compare with Fig. 13.

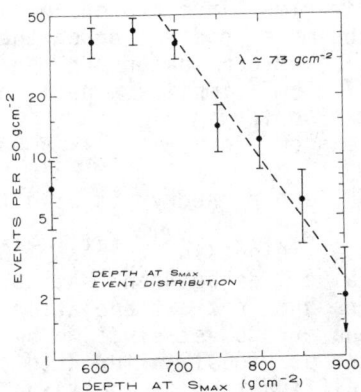

Fig. 15. Differential distribution of 90 events (at E~10^{18})vs. depth of shower. Attentuation slope Λ~73g cm^{-2} implies σ_{pp}^{tot} ~ 120mb.

Measurement of σ_{pp}. In Fig. 15 we show the distribution of shower maxima vs depth of maximum for a select sample of about 90 events with energies near 10^9 GeV($S^{1/2}$~$4 \cdot 10^4$ GeV). The event sample was selected by demanding that the estimated error in shower maximum location be within ± 50 g cm^{-2}. The slope of this distribution (Λ_m) at large shower depths should relate to the nucleon-air interaction length λ_n. This relationship has been investigated in detail by Gaisser et al[5]. They conclude that: $\Lambda_m \sim 1.6 \lambda_n$ and that the distribution of shower maxima is, in fact, as sensitive to the value of λ_n as is the distribution of even "earlier" observed points along the shower profile such as $\Lambda_{1/4\,max}$. Furthermore, we should note that the slope of this distribution at large depths should be determined preferentially by protons as opposed to heavier cosmic ray primaries since protons presumably would be more penetrating on the average. We note that our measured slope
$\Lambda_m \sim 73$ g cm^{-2} implies a nucleon interaction length of
$\lambda_n \sim 48$ g cm^{-2} or $\sigma_{p\text{-air}} \simeq 500$ mb

and if Glauber theory[6] is used to estimate σ_{pp}, we obtain $\sigma_{pp}^{tot} \simeq 120$ mb. This value lies between that obtained by a $\ln s$ and $\ln^2 s$ extrapolation.[5] (We quote no errors yet since we believe that the previously alluded to systematic difficulties may outweigh statistical problems.) Such a value for the cross section implies that a significant fraction of the cosmic rays contained in this data sample are protons. (If they were mostly Fe nuclei their behavior is quite remarkable; however the presence of lighter nuclei, such as alphas, can certainly not be ruled out.) We point out that inaccuracies in locating the depth at maximum would probably decrease our estimated value of Λ_m. Hence, our estimate of

σ_{pp}^{tot} probably represents a

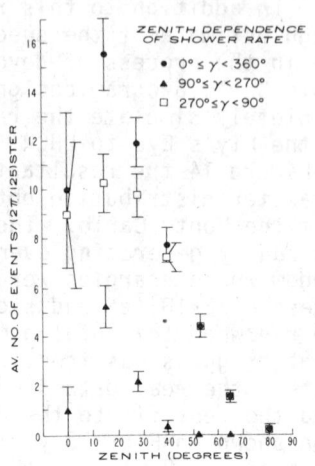

Fig. 16. Zenith angle event distribution for about 600 showers with energies >10^{18} eV. γ-angle intervals refer to showers impacting either behind or in front of Fly's Eye.

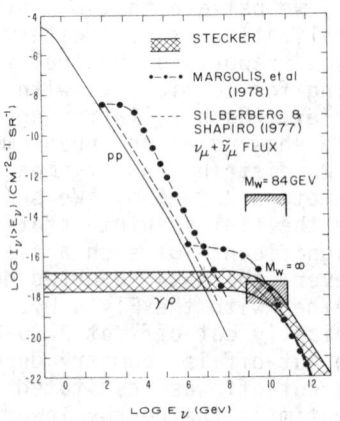

Fig. 17. Estimated limits on extragalactic neutrino flux near $E \sim 10^{20}$ eV given 84 GeV $\lesssim M_W \lesssim \infty$.

lower limit. Clearly, a larger number of more accurately measured events is necessary to (1) more accurately determine Λ_m and (2) look for changes in Λ_m indicative of compositional effects. A final value for σ_{pp} may await data taken with both Fly's Eyes operational.

Extragalactic Neutrino Flux

A search for the extragalactic neutrino flux ($E\nu \sim 10^{20}$ eV) is necessarily dependent on the value of the neutrino cross-section at these energies. We consider two limiting cases: $M_W = \infty$ (point interaction) and $M_W = 84$ GeV/C^2 (Weinberg-Salam choice). If $M_W = \infty$, the earth is opaque to neutrinos of this energy and we search for them by looking for events with zenith angles between 80°-90°(the atmosphere is ≥ 5000 gm cm^{-2} so that hadrons will not penetrate into the fiducial volume). Fig. 16 shows the zenith angle distribution of ~ 600 events obtained with 2/3 of Fly's Eye I operational for about 6 months. We see no events with $80° < \theta_z < 95°$ and set a limit on the $\nu_\mu + \nu_e$ flux shown in Fig. 17 ($M_W = \infty$). If $M_W = 84$ GeV/C^2, the earth is an $\sim 10\%$ transmitter and we can search for neutrinos by looking for upward-going (flasher-like) events. The best sensitivity is achieved for the ν_e flux, since at these energies the Landau-Pomeranchuk-Migdal effect[7] is operative and gives a radiation length in earth of ~ 100m. The result for this assumption is also shown in Fig. 17 ($M_W = 84$ GeV/C^2).
These limits should improve by factors of 20-100 as the experiment progresses and more realistic acceptance calculations are done. If and when the W boson is found and its mass determined, more definitive limits on the extragalactic flux can be set. We note that since the most obvious source of such high energy neutrinos is the interaction of the primary cosmic ray flux with the 3°K black body radiation, such measurements could, in principle, confirm the universality of the 3°K radiation in extragalactic space.

Future Prospects:

There is a two prong thrust to improve the Fly's Eye: noise reduction and spatial resolution improvement. To reduce the night sky background radiation from stars, planets and street lights, optical filters (UG-1) are being installed on 14 phototubes as a small scale test. We expect the filters will enable us to expand the visible volume of the detector as well as extend the observation time into nights when a small fraction of the moon is visible. We are also designing a fine resolution Fly's Eye to supplement our existing detectors. Optimization of the design based on our experience with the existing eye and new phototubes and electronics has begun this summer. The high resolution eye will enable us to extend the visible fiducial area beyond our present radius as well as give us improved shower profile measurement.

Whether we will proceed to fully instrument the second eye, develop a single high resolution eye, or both, will depend to a large extent on understanding the data which we will take with both eyes in the following year.

REFERENCES

1. K. Greisen, Phys. Rev. Let. 16, 748 (1966).
2. T. K. Gaisser and A. M. Hillas, 15th Int. Cosmic Ray Conf. 8. 353 (1977).
3. F. W. Stecker, NASA Tech. Memorandum 79609, GSFC, Greenbelt, MD. (1978).
4. A. A. Watson, 16th Moriond Astrophysics Meeting "Cosmology and Particles" Les Arcs, France, edited by J. Audouze et al, editions Frontieres March 15-21, 1981.
5. T. K. Gaisser (see proceedings--Madison Conference on High Energy Collider Physics, Dec. 1981.)
6. R. J. Glauber and G. Matthiae, Nucl. Phys. B21, 135 (1970).
7. E. Konishi, et. al., Nuovo Cimento, 44A, 509 (1978)

PHYSICS WITH UNDERGROUND LEPTONS OF ATMOSPHERIC ORIGIN

T. K. Gaisser

Bartol Research Foundation of The Franklin Institute
University of Delaware, Newark, DE 19711

ABSTRACT

Muons of great energy and neutrinos of all energies produced in the atmosphere by cosmic rays can penetrate to deep underground detectors. With existing and proposed proton decay detectors in mind, we review the physics accessible to underground cosmic ray experiments. Primary emphasis is on use of multiple coincident muons to study composition of primary cosmic rays at energies inaccessible to direct experiment

INTRODUCTION

Cosmic ray muons and neutrinos originate from the cascades produced in the atmosphere by primary nucleons:

$$N \longrightarrow \begin{cases} \pi^{\pm}, K \longrightarrow \begin{cases} \mu \longrightarrow \begin{cases} e \\ \nu_\mu \; \nu_e \end{cases} \\ \nu_\mu \end{cases} \\ \pi^0 \\ \hookrightarrow 2\gamma \rightarrow \text{electromagnetic cascades} \end{cases} \quad (1)$$

The flux of primary nucleons is a steeply falling function of energy, approximately given by

$$\frac{dn}{dE} \propto E^{-\gamma} \quad (2)$$

with $\gamma \sim 2.6$, the exact value depending somewhat on the energy. Energy spectra of secondary hadrons in the atmosphere are similar, and spectra of muons and neutrinos are even steeper because of the energy dependence of the dilation of the lifetimes of the parent mesons. Only the muons and neutrinos from the atmospheric cosmic ray cascades can penetrate to deep underground detectors. The muons must have very high energy at production to penetrate to the detector (e.g. ~600 GeV to reach 2000 m.w.e.). On the other hand the Earth is transparent to virtually all cosmic ray neutrinos, most of which have energies $\lesssim$ 1 GeV.

NEUTRINOS

One can use underground detectors to compare rates of upward and downward going neutrinos and search for effects of neutrino oscillations (e.g. $\nu_\mu \leftrightarrow \nu_e$) which may appear if the oscillation length is less than the diameter of the Earth.[1-5] The goal here is to take advantage of the large difference in path length of neutrinos produced in the atmosphere immediately above the detector as compared to those produced in the atmosphere on the other side of the Earth. Success depends on having a large enough detector to get a useful rate of neutrino interactions. This is in the 10 kiloton range for contained interactions.[1,3] One can attempt to overcome this limitation by measuring the angular dependence of horizontal and upward going muons produced by ν_μ in a large volume of rock surrounding the detector.[6] The price is that the neutrinos have higher energy in this case so that the sensitivity to neutrino mass difference is somewhat less than in case of direct observation (because the oscillation length increases with energy).

In all cases involving neutrinos in the GeV range, effects of the Earth's magnetic field are likely to be important since the flux of upward and downward neutrinos will depend on integrals over different geomagnetic field configurations. Thus, despite the fact[3] that upward and downward geometries are symmetric, a difference in effective geomagnetic cutoffs for the corresponding primaries will in general lead to differences in the fluxes of produced neutrinos. Moreover, this effect cannot entirely be removed by measuring the ratio ν_e/ν_μ. Because of the difference in their genealogies, the relation between E_ν and the primary energy E_0 will be different for the two types of neutrinos, so that a difference in energy of the geomagnetic cutoff can affect the ratio as well as the overall flux.

For these reasons, and also because neutrino interactions are a major background for proton decay searches, a full calculation of the expected neutrino flux, taking account of effects of the geomagnetic field, is necessary. The calculation can be normalized by comparison to measurements of low energy atmospheric muons, particularly at high altitudes.[7] Existing calculations[8,9] differ by a factor of two at 1 GeV where they overlap.

MUONS

The event rate in deep underground detectors is dominated by muons produced high in the atmosphere with sufficient energy to penetrate the overlying rock. Study of underground muons is an old subject.[10,11] Motivation for renewed study of the subject comes from the prospect of the new proton decay detectors that are significantly larger and/or of significantly higher resolution than previous deep detectors. The basic observables are the rates of events with exactly N_μ coincident muons (from the same primary cosmic ray) and the muon separations. These quantities reflect the primary composition and energy spectra as well as the physics of muon production in hadron collisions. Because high energy muon physics (e.g. study of prompt muons) will be accessible to new $\bar{p}p$ colliding beam experiments, it would appear that the most important new physics accessible to measurements of underground muons will be the cosmic ray astrophysics of the primary composition and energy spectra. (Miyake[10]) points out, however, that observation of "muon bundles", groups with internal spacing of a few centimeters, requires new muon physics. Very closely spaced coincident muons also appear in the Mont Blanc[12] and Soudan I detectors,[13] as reported at this meeting, and it remains to be seen through simulations whether these can be fluctuations of ordinary events or not.)

Details of the simulations required to interpret measurements of multiple muons underground have been described in recent papers.[14,15] Here we only review some of the main features in order to discuss the sensitivity of detectors at various depths. We also give a very brief overview of astrophysical questions that the measurement of muons may illuminate.

Full Monte Carlo simulations show[15] that the mean number of muons at the depth of the detector per primary proton is well described by Elbert's parametrization[16]:

$$< N_\mu > = \left(\frac{14.5 \text{ GeV}}{E_\mu \cos\Theta} \right) \left(\frac{E_o}{E_\mu} \right)^{p_1} \left(1 - \frac{E_\mu}{E_o} \right)^{p_2} \quad (3)$$

The first factor is the explicit time dilation factor (sec Θ/E_μ) that arises because of the competition between decay and interaction of parent pions and kaons at any particular atmospheric depth. Relative probability of decay increases with increasing zenith angle Θ because the density to a fixed slant depth is decreasing. The second factor would have $p_1=1$ for a pure multiplicative cascade (e.g. approximation A of E-M cascade theory). The power is less than unity here ($p_1=.757$) because the relevant parent mesons are produced deeper in dense atmosphere as E_o increases for fixed E_μ. The last factor is a threshold factor that arises ultimately from quark counting rules that govern inclusive meson production near Feynman x=1 in hadronic collisions. Here $p_2 \sim 5.25$.

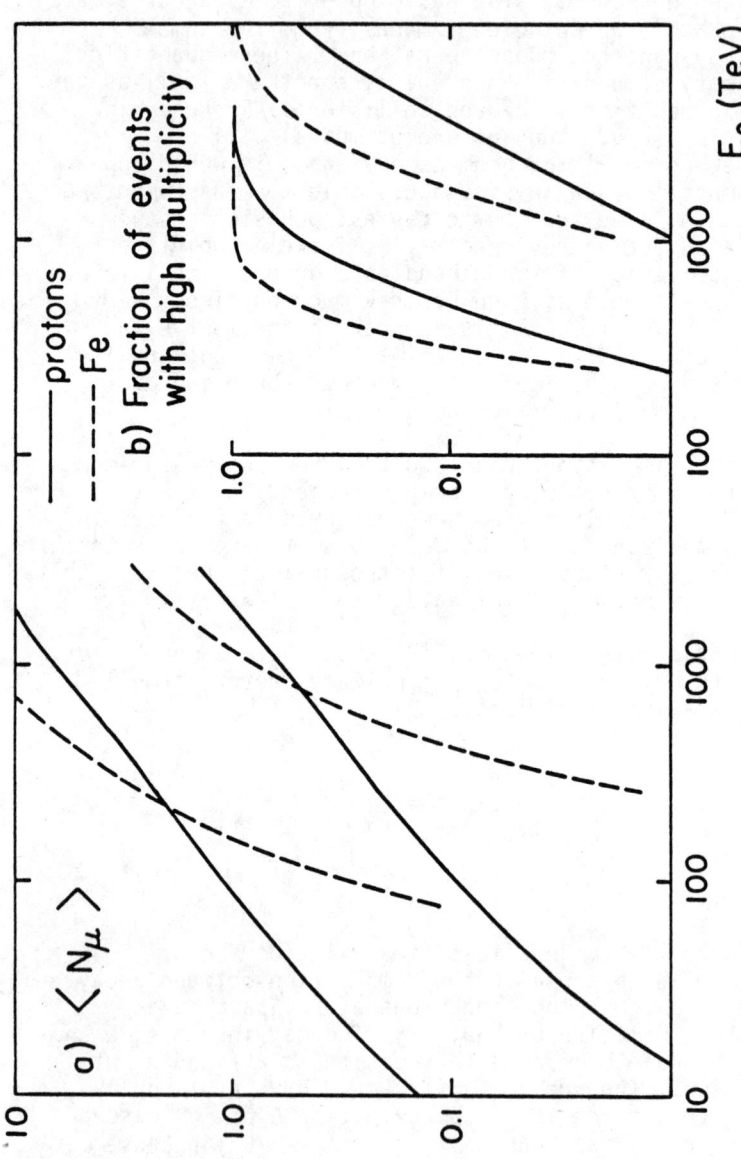

Figure 1 - a) Average number of muons per proton primary of energy E_0. The upper left hand set of curves is for a detector at 1750 m.w.e. vertical depth assuming a zenith angle of 30°. The lower right hand set is for 4200 m.w.e. and for vertical muons.
b) Fraction of high multiplicity events among events with at least one muon at the detector level. The two sets of curves are as in part (a). The shallow set is for $N_\mu \geq 7$ and the deeper set is for $N_\mu \geq 4$.

For a primary of mass A and total energy E_0, $E_0/E_\mu \to E_0/(AE_\mu)$ in the second and third factors of Equation (3) and the whole is to be multiplied by the number of nucleons A. Thus

$$<N_\mu>_A / <N_\mu>_{A=1} \to A^{.243} \tag{4}$$

as $E_0 \to \infty$ with E_μ fixed.

On the other hand, for $E_\mu/E_0 < 1 < AE_\mu/E_0$, the ratio in Equation (4) vanishes because of the threshold factor. This leads to the characteristic behavior of $<N_\mu>$ shown in Figure 1a. The multiplicity distribution at the detector depth X (measured in km. of water equivalent) is Poisson with mean $<N_\mu>$ and (very roughly):

$$E_\mu (TeV) \sim .53 \, (e^{.4X} - 1). \tag{5}$$

Figure 1b shows the corresponding probability per event of having > N_μ muons at each of two detector depths for two primary masses. (Note that at the shallower depth it is necessary to go to higher multiplicity to get as great a sensitivity as at the deeper depth. This reflects the fact that it is relatively easy to produce multiples with protons at the shallower depth.) The difference in the ratio of multiples to singles for Fe and p as a function of E_0 illustrates the potential sensitivity of this type of experiment to primary composition. A discussion of inevitable complexities, such as effect of a detector of finite area, lack of primary energy resolution and low flux at high energy is postponed to a later section.

RELATED ASTROPHYSICS

The fundamental observables of cosmic ray astrophysics are the chemical composition and energy spectra of each component of the primary cosmic rays. Measurements of these quantities constitute a large part of the experimental basis of theories of the origin, acceleration and propagation of cosmic rays. The general questions include:

o **Origin**: What is the nature of the source region and source material? e.g. is the source material primordial (p, α) or processed (C, O, Si, Fe ...)? (One must keep in mind that an injection mechanism may distort the answer.)

o **Acceleration**: What kinds of electromagnetic-plasma processes can achieve > 10^8 x present machine energies of 1 TeV? e.g. learn scales and velocities of shock waves, turbulence and magnetic irregularities.

o <u>Propagation</u>: What is the medium like in which cosmic rays propagate? What are the ambient magnetic fields and particle densities? How long do cosmic rays stay in a local region, in the galaxy, in the local supercluster? What is the nature of the diffusion process and are there any extragalactic cosmic rays?

These are the classic questions of cosmic ray astrophysics that have been studied intensively at low energies ($< 10^{12}$ eV) where direct measurements of the primaries have been made. The flux of primaries above 10^{14} eV is less than 1 particle per m^2. sr. hour, so the energy range around 10^{15} eV and above is likely to remain inaccessible to direct measurements for a long time. At these energies, one must rely on interpretation of indirect measurements of atmospheric cascades to determine the properties of the primaries. Such experiments overcome the problem of low flux by using large detectors exposed for long periods deep in the atmosphere. The price is high since interpretation of the cascades is difficult, depending on extensive computer simulation. On the other hand, the astrophysics around 10^{15} eV is particularly interesting[17] because the spectral index, γ, of the overall energy spectrum is known to change around this energy.

Measurements of coincident multiple muons underground is one type of air shower experiment that is relevant to cosmic ray astrophysics around 10^{15} eV. Its relation to other cosmic ray experiments and the status of the field in general is treated in several recent reviews.[18-20] The Proceedings of a recent conference on the subject[21] contains several relevant papers. Prospects for significant progress in interpretation of cosmic ray cascade experiments are greatly enhanced by two experimental developments: First, the advent of colliding beam interactions at $\sqrt{s}$ = 540 GeV (which corresponds to $E_0 \sim 1.6 \times 10^{14}$ eV protons on a stationary target), and second, construction of an experiment[22] to be flown on space shuttle in two or three years (as well as new balloon flights of large exposure[23]) that will measure primary composition directly up to around 10^{14} eV. These new experiments will serve to normalize and calibrate the indirect methods, which can then be used to extend our knowledge of primary composition and spectra into the $10^{15} - 10^{16}$ eV range.

UNDERGROUND MUONS: DETAILED CONSIDERATIONS

The maximum energy accessible to underground experiments is determined by flux. Table I summarizes the fluxes of events of several multiplicities at the two detector depths and angles corresponding to Figure 1. These fluxes are obtained by convoluting the probability per primary (Figure 1b) with the primary spectrum. The range of primary energies that gives the dominant contribution in each case is also shown. The Table is still artificial in that the effect of a finite detector has not been included; i.e., N_μ here means multiplicity at the depth of the detectors, regardless of lateral separation of the muons. Nevertheless, the essential features of an experiment with an underground detector alone are illustrated:

o Most events are single muons from low energy proton primaries.

o At the shallower depth one must go to higher multiplicity to reach a situation where most multiples come from heavies.

o Primary energies $\gtrsim 10^{15}$ eV are accessible.

o Ratio of multiples/singles integrated over the primary spectrum will be very sensitive to absolute normalization of the primary spectrum in two different energy ranges.

o Ratio of high multiples to low multiples ($N_\mu > 1$) is somewhat less sensitive to the normalization problem but will be very sensitive to the effects of a finite detector.

Experiments with an underground detector alone have a sensitivity of roughly a factor of two[14,15] in measurable quantities in distinguishing between two quite different primary compositions: one with 8 per cent Fe and the other with 50 per cent Fe primaries around 10^{15} eV. Interpretation of shallow experiments is further complicated by distortion of the lateral distribution due to the Earth's magnetic field.[15]

There are two experimental techniques that can improve the sensitivity to primary composition:

1) Use of a surface array in coincidence with the underground detector to estimate the primary energy. The goal is to compare multiples/singles at nearly the same energy to increase the sensitivity to heavies and to remove uncertainties of normalization.

2) Use of an outrigger detector to measure the lateral distribution. This will remove an important source of uncertainty from the calculation since the rates of N_μ contained in a finite detector are very sensitive to the actual lateral distribution of the muons.

TABLE I

Flux (a) and primary energy range (b) of events with at least N_μ muons at the depth of the detector

		$N_\mu > 1$	$N_\mu > 4$	$N_\mu > 7$
X = 1.75 km.w.e.	p	125,000	89,000	170
		2–30	50–700	300–2500
θ = 30°	Fe	3,400	900	400
		70–400	170–1000	300–1400
X = 4.2 km.w.e.	p	4500	~10	~1
		6–130	> 1000	> 5000
Vertical	Fe	200	20	~10
		300–2000	> 1500	>3000

a) Flux is quoted in events per 100m² per week per sr. The normalization is approximate and artificially assumes that primaries in the stated range are either all protons or all Fe. A spectral index of 2.6 is used to make these estimates.
b) Primaries in the stated energy range (in TeV per nucleus) produce ~80 per cent of the total flux quoted; i.e., about 10 per cent of the flux is produced by primaries with energies below and about 10 per cent with energies above the stated range.

There are problems with the use of a surface array that require further consideration.[15] For an array of particle detectors there are large fluctuations in N_e (the measured particle number in the accompanying surface air shower) for given E_0, and the relation between E_0 and N_e depends on primary mass. These problems become more severe at lower energy and deeper in the atmosphere where one is working more on the tail of the shower. A surface array that detects atmospheric Cherenkov light would provide a less fluctuating measure of E_0,[24] but requires clear, dark nights for operation. It could nevertheless turn out to be advantageous for a shallow detector (with relatively high rate) and a surface deep in the atmosphere, as pointed out by Shupe during the Conference.

To illustrate the potential of a surface-underground combination, I show in Table II an estimate of the sensitivity of the experiment being constructed at Homestake[25, 26] The numbers assume complete containment of all muons in the underground detector and so are artificial. They must still be corrected for "feed down" (i.e., N_μ at the depth $\to N'_\mu < N_\mu$ in a finite detector with large fluctuations in N'_μ for fixed N_μ). The sensitivity is now about a factor of 5.

TABLE II

Example for 200 m^2 detector under 4.2 km.w.e. with surface array at 1.6 km. above sea level for $3.10^5 < N_e < 10^6$ or $E_0 \sim 2.10^{15}$ eV

	$N_\mu > 2/N_\mu > 1$	$N_\mu > 3/N_\mu > 1$
All Protons	.35	.06
50 Per Cent Fe	.61	.35

Finally, in Table III I have summarized rates and other relevant parameters for a surface-underground combination for various vertical distances, d, between the surface and underground components. $\pi <r>^2$ is the effective area of the muon shower front. I assume the surface array to have an effective area of $A_s = 0.1$ km^2 for showers with $E_0 > 10^{15}$ eV, and the underground detector is assumed to have $A_d = 200$ m^2.

TABLE III

Comparison of depths for surface-underground combination

Depth		$E_\mu(\theta=30°)$	$<r>$	$\pi<r>^2$	$\dfrac{A_s A_d}{d^2}$	Coincidence Rates per yr.
km	km.w.e.	(TeV)	(m)	(m^2)	(m^{2}sr)	
0.6	1.8	0.7	8	200	56	5000
1.1	3.4	2.0	4	50	16	1400
1.6	4.2	3.1	3	28	8	490
2	5.2	5.3	2	12	5	150

At shallow depths $\pi < r >^2 \sim A_d$ and containment is poor. Moreover, as mentioned above, one must go to high multiplicity to get sensitivity to heavy primaries. One cannot compensate by looking at very large zenith angle because containment becomes worse. On the other hand, for great depths the rate is very low. However, the practical range would seem to span > 0.5 to < 2 Km of rock and thus to include all but the deepest proton decay detectors.

SUMMARY

Several factors lead to renewed interest at this time in use of underground muons to study primary composition in the $10^{15} - 10^{16}$ eV range, which is a particularly interesting region astrophysically:

o There are new and planned large, deep detectors of high resolution designed for study of proton decay;

o Direct measurements of primary composition with high statistics to 10^{14} eV will provide a calibration by 1984-85.

o Experiments at $\bar{p}p$ colliders are helping to determine hadronic physics to $\sqrt{s} \sim 1$ TeV or $E_0 \sim 10^{15}$ eV, removing a source of ambiguity in the calculations needed to interpret measurements of coincident multiple muons underground.

o Ancillary surface arrays and outrigger detectors to improve the sensitivity of the experiment are relatively inexpensive compared to the basic proton decay detector.

Further detailed studies are urgently required to verify the sensitivity to primary composition of real detectors with a realistic mixture of all primary nuclei and to optimize the design of the surface-underground combination.

Acknowledgment: This work is supported in part by the United States Department of Energy under Contract No. DE-AC02-78ER05007.

REFERENCES

1. B. Cortez and L. Sulak in Unification of Fundamental Particle Interactions (Plenum, New York) ed. S. Ferrara, J. Ellis, P. van Nieuwenhuizen, p. 661 (1980).
2. P. H. Frampton and S. Glashow, Phys. Rev. D25, 1982 (1982).
3. D. S. Ayres, T. K. Gaisser, A. K. Mann and R. E. Shrock, to be published in Proc. DPF Summer Study, Snowmass (1982). See also A. K. Mann, this conference.

4. V. J. Stenger, Proc. DUMAND Symposium, Vol. II, p. 37 (1980), S. Pakvasa, Ibid., p. 45 and R. Silberberg and M. M. Shapiro, Ibid., p. 59.
5. D. Cline, this workshop.
6. K. Lande, this workshop.
7. D. H. Perkins, Oxford University Preprint 60/82, presented at 21st Int. Conf. on High Energy Physics, Paris, 1982.
8. A. C. Tam and E. C. M. Young, Proc. 11th Int. Cosmic Ray Conf. (Budapest) p. 307 (1969) compute the energy spectra of muon neutrinos from 200 MeV to several GeV. See also E. C. M. Young, Cosmic Rays at Ground Level, Institute of Physics Press p. 105, A. W. Wolfendale, ed., London and Bristol (1973).
9. L. Volkova, Sov. J. Nucl. Phys. 31, 1510 (1980) and Proc. DUMAND Summer Study Vol. 1, p. 75 (1978) computes fluxes of neutrinos above 1 GeV.
10. S. Miyake, this workshop.
11. G. H. Lowe et al., Phys. Rev. D13, 2925 (1976) and D12, 651 (1975).
12. P. Picci et al., this workshop.
13. L. Price et al., this workshop.
14. J. W. Elbert, T. K. Gaisser and Todor Stanev, submitted to Phys. Rev. D (1982).
15. T. K. Gaisser and Todor Stanev, to be published in Proc. Argonne Workshop on Proton Decay (1982).
16. J. W. Elbert, Proc. DUMAND Summer Workshop, ed. A. Roberts, Vol. 2, p. 101 (1978).
17. See the paper by J. Ormes in Proc. Penn Workshop on Very High Energy Cosmic Ray Interactions, April, 1982, for a discussion of Cosmic Ray Astrophysics around 10^{15} eV.
18. T. K. Gaisser, Comments Nucl. Part. Phys. 11, 25, (1982).
19. T. K. Gaisser and L. W. Jones, to be published in Proc. DPF Summer Study at Snowmass, 1982.
20. T. K. Gaisser and G. B. Yodh, Ann. Revs. Nucl. and Particle Science 30, 475 (1980).
21. Proc. Workshop on Very High Energy Cosmic Ray Interactions, University of Pennsylvania, 22-24 April, 1982, ed. M. Cherry, K. Lande and R. Steinberg.
22. D. Müller, Proc. Penn Workshop, Ibid., 1982.
23. W. V. Jones et al., Proc. Penn Workshop, Ibid., 1982.
24. A. A. Andam et al., Phys. Rev. D26, 23 (1982) and references therein.
25. M. Cherry et al., Proc. Penn Workshop, Ibid., 1982.
26. E. Fenyves et al., Proc. Penn Workshop, Ibid., 1982.

DETECTION OF GRAVITATIONAL COLLAPSE

J. Craig Wheeler and John A. Wheeler
University of Texas, Austin, TX 78712

ABSTRACT

At least one kind of supernova is expected to emit a large flux of neutrinos and gravitational radiation because of the collapse of a core to form a neutron star. Such collapse events may in addition occur in the absence of any optical display. The corresponding neutrino bursts can be detected via Cerenkov events in the same water used in proton decay experiments. Dedicated equipment is under construction to detect the gravitational radiation. Events throughout the Galaxy could be detectable, but are expected only at intervals exceeding a decade. Nevertheless, the next event could come tomorrow, so every attempt should be made to make the monitoring for such events routine.

SUPERNOVAE AND GRAVITATIONAL COLLAPSE

Bursts of neutrinos and gravity waves are expected from the gravitational collapse of a stellar core to form a neutron star. Such collapse is expected for the cores of massive stars and is thought to be connected with the explosion mechanism of supernovae of Type II. On the other hand, some collapse events may give no burst of light, only neutrinos and gravity waves.

Type II supernovae are identified by their normal hydrogen-rich spectra. Studies of their spectra, light curves and their correlation with spiral arms all point to the precursor being a massive star. The rate of explosion of Type II supernovae in the Galaxy suggests that they come from stars of about 10 to 20 $M_\odot$. Theoretical studies of the evolution of stars of such mass show that the formation of an iron core of about 1.5 $M_\odot$ and its subsequent collapse is very likely[1]. (Severe internal rotation might alter this conclusion.) The rates of explosion of Type II supernovae are roughly equal to the rate of formation of pulsars, rotating magnetized neutron stars. Thus with canonical numbers there is some reason to think that the supernova explosion in a star of 10-20 $M_\odot$ is associated with gravitational collapse. This association with pulsars may be a bit simplistic, as will be argued below.

Above 20 $M_\odot$ some stars may make neutron stars. Others may undergo total collapse to give black holes. There is no present certainty as to which stars do which at any particular mass. The rate of events from such massive stars is very small, less than one per century in the Galaxy. There may be a different neutrino signature for the two events. A neutron star may have a rather long deleptonization phase in which the binding energy of the neutron star, $\sim 10^{53}$ ergs, is radiated. The neutrino burst from the formation of a black hole is likely to be truncated after $\sim$ 10 msec. Thus, the detection of one of these rare events would be very useful.

Type I Supernovae are hydrogen deficient and uncorrelated with the arms of spiral galaxies. They do seem to be correlated with regions of active star formation in small irregular galaxies. The precursor stars are presumably less massive than those of Type II supernovae, but the nature of these precursors is uncertain.

A currently popular model for Type I supernovae is one in which a thermonuclear explosion results in total disruption, and no collapse[2]. Such an explosion would produce neutrinos but they would be few in number and have an energy of order one MeV which would make them virtually impossible to detect.

There is some evidence in favor of the prediction that Type I events do not leave neutron stars. Observations with the Einstein Orbiting X-ray Observatory have failed to detect thermal X-ray emission from the hot surface of neutron stars which might be present in the putative Type I remnants of SN 1006, SN 1572, and SN 1604[3]. An independent constraint is the failure to see an X-ray synchrotron nebula in these remnants. This is an important constraint because such a nebula is not subject to pulsar beaming which could hide a radio pulsar. Such a nebula is obvious in the Crab Nebula and detected for the Vela supernova remnant as well as around some pulsars[4].

If the thermonuclear explosion models are correct, then Type I supernovae should be discounted when trying to estimate collapse rates from supernova rates. On the other hand, if collapse is involved and a neutrino or gravity wave burst could be detected, then the whole class of thermonuclear models could be discarded. A definite statement about the simple existence of a neutrino or gravity wave burst or lack thereof associated with a Type I event would be very useful.

EVENT RATES

The total rate of supernova events in the Galaxy is roughly one per 25 years[5]. This rate results from estimates based on historical rates corrected for Galactic extinction, or by interpolation from extragalactic rates in different types of galaxies. Pulsars are formed at a rate of about one per 35 ± 15 years[6]. There is no useful estimate of the rate of non-magnetic non-optical events.

The total supernova rate in the Galaxy will not be the relevant number to estimate gravitational collapse if Type I events do not produce collapse. Unfortunately, the ratio of Type I to Type II events in the Galaxy is very uncertain. Of the seven or so historical events, three are putatively of Type I. The light curves of SN 1572 and SN 1604 by Tycho and Kepler[7] are roughly more consistent with Type I than Type II. SN 1006 was very bright and so may have been a Type I which are typically about three times brighter than Type II. Assigning a type to SN 1054 is very difficult despite the historical record. It clearly left very different remnants, both compact and extended, than SN 1006, SN 1572 or SN 1604. Arguments that it may have been Type II depend on assigning the first Chinese record to maximum light. The Japanese, however, apparently saw it earlier, and the a priori chance that they saw it on the rapid rise is small. Without knowing the phase of the early observations, no useful constraint on the light curve can be assigned. Cas A may not have been observed at all and in any case was too dim to be either classical type of supernova. Thus the only identifiable supernovae in the Galaxy were Type I. None can be assigned as Type II.

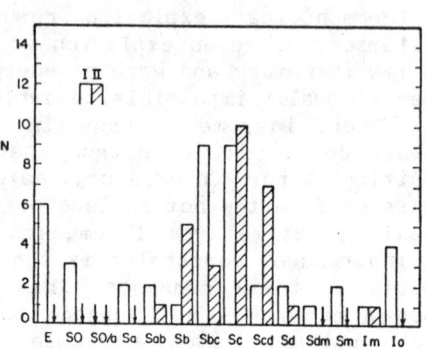

Figure 1 - The number of supernovae definitely identified to be of Type I or Type II is presented as a function of the type of the host galaxy in the sample of Oemler and Tinsley[7].

In other galaxies the ratio of Type I to Type II varies with the type of galaxy as shown in Figure 1. No Type II are seen in gas poor elliptical galaxies nor in I0 galaxies which are small irregular galaxies composed mostly of old stars but interspersed with patches of active star formation. This suggests that Type I events are and are not associated with moderately massive stars, adding to the confusion concerning their precursors. Spiral galaxies are classified by the degree of tightness with which the arms are wrapped, tightly for Sa, loosely for Sc. Sc galaxies are prodigous supernova producers and the numbers of Type I and Type II are roughly equal. In the sample of Sa and Sb galaxies the ratio is not at all clear. The probability of seeing one Type I and five Type II in Sb galaxies when the true rates are equal is only a few percent. The ratio of Type I to Type II may vary rapidly with galactic type. The Galaxy is roughly an Sbc, right in the range of high uncertainty.

The standard assumption is that the rates of Type I and Type II are equal. Then Type II occur at roughly one per 50 years. This is roughly comparable to the pulsar rate as mentioned previously.

On the other hand, the historical rates may show a bias toward Type I. In addition, the dearth of pulsars is not confined to the three historical remnants associated with Type I. The search for X-ray synchrotron nebulae has revealed nothing detectable in the majority of older supernova remnants surveyed. If these events made pulsars they should be detectable because pulsars live $\sim 10^7$ years versus $\sim 10^5$ years for the remnants. If there is evidence for pulsars in only roughly 20 percent of the extended remnants, one might argue that only 20 percent of the supernovae were Type II, given the uncertainty in the Type I/Type II ratio. Then the rate of formation of pulsars in supernovae would be about one in 125 years. This can not be the only pulsar production mechanism since such a rate disagrees seriously with the observed pulsar formation rate.

The fact that most remnants show no evidence for pulsars implies that there are ways to make pulsars without producing an extended remnant. Therefore, one should scale estimates of collapse rates with the supernova rates only with great caution.

If there are non-magnetic collapse events in Type I and Type II supernovae or events without any optical display, then the rate of collapse would be higher than either the pulsar rate or the supernova rates imply. A very optimistic estimate for the rate of collapse in the Galaxy would be about one every ten years. How does this compare with the death rate expected from the known number of stars and their astrophysically calculated evolution rates? Comparable if collapse occurs in all stars with $M > 4\ M_\odot$. A $4\ M_\odot$ limit is a low, but not impossible number. Collapse may be triggered in old white dwarfs by mass transfer in binary systems or slow changes in composition at a rate unrelated to the current star formation rate so no specific mass limit can be assigned. A rather pessimistic limit to the collapse rate would be the upper limit for pulsars, about 1 per 50 years, corresponding to all stars above about $10\ M_\odot$.

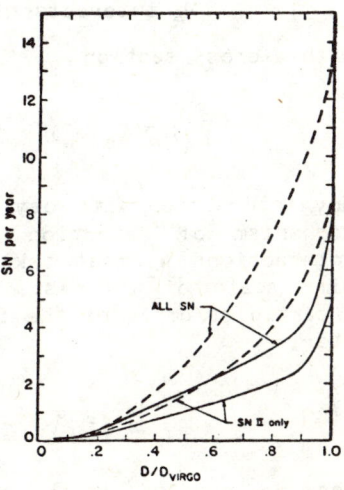

Figure 2 - The rate at which supernovae are expected to occur in external galaxies is given as a function of the fractional distance from our Galaxy to the center of the nearby rich Virgo cluster of galaxies. Estimates for this distance range from 15 to 25 Megaparsecs. The solid lines indicate the rate of supernova events between r and r + dr. The dashed lines give the integrated rate out to the distance r. In each case, the upper curve gives the expected rate for all supernova types, and the lower curve gives the rate for Type II supernovae only.

Extragalactic supernova rates are much higher at sufficient distance. Figure 2 shows the rates of all supernovae, and the rates of Type II only, in successive spherical shells out to the Virgo cluster. Also shown is the integrated rate. These numbers are based on an analysis by Tammann[8], but with care taken not to apply correction factors intended for spiral galaxies and for Type II to all galaxy types and all supernovae. The distance to Virgo and intermediate points is as uncertain as the Hubble constant and the age of the Universe. Estimates for the distance to Virgo center therefore range from 15 to 25 Mpc. The rates for supernovae approach one per year for distances of ~ 3-5 Mpc. Unfortunately, the neutrino and gravity wave signals from such a distance will remain impractically small for the foreseeable future.

DETECTING THE NEUTRINO BURST FROM COLLAPSE

The solar neutrino experiment operated by Ray Davis and his colleagues (most recently reviewed in these proceedings) could detect a nearby collapse. For larger, more sensitive experiments, simple water may be the most effective neutrino detector. The detection mechanisms are

$$\bar{\nu}_e \text{ interacts with a proton } (n = 2 \text{ per } H_2O)$$

with a cross section

$$\sigma(\bar{\nu}_e, p) = 8.5 \times 10^{-42} \text{ cm}^2 \; (\varepsilon_{\bar{\nu}_e}/10 \text{ MeV})^2,$$

and

$$\nu_e \text{ interacts with an electron } (n = 10 \text{ per } H_2O)$$

with a cross section

$$\sigma(\nu_e, e^-) = 1.7 \times 10^{-43} \text{ cm}^2 \; (\varepsilon_{\nu_e}/10 \text{ MeV}).$$

Any interaction with oxygen is negligible by comparison. For each mechanism of detection one counts on measuring the number of interactions N_d that take place in a mass of water M in the total time scale of interest. That is governed by the time integral, $F(\text{particles/cm}^2)$, of the flux at the earth of the neutrinos from the star,

$$F = 8.3 \times 10^9 \text{ cm}^{-2} \; N_{56} \; (D/10 \text{ kpc})^2.$$

Here $N_{56} = N/10^{56}$ is the neutrino emission from the star totaled over the time of interest and D is the distance from the star. The expected number of events in the detector via a given mechanism is

$$N_d = n N_0 F \, M/\mu$$

Current calculations of gravitational collapse differ in detail but agree in general outline. After neutrinos are trapped at about 10^{12} g cm^{-3} the central portions of the star collapse in a low entropy, homologous ($v \propto r$) fashion. The collapse of this inner core halts when it reaches nuclear density and the equation of state stiffens. A shock forms as material rains in on the halted inner core.

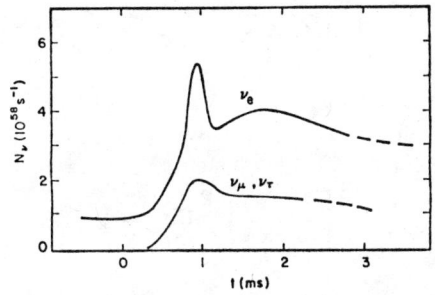 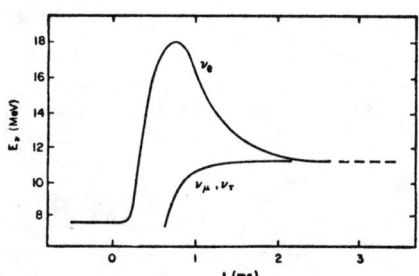

Figure 3 - The rate of emission of neutrinos during the gravitational collapse of a stellar core is given as a function of the time since the formation of the shock when the core bounces at nuclear densities. The upper curve presents the rate of electron neutrinos which result from electron capture. The lower curve gives the rate of other neutrinos from thermal processes. The curves shown are a somewhat schematic representation of the results of Mazurek, Cooperstein and Kahana[9] and those of Van Riper[10].

Figure 4 - Representative energies are given for the neutrinos emitted during gravitational collapse.

Neutrino emission begins with electron capture in the collapsing material in the regions beyond the neutrino trapping radius. A short intense relatively energetic burst accompanies the arrival of the shock at the "neutrinosphere", $\tau_{\nu_e} \sim 1$ as shown in Figures 3 and 4. Energy can rise to ~ 20 MeV per neutrino for ~ 1 msec. A plateau of ~ 10 msec then follows in which electron capture proceeds in the post-shock material. Energy drops to ~ 10 MeV, but the pair formation process can begin to provide a flux of $\bar{\nu}_e$ comparable to ν_e.

Over longer timescales the situation is uncertain. A period of "deleptonization" ensues in which the binding energy of the neutron star is emitted as neutrinos. This results in the release of ~10^{53} ergs worth of neutrinos but the timescale and energy spectrum are uncertain, depending, for instance, on whether the core is quasi-static or subject to strong hydrodynamic motions. The neutrino energy is also uncertain.

The original burst is predicted to give ~ 10^{56} ν_e of ε_{ν_e} ~ 10 - 20 MeV over a period of ~ 10 msec. The next 10 msec could result in ~ 10^{55} ν_e and $\bar{\nu}_e$ with ε_ν ~ 10 MeV. The total energy involved is ~ 10^{51} ergs. If the neutrinos continue to come out at ~ 10 MeV during deleptonization then another ~ 10^{58} neutrinos could emerge. If the collapse leads to the formation of a black hole then perhaps only the first phase will be observed.

The mass of the detector in metric tons per detected neutrino can then be written as:

$$\frac{M_{\nu_e}}{N_{d,\nu_e}} = 2.2 \times 10^3 \text{ m tons} \left[\frac{D}{10 \text{ kpc}}\right]^2 \left[\frac{\varepsilon_{\nu_e}}{10 \text{ MeV}}\right]^{-1} N_{\nu_e,56}^{-1}$$

and

$$\frac{M_{\bar{\nu}_e}}{N_{d,\bar{\nu}_e}} = 2.1 \times 10^3 \text{ m tons} \left[\frac{D}{10 \text{ kpc}}\right]^2 \left[\frac{\varepsilon_{\bar{\nu}_e}}{10 \text{ MeV}}\right]^{-2} N_{\bar{\nu}_e,55}^{-1}$$

These expressions give a reasonable portrayal of the expected detection rate in the first several tens of milliseconds, the period covered by current calculations. If a simple detection is all that is required and deleptonization produces neutrinos of ~ 10 MeV then a kiloton detector is probably adequate to "see" the 10^{58} neutrinos from a collapse anywhere in the Galaxy. Clearly, however, the case is very marginal for resolving the original burst to compare observations with the theory of core bounce and shock formation. Detection of black hole formation at 10 kpc would be difficult.

GRAVITY WAVES FROM GRAVITATIONAL COLLAPSE

Gravitational radiation, even more surely than neutrinos, emerges from the collapse of a stellar core to form a neutron star. The mechanism for gravity wave production depends on unknown physical factors, however, and detection requires pioneering technology. Despite these difficulties, it is a remarkable testimony to the advance of astrophysics in the last three decades that hopes and efforts have risen hand in hand for detection of gravity waves and of neutrinos.

The contrast is clear between a single pulse of gravitational radiation, associated with a single sudden change in the mass quadrupole moment of the compact object and the continuous train of waves associated with a quadrupole altering periodically with time by reason of vibration or rotation. In either case, the relevant measure of the strength of the gravity waves at the detector is the fractional extension, h, of the length transverse to the direction of the travel of the wave. This quantity is given approximately by

$$h \sim 2GI/c^4 Dt^2$$

The reduced quadrupole moment $I \sim 0.1 MR^2 \sim 10^{44}$ g cm^{-2}, altering by a large fraction of itself in a time $t \sim 0.1 t_{dyn} \sim 10^{-4}$ s at a distance, D, of the order the Galactic center gives $h \sim 10^{-19}$; for the Virgo cluster at about 20 Mpc, $h \sim 10^{-22}$. This estimate is consistent with the detailed calculations of Saenz and Shapiro[11].

Most stars are believed to have an appreciable amount of rotation already before collapse and consequently may rotate very fast indeed after a shrinkage in dimensions of two or three orders of magnitude. At one extreme is a collapse very nearly spherically symmetric, with the rotation of the neutron star at the end so modest that the rotation makes only a small perturbation on the idealized spherical form.

At the other extreme is the "collapse, pursuit, and plunge" scenario envisaged by Ruffini and Wheeler[12]. Here the angular momentum is so great that the neutron fluid assumes a pancake form immediately after collapse. This pancake is unstable and fragments into separate neutron stars that revolve about their common center of gravity, radiating both energy and angular momentum in gravity waves. The pulse generated in such an act of fragmentation should be comparable to, and perhaps even greater than, the pulse generated in the original act of collapse itself. Most estimates of the gravitational radiation to be expected from collapse neglect all but the original pulse. The reason is not lack of interest in those events. It is the difficulty in giving them a detailed hydrodynamic analysis[11].

November 1982 saw the discovery of a pulsar associated with a neutron star spinning around its axis 641 times per second[13]. No object has focussed attention more dramatically on the idea of detecting gravitational waves of such a sharply defined and directly measured frequency. Advance knowledge of the amplitude of the rotating quadrupole will be difficult to acquire despite the accurate period. "Mountains" on the surface of the neutron star less than a centimeter high would suffice to make the Crab pulsar a quite significant source of continuous gravitational radiation, and the new pulsar even more so.

The search for continuous gravitational radiation may tell more about neutron star geology than any present knowledge of that geology can teach about gravitational radiation. No object would seem better suited as a test case than a very fast pulsar such as that picked up by the newly employed fast timing technology at the Arecibo radio telescope.

For the detection of gravitational radiation many ideas have been proposed. Detectors have been built at more than a dozen centers. Descriptions of ideas and devices are available in several books including Misner, Thorne and Wheeler[14] and Smarr[15]. There is always the possibility that some ingenious and novel detector can be invented superior to anything now conceived. An interesting but so far unrealized try in this direction is the proposed device described in CERN reports by Emilio Picasso and Luigi Radicati. Two superconducting cavities each carry an electromagnetic wave. A hole couples them. They respond differently to the gravitational wave by reason of their different orientation to its polarization. In consequence, a little energy is taken from one mode and given to the other. Measurement of this energy transfer is the tool for detection.

Interesting though this and other novel ideas are, present efforts at detection concentrate on two simpler and better known proposals, the Weber bar and the Michelson interferometer.

The bar, with a mass of the order of a ton, has a quadrupole moment and therefore responds to a gravitational wave that travels at any angle to its axis. The expected extension of a bar of length

$L = 1$ m under the influence of a gravitational wave of amplitude $h = 10^{-22}$ is only of the order of 10^{-20} cm, unbelievably small compared even to the dimensions of an atomic nucleus, and on this account at first sight utterly beyond measurement. However, the quantity one determines experimentally is, in principle, related to an average over all the atoms of the bar, not the position of any individual atom. More concretely, one is concerned with the amplitude of the lowest mode of longitudinal vibration of the bar, typically endowed with a frequency of the order of a 1000 cycle/s. At room temperature this mode carries on the average 4×10^9 quanta; at liquid helium temperature, 4×10^7 quanta. Granted the most favorable phase relation, a sudden gravitational wave with $h \sim 10^{-22}$ will suddenly increase or decrease this number by ~ 100. That is the detection problem. The analysis of the oscillator has to be conducted at the quantum level. By contrast, the gravitational wave, weak though it is, contains such an enormous number of quanta that it can be envisaged for detection purposes in purely classical terms. The frequency of the bar is selected so that it will respond to the expected short pulse, $\sim 10^{-3}$ s, from a collapse event. Current Weber bars operated at room temperature require a wave with $h \sim 10^{-16}$ and operated at 4K, a wave of $h \sim 10^{-17}$, and hence are sufficient to detect only relatively nearby collapse events[16].

The Michelson interferometer, whether on the ground or in space, is conceived as having each mirror, half silvered or fully silvered, stationed on a mass that is free (in space) or effectively free (earth bound but suspended like a pendulum). Ronald Drever of Glasgow and Caltech and his Caltech colleagues have under construction such an interferometer with a forty meter base line. With a mirror reflectivity of $R = 0.997$ this device would give, Drever estimates, a gravity wave sensitivity of $h \sim 10^{-19}$. His group estimates that improvements are conceivable which would bring this figure to something of the order of 10^{-21}.

CONCLUSIONS

There are many uncertainties involved in the estimate of the detectability of collapse events in the Galaxy. We do not know which supernovae involve collapse. We do not know how often collapse occurs without associated optical display or extended remnant or without either. We do not know how often collapse occurs with negligible magnetic field, with or without a supernova. The details of the neutrino spectrum and time history are also uncertain and subject to revision. We do not know the mass quadrupole moment.

The scientific payoff of the detection of a gravitational collapse event depends on a number of factors. The event must be <u>confirmed</u>. Even then, the simple statement that a confirmed collapse had occurred, while intensely interesting, might be of little benefit if the data is too sparse. The more temporal and spectral information, the more valuable the event. In addition, an event which was susceptible to correlative studies with all the armaments of optical, radio and X-ray astronomy would be of most use.

Neutrino detectors should have thresholds < 10 MeV and time resolution on scales from ~ 0.1 to 1000 msec. In order to confirm an event absolute timing capability for individual detectors is mandatory. Such absolute timing capability would also enable position location through triangulation. If several kiloton detectors were placed about the earth with spacing of order 1000 km the angular resolution would be ~ 1' Δt (μsec) where Δt is the absolute timing resolution in μsec. Microsecond accuracy would enable a search for an optical counterpart. Timing to ~ 10 nsec could locate a source to an accuracy of order a second of arc.

Discovering an optical counterpart at great distance is not out of the question. Supernovae are estimated to range in peak brightness from -17^m to -20^m depending on the type and distance estimates. The average obscuration in the Galaxy is about 2^m per kpc. Thus a supernova at 10 kpc would appear at m ~ 15 to 18. This is bright enough not only for detection, but for detailed spectral analysis which would determine the type. The actual optical detectability would depend sensitively on the supernova type and the direction in the Galaxy. In some directions the obscuration is very heavy. It is somewhat lower toward the Galactic center, increasing the liklihood of seeing a supernova there.

Type II supernovae are bright and are more likely to be confined to the high obscuration zones of spiral arms. Type I events are brighter and tend to avoid the regions of large obscuration. Simple association of a collapse event with a Type II would not be very useful because it is expected. Lack of collapse would be very surprising but difficult to confirm given the weakness of the signals and the likely obscuration of an optical Type II event. Detection of a Type I and a basic yes/no statement concerning an associable neutrino or gravity waves pulse would be immensely useful in selecting among competing classes of theories.

The chance of an optical, radio, or X-ray correlation (or definite lack thereof) and detailed neutrino or gravity wave temporal and spectral information increases greatly as the distance is reduced. The price, of course, is that the probability of occurrence is lessened approximately as the area of the Galactic plane surveyed.

The low rate of occurrence of supernovae and collapse events in the Galaxy and particularly in the solar neighborhood should not be used as an argument against the construction of appropriate detection devices. Statistics aside, the next event could occur tomorrow. Detection of neutrinos from gravitational collapse should probably not be the first priority, given the low rate, but it would be a tragedy if an event occurred in the near future and the various proton decay experiments were not instrumented in an appropriate manner to detect it.

We are grateful to Kip Thorne, Stuart Shapiro and William Dean for helpful conversations. This work was partially supported by NSF grants AST 8201210 and PHY 7826592.

REFERENCES

1. J. C. Wheeler, Rep. on Prog. in Phys. $\underline{44}$, 135 (1981), and references therein.
2. J. C. Wheeler, in Supernovae ed. M. J. Rees and R. J. Stoneham (Reidel, Dordrecht, 1982).
3. K. Nomoto, and S. Tsuruta, Ap. J. $\underline{250}$, L19 (1981).
4. D. Helfand, in Supernovae ed. M. J. Rees and R. J. Stoneham (Reidel, Dordrecht, 1982).
5. G. A. Tammann, in Supernovae ed. M. J. Rees and R. J. Stoneham (Reidel, Dordrecht, 1982).
6. A. Lyne, in Supernovae ed. M. J. Rees and R. J. Stoneham (Reidel, Dordrecht, 1982).
7. A. Oemler, and B. M. Tinsley, Astron. J. $\underline{84}$, 985 (1979).
8. G. A. Tammann, in Proceedings of the 1976 DUMAND Summer Workshop ed. A. Roberts (Fermi Laboratory, Batavia, IL, 1976).
9. T. J. Mazurek, J. Cooperstein, and S. Kahana, in Proceedings of DUMAND (1980).
10. K. A. Van Riper, Ap. J. $\underline{257}$, 793 (1982), and private communication.
11. R. A. Saenz, and S. L. Shapiro, Ap. J. $\underline{244}$, 1033 (1981)
12. R. Ruffini, and J. A. Wheeler, in Proceedings of the Conference on Space Physics (European Space Research Organization, Paris, 1971); updated in Black Holes, Gravitational Waves and Cosmology ed. M. J. Rees, R. Ruffini and J. A. Wheeler (Gordon and Breach, New York, 1974).
13. D. Backer, private communication (1982).
14. C. W. Misner, K. S. Thorne, and J. A. Wheeler, Gravitation (W. H. Freeman and Company, San Francisco, 1973).
15. L. Smarr, (ed) Sources of Gravitational Radiation, (Cambridge University Press, Cambridge, England, 1979).
16. T. Tyson, and R. P. Giffard, Ann. Rev. Astr. and Astrophys. $\underline{16}$, (1978), p. 521.

SIGNATURES FOR UNDERGROUND NEUTRINOS

Marshall Crouch
Case Western Reserve University, Cleveland, Oh. 44106

ABSTRACT

Careful attention to the concept of the signature is especially important in neutrino experiments where the background usually predominates strongly over the signal. Examples are given of signature techniques that have been successful, as well as cases where signatures have proved to have shortcomings.

THE CONCEPT OF SIGNATURES

The classic experiment of Chamberlain, Segre[1] et al. in 1955 to study antiprotons in a flux of pions greater by a factor of 30,000 represented one of the first instances where the powerful technique of prescribing a detailed signature was exploited. Their experimental arrangement (Fig. 1) used a complex array of state-of-the-art bending magnets, focussing magnets, scintillation detectors, threshold Cerenkov detectors, window Cerenkov detectors, and time of flight techniques to select a beam of charged particles in a narrow range of momentum and then to recognize the lower velocity component present in the beam. At about the same time, Reines and Cowan[2] were able to observe inverse beta decay interactions produced by reactor antineutrinos in the overwhelming background environment of the reactor, despite the extremely small cross section for the interaction, by insisting that each of the two reaction products authenticate itself with coincidence signals separated by a characteristic delay time associated with the thermalization, diffusion and capture of the product

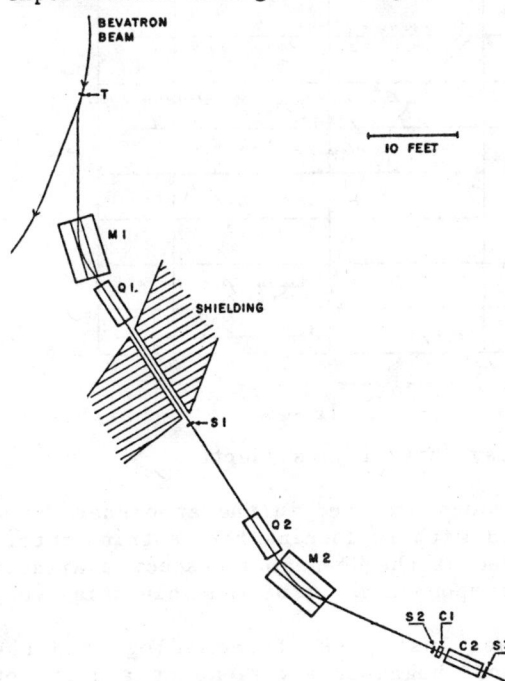

Fig. 1. The antiproton experiment

neutron. The concept of signature introduced in these landmark experiments has much of the flavor of the design of an obstacle course. The experimenter must carefully devise an obstacle course which the particles or interaction of interest are able to negotiate, but which provides insuperable obstacles for the background.

COSMIC RAY NEUTRINOS

After the field of experimental neutrino interaction physics was inaugurated with the experiments of Reines and Cowan[2], it was inevitable that physicists would consider next whether natural sources of neutrinos could be observed. The most obvious possibilities are the low energy electron neutrinos from the sun, and the muon neutrinos expected to be present as a component of the cosmic radiation.

Regarding the latter, the curve of cosmic ray intensity vs. depth in the earth measured during the period 1925 - 1964 (Fig. 2)

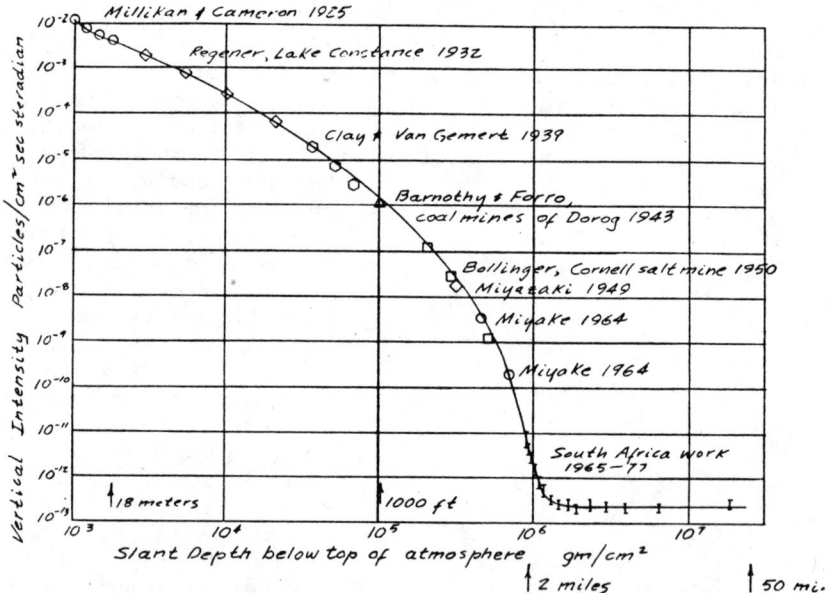

Fig. 2. Cosmic ray intensity vs. depth

showed that the intensity of muons produced in the atmosphere, the principal background to contend with in looking for neutrino interactions, is sufficiently reduced at the deepest sites now available that a study of the neutrino component might be feasible using very large underground detectors.

In 1964 a group from Case Institute of Technology and the University of the Witwatersrand[3] began observations at a depth of about 2 miles in a South African Gold Mine (Fig. 3). At such a site, atmospheric muons are still observed to arrive in the vertical direction, but apparatus which observes in a sufficiently oblique direction will be shielded by a slant depth of rock absorber which screens out virtually all of the charged cosmic rays.

Thus the signature for cosmic ray muon neutrinos is the arrival of
(locally produced secondary) muons in directions where the slant
thickness of the rock layer is more than three miles. The segmented
scintillation detector array shown had dimensions carefully chosen
so that some of the residual atmospheric muons would be observed,
giving cross-tunnel Upper-Lower signals. However other cross tunnel
signals are at zenith angles such that they can only arise from
nearby neutrino interactions. 35 neutrino interaction events were
observed in a 2-year period.

The figure shows another aspect of the neutrino signature as

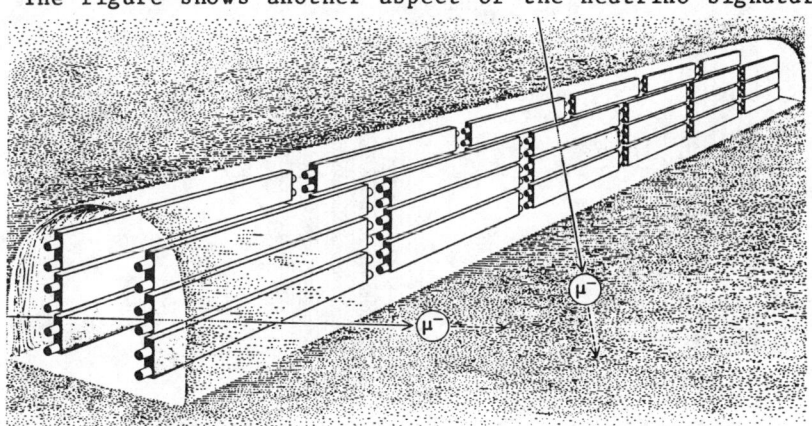

Fig. 3. The first South Africa array

well. These unshielded scintillation detectors also have some
potentially troublesome background from
juxtaposition of pulses due to natural
radioactivity. However the cross tunnel
signals (Fig. 4) constitute a clean 8-
fold coincidence record for which
spurious background rates are essentially zero. In this photograph, the
apparent sequential nature of the two
trains of 4 photomultiplier pulses is
produced by a delay line network.

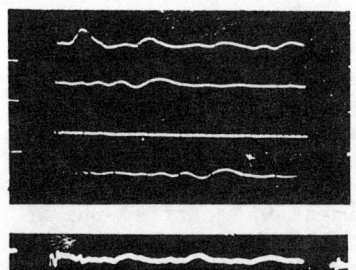
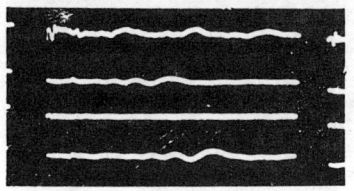

At about the same time a collaboration of physicists from Tata Institute, Osaka City University, and Durham
University undertook similar measurements in the Kolar Gold Fields in
India[4]. Their apparatus (Fig. 5.) also
featured scintillation counter telescopes aimed generally in the horizontal direction. This array had somewhat
smaller aperture, but the counters were

Fig. 4. Neutrino
interaction signal.

shielded with lead and were used to trigger arrays of neon flash
tubes to display the tracks of the detected charged particles. A
a smaller number of events was recorded, but they were able to

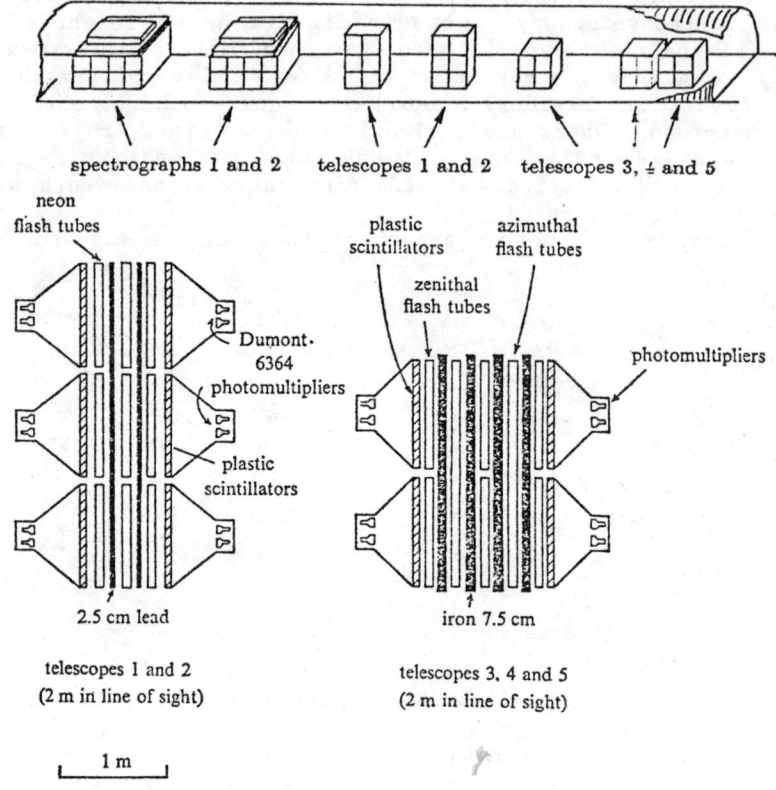

Fig. 5. The KGF atmospheric neutrino detector

measure the angular distribution of detected particles (Fig. 6) which showed very clearly the composite nature of the cosmic rays at that depth. Atmospheric muons, of course, are observed at small zenith angles, whereas the neutrino-induced signals are actually peaked in the horizontal direction.

Between 1967 and 1971 a second experiment was carried out in the South African mine at a depth of slightly over two miles. The Case Western Reserve University - University of the Witwatersrand - University of California, Irvine group[5] operated a 20 ton scintillation detector in a geometry giving a larger aperture, with a 32 ton array of neon flash tubes to provide information on particle trajectories. 132 neutrino events were recorded, giving a more detailed angular distribution (Fig. 7). These measurements permitted the cosmic ray intensity vs. depth curve to be extended to slant depths of about 50 miles. Thus the 1971 measurements shown on Fig. 1 display unmistakeably the expected behavior at great depths, where a constant flux of neutrinos is the only surviving component of the cosmic radiation. Thus Fig. 1 shows a 50 year odessy of cosmic ray physicists over the world which might be called a real life "journey to the center of the earth".

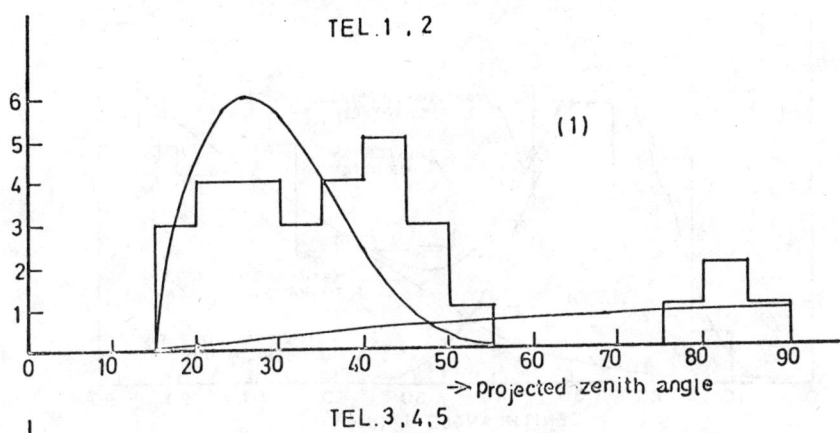

Fig. 6. The KGF composite angular distribution

It should be mentioned that University of Utah[6] also placed a large detector at an underground site to study atmospheric neutrinos. As a signature to differentiate between neutrino induced events and atmospheric muons they incorporated fast timing capability in an array of track detectors to recognize near-horizontal upward travelling muons. However at their relatively shallow depth the large background of charged cosmic rays proved to make it difficult to exploit this signature. During the period of operation 10^6 muons were detected, among which there were 5 neutrino candidates. The principal thrust of the measurements actually made with this array proved to be studies of atmospheric muons, initially planned as a secondary objective.

As for the future of further studies of atmospheric muons, it is difficult to say. The measurements described have merely confirmed the existence of a component of the cosmic radiation which had been expected. However the several preliminary studies of atmospheric neutrinos leave us at about the stage where studies of cosmic ray muons stood after the classic 1939 experiment of Rossi, Hillberry and Hoag[7]. Further studies of flux and angular distribution, energy spectra, details of the interactions, and possible correlations of events with signals in shallower detectors could yield interesting information. Measurements made to date include only a handful of interesting decay events, some being stopped muons with others having some alternate explanation. Unique effects of possible neutrino oscillation processes might be observed, complementing other kinds of measurements. What might appear to be a sub-field already neatly wrapped up may yet contain many surprising facets as unexpected as in many similar areas in the past. It is of course also true that a detector of sufficient size and resolution to study atmospheric neutrinos has the capability to recognize other signals of interest, both predicted and unexpected. Neutrino signals from beyond the earth's atmosphere are not inconceivable and would be an exciting added bonus.

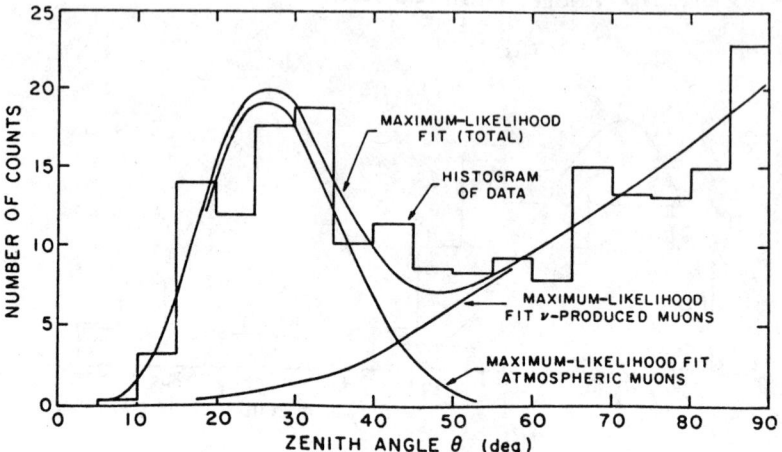

Fig. 7. Angular distribution from the second South Africa experiment.

Regrettably the signature used in triggering the South Africa detector was so rigorously dedicated to atmospheric neutrino observation that the array was not sensitive to a sequence of signals on a millisecond time scale. Future experiments should avoid any restrictive logic in recording systems, in matters such as timing of signals, amplitude, or multiplicity.

There are two features of the atmospheric neutrino experiments which warrant some further comment. The first is that the effective target mass for the neutrino interactions is very much larger than the actual detector mass. The South Africa detector was a liquid scintillation counter aggregating 20 tons. The neon flash tube array represented another 32 tons, but only signals in the scintillation detectors were capable of producing triggers. Eq. 1

$$N = M \, t \, \Omega \, \epsilon \, N_A \int \phi(E) \sigma(E) \, dE \quad (1)$$

20 ton detector
N = 132 neutrino interactions in t = 2 years
what effective target mass M ?

assume $\epsilon = 100\%$
σ = Fermilab cross sec.
ϕ = calculated flux from atmospheric cascade
N_A = Avagadro's no. (nucleons/gram)
$\Omega = 4\pi$

evaluate integral as $\int \sigma \phi \, dE = 3.0 \times 10^{-16}$ g^{-1} sec^{-1} sr^{-1}

solve for target mass   M = 800 tons !

gives the relation between number of events recorded, target mass, running time, cross sections, etc. We find the remarkable result that the 132 neutrino interaction events recorded corresponds to a target mass of 800 tons, throughout which events are recorded with 100% efficiency. Of course the "mass amplification" effect characteristic of this kind of experiment necessarily implies that only a small fraction of the interaction vertices occur in the detector. Nevertheless in both the KGF and South Africa experiments a number of events were recorded in which the vertices occurred sufficiently close to the detector array that as many as 7 secondary particle tracks were recorded.

The other feature is the extreme sensitivity of the composite angular distribution to the depth. Fig. 8 shows the angular distribution of neutrino induced muons, constant independent of depth, together with the distributions of atmospheric muons at three depths. The KGF experiment was located at a relatively deep site, but the additional depth of the South Africa experiment makes a drastic improvement in the prominence of the neutrino peak relative to the atmospheric muon distribution. The third curve shows the distribution at a depth of 5 km under water, the proposed depth for undersea arrays. The curve has been drawn with ordinates multiplied by a factor 1/10. The intensity of the atmospheric muons is so great that it is impossible to distinguish the neutrino component confidently using the shape of the distribution. At the depths of the University of Utah experiments and the salt mines that have been used for underground experiments, the predominance of the atmospheric muon peak is enormously greater yet. Accordingly if neutrino measurements are to be successfully carried out at such depths some very clever signature must be insisted upon, probably involving time domain features.

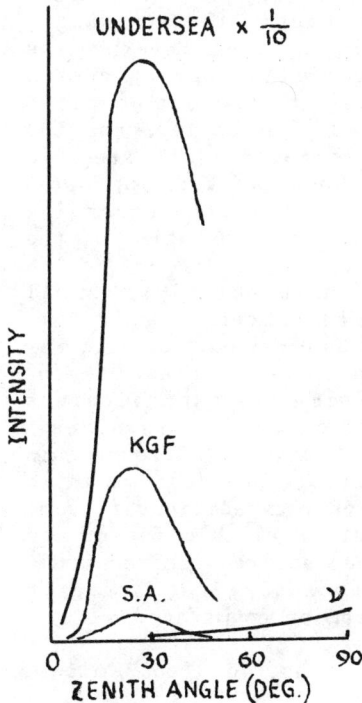

Fig. 8. Effect of depth on angular distribution.

A number of large underground detectors currently being operated are dedicated to the search for proton decay, but are related to experiments previously described in that the atmospheric neutrino interactions represent the principal background for the measurements. Two comments regarding the signatures being used may be pertinent because of this relationship.

In the IMB proton decay experiment[8] a new dimension in signature utilization is being attempted. Recognition of a characteristic pattern of back-to-back Cerenkov cones is to be used, aided by anticoincidence signals, to observe events where the background of charged cosmic ray particles (3/second) is larger than the expected signal by a factor that may be as great as 10^7. Of course the more troublesome background due to energetic neutrinos is many orders of magnitude smaller. Results should be forthcoming shortly to reveal the effectiveness of this totally new concept in the character of signatures.

In the KGF proton decay experiment of Prof. Miyake[9], the trigger rate is only 2/day. This is partly due to the much greater depth of this experiment (7600 mwe vs. 1570 mwe) and the smaller size (10^2 tons vs. 10^4 tons). However it appears to be partly the result of establishing a triggering signature and data recording system so strongly dedicated to the objective that the array may be blind to other possible interesting classes of signals. The design of the experiment may have been partly motivated by budget considerations or efforts to minimize setting up time. Nevertheless it would be unfortunate if this unique facility, having such a large mass at such great depth, were to fail to exploit possible secondary objectives. From the standpoint of flexibility, the IMB data acquisition system using an on-line computer with extensive pattern recognition software seems to be a better approach. Possibly a way could be found to expand the KGF data recording system without interrupting the important primary operation of the array.

The curve shown in Fig. 1, which includes only a small sampling of underground measurements that have been carried out, deserves some comment on the sites used. Underground cosmic ray experiments have been conducted in many, many mines, caves, tunnels, etc. throughout the world. Each time the experimenters have had to conduct a search for a suitable site which involves a considerable expenditure of time, effort and expense. Although this may take the form of pleasant junkets, jetting about the world to negotiate with mining moguls, it would be very expeditious if some sort of national underground laboratory were available for use on such occasions. Unfortunately to provide the features desired would be so expensive that it may still be necessary to locate in existing mines. Nevertheless the features can be enumerated.

1. Generous space provided at various depths, down to 3 miles if possible.
2. Preferably located under level terrain, with "standard rock". A site in a region such as the Sangre de Cristo mountain range would be distinctly inferior.
3. Capability of moving large and heavy equipment to the various depths.
4. Ample electrical power, water supply, etc.
5. Adequate cooling and ventilation.
6. Reasonable safety features, such as 2 access shafts for the shallower areas.

It is still true that gravity wave observations and searches for cosmic neutrinos are most effective with several widely separated observing stations. Coincident observations can serve to distinguish true events from spurious ones, while measurements of time differences may give information on direction of arrival of signals.

NEUTRINOS FROM BEYOND THE EARTH'S ATMOSPHERE

The sun, 8 light minutes further in space, is expected to emit neutrinos from its core, giving important information on the solar energy generation process. The famous Brookhaven radiochemical experiment headed by Ray Davis[10] has, after a great effort, succeeded in observing a signal of about the expected intensity. Measurements are not able to give information on either distance of the source or direction of arrival. There is some disagreement with the prevailing theoretical prediction of expected neutrino flux, but this appears to be due to inadequacy of the theory.

For cosmic neutrinos from more distant sources, the outlook for experimental neutrino astronomy is harder to assess. Probably had large underground detectors been operating in 1054, 1572, or 1604 A.D. extremely interesting signals related to the bright supernova observed in those years would have been recorded. However the expected signal rate is difficult to judge. A brief discussion following a symposium held in Tokyo where A. Chudakov described his 100 ton underground scintillation detector and expectations for it illuminates the point. T. Suda (Tokyo University): "Then, in your lifetime you probably expect to observe one event". Chudakov: "Ha! You think I am going to die!".

The experimentalist is obliged to be guided by predictions of theorists working in the field of astrophysics, which is a difficult field because of the need to extrapolate so many orders of magnitude beyond the realm of laboratory experience. Predictions have been made of possible measureable neutrino fluxes associated with certain processes in stellar evolution, as well as characteristic time structures in the radiations. There are estimates of the frequency of occurrence of such events in our galaxy, or inside a region bounded by a given radius. One can only start with these predictions, augment the stated error bars in accordance with past experience, and decide whether there is a reasonable likelihood of obtaining rewarding results from detector arrays that can be devised. In a field like this, just as an experienced aircraft pilot should continuously have in mind possible emergency landing sites during a flight, the experimenter should be alert to all possible secondary objectives that might be attainable as consolation prizes.

For several years a multilayer 7 ton liquid scintillation detector has been operated underground in the salt mine at Cleveland by W.R. Kropp, J.S. Hansen, and F. Reines of University of California, Irvine, C.B. Bratton of Cleveland State University, and myself. The primary objective was to set a limit on the steady antineutrino flux arriving at the earth (and/or emanating therefrom). A 1.4 ton detector was also operated in coincidence at a distance such that the decoherence curve for charged cosmic rays had fallen to zero. A rather restrictive signature was established, viz. a signal greater than 8 MeV in one slab of the detector, unaccompanied by signals in neighboring slabs. In addition, a delayed coincidence signal due to the 2.2 MeV gamma ray from capture of the product neutron in the scintillator was required. A large area "umbrella" and "wall" proportional counter surrounded the scintillator as well as a 20 ton lead absorber. Signals having this signature were recorded at a rate which seemed unreasonably high, indicating that the signature still did not constitute an "obstacle course" capable of excluding the backgrounds which are present.

A small number of multiple events of some interest were recorded. They were almost certainly too numerous to be of astrophysical significance, yet they were not spurious electrical signals, having very clean waveforms. The energies and timing rule out interpretation as neutron capture events, certainly in the hydrogenous scintillator. Fig. 9 shows the event of highest multiplicity which was recorded. We still find it inexplicable.

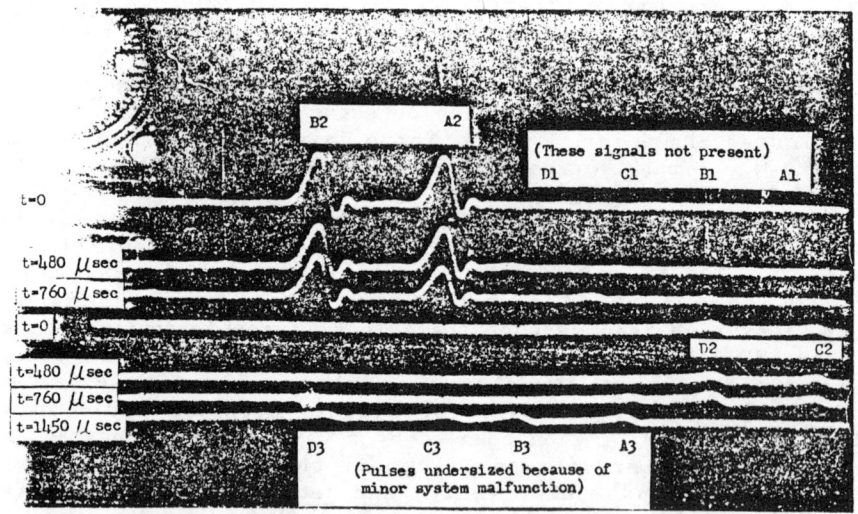

Fig. 9. A train of 4 clean ~10 MeV signals recorded June 14, 1975 in the 7 m^3 underground detector, with delay times indicated.

In closing, I would like to make a plea for cooperation between groups operating large underground detectors. The field of radioastronomy matured from an early period where angular resolution was initially painfully crude to one where cooperative very long baseline interferometer measurements are superior to the best optical observations. If groups operating large underground detectors take care to measure as precisely as possible the time of occurrence of events, and share results that are of potential astrophysical significance, it is possible that here too the importance of the whole observations may be of greater value to all concerned than the sum of the parts.

REFERENCES

1. O. Chamberlain, E. Segre, C. Wiegand, and T. Ypsilantis, Phys. Rev. 100, 947 (1955).
2. F. Reines, C.L. Cowan, F.B. Harrison, A.D. McGuire, and H.W. Kruse, Phys. Rev. 117, 159 (1960).
3. F. Reines, W.R. Kropp, H.W. Sobel, H.S. Gurr, J. Lathrop, M.F. Crouch, J.P.F. Sellschop, and B.S. Meyer, Phys. Rev. D4, 80 (1971).
4. M.R. Krishnaswamy, M.G.K. Menon, V.S. Narasimham, K. Hinotani, N. Ito, S. Miyake, J.L. Osborne, A.J. Parsons, and A. W. Wolfendale, Proc. R. Soc. (London) A323, 489 (1971).
5. M.F. Crouch, P.B. Landecker, J.F. Lathrop, F. Reines, W.G. Sandie, H.W. Sobel, H. Coxell, and J.P.F. Sellschop, Phys. Rev. D18, 2239 (1978).
6. H.E. Bergeson, G.L. Cassiday, J.W. Keuffel, and J.A. Thompson, Phys. Rev. Lett. 31, 1091 (1973).
7. B. Rossi, N. Hilberry, J.B. Hoag, Phys. Rev. 56, 837 (1939).
8. D. Sinclair, Session III paper, this conference.
9. S. Miyake, Session IV paper, this conference.
10. W.A. Fowler, Session II paper, this conference.

Particle Physics Below the Earth's Surface, An Overview of Possibilities

J.G.Learned
Univ. of Hawaii, Manoa, Honolulu, HI 96822

ABSTRACT

Potential subterranian cosmic ray and high energy particle physics experiments of current interest are reviewed. It is argued that most of the experiments may be attempted with instruments that fall generally into two classes: 1) detectors with greater than one thousand ton mass and MeV energy sensitivity, and 2) those with far greater mass, megatons or more, but much higher energy threshold, TeV. Those in class 1 are useful for nucleon decay searches, solar neutrino observations, and monitoring for supernovae in our galaxy. Class 2 detectors are aimed at high energy neutrino astronomy and various cosmic ray muon studies and the nature of interactions at ultra high energies. Both types can do useful muon studies, depending particularly upon depth. Finally the prospects for second generation nucleon decay searches are discussed and it is concluded that a national underground laboratory may be justified sometime in the future, after the results of contemporary endeavors are available, but that at present it is not possible to define the characteristics required of such a national facility. However, it is suggested that studies be undertaken to evaluate the various options, including novel technology, such as TPC-like devices, and to evaluate location options, such as being near a surface cosmic ray installation, or being located in the ocean.

INTRODUCTION

In the context of the topic of this symposium it seems appropriate to review the various types of high energy particle physics and cosmic ray experiments which are currently of interest and which are best carried out beneath the earth's surface. This discussion will not include such research as double beta decay studies and radiochemical neutrino astronomy because they involve unique technology, very different from the large tracking detectors which we want to consider here. The intent of this paper was initially to delve into the question of what would be required of a national scale detector to make best multiple use for nucleon decay search and other investigations. It is realized that we do not yet have the necessary scientific data to make a sensible start on defining the requirements of a second generation nucleon decay detector. That information will soon be available from first round dedicated nucleon decay searches, hopefully. Neither do we have adequate technological background from design studies and laboratory tests of novel techniques which may be necessary to make such an enormous device economically feasable.

The talk given at Los Alamos included a review of the current status of the DUMAND program, aimed at producing a deep ocean counter for TeV phenomena. The project has been extensively reported on elsewhere. Reference is made particularly to the "Proposal to Construct a Deep-Ocean Laboratory for the Study of High Energy Astrophysics, Cosmic Rays, and Neutrino Interactions"[1].

REVIEW OF SUBTERRANIAN EXPERIMENTS

In order to bring the discussion into focus we present Figure 1, illustrating the relationship between energy sensitivity and size for some of the most actively discussed cosmic ray and high energy physics experiments that can be carried out with large subterranian detectors. The detectors appropriate to these studies should be able to track relativistic particles. The matter for discussion is how large they should be, how much spatial, temporal, and energy resolution is needed, where must they be located, and what is the best choice for technique (as mainly governed by economics). We will now discuss, briefly, the various experiments.

Solar Neutrinos

Beginning at the low energy end, the requirements for a counting (as opposed to an integrating radiochemical) solar neutrino experiment are difficult: a sensitivity to interactions in the 1-8 MeV range and a mass of at least several hundred tons[2]. Detectors of this type must be designed with proper concern for local radioactive background, which extends into this energy regime, including the radioactivity of the detector itself. The goal of such an experiment

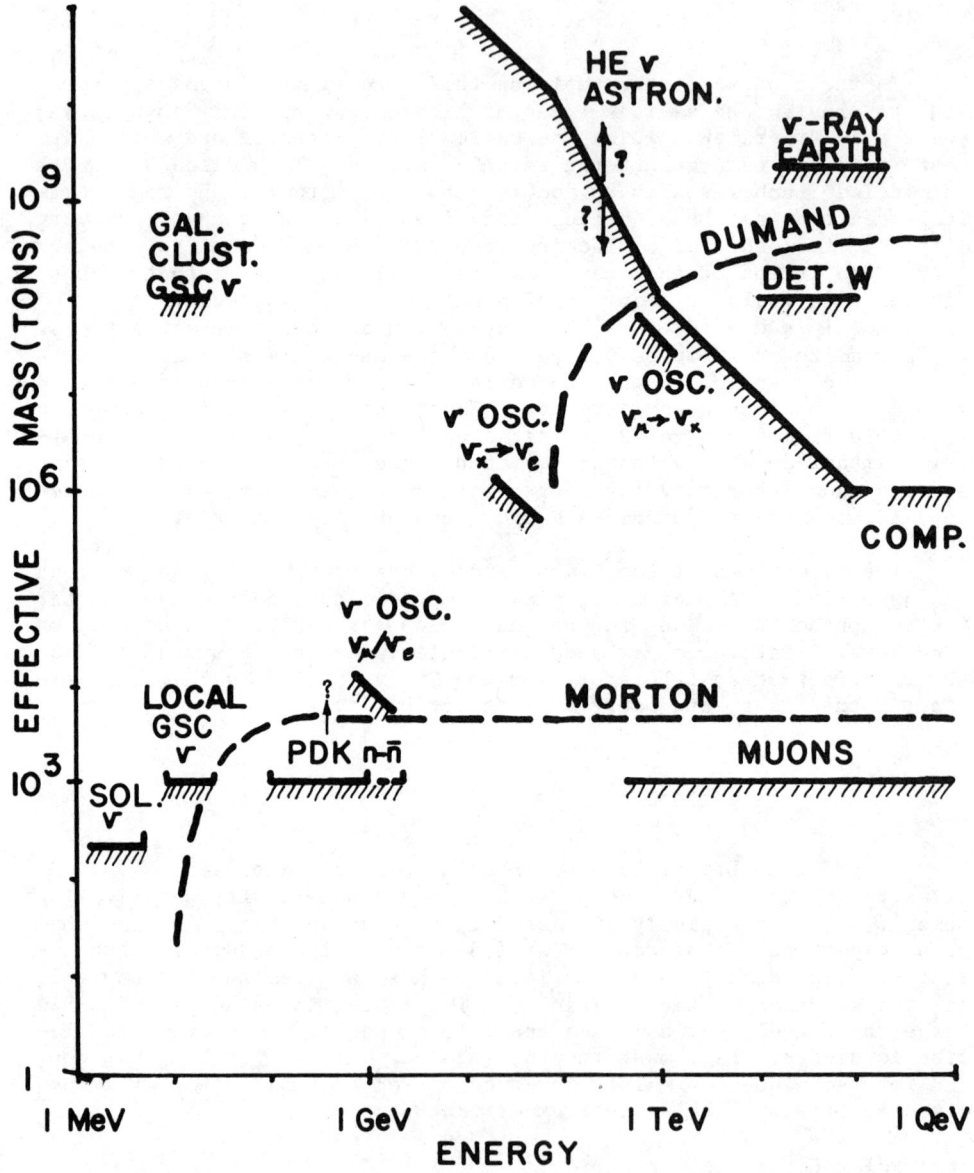

Figure 1. Relationship between energy sensitivity and size of detector required to explore various high energy particle and cosmic ray phenomena with underearth counting detectors. See text for discussion. All entries require significant qualification and the graph should be taken as pictorial only.

is to detect the solar neutrinos with some directivity from the elastic interaction of the (anti-) neutrinos with electrons in the target material. This measurement would permit associating the observed signal with the apparent solar motion, which is not possible with radiochemical detectors. It does not seem practical at present, however, to design a counting instrument that can observe the 1 MeV PEP neutrino flux. Given the continuing conflict between the observations at Homestake and the theoretical calculations for the ^{8}B solar flux[3], a direct observation is all the more important. Note that the detectability of signals in this range is strongly energy dependant due to weak interaction cross section (E^2) and detector sensitivity (typically another E^1 or more), as opposed to backgrounds that fall steeply with energy.

Stellar Collapse and Other Sources of Neutrino Pulses

Neutrinos from gravitational stellar collapse are expected to have energies in the neighborhood of 10 MeV[4,5]. The length of the pulse is not universally agreed upon and depends upon calculational details because the neutrinos fill phase space in the collapsing star and diffuse out over a period of time large compared to the infall time (~ 10 ms), with estimates ranging from 100 ms to a few seconds. The total power in neutrinos is much less ambiguous, neutrinos being the dominant mode of radiation for the gravitational potential energy, and the stellar mass range being well constrained (to about 1 solar mass). We can calculate the range of observability of a GSC in a given detector, though the signal-to-noise will depend upon the neutrino spill time. The results are that with a detector of a thousand tons we should be able to "see" much of our galaxy[6]. The highly desireable goal of building a device that would be capable of observing GSC from throughout our super cluster out to Virgo, and which would yield an estimated rate of one per day, is not within reach. Such a detector would need to have 10 MeV sensitivity throughout at least 10^8 tons[4]. Once having made a detector to watch our galaxy, the gain in expected GSC rate would be painfully slow with increasing size. The rate of supernovae in our galaxy is only once every several hundred years by direct observation. But, from observations of external galaxies it is expected to be once in 10-50 years, depending upon which estimate one believes. The actual GSC rate (if it is possible to have GSC without mantle blowoff, though most calculations are not able to produce ejection) could be quite a bit higher, perhaps once every several years, though most experts would say that such a high rate is unlikely[4].

There are no other sources of neutrino bursts that are expected to come close to the power of GSC (eg. from Novae, solar flares, carbon flash, and x-ray and γ-ray bursts)[4]. Building a dedicated instrument to search solely for extraterrestrial neutrino bursts is not a viable enterprise and the exciting business of searching for neutrino pulses from GSC will continue to ride with the fortunes of other primary endeavors. Conversley, it would be a pity if the nucleon decay detectors did not have adequate sensitivity or appropriate instrumentation to watch for neutrino bursts of 10 MeV or so. A minor but important issue is that designers of such detectors make provision for absolute time recording to a precision of ~1 ms in order to use the relative timing between detectors for calculating source direction.

It has occassionaly been suggested to make a detector capable of detecting terrestrial low energy sources, pulsed (bombs) or otherwise (reactors). Calculations indicate the range of detectability of either type of source to be limited to ranges of at most a kilometer, for typical instruments under discussion[4].

Nucleon Decay Searches

Next in increasing energy we come to new nucleon decay detectors, the requirements for which are a mass of a thousand tons or more (discussed in more detail below) and an energy sensitivity down to about 100 MeV. Having a sensitivity to any muon decay subsequent to a nucleon decay, or background cosmic ray neutrino interaction, is highly desireable (that is energy down to ~10 MeV over a period of ~10 μsec). The same characteristics will largely suffice to search for neutron-antineutron oscillations in a nucleus. The observation of such a signal would be spectacular, but the absence of a signal is difficult to interpret.

Magnetic Monopole Searches Underground

Nucleon decay detectors naturally are able to search for monopoles that catalyze nucleon decay, though there is concern about dead time and events being reflected if a monopole made an initial interaction near the edge of the fiducial volume. Searching for monopoles that do not catalyze nucleon decay, and other exotica with unusual velocities and/or levels of ionization should be considered, but may require some special additional instrumentation[8].

Cosmic Ray Neutrinos and Neutrino Oscillations

A number of studies may be performed using the cosmic ray neutrinos and muons. The neutrino studies can be divided into two types: those utilizing neutrino interactions inside the detector and those utilizing the muons passing through the detector from interactions some distance away. The total neutrino interaction rate from neutrinos due to cosmic rays hitting the atmosphere is about one event per 8 tons of matter per year with energy above 100 MeV. The spectrum of the neutrinos falls off as the 2.5 power of the energy in the few GeV region, steepening to 3.8 above a TeV[1]. Thus for detectors of the nucleon decay variety, the neutrino events will mostly have energy of <1 GeV, and there will be very few with more than 10 GeV. Since we know the interaction characteristics well in this range (from accelerator studies) and since the neutrino flux is assumed to be well calculable (from our knowledge of the primary cosmic rays), studies of neutrino oscillations become possible[9]. The energy is relatively low and the flight distance is very large (L/E of

10^4 m/MeV, as compared with ~10 m/MeV for reactor experiments). However, because the neutrino direction is poorly determined (30^0) and the statistics will be small, the comparisom will have to be not much more than up-to-down ratios, including the effect of downgoing muons possibly reducing or eliminating part of the solid angle. The net result is that such measurements of neutrino oscillations for contained neutrino interactions will likely be of marginal use. Also, because the ratio of muon neutrinos to electron neutrinos is 2:1 at low energies, the measurement of oscillations will only be sensitive to fairly large mixing angles, probably $\sin^2\theta > 0.2$. Another way to seek effects of neutrino oscillations is via the angular distribution of throughgoing muons produced by charged current muon neutrino and antineutrino interactions near the detector[1]. One can thus look for the disappearance of such neutrinos as indicated by an asymmetry in the otherwise exactly up-down symmetric flux (peaked by 10:1 about the horizontal plane). The increasing cross section and muon range with energy push the region of sensitivity for this measurement up to the several hundred GeV range. Again statistics are the limitation but a detector of $>10^4$ m^2 will be able to achieve a sensitivity to a $\delta m^2 > 0.1$ eV^2 for large mixing angles[1]. Another possible experiment would use the appearance of an electron neutrino flux in the regime where it is expected to be very small (above 100GeV where few muons decay and below 10 TeV where directly produced electron neutrinos begin to be important)[10]. This experiment would require the good resolution more typical of nucleon detectors but a size more characteristic of DUMAND, and will not be practical soon. Another potentially nice experiment would be to look for τ's produced in the detector, but that is obviuosly even more difficult.

The suggestion has often been made to locate an underearth detector so as to be able to point an accelerator neutrino beam toward it and to make a long baseline neutrino oscillation test thereby[11]. The numbers are not very encouraging, however, and pointing the neutrino beam downwards is very expensive.

Weak Interactions above W Threshold

At very high energies, above 10 TeV, the propogator effects in the weak interaction will have two potentially observable effects, in the energy transfer distribution (y distribution) and in saturation of the total cross section. Since it is expected that the W and Z will be found at colliding beam machines, the important study is to confirm our theory of the electroweak interaction at center of mass energies above the mass of the W and Z. This will not be possible at fixed target machines until a neutrino beam is available from something like the giant 40 km diameter machines now discussed, but certainly not built for a decade or more. DUMAND is the only experiment that has a chance to make this observation, but even it is marginal.

Geophysics

An amusing geophysical observation becomes possible as the attenuation of the earth becomes significant, one may "neutrino-ray" the earth to obtain the density versus radius profile[1]. This has been measured only indirectly with seismic studies that require knowledge of the composition of the core and equation of state (not accessable in the laboratory). As usual the predicted statistical precision is such that only a very crude measurement can be expected using a detector of the proposed DUMAND size and neutrinos from ordinary π and K decay. Large direct production or extraterrestrial fluxes could make a significant difference, though.

Extraterrestrial Neutrinos

Finally, one can search for extraterrestrial sources of neutrinos. There is no question that such sources exist[1]. We know that there are cosmic ray nucleons of energies up to $\sim 10^{20}$ eV and these surely lead to neutrinos of similar energies. We also know that there are many sites of strong non-thermal radiation in the universe, associated with the most luminous and peculiar objects. There are similarities between many of these, from SS-433 in our galaxy to distant Quasars, involving apparent particle beaming. The big question for neutrino astronomy is whether such sources are purely electronic or involve protons or nuclei. The best, but still not definitive evidence we now have is the direct observation of TeV γ-rays from a handfull of sources, revealing the peculiar fact that some energetic astronomical objects radiate the bulk of their power at the high end of the spectrum! There is good reason to believe in high energy neutrinos from otherwise directly unobservable sources, but when pinned down to making numerical predictions we are caught in somewhat of a "Catch-22", that the very sources best observed with high energy neutrinos are those about which we know least from photon astronomy. Even the upper limits on neutrino fluxes that one can calculate from energy bounds are not very useful. It is a frustrating situation and the best we can do is to work at it incrementally until a signal is found. The new round of nucleon decay detectors represent a significant improvement over earlier detectors in neutrino sensitivity so we might be lucky and find extraterrestrial signals with them, though by most estimates that prospect must be judged unlikely.

It is somewhat surprising to many people that the best energy band to begin searching for extraterrestrial neutrino signals is between 100 GeV and 10 TeV[1]. This result depends upon the assumption of power law spectra (but is not very sensitive to value of the spectral index). It comes about because of the rising cross section and muon range on the low side, and the saturation of both at high energies along with decreasing fluxes on the high side. Nucleon decay detectors suffer in comparisom partly due to angular precision (both experimental and from the interaction itself) and due to the high atmospheric neutrino background. DUMAND is calculated to have a flux sensitivity comparable to γ-ray detectors in the same TeV energy range ($1/m^2$ month) while the nucleon decay detectors are expected to be somewhat less sensitive (several orders of magnitude)[1].

Cosmic Ray Muons

Next we discuss the prospects for subterranian studies utilizing cosmic ray muons[1,12]. Underground muons have been studied for 30 years or more and the depth intensity and angular variation are known and understood to an absolute precision of roughly 20%, for single muons. Multiple muons have been observed, as well as the "decoherence" function, expressing the density of muon pairs as a function of separation. There have been hints of peculiarities in the frequency of tight groups or "muon bundles" and the distribution of muons at large separations, >10 m, as well as recent hints of anisotropy in the rate of high multiplicity events[12]. The new generation of nucleon detectors should should clear this situation up. No one instrument is suited to investigate all these problems because the bundles will need a high resolution instrument and the large separation phenomena a very large aperture detector. Also we really need to explore these phenomena at a variety of depths in order to understand the energy dependance.

In general the study of underground muons is useful for comparisom with predictions arrived at through rather complicated Monte Carlo simulations beginning with a model for the composition and interaction of the primary cosmic rays hitting the earth's atmosphere. As demonstrated by published data and interpretations, once it is observed that there is some disagreement it is very difficult to pin down the source of the problem[12]. One area though that shows a rather clean effect is the production of multiple muons by Fe nuclei compared to protons as primaries. For the same total energy, Fe nuclei are much more efficient in producing muons because the nucleus fragments high in the atmosphere producing a large number of lower individual energy π's. The picture is somewhat complicated by the depth dependance: for a given energy per nucleus the proton will produce muons reaching greater depths, but fewer of them. There is thus a cross over depth at which the multiplicity will be equal for a given total incoming energy. This crossing occurs at higher energy with greater depth, and at smaller multiplicity as well. For a given underground installation we must convolute this picture with the steeply falling spectrum of cosmic rays. Using Elbert's formula[12] we can estimate that the cross over energy for primaries of iron and protons to produce the same muon multiplicity is about 325 times the energy required to penetrate to that depth. Now one of the most interesting questions in cosmic rays at present is the controversy over the composition of the primary flux at about 10^{15}eV/nucleus. It appears quite possible that the flux is dominated by iron near this energy, plus or minus a decade. The energy ratio given above then tells us an upper limit on the depth at which we can explore the composition of 10^{15}eV primaries, namely 4800 mwe in vertical depth. The multipicity distributions at depth are predicted to be (reasonably) Poisson-like. The ratio of muon multiplicity from iron to protons shows a saturation value, using the same formula, of 2.7 for shallow depths. The optimum separation of multiplicities will occur when the multiplicity due to iron has a mean of about 1 (so that multiple muon events will be mostly due to iron). This depth is around 4 kmwe vertically. The conclusion to be drawn is that the composition question will be best explored by relatively deep detectors.(Homestake and DUMAND are in the right range, while the Park City, Soudan and Morton detectors are too shallow, and the Mont Blanc and Kolar detectors are too deep.) It is also clear that this exploration of composition will be greatly facilitated if the primary energy can be simultaneously observed, as with an Extensive Air Shower array at the surface. Showers, consisting mostly of electrons, do not penetrate to sea level well at this energy and hence a hybrid detector linking both EAS Array and underground instrument would be best placed in the mountains, as high as possible. An alternative to this is to utilize an air Cerenkov detector to "see" the shower at its maximum, though then one pays a large price in dead time awaiting good dark seeing conditions (<10% time).

The size of a detector must also be considered for these muon multiplicity studies. The characteristic muon separation is 2-5 m in the range of depths and energies under discussion and hence in order not to distort the multiplicity distribution one would like a detector with dimensions large in comparisom, >20 m. Calculations for the $9 m^2$ Soudan detector show that it severly truncates the distribution above a multiplicity of 5 muons. Moreover the earth's magnetic field will give a North-South elongation to the distribution, an important effect for the Soudan detector. It is less important at great depths (higher muon energies) such as Homestake, or in larger detectors such as Morton.

Detectors of all sizes, from Soudan I to DUMAND, have and will continue to produce useful cosmic ray muon studies as dictated by their particular attributes. There are a number of other cosmic ray muon studies possible, (particularly with DUMAND, because of its size) but we shall not discuss them here.

Classes of Detectors

Summarizing this section, we have superficially reviewed the most prominently discussed experiments for deep underearth detectors. They seem to naturally divide into two classes: 1) detectors of the large nucleon decay search type, which have masses of a thousand tons or more and energy sensitivity down to a few MeV and are located at depths of 2-4 kmwe, and 2) detectors of more than $10^6 m^3$ at depths of 4 kmwe or more and energy sensitivity down to about 100 GeV. Because of the volume the only proposal for a detector of the second class is for DUMAND, in the ocean. There seem to be many possibilities for detectors in the first class. We shall discuss those below.

NEXT GENERATION NUCLEON DECAY DETECTORS

Present limits on the lifetime for protons present a formidable challenge for the design of a next generation instrument. Unfortunately one solution, independant of the results of the first round of dedicated nucleon searches will not suffice. For example, if the decay is found soon with a lifetime in the 10^{31} year range then a ten thousand ton detector (6×10^{33} nucleons) can explore the decay modes. This instrument should be sensitive to the modes under investigation (K's?, multi-π's?, μ's?, multi-ν's?) and we will need the hints from the first round since one cannot afford to optimize all resolutions. I don't propose to present a design for such a detector here, but only point out that extrapolations from traditional accelerator neutrino detectors would lead to outrageous costs for a 10 kiloton detector. We must contemplate new techniques that scale well to great sizes. It is not clear to me that large water Cherenkov detectors might not suffice. An improved water Cherenkov detector might use imaging optical detectors for better track resolution. Another type of detector that scales well is of the time projection chamber type, either liquid or gas.

On the other hand, if we do not find the decay in the first round with a lifetime limit between 10^{32} and 10^{33} years, other action will be appropriate. First one will have to ask if it is worth another round considering the open ended predictions, given the present apparent demise of simple SU(5). The detector mass required to make a factor of ten improvement in lifetime sensitivity will probably have to be two decades larger in mass because of the necessity of subtracting indistinguishable neutrino interactions. We are suddenly faced with the prospect of building a 100 thousand to million ton detector, and it must have fairly good resolution.

Neither is the type of detector clear, nor is the best location obvious to me. At some sufficiently large size it will not be practical to place such a detector underground. Although placing a complex instrument at sea is surely not trivial, we who have been studying DUMAND have learned much, including some of the advantages of working at sea. There is a wealth of experience in the oceanographic field, the Navy and the oil industry, of which we have been able to take advantage. Working in mines has become more familiar to high energy physicists of late, but still is not a trivial or comfortable business compared to the accelerator laboratory. Table I presents a list of some of the tradeoffs between working in deep mines versus the deep ocean. A choice cannot be made at this time because of lack of information. I only claim that a wide range of options should be considered before we undertake the next step in nucleon decay searches.

Table I: Some Tradeoffs between mine and ocean deployment.

Characteristic	Mine	Ocean
Cost of Cavity	Very High	Container Required
Installation Cost	High	High--Preassemble
Safety	Concern	No Prob. after Inst.
Reliability Requirements	High	Very High
Recoverability	Easy	With Difficulty
Reparability	Easy, Probably	Very Difficult
Remote Operation	Desireable	Required
Experience	Fair	Little (HEP)
Risk	Some	Yes--Segment Detector
Know Overburden	Fair	Excellent
Vary Depth	No	Possible
Bury in Large Array	No	Possible
Scale Up Later	Difficult	Easy
Surface Array	Possible	Difficult, but Possible

CONCLUSION

In order to be ready to make an informed choice, I believe we should start now to evaluate some of the novel technologies, both on paper and in the laboratory. We should also seriously set about evaluating the options in size, depth, and location for a second generation nucleon decay detector. It is probably wise for the U.S. to build at least one fine grain detector in the thousand ton class to complement the present large water Cherenkov detectors and to gain the experience with that type of instrument. It seems to me that the need for a National Underground Laboratory has not yet become manifest.

REFERENCES

1. International DUMAND Collaboration, Hawaii DUMAND Center, Oct. 15, 1982.
2. H. Chen, Proceedings of the 1982 Summer Workshop on Proton Decay Experiments, ANL-HEP-PR-82-24, 274 (1982).
3. R. Davis, Op. Cit., 399.
4. Proceedings of the 1976 DUMAND Summer Workshop, A. Roberts, ed., HDC (1977). See particularly the papers by Tammann and by J. C. Wheeler.
5. Proceedings of the 1980 DUMAND Summer Workshop, V. Stenger, ed., HDC (1981).
6. M. L. Cherry, et al., ANL-HEP-PR-82-24, 300 (1982).
7. S. Errede, et al., Proceedings of the Wingspread Conference on Monopoles, (1982).
8. P. Bosetti, et al., this conference.
9. P. V. R. Murthy, Proc. 17th Int. Cos. Ray Conf., Paris, (1981).
10. D. Cline, ANL-HEP-PR-82-24, 224 (1982).
11. A. Mann, et al., Op. Cit., 38 (1982).
12. See various papers in the Proc. Workshop on Very High Energy Cosmic Ray Interactions, ed. Cherry, Lande, and Steinberg, U. of Penn. (1982).

BEYOND PROTON DECAY:
OTHER PHYSICS POSSIBILITIES WITH THE HOMESTAKE
NUCLEON DECAY DETECTOR

M.L. Cherry, I. Davidson, K. Lande, C.K. Lee, E. Marshall,
and R.I. Steinberg
Dept. of Physics, Univ. of Pennsylvania, Philadelphia, PA 19104

The Homestake Gold Mine presently houses the
Brookhaven solar neutrino experiment and a
300-ton water Cerenkov detector at a depth of
4200 meters water equivalent. The Cerenkov
detector has been used to study nucleon decay,
multiple muons, and neutrino bursts. An array
of liquid scintillator, with surface area of
130 m^2, is presently being constructed to
measure magnetic monopoles, neutrino oscillations,
underground muons, and neutrino bursts. At the
same time, a 1 km^2 extensive air shower array
is being built on the surface in order to measure
the high energy cosmic ray composition with
simultaneous surface and underground shower
measurements. Future plans call for a 1406-ton
liquid scintillator Tracking Spectrometer to
measure nucleon decay, n-$\bar{\text{n}}$ transitions, and
the low energy cosmic ray neutrino spectrum.
We describe the present results and the possibilities for physics other than nucleon decay
in the nucleon decay detectors.

I. INTRODUCTION.

The growing interest in searches for nucleon decay has also
focused attention on a series of other important particle physics
questions that can best be answered in a non-accelerator, low-
background environment. The nucleon decay detectors themselves can
carry out several of these experiments, including:

1) the search for massive magnetic monopoles;
2) a search for neutron-antineutron annihilation in nuclei;
3) an investigation of neutrino transmission through the earth
   and the possibility of either vacuum or matter oscillations;
4) a study of neutrinos from extraterrestrial sources, in
   particular collapsing stars; and
5) the measurement of the multiplicity and transverse
   momentum distributions of cosmic ray muons. This study
   in conjunction with a surface extensive air shower array
   (to determine the energy of the cosmic ray primaries)
   permits a measurement of the primary mass composition.

0094-243X/83/960248-17 $3.00 Copyright 1983 American Institute of PHysics

The existing 300-ton water Cerenkov detector at Homestake and the proposed 1406-ton Tracking Spectrometer nucleon decay experiment are capable of carrying out all of these measurements. In this report, we describe the Homestake program, summarize the non-proton decay results from the water Cerenkov detector, and describe the studies to be carried out in the future. We limit ourselves, however, to the physics obtainable with the nucleon decay detectors; the ^{37}Cl and ^{71}Ga solar neutrino detectors have been described elsewhere[1].

The 1406-ton Homestake Tracking Spectrometer[2] will make possible a search for nucleon decay up to a lifetime of 5×10^{32} years. The first set of 200 scintillator elements, covering the outside of a hollow 8m x 8m x 16m box, will be installed in the mine shortly, and will be used as a large-area detector for GUTs monopoles (Section II) and neutrino oscillations (Section III). In Section IV, we describe the possibility of searching for n-n̄ oscillations in the full Tracking Spectrometer. The present Homestake detector has been used to study underground muons[3] and to search for neutrino bursts from collapsing stars[4], as well as to place a lower limit of $1.5 - 3 \times 10^{30}$ years on the proton decay lifetime[5]. We summarize our present muon results in Section V. The composition of the cosmic ray beam near 10^{15} eV can be determined by combining the underground muon observations with simultaneous measurements of the showers at the earth's surface. A surface air shower array is presently being constructed; the composition measurement with the combined surface array-underground telescope is described in Section VI. In Section VII we summarize the present results of our search for neutrino bursts from collapsing stars.

II. MAGNETIC MONOPOLES & THE HOMESTAKE LIQUID SCINTILLATOR TELESCOPE.

The Grand Unified Theories that predict nucleon decay also predict the existence of magnetic monopoles with mass $\sim 10^{16}$ GeV. These 't Hooft-Polyakov monopoles would have been produced in the early stages of the universe with a flux density comparable to that of nucleons. Although there are no direct estimates of the density of monopoles remaining today, upper limits can be established from considerations of closure of the universe and stability of galactic magnetic fields, which suggest that on a large scale the mean flux of monopoles can be no greater than 10^{-4} m^{-2} yr^{-1} sr^{-1}. These limits do not apply to monopoles trapped in local substructure of the Galaxy -- for example, monopoles orbiting like micrometeoroids in the Sun's gravitational field[6]. The expected velocities of such local monopoles would then be $\beta \sim 10^{-4}$, comparable to the earth's orbital velocity.

Cabrera[7] has recently reported the possible detection of a magnetic monopole with magnetic charge 137e/2 in a 20 cm^2 superconducting loop. The flux corresponding to one event in six months is 0.6 cm^{-2} yr^{-1} sr^{-1}, much higher than the maximum fluxes allowed by the galactic field and closure constraints. The only

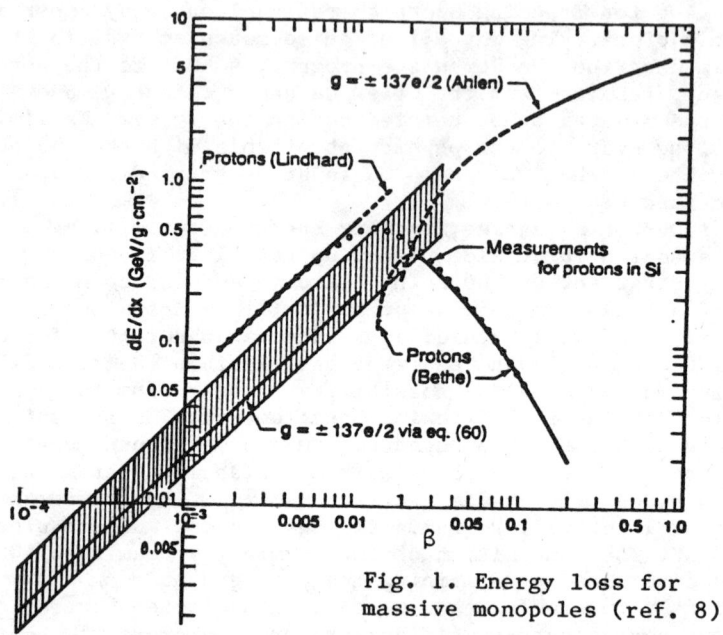

Fig. 1. Energy loss for massive monopoles (ref. 8)

way to explain such a high apparent flux (assuming it to be real) is to postulate some sort of local enhancement like the gravitational trapping mechanism. The low monopole velocity in the gravitational trapping model is also handy to explain away the fact that several larger detectors have reported only upper limits to the flux: the velocity threshold of other monopole searches was typically 10^{-2}, so only Cabrera's device was sensitive to slow monopoles.

Ahlen and Kinoshita[8] suggest that monopoles lose 2 MeV cm^2 g^{-1} at $\beta = 10^{-4}$, the same as minimum ionizing particles, and that dE/dx increases linearly with velocity up to $\beta = 10^{-2}$ (Fig. 1). Since the ionization efficiency of such slow monopoles is likely to be on the order of or less than 10%, a detector sensitive to slow monopoles must produce hundreds of detectable ion pairs for a throughgoing minimum ionizing particle. One example of such a detector is a thick liquid scintillator. With a scintillator thickness of 15-20 g cm^{-2}, a monopole threshold at 10% of minimum ionizing corresponds to a 3-4 MeV energy deposit, above the level of typical energy deposits from background radioactivity. In addition, in a 30 cm thick scintillator, the pulse width is 1/β ns, thus providing an appreciable broadening of the detector pulse for slow monopoles and thus another constraint on the monopole signal.

A meaningful improvement over present results thus requires a search with a very large area detector at a very low threshold. In order to suppress cosmic ray-induced background, a deep mine site is desirable. Experiments performed at the surface of the earth are

troubled not only by muons, which can be partially shielded by
appropriate anticoincidence counters, but by neutral components of
cosmic rays which can penetrate veto counters and can simulate both
fast and slow monopoles. Slow monopole simulation by these cosmic
ray secondaries is particularly serious since the typically small
energy deposits will match those expected from monopoles, but
heavily ionizing fast monopoles can also be mimicked by muon
bundles. This latter problem will be compounded if there is
saturation of the readout electronics and the full ionization
cannot be measured. In a deep underground location not only are
there very few muons to confuse the interpretation, but the muons
that exist are strongly peaked in the vertical direction and are
thus readily distinguishable from monopoles which are likely to
be isotropic.

In Fig. 2 we show the monopole telescope to be constructed
at Homestake in collaboration with B. Cleveland, R. Davis, and D.
Lowenstein at Brookhaven. It consists of a hollow 8m x 8m x 16m
box of 30cm x 30cm x 8m liquid scintillation detectors surrounding
the existing ^{37}Cl solar neutrino tank. Each of the 200 scintillator
elements is a teflon-lined PVC box containing a low-cost liquid
scintillator developed to have excellent light collection and
transmission characteristics, a light attenuation length greater
than 8 m, excellent long-term stability, a high flash point, and
low toxicity[2]. Each detector element is viewed by two 5-inch
photomultiplier tubes in coincidence, one at each end. The detector
elements have been developed for the proposed 1406-ton Homestake
Tracking Spectrometer nucleon decay experiment, which will consist
of three 8m x 8m x 8m stacks with a sensitivity corresponding to a

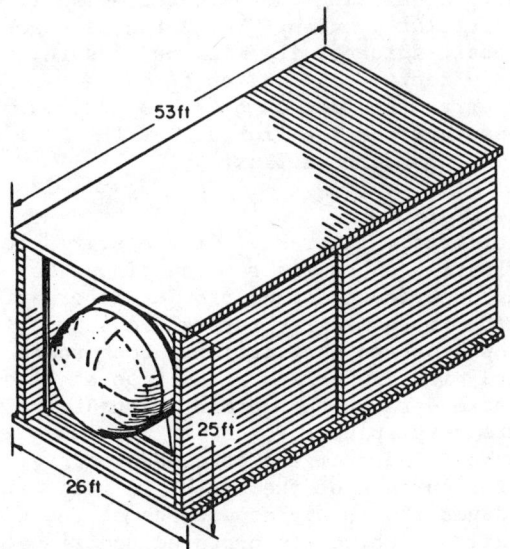

Fig. 2. Homestake Large Area Liquid Scintillation Detector

total nucleon lifetime of 5×10^{32} yrs. This sensitivity is made possible by a combination of excellent energy resolution ($\pm$ 4.2%, more than sufficient to produce a sharp total energy peak at 938 MeV for neutrinoless nucleon decay modes), $\pm$ 1.3ns time resolution, spatial resolution of $\pm$ 15 cm (0.32 radiation lengths), and a very low muon background flux (1100 m^{-2} yr^{-1}, a factor of 10^7 below the surface muon flux). Fast muons passing through the midpoint of one of our modules produce an average of 350 photoelectrons at each photomultiplier. A particle ionizing at 1/100 of minimum would thus produce 3-4 photoelectrons at each photomultiplier and so would be easily visible. Accidentals from ambient radioactivity will probably limit the two-layer detector to an effective threshold of 5 MeV, or about 10% of minimum ionizing, however.

We will have available several pieces of information allowing us to identify a traversing monopole depending on its velocity and ionization: 1) the pulse widths in the entering and exiting counters should each be 1 ns/β; 2) the delay between the pulses in the entering and exiting counters should be 25 ns/β; and 3) both pulse heights should be consistent with expectations for a monopole at the velocity determined by 2). For the velocities of interest, the delay time between the two counter pulses will be 25-250μsec. Such long delays can only be correlated in a very low background environment, such as that available in a deep mine. The most severe background will be due to two independent, traversing muons, each of which is detected in only one of the two counters through which they pass. The accidental coincidence rate for independent muons is 4×10^{-5} yr^{-1}. This accidental rate will be further reduced by consideration of the pulse heights and widths in the two counters and by the necessity of missing the outgoing pulse from each of the two accidentally correlated muons. The accidental backgrounds associated with local radioactivity will be measured in situ and will determine our effective ionization threshold.

The detector array provides an aperture of 1500 m^2 sr. One event per year thus represents a flux of 2×10^{-15} cm^{-2} sec^{-1}sr^{-1}, about 3×10^{-6} times the Cabrera flux.

III. ATMOSPHERIC NEUTRINOS.

There is considerable interest in the energy spectrum, flux, and angular distribution of cosmic ray neutrinos. Not only does a measurement of the neutrino spectrum provide information about the production mechanism in cosmic ray interactions, but it also makes possible studies of neutrino oscillations, backgrounds for nucleon decay searches, and direct neutrino production at high energies.

The experimental data on the cosmic ray neutrino flux and energy spectrum are very sparce. For neutrinos above 1 GeV the spectrum can be calculated from the observed spectrum of cosmic ray muons that reach the surface of the earth. In this calculation it is necessary to assume the energy dependence of the pion-to-kaon ratio and the fraction of directly produced neutrinos. Since the muon intensity is measured only in the energy range 2 GeV $\lesssim E_\mu \lesssim$ 10

TeV, the neutrino spectrum below and above this range can only be determined by extrapolation or modeling. The neutrino flux below 1 GeV is particularly important as the background for nucleon decay searches. Using the calculations of Osborne and Young[9], we find that the interaction rate per nucleon corresponds to an apparent nucleon lifetime of 3×10^{30} yrs. This rate would give rise to 100 interactions per year in a 500-ton detector. The 140-ton Kolar detector has reported several fully-contained candidate proton decay events. The identification of these events as proton decays depends critically, however, on statistical arguments based on calculations of neutrino interaction rates which indicate that the events are unlikely to be caused by masquerading neutrinos. A direct measurement of the neutrino background is therefore crucial to check the background for the KGF and other large presently operating nucleon decay detectors.

A neutrino oscillation measurement is also extremely exciting. By measuring the rate of neutrino-induced muons as a function of zenith angle, the Homestake scintillator array can study neutrino flight paths over the range $100 - 10^4$ km, and thus permits a search for $10^{-1} eV^2 > \delta m^2 > 10^{-3} eV^2$. This is among the most sensitive mass difference investigations possible with terrestrial neutrino sources and is inferior only to the mass difference sensitivity of the Gallium solar neutrino experiment. High energy cosmic ray muon neutrinos interact in the rock around the detector, producing muons which then traverse the detector. Neutrino oscillations would cause the muon neutrinos to convert into electron or tau neutrinos, both of which are unlikely to produce signals in our muon detector box. Electron neutrinos interacting in the rock will give rise to electrons which shower in the rock and thus have secondaries of very limited range. At low energies, tau neutrinos can undergo only neutral current scattering, while at higher energies the produced taus will decay almost immediately into short-range secondaries. The ^{37}Cl tank in the detector core provides a muon energy threshold $E_\mu \gtrsim 2$ GeV, thereby suppressing the contribution of muon-decay neutrinos (and therefore the effect of an admixture of electron neutrinos). The absence of neutrino oscillations is characterized by a flux of muons per unit solid angle that is independent of zenith angle. The oscillation of muon neutrinos into either electron or tau neutrinos will result in a reduction of muons as the zenith angle and thus the neutrino flight path is increased. In the absence of oscillations, we expect about 200 neutrino-induced muons per year. The reduction in intensity and the angular range in which this reduction begins will determine both the mass difference and the mixing angle (Fig. 3).

The fine angular resolution of our detector ($< 3°$) will also permit a search for the matter oscillations predicted by Wolfenstein several years ago. His suggestion is that, since electron and muon neutrinos see different potentials in traversing matter, they will build up a phase shift and thus oscillate. In order to give rise to measurable oscillation it would be necessary

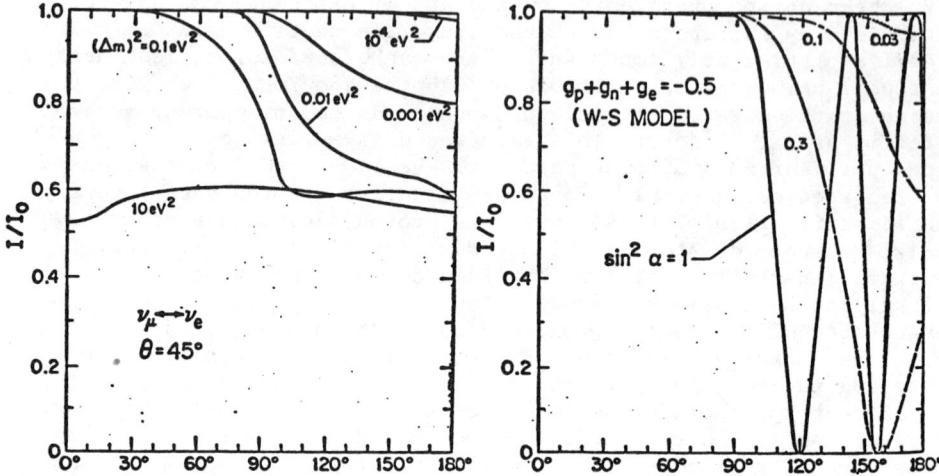

Fig. 3. Vacuum oscillations.   Fig. 4. Matter oscillations.

to traverse a significant fraction of the earth's diameter. These matter oscillations will occur for zenith angles $\theta > 145°$ and will be characterized by violent oscillations with periods of several degrees in zenith angle. This phenomenon should not only be detectable in our experiment but should be distinguishable from vacuum oscillations (Fig. 4).

IV. NEUTRON-ANTINEUTRON CONVERSION.

In addition to $\Delta B = 1$ violations of baryon number, some theories predict $\Delta B = 2$ processes. With $\Delta B = 2$, $n \to \bar{n}$ conversion becomes possible. The experimental signature of such an event is clear--the $\bar{n}$ immediately annihilates with another nucleon in the same nucleus and releases 2 GeV of rest mass energy, mainly in the form of low energy pions. (The average number of pions is about 5, roughly evenly distributed among π^+, π^-, and π^0.) Such an event should be clearly visible in a nucleon decay detector such as the Homestake Tracking Spectrometer. H. Anderson has suggested doing an n-$\bar{n}$ experiment by installing duplicates of the Tracking Spectrometer liquid scintillator elements in the neutron beam at the Los Alamos Omega West reactor[10]. The experiment can also be performed underground at Homestake simultaneously with the nucleon decay search.

The best existing limit for this sort of event comes from the Homestake water Cerenkov data, where we searched for $\mu \to e$ decays in 150 tons of water. In 407 days, we observed 5 2-module events, moving up or to the side, unaccompanied by a veto from the top scintillator or the bottom Cerenkov tank, with observed energies less than 2 GeV. There were 7 comparable events with a veto, so that we take an upper limit to our observed n-$\bar{n}$ candidates to be 2. The total exposure time was 6.5×10^{31} nucleon-yrs, so we find a lower limit to the n-n̄ lifetime to be

$$\Gamma^{-1}_{n-\bar{n}} > \frac{(6.5 \times 10^{31} \text{ nucleon-yrs}) \times B}{2 \text{ candidate events}} = 9 \times 10^{30} \text{ yr}$$

where B ~ 0.3 is the probability of producing and detecting a decay muon following a neutron oscillation event. Following the conservative Mohapatra-Marshak estimate[11], we take Γ to be

$$\Gamma = \frac{2\pi}{\hbar} |H|^2 \frac{dN}{dE} ,$$

where the matrix element H = δm x an overlap integral between two nucleons is H ~ 0.01 δm, and dN/dE ~ (1 GeV)$^{-1}$. The experimental limit then gives a mixing time

$$\tau = \hbar/\delta m > 3 \times 10^5 \text{ sec.}$$

This is already larger than several theoretical estimates[11].

The full 1406-ton Tracking Spectrometer will have energy resolution good enough to check that the total energy released is close to 2 GeV, and (unlike a Cerenkov counter) will have an energy threshold low enough (E_{min} ~ 1 MeV in a single element) to see nonrelativistic particles and hence measure the full energy loss. Accurate timing resolution will make it possible to determine that particles are flowing away from a vertex, and to verify momentum conservation. Spatial resolution of ± 0.3 radiation lengths makes it possible to identify tracks flowing out from a vertex, and determine the multiplicity unambiguously. With a slow neutron flux of 3×10^{11} sec^{-1} in the Omega West experiment, Anderson expects a sensitivity of $\tau > 10^7$ sec in a 200 day run. A similar sensitivity will be attainable at Homestake.

V. UNDERGROUND MUONS & THE HOMESTAKE WATER CERENKOV DETECTOR.

High energy cosmic rays are a unique tool with which to probe simultaneously basic phenomena in astrophysics and particle physics. There is great astrophysical interest in the region around 10^{15} eV, an energy range too high for direct observations and below most air shower experiments. There appears to be agreement that there is a change in slope in the primary cosmic ray spectrum. The steepening in the spectrum may reflect the upper energy limit of the acceleration process, a changing source mechanism, a new class of sources, or energy- or charge-dependent propagation effects. At energies up to 100 GeV/nucleon, cosmic ray path lengths in interstellar space are clearly observed to decrease with increasing energy. If this trend continues, then at energies near 10^{15} eV, where the gyroradius becomes comparable to the thickness of the galactic disk, the composition measured at earth should approach closely to the original source composition. In addition, recent analyses of air shower data indicate the onset of sizeable anisotropies near 10^{15} eV. Statistically significant direct measurements of the composition will soon be extended to energies well above 1 TeV/

nucleon with JACEE and Spacelab, and many of the uncertainties in the particle physics and interaction models will be resolved up to laboratory energies above 10^{15} eV at the new CERN and FNAL colliders, so it appears both interesting and timely to study the region of the cosmic ray spectrum between 10^{14} and 10^{16} eV[12,13].

The existing Homestake water Cerenkov detector (Fig. 5) consists of a water Cerenkov counter and liquid scintillation detector, with upper surface area of 113 m^2, surrounding the ^{37}Cl Brookhaven solar neutrino detector. Located at a depth of 1480 meters of rock (4200 meters of water equivalent) in the Homestake Gold Mine, Lead, South Dakota, the device has been used to study the high-energy muon component of showers generated by 10^{14}-10^{15} eV/nucleon cosmic ray primaries. Muons observed deep underground are unique among cosmic ray components in

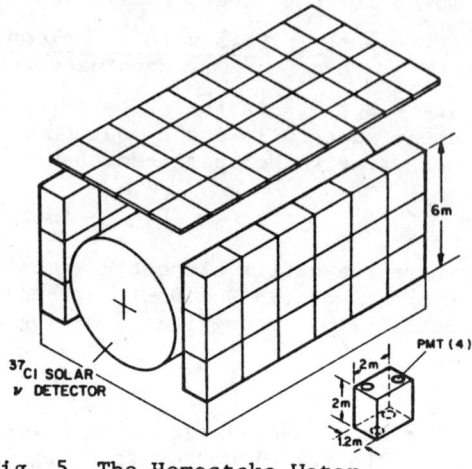

Fig. 5. The Homestake Water Cerenkov Detector.

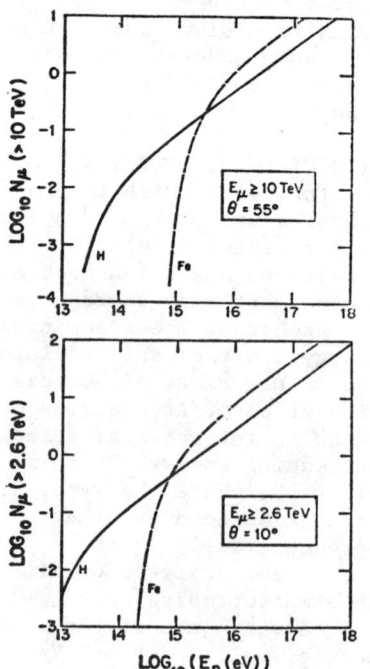

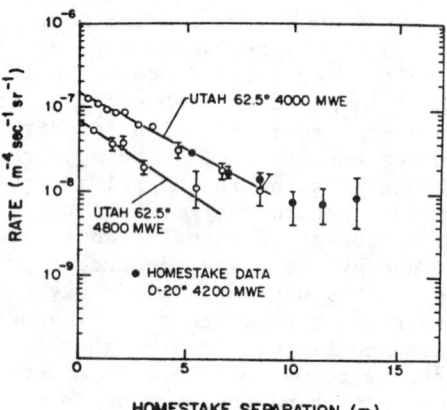

Fig. 7. Decoherence curve. Solid points are the Homestake results; open circles are the Utah results normalized to Homestake.

Fig. 6. Number of underground muons vs. primary energy for $10°$ and $55°$ zenith angles.

that they carry direct information about the initial stages of the shower. In order to reach our depth, muons must have roughly 2.7 TeV at the surface of the earth. Such muons can be produced by primary protons with energies in excess of 10^{13} eV or by primary iron with energies above a few times 10^{14} eV (Fig. 6). A proton generally gives rise to a single high energy muon while an iron, consisting of a superposition of 56 separate nucleons, has a large probability of multiple muon production, particularly above 10^{15} eV. For small showers (E $\lesssim 10^{15}$ eV) we expect to observe single muons primarily from cosmic ray protons, while for large showers (E $> 10^{15}$ eV) we expect a mix of single and multiple muons from protons and heavy (nominally iron) primaries.

In eighteen months of running with the present detector, we have observed 7124 vertical muon events. The multiple muon rates and lateral distribution have been presented and discussed in ref. 3. The pion transverse momentum can be derived from the observed number N(r) of underground muon pairs as a function of the separation r. If dN/dr is the number of such events/meter of separation in an observation time t, solid angle Ω, and total detector area A, then the separation distribution can be expressed in terms of the decoherence curve

Table 1. Rate of multiple muons

Multiplicity	Number of events
1	6814
2	275
3	31
4	4
>4	0

$$R(r) = \frac{1}{2\pi\, r\, \varepsilon(r) A \Omega t} \frac{dN}{dr} \qquad (1)$$

where $\varepsilon(r)$ is the geometric efficiency with which a muon pair can be detected at a given separation r. The measured decoherence curve is shown as the solid points in Fig. 7, together with the earlier Utah results[14] at 4000 and 4800 m.w.e. The Utah points are corrected for the difference in altitude and zenith angle between Utah and Homestake.

The results of several accelerator and cosmic ray experiments indicate a slow increase of $\langle p_t \rangle$ with $\sqrt{s}$, as shown in Table 2.

Table 2. Summary of $\langle p_t \rangle$ for recent high energy experiments.

$\sqrt{s}$ (GeV)	$\langle p_t \rangle$ (MeV/c)	
14	325 $\pm$ 2	FNAL
20	340	FNAL
53	350	ISR
180	400 $\pm$ 20	balloon
200	440 $\pm$ 20	balloon
170 - 400	500 - 670	Utah underground muons
200 - 400	570	air showers
200 - 1200	140 - 500	Chacaltaya
540	500	CERN Collider
400 - 600	500	present experiment

In particular, the Utah results, based on Monte Carlo calculations assuming a mixture of protons and iron in the primary beam, give $\langle p_t \rangle$ ranging from 500 to 670 MeV/c, depending on the details of their models. The CERN $\bar{p}p$ results[15] at $\sqrt{s}$ = 540 GeV give $\langle p_t \rangle$ = 500 MeV/c. Elbert et al.[3] have performed a full Monte Carlo simulation of our results, and have obtained reasonable agreement with a model assuming $\langle p_t \rangle$ = 500 MeV/c and an iron-to-proton ratio increasing with energy and reaching unity at a primary energy near 100 TeV/nucleon.

The primary composition is most sensitive to the rates of multiple muons. Elbert et al.[3] find that the results may be consistent with either a proton-rich spectrum based on an extrapolation of direct particle-by-particle balloon and satellite measurements at low energies (below 100 GeV), or an iron-rich spectrum based on the indirect measurements of hadrons in air shower cores by the Maryland group[16] at 10 - 1000 TeV. Calculations have been performed for a variety of different spectra and compositions, but no combination has been found which reproduces the full set of data in detail. The primary energy range of the present experiment is not a measured quantity, and can be determined only with the detailed Monte Carlo shower studies of ref. 3. The experiment is sensitive to primary cosmic rays of 10^{14}-10^{15} eV, although the typical energy of primaries producing underground muons is 100 - 200 TeV/nucleon. We are presently installing an air shower array on the surface above the underground detector in order to provide this measurement directly.[17]

VI. MEASUREMENTS OF COSMIC RAY COMPOSITION WITH A COMBINED SURFACE AIR SHOWER ARRAY AND DEEP UNDERGROUND DETECTOR.

Previous experiments at air shower energies (including our own) have used <u>either</u> the surface measurements or the underground muon data to extract the composition information, and have produced results which are highly model-dependent and plagued by the unavoidable difficulties attendant upon separating out the imperfectly-known particle physics effects from the astrophysical effects of composition and energy spectrum. With the combination of a surface array and a deep underground detector, we will have <u>both</u> types of data available simultaneously. By measuring the total energy/nucleus on the surface and the energy/nucleon underground we will determine the composition in a way which depends essentially on energetics, and very little on the detailed calculations of specific interaction and propagation models.

In attempting to probe the composition using the underground data alone, one can compare the rate of multiple underground muons to the rate of singles. This amounts to comparing the integral flux of iron above 10^{15} eV (together with the contribution of still higher energy protons) with the much greater integral flux of protons below 10^{15} eV. Small inaccuracies can then be inordinately important. On the other hand, one enhances the composition information by tagging the underground muon data with the shower size measured on the surface, so that one is measuring muon multiplicities in fixed shower size windows, and relating protons to iron at comparable total energies.

Of the five deep detectors, only Kolar and Homestake have acceptably smooth and accessible surface terrain for a shower array. The Kolar group has in fact already operated an air shower array over a shallow (220 GeV threshold) underground detector, and presently plan an air shower array above a deeper (1 km) underground detector in order to increase their muon threshold and make it possible to study the composition in the 10^{15} eV range.

We are currently adding an extensive air shower array on the earth's surface above the Homestake deep underground detector. The array will consist of approximately 100 particle detectors, each 3 m^2, deployed over an area of roughly 1 km^2 at a depth in the atmosphere of 850 g. Together with a 200 m^2 underground detector, these surface detectors will constitute a large solid angle (0.5 sr), large aperture, high energy threshold cosmic ray telescope. The experiment will be carried out as a collaborative effort with the University of Leeds/ Haverah Park air shower group (A. Watson, J. Lloyd-Evans, R. Reid), the Tata Institute/Kolar Gold Fields group (R. Sivaprasad), the University of Texas (E. Fenyves), and the South Dakota School of Mines and Technology (T. Ashworth).

In order to predict the detector performance, we have performed Monte Carlo calculations simulating one year's worth of cosmic ray events. The calculations have been based on the Monte Carlo code developed at Bartol and Utah by Gaisser, Elbert, Stanev, and their collaborators. In Figs. 8a and 8b, we show the distributions of observed muon number vs observed electron number for one

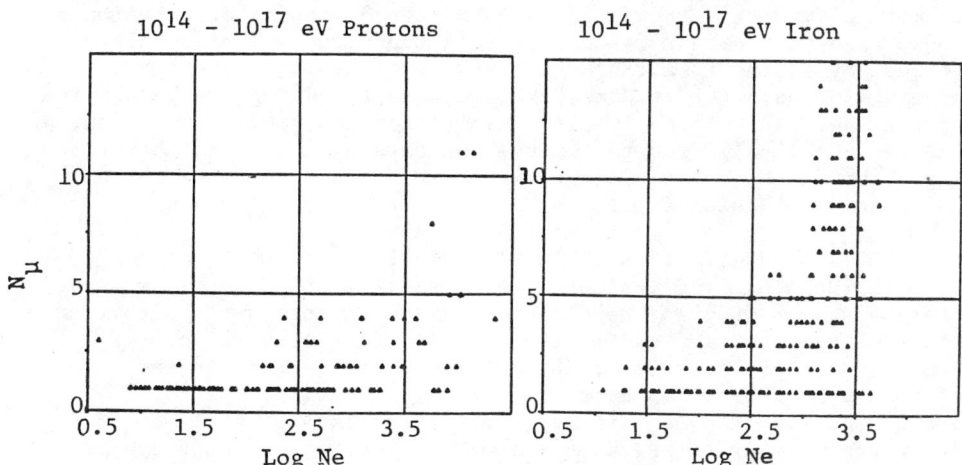

Fig. 8. Expected distributions of observed N_μ vs. N_e for one year's worth of protons (left) and iron (right).

year's worth of protons and iron, respectively. Only events which trigger the underground detector are shown. As expected, the proton distribution is strongly concentrated at the 1 muon level. At observed shower sizes above about $N_e = 100$ (corresponding to slightly above 10^{15} eV), iron nuclei begin generating copious numbers of detected muons. The distributions are dramatically different. By choosing both N_e and N_μ, one can clearly determine whether a particular event is more likely to be derived from a parent sample of protons or iron.

It is impossible to perform this sort of discrimination using just the underground muon number. For example, $N_\mu = 4$ may come from an iron with $N_e = 100$ (2×10^{15} eV), or it may come from a proton with $N_e = 1000$ (3×10^{15} eV). For some compositions and spectral mixes, such protons will be much more abundant than the iron, while for other spectral mixes, the reverse is true. Likewise, $N_\mu = 11$ may be due equally well a priori to iron with $N_e = 10^3$ (6×10^{15} eV) or to protons with observed $N_e = 10^4$ (2×10^{16} eV). Again, a flat iron spectrum would have to be suppressed at an energy near 6×10^{15} eV, so that it is quite possible that the flux of 2×10^{16} eV protons might be large compared to the 6×10^{15} eV iron flux, and the underground muon observation alone could not discriminate between iron and protons.

At energies below about 1 TeV, observations on balloons and spacecraft have measured the cosmic ray spectrum and composition directly with emulsions and counters. At higher energies, experiments like the Utah and Homestake underground muon studies and the Maryland air shower core measurements have provided data which can be fit to Monte Carlo calculations assuming specific models of the primary spectrum and composition. These latter experiments have provided tantalizing suggestions of a large abundance of heavy primary cosmic rays. But, as shown in ref. 3, the interpretation of these experiments is not at all straightforward, and certainly not unique. By adding an air shower array on the surface above the underground detector at Homestake, we will be adding the capability to take data which clearly discriminate between protons and iron, and above all permit us to analyze our data using a straightforward algorithm which does not depend on the details of a Monte Carlo calculation of assumed primary spectrum and composition.

VII. NEUTRINO BURSTS FROM COLLAPSING STARS.

In the final gravitational collapse of a sufficiently massive star, a sudden drop in the electron degeneracy pressure supporting the stellar core can lead to the release of a large pulse of electron capture neutrinos, thermal $\nu - \bar{\nu}$ pairs, and (in the case of an asymmetric collapse) gravitational radiation. The total energy radiated may be 10^{53} ergs, corresponding to nearly 10^{58} neutrinos with energies of 10 - 100 MeV. The rate of such events per galaxy is highly uncertain, but rates of once every 10 - 40 years have been suggested based on optical and radio surveys of supernovae in external galaxies and supernova remnants in our own galaxy. Pulsar birth rates have indicated collapse rates as high as once every 4 - 6 years.

In order to detect neutrino bursts, a lengthy search is required with massive, well-shielded detectors located deep underground or underwater, having a low background counting rate and a high efficiency for detecting low-energy ν and $\bar{\nu}$. For the particular case of the model described by Freedman et al.[18], a burst of 10^{58} 10 - 20 MeV ν_e and $\bar{\nu}_e$ at the Galactic Center would result in 10 - 20 interactions in the Homestake Cerenkov detector in less than a second, with roughly equal contributions from ν_e and $\bar{\nu}_e$.

The main detector background is due to Compton scattering of gamma rays from local radioactivity. In order to separate clusters of counts due to real neutrino bursts from this steady radioactivity-induced background, we study the distribution of events in time -- i.e., we look for bursts of N events occurring within a time interval t_N shorter than would be expected from random statistics. The relation between N and t_N such that we would expect one random background burst per year is shown as the smooth curve in Fig. 9. Also shown is the experimentally observed distribution for 384 effective days of "on time". The points in the lower right-hand region of the figure are uninteresting, and have been artificially suppressed for clarity. There is excellent agreement between the predicted curve and the observed upper boundary of our events, indicating that there is no unanticipated background or source giving rise to a burst rate appreciably greater than the maximum predicted collapsing star rates. There is a clearly defined region above the distribution in Fig. 9 in which a potential astronomical signal could be unambiguously identified; this region of the graph

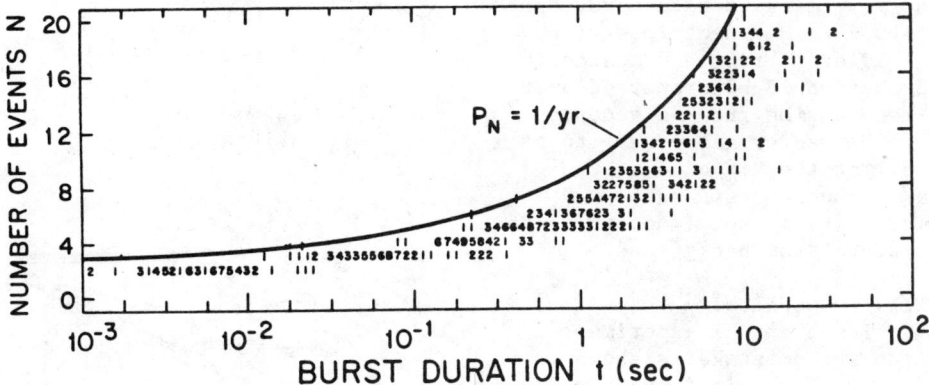

Fig. 9. Measured distribution of the number of events N in a burst vs burst duration t for 384 days of running time with the water Cerenkov detector. The smooth curve shows the expected relation for random background at a rate of 1 burst per year.

is empty. Although our data indicate no evidence of neutrino bursts over a period of about a year, the results clearly demonstrate that the background rate is sufficiently low to permit the detection of events consisting of 5 or more neutrino counts within 0.1 sec or less, or more than 10 counts in several seconds, corresponding to a burst of $\sim 10^{58}$ neutrinos with $E \geq 10$ MeV at the Galactic Center.

It is interesting to consider the effect of massive neutrinos on a burst search. For a source at a distance d emitting neutrinos with energies E_1 and E_2, the effect of a non-zero mass m is to introduce a dispersion in the flight time from the source to the earth, given by

$$\Delta t = \frac{m^2 d}{2c} \left(\frac{1}{E_1^2} - \frac{1}{E_2^2} \right).$$

If $m = 10$ eV/c^2, then a source at the Galactic Center emitting a narrow pulse of neutrinos with energies ranging from 10 to 20 MeV will produce a pulse width at the earth of 375 ms. If, however, a burst were observed with a time structure as narrow as 10 ms (comparable to the burst rise time suggested by the model of Wilson[19], for example), then one could place an upper limit on the neutrino mass of about

$$1.5 \, (d/10 \text{ kpc})^{-1/2} \text{ eV}/c^2,$$

about an order of magnitude lower than the limits obtainable with proposed terrestrial experiments. Another effect of a non-zero mass is to effectively limit neutrino burst searches to our Galaxy, even for very massive detectors. Bursts of 10 - 20 MeV neutrinos with mass $m = 10$ eV/c^2 from extragalactic sources would be dispersed over times ranging from 30 seconds for the nearby galaxy M31 to an hour for the Virgo cluster, making such bursts very difficult, if not impossible, to separate from background.

VIII. CONCLUSION.

The eventual experimental setup at Homestake is shown in Fig. 10. The scintillator array is presently being constructed to replace the water Cerenkov detector. The mass of the 200 new scintillator elements will be 140 tons, the same as the side modules of the Cerenkov

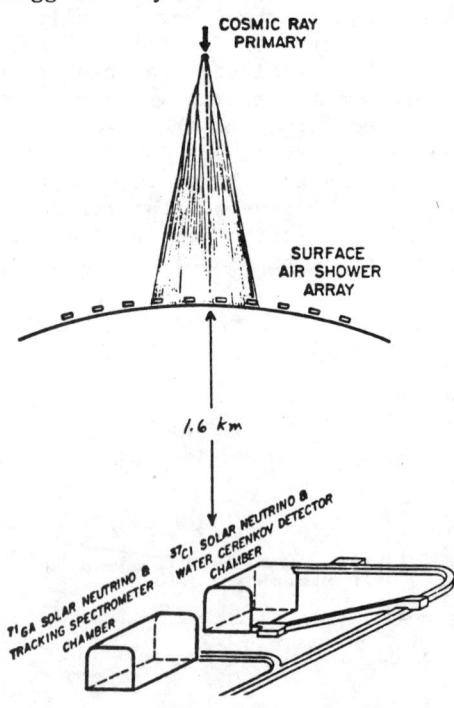

Fig. 10. Planned experimental areas at Homestake, showing the surface array, the existing solar ν-nucleon decay room, and the planned Gallium - Tracking Spectrometer chamber.

detector. They will be used to continue the neutrino burst search
with better resolution than at present, to search for GUTs
monopoles and neutrino oscillations, and to study high energy muons
in conjunction with the surface array, which is scheduled to be
completed during 1983. The Gallium/Tracking Spectrometer room is
shown in more detail in Fig. 11. The full nucleon decay detector
will consist of three stacks, each 8m x 8m x 8m, of scintillator
elements in crossed x- and y-arrays. The key feature of the
detector is the ability to measure the total energy in an event
(either a nucleon decay or a background event) with a resolution of
4%. Fully contained events (approximately 50% of the total number
of events, with no energy deposited in the outer detector layers)
should give a clear 938 MeV peak standing out above the neutrino-
induced background. At the rate suggested by the events seen at
Kolar and Mt. Blanc, we expect 70 real events and 1ν event in our
energy window in a 1-year run. Background is further suppressed by
insisting on momentum balance (within the constraints imposed by
Fermi motion inside the nucleus). In addition to the two large
experiments, there will be space in the room for additional
experiments such as double beta decay. Construction of the new room
will require approximately nine months, during which data-taking in
the original room will continue uninterrupted.

Funding for these experiments has been provided in part by DOE
and by NSF. We are grateful for the extensive and generous

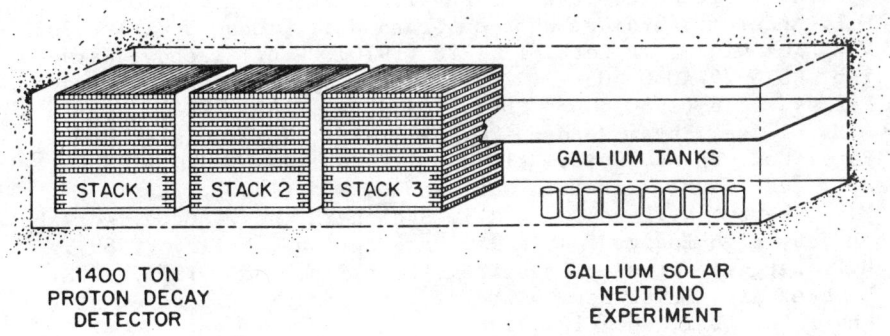

Fig. 11. Proposed new 5000 m^3 Homestake experimental room, with
the full three stacks of the Tracking Spectrometer nucleon decay
detector and the gallium solar neutrino tanks. The new room is to be
built adjacent to the present 2000 m^3 room at a depth of 4200 m.w.e.

support of the Homestake Mining Company management and personnel, who maintain the facilities and freely offer their mining expertise. We also appreciate the advice and essential contributions of numerous people, in particular R. Davis, Jr., B. Cleveland, W. Frati, R. Punkinnen, E.J. Fenyves, A.A. Watson, J. Lloyd-Evans, T.K. Gaisser, T. Stanev, J. Elbert, T. Ashworth, and M. Deakyne.

REFERENCES

1) J.N. Bahcall and R. Davis, Jr., Science 191, 264 (1976); R. Davis, Jr., Summer Workshop on Proton Decay Experiments, Argonne (1982); W. Hampel, Workshop on Science Underground, Los Alamos (1982).
2) M.L. Cherry, I. Davidson, K. Lande, C.K. Lee, E. Marshall, R.I. Steinberg, B. Cleveland, R. Davis, Jr., and D. Lowenstein, Summer Workshop on Proton Decay Experiments, Argonne (1982); and 3rd Workshop on Grand Unification, Chapel Hill (1982).
3) M.L. Cherry, M. Deakyne, K. Lande, C.K. Lee, R.I. Steinberg, B. Cleveland, E.J. Fenyves, submitted to Phys. Rev. D (1982), and ref. 13, p.278; J.W. Elbert, T.K. Gaisser, T. Stanev, submitted to Phys. Rev. D (1982); J.W. Elbert, ref. 13, p.312.
4) M.L. Cherry, M. Deakyne, T. Daily, K. Lande, C.K. Lee, R.I. Steinberg, and E.J. Fenyves, J. Phys. G 8, 879 (1982).
5) M.L. Cherry, M. Deakyne, K. Lande, C.K. Lee, R.I. Steinberg, and B. Cleveland, Phys. Rev. Lett. 47, 1507 (1981).
6) S. Dimopoulos et al., Harvard preprint HUTP-82/A016 (1982); S.L. Glashow, 3rd Workshop on Grand Unification, Chapel Hill (1982).
7) B. Cabrera, Phys. Rev. Lett. 48, 1378 (1982).
8) S.P. Ahlen and K. Kinoshita, submitted to Phys. Rev. D (1982).
9) J.L. Osborne and E.C.M. Young, in Cosmic Rays at Ground Level, ed. by A.W. Wolfendale, Inst. of Physics, London, p.85, 105 (1973).
10) H.L. Anderson, 3rd Workshop on Grand Unif., Chapel Hill (1982).
11) V.A. Kuz'min, JETP Lett. 12, 228 (1970); S.L. Glashow, Harvard rep. HUTO-79/A040,A059 (1979); R.N. Mohapatra and R.E. Marshak, PRL 44, 1316 (1980), and Phys. Lett. 94B, 183 (1980).
12) A.M. Hillas, Phys. Reports 20, 59 (1975); T.K. Gaisser et al., Revs. Mod. Phys. 50, 859 (1978); T.K. Gaisser and G.B. Yodh, Ann. Revs. Nucl. Particle Science 30, 475 (1980).
13) M.L. Cherry, K. Lande, and R.I. Steinberg, eds., Proc. Workshop on Very High Energy Cosmic Ray Interactions, Phila. (1982).
14) H.E. Bergeson et al., Phys. Rev. Lett. 35, 1681 (1975); G.H. Lowe et al., Phys. Rev. D12, 651 (1975); J.W. Elbert et al., Phys. Rev. D12, 660 (1975).
15) K. Alpgard et al., CERN preprint EP 81-152 (1981).
16) J.A. Goodman et al., ref. 13, p. 174.
17) M.L. Cherry, I. Davidson, K. Lande, C.K. Lee, E. Marshall, and R.I. Steinberg, ref. 13, p.356.
18) D.Z. Freedman, D.N. Schramm, and D.L. Tubbs, Ann. Rev. Nucl. Sci. 27, 167 (1977).
19) J.R. Wilson, Phys. Rev. Lett. 32, 849 (1974).

COSMIC RAY PHYSICS UNDERGROUND: SOME PUZZLES

M.R.Krishnaswamy, M.G.K.Menon, N.K.Mondal
V.S.Narasimham and B.V.Sreekantan
Tata Institute of Fundamental Research, Bombay

Y.Hayashi, N.Ito and S.Kawakami
Osaka City University, Osaka
and
S.Miyake
University of Tokyo, Tokyo

ABSTRACT

A series of experiments has been carried out since 1961 at depths of 750 to 8400 hg/cm^2 in the Kolar Gold Mines and the general features of atmospheric muons and neutrinos have been studied. However, there are some problems which are incompletely understood and some of the events observed suggest the existence of new phenomena caused by unknown particles or processes. We point out some problems observed at these great depths which are relevant to the prompt muons in hadronic collisions; the increase of average transverse momentum in the very high energy region, muon bundles, Kolar events, large electromagnetic cascades, and so on.

INTRODUCTION

A series of experiments performed in the Kolar Gold Field (KGF) over the past two decades was a systematic study of cosmic ray muons and neutrinos, and other exotic phenomena and processes at great depths underground. The experimental condition at KGF and a brief sketch of the earlier work follows.

The surface of the Kolar Gold Mines is at a height of ~900 m above sea level at 12.9° N latitude, close to the geomagnetic equator. The composition of Kolar rock known as "hornblende" is somewhat special, it has a mean density of 3.03 g/cm^3, mean Z/A = 0.495, and mean Z^2/A = 6.4. The surface is flat within ~ 20 m over an area several kilometers in extent. The quality of rock is also reasonably uniform over the wide range of area and depth except for a very thin layer at the surface. Owing to the long history of mining at Kolar, many suitable places for cosmic ray observations are available. (see Fig.1)

The earlier experiments (1961 - 64) at 6 depths underground ranging from 816 to 8400 hg/cm^2, measured the intensity of atmospheric muons and their zenith angular distributions at these depths. During 1965 - 70, a larger scale experiment to detect cosmic ray neutrino interactions was carried out at 7000 hg/cm^2, in collaboration with the Durham group, U.K. As a consequence, the general features of muons and neutrinos of cosmic ray origin became fairly clear to a depth of about 10,000 hg/cm^2. During the period 1968 -1973, more precise measurements of intensity and zenith angular distribution of atmospheric muons were carried out at the depths of 754,

1500, 3375 and 6045 hg/cm^2 using neon flash tubes. These experiments showed clearly that muons are produced predominatly through the well-known channels of pion and kaon decay up to primary hadron energies of the order of 10^5 GeV and the contribution due to the decay of every short lived particles is negligible. These experiments have not only provided the variation of muon intensity with depth, but also provided information on multiple muons, rock showers and exotic events of various types. Some anomalous events that cannot be understood in terms of normal processes are called "Kolar events". A search for this type of event was continued from 1975 to 1981 at the depth of 3375 hg/cm^2, with a specially designed detector comprising proportional counters to provide a trigger, and 9 vertical planes of neon flash tubes, interposed with about 10 cm thick iron walls. At the end of 1980, a new detector, shown in Fig.2, started operation. The detector is made up of iron pipes of square cross section, which are arranged in an orthogonal configuration with alternate layers made up of 4 m and 6 m long counters. Each of 34 layers of proportional counters used has a 1.2 cm thick iron plate. Details are shown in Fig.2.

The results of these experiments have already been reported but some of the topics will be discussed in this paper to assess researches which might be appropriate for future experiments deep underground.

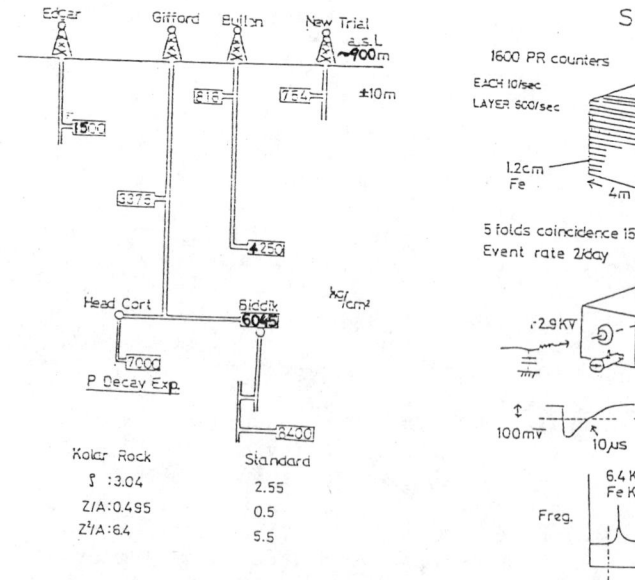

Fig. 1. Schematic diagram of KGF.

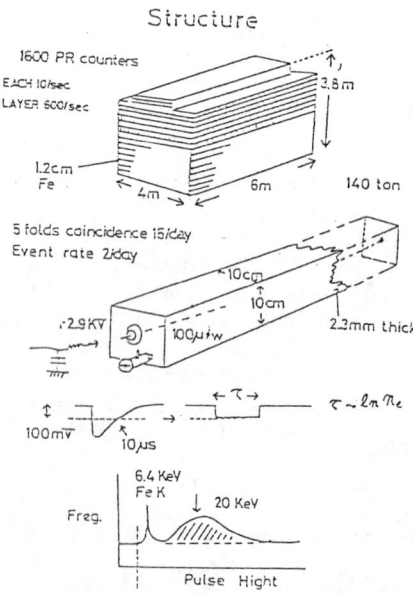

Fig. 2. Detector of proton decay experiment at 7000 hg/cm^2.

1) Prompt Muons

The search in cosmic rays for prompt muons, produced directly in hadronic collisions or through very short lived massive particles is based on the effect such prompt muons would have on the energy spectrum and zenith angle dependence. Although the ordinary cosmic ray muons through well-known channels of pion and kaon decay have an exponent in the energy spectrum steeper than that of primaries by 1 and a sec θ effect in their zenith angular distribution, the prompt muons are supposed to have an energy spectrum which is parallel to that of the primaries and an isotropic intensity independent of their zenith angles.

Fig.3 shows zenith angular distribution observed by the detector in Fig. 2 at 7000 hg/cm².

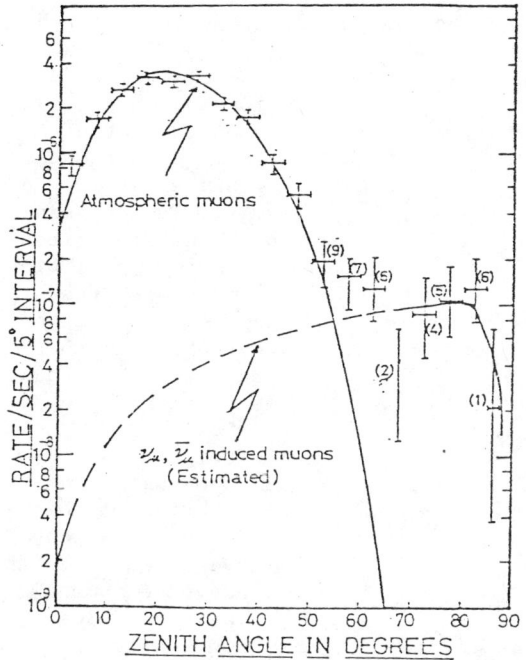

Fig. 3. Zenith angular distribution observed at 7000 hg/cm².

The muons are of two components; one is of atmospheric origin which is distributed over zenith angles $< 55°$ and the other is from neutrino induced muons in the rock. The solid curve in Fig.3 is that expected from ordinary muons taking into account the aperture of the detector. It shows good agreement with observed data up to 55° which corresponds to a slant depth of 12000 hg/cm². It means the empirical formula for the depth-intensity relation for atmospheric muons,

$$I(h,\theta) = \frac{174}{h + 400}[h\sec\theta + 11]^{-1.53}\exp-(8.0\times10^{-4}h\sec\theta), \quad (1)$$

where h, the depth in units of hg/cm², is still applicable at these great depths. A 12000 hg/cm² depth corresponds to 200 TeV if one uses an average energy loss for energetic muons, but the effective threshold energy considering bremmsstrahlung fluctuations is estimated as about 60 TeV to penetrate this thickness of rock.

Since the decay probability of pions at 60 TeV is about 0.2 % and the effect of prompt muons is not significant in this energy region, one can set the upper limit on the rate of prompt muons to pions as less than 0.1 % at an energy of primary cosmic rays of 400 TeV.

The result, which is relevant to the production cross section of charmed mesons, indicates that the dependence on energy is small.

2) Parallel Muons

We measured parallel muons using visual detectors (neon flash tubes) at depths 754,1500,3375,6045 and 7000 hg/cm^2. They have a zenith angular distribution which is similar to that of single muons and, because of higher primary energy, a larger number of muons from first and second collisions reach these great depths. Since the scattering in the penetration of rock is small and geomagnetic effects do not change the opening angle at production, one can use the decoherence curve to estimate the transverse momentum, Pt, at production,

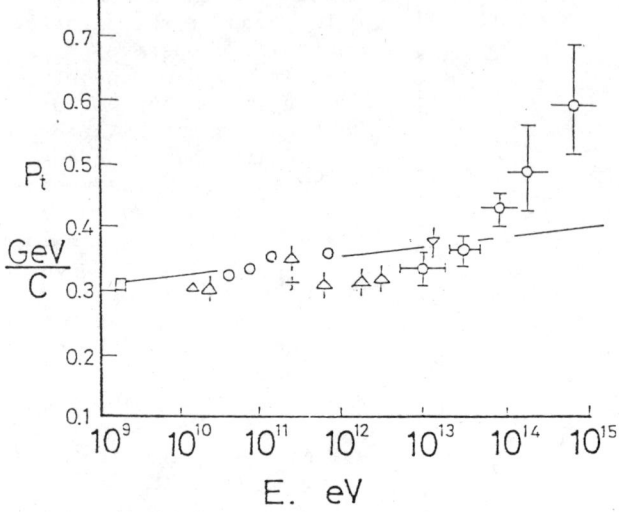

Fig.4. Average Pt vs. primary energy.

assuming their production height is about 20 km above sea level.
The procedure to get Pt from the observed data is as follows ;
a) Observed data, frequency vs separation between two muons, should be corrected according to the detection efficiencies for individual bins.
b) The corrected frequencies are divided by separation,r,in order to get a decoherence curve.
c) A distribution function of the form of $\exp(-r/r_0)$ is deduced from the decoherence curve. Then, Pt will be obtained from the equation ;

$$\langle Pt \rangle = \frac{2\ r_0 \times P}{20\ \mathrm{km}} \qquad (2)$$

where P is an average momentum of parents of muons.

Table I

depth hg/cm^2	ratio to single muons	r_0 (m)	$\langle Pt \rangle$ (MeV/c)	P (TeV/c)
754	2.11 x 10^{-3}	13	390	0.3
1500	4.00 "	6.5	422	0.65
3375	6.40 "	2.4	520	2.2
6045	5.93 "	1.0	600	6
7000	10^{-2} *	1.0	800	8

* area of detector was 24 m^2 (all others are from two sets of 4 m^2)

As shown in Fig. 4, it appears that Pt is rapidly increasing with the energy of the primary particle above 10^{14} eV. It may indicate some change in the interaction characteristics.

3) Muon Bundles

As mentioned, parallel muons were analyzed in the form $\exp(-r/r_0)$ assuming the multiple production of pions to be normal. However, there are special types of muon bundles observed at 3375 hg/cm^2. The events have two anomalies; (1) the event rate is unexpectedly high compared with an extrapolation of low multiplicity events (Fig. 5), and (2) the spread of the essential part of the events is limited to an area of about 2 m diameter. An example is shown in Fig. 6. While the event has well defined single tracks (13), they are only in one half of the detector volume. Therefore, it is difficult to explain them in terms of a conventional model of muon production even if an extreme case of heavy chemical composition of primaries is assumed.

It can be easily shown that large multiplicity events involving high energy muons are produced predominantly by heavier primaries instead of protons. However, the main features of this phenomenon are quite clear in a qualitative sense.

A common feature of these events is the bunching of a majority of muons in a small area, whereas small multiplicity events have a typical separation of 3 - 5 m consistent with a mean Pt of about 500 MeV/c. The small Pt coupled with high multiplicity of these events are the anomalous features that should be understood either in terms of normal primary cosmic ray interactions, or some unknown phenomena.

There are no reports of this type of phenomenon at either shallower or greater depths. It may be best studied at a depth around 3000 - 4000 hg/cm^2 because it may be obscured by other events at shallow depth and absorbed at great depth.

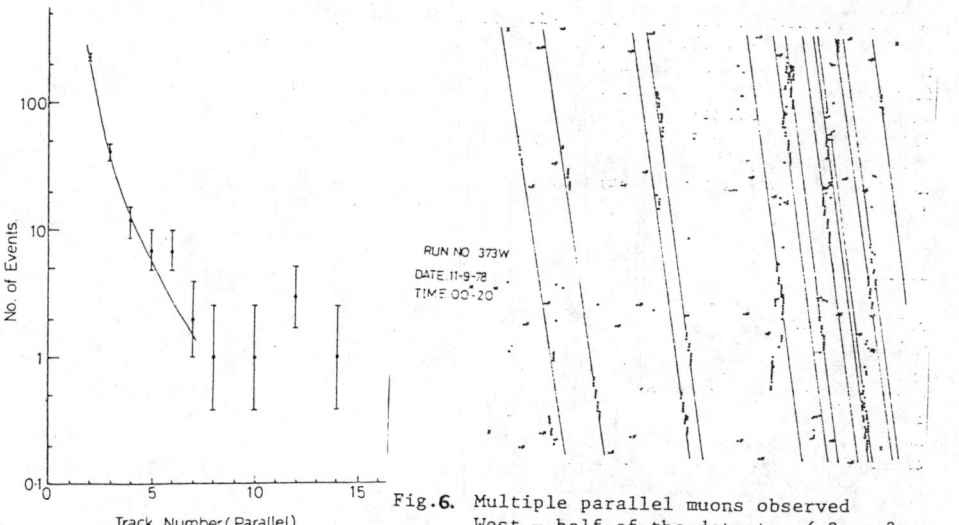

Fig. 5. Number spectrum of muons.

Fig. 6. Multiple parallel muons observed West - half of the detector (2m x 3m, area at top) East - half does not show muons.

4) Kolar Events

In the experiments deep underground, a special class of events has been observed, characterized, in general, by 3 charged particles arising from a vertex in air with large opening angles (Fig. 7). The rate of occurence seems to be independent of depth. Therefore, the most plausible interpretation of these events is that they are due to the decay in flight of new massive and long lived particles produced in neutrino interactions in rock. The new particles have to be massive 2 - 5 GeV, and have long lifetimes of order 10^{-9} sec. There is one special event which shows possible successive decays through two kinds of new particles as indicated in Fig. 8. The particle of the first stage may have a mass of 10 GeV or more and a rather shorter life, of the order of 10^{-10} sec.

There is still no information from accellerater experiments regarding this type of event, but we may be dealing with new phenomena or a completely new situation.

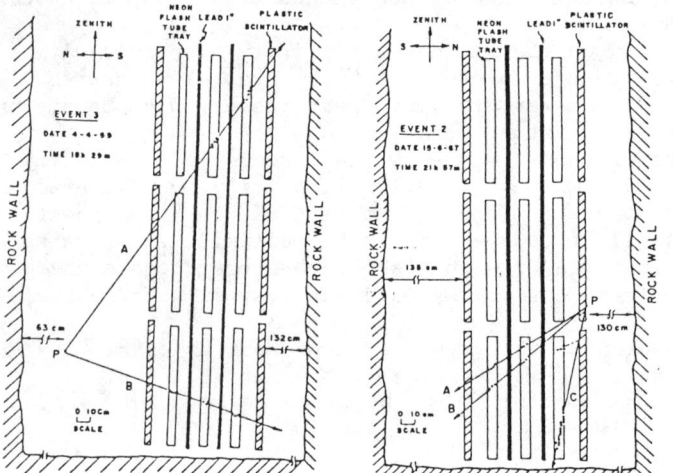

Fig. 7. Two examples of Kolar event at 7000hg/cm².

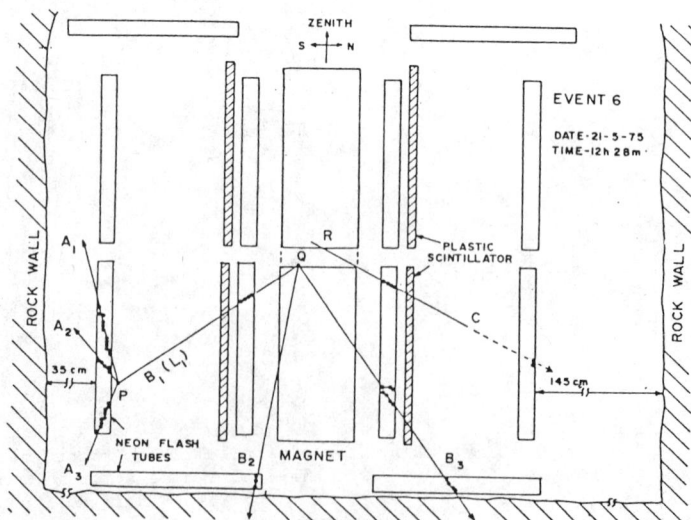

Fig. 8. An example of possible successive decays at 3375 hg/cm².

5) Large Electromagnetic Cascade

In the experiments conducted at a variety of depths, in addition to single muon events, cascades of different sizes were recorded in telescopes comprising lead absorbers, scintillators, and neon flash tubes. Most of these cascades are clearly due to electromagnetic as well as photo-nuclear processes involving muons, but there are some large cascade showers with an energy content more than a few hundred GeV. Although these showers are rare, their frequency is much higher than predicted, and almost independent of depth between 3000 – 8000 hg/cm^2. The angular distribution of the shower is also quite broad, coming even from nearly horizontal directions. These features are difficult to understand in terms of conventional knowledge of the energy spectrum and interaction properties of atmospheric muons as well as neutrinos. However, the exact value of the shower energy was not known because of saturation of the detector.

In a new series of experiments, employing a detector comprising proportional counters of area 24 m^2 at 1840 hg/cm^2 in depth, an unusual cascade has been recorded which may belong to the same category

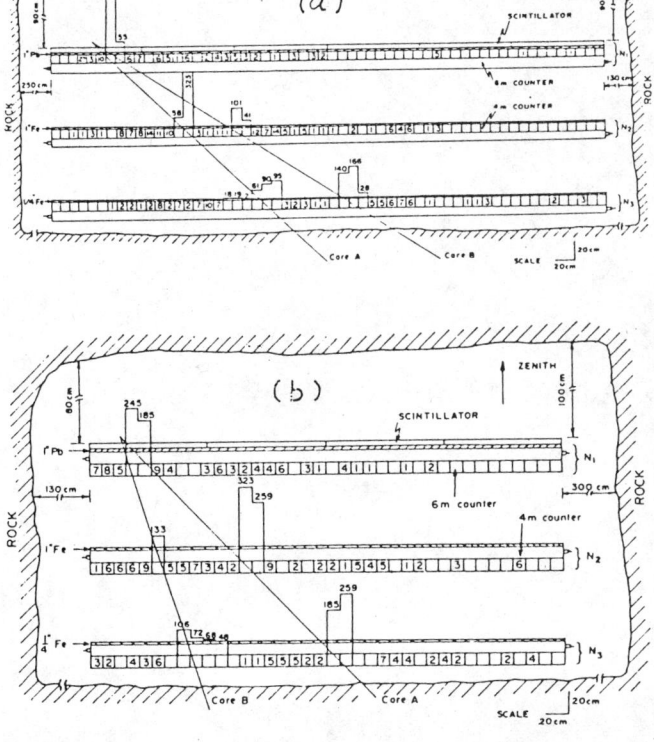

Fig. 9: The orthogonal views of the event in the 4 m and 6 m counters are shown in 'a' and 'b' respectively. The numbers within the squares are the measured particle numbers and the histograms are density profiles.

(Fig. 9). The event has a double core structure of cascade nature with an opening angle of 26°, with energies of about 100 GeV and 80 GeV respectively. Therefore, the relative Pt is about 40 GeV/c.

This unusual event cannot be understood in the framework of any known processes. It is again indicative of the production of very massive particles in lepton-hadron collision.

There is no fixed conclusion in this paper, but it is hoped that this review of current unsolved problems may be useful in discussion of future plans for "Science Underground".

REFERENCES

The following reference numbers correspond to the topic numbers.
2) Krishnaswamy, et al., 15th ICRC Plovdiv Vol. 6 161 (1977).
3) Krishnaswamy, et al., 16th ICRC Kyoto Vol. 13 378 (1979).
4) Krishnaswamy, et al., PRAMANA Vol. 5 211 (1975).
5) Krishnaswamy, et al., 15th ICRC Plovdiv Vol. 6 137 (1977).
   ibid,   16th ICRC Kyoto Vol. 13 14 (1979).
   ibid,   16th ICRC Kyoto Vol. 6 128 (1979).
ICRC - - - International Cosmic Ray Conference

A SOLAR BREEDER TO EXPLAIN CONSTANT LUMINOSITY AND THE LACK OF NEUTRINOS

C. Alexander and A. M. Perry
Oak Ridge National Laboratory, Tennessee 37830

L. H. Levine and L. M. Libby
University of California at Los Angeles, CA 90024

ABSTRACT

A breeder supplying energy in the sun runs for billions of years. Instead of solar neutrinos, it produces antineutrinos from fission products and actinide decays. It gives the sun constant luminosity as required by archaen biology. This follows because reactors are negatively thermostatted.

INTRODUCTION

To keep the sun's energy constant over the last 3.5 b.y., as required by the archaen biological evidence and to explain the observed insufficient flux of solar neutrinos, we suggest that the sun's power is produced by a breeder. Namely, the thick H and He shell formed by neutron and alpha decay covers a low level neutron chain reaction in the solar core. Such a chain reacting core would have a negative temperature coefficient, so that if cooling occurs, the chain reaction increases its rate, keeping the temperature constant. That is, if heating occurs, the core expands and the rate of the chain reaction slows down; and conversely if the system cools, the core contracts and the rate of the chain reaction speeds up, as is well known in neutron reactors.

Further, we offer an explanation why the expected neutrino flux from fusion in the sun is not emitted. In our breeder model for the sun, fission products are formed which emit anti-neutrinos instead of neutrinos, none of which would have been recorded by the experiment of Davis et al. Superheavy elements decay to the breeder.

We propose that light elements have been produced by conventional nucleogenesis mainly on the outsides of stars, and thus they emit very few neutrinos. Fusion may occur in the shell of hydrogen and helium that surrounds the chain reacting core, the hydrogen being formed from neutron decay and the helium from alpha decay as time goes on, so that fission and fusion both contribute to the solar luminosity. We assume a sun of central temperature about 11,000,000 degrees, e.g., Schwarzschild 1958, equivalent to about 1 KeV. The neutron temperatures however are higher on the average, as they slow down from fission energies, having a spectrum equivalent to that in a breeder. We choose the spectrum as that of an LMFBR, referred to in ACE parlance (Circa 1970) as the "Advanced Oxide Breeder Reaction."

THE COMPUTATIONS

1. Notation:

$$P_s = \text{sun's power} \approx 3.9 \times 10^{23} \text{ kw}$$
$$M_s = \text{sun's mass} \approx 2 \times 10^{30} \text{ kg}$$

where subscripts 2,5,8 refer respectively to ^{232}Th, ^{235}U, ^{238}U

$N_8 \equiv \dfrac{M_8}{M_s} \sim$ relative abundance of ^{238}U atoms

$N_5 \equiv \dfrac{M_5}{M_s} \sim$ relative abundance of ^{235}U atoms

$N_2 \equiv \dfrac{M_2}{M_s} \sim$ relative abundance of ^{232}Th atoms

---

$\lambda_8 = 0.154/10^9$ yr = decay probability of ^{238}U
$\lambda_5 = 0.972/10^9$ yr = decay probability of ^{235}U
$\lambda_2 = 0.05/10^9$ yr = decay probability of ^{232}Th

$\mu_i = \sigma_i \phi = \int_0^\infty \sigma_i(E) \phi(E) \, dE$ where σ_i is the neutron absorption cross section of nuclide i (fission + capture)

$\mu_5 = 0.1/N_5$ per b.y.

$\mu_8 = 0.162 \, \mu_5$

$\mu_2 = 0.154 \, \mu_5$

$\mu_5' = 0.224 \, \mu_5$

		barns
^{235}U	σ_f	1.876
	σ_c	0.543
	σ_a	2.419
^{238}U	σ_f	0.047
	σ_c	0.346
	σ_a	0.393
^{232}Th	σ_f	0.0113
	σ_c	0.360
	σ_a	0.371

2. Equations:

$$\frac{dN_8}{dt} = -(\lambda_8 + \mu_8) N_8$$

$$\frac{dN_5}{dt} = -(\lambda_5 + \mu_5) N_5 + \mu_8 N_8$$

$$\frac{dN_2}{dt} = -(\lambda_2 + \mu_2) N_2 + \mu_5' N_5$$

The reactivity constant is:

$$k = \left\{ \frac{2.9 \, N_5}{1.289 \, N_5 + 0.209 \, N_8 + 0.198 \, N_2} \right\}$$

and k must be $\geq$ 1 in order for a chain reaction to occur.

We find that lead and fission products contribute only negligibly to k; Lead has a very small cross section for neutrons and fission products, having a smaller weight, float up out of the core.

The graph, Fig. 1, shows how N_2, N_5, N_8, and k vary with time for a mixture initially consisting of 70% N_8, and 30% N_5. This is the mixture corresponding to extrapolating back today's uranium consisting of 99.3% N_8 and 0.7% N_5.

The critical time, namely the time, T_{crit}, at which k falls below 1, varies with $N_8(t=0)$. For N_8/N_5 = .7/.3 at t=0, the critical time is 4.6 b.y. At higher values of N_8/N_5, T_{crit} increases to as much as 4.8 b.y. In our hypothesis after a couple of billion years a shell of hydrogen and helium forms around the uranium core and begins to contribute to the solar luminosity by fusion. The fusion reactions then contribute the neutrinos found by Ray Davis et al.

UPWARD FLOAT OF FISSION PRODUCTS

The plasma of ~1 KeV acts like a mixture of perfect gases,[2] namely gases of uranium ions, fission product ions, and electrons. Uranium has 12 bound electrons and 80 free electrons; "fissionium" has 4 bound electrons and 40 free electrons. Fissionium gas floats up in the uranium gas much as helium gas floats up in the earth's atmospheric gases, because each has a smaller atomic mass than the masses of the surrounding particles. The effective particle mass of fissionium is 118/(40 + 1) = 2.88 and that of uranium is 238/(80 + 1) = 2.94.

The solar profiles computed by Schwarzschild[1] show that 90% of the energy is produced in the core of $0 \leq R/R_\theta \leq 0.2$. Therefore to the float rate of fissionium we have used conditions at R/R_θ = 0.1 as representative of the core. These are:

ϕ = 76 gm/cm^3; M = 1.99 x 10^{32} gm

n = 1.55 x 10^{25} particles/cm^3 of uranium ions and electrons

R = 6.96 x 10^9 cm

The vertical velocity ω for steady motion is given by eqn (22) of Ref. 3. Taking only the first term, eqn (25), we find a lower limit on the vertical velocity, as follows:

$$\omega \geq (q/C_p)/(dT/dR) : \quad n = 1.6 \times 10^{25} \text{ particle/cm}^3$$

$$C_p = \frac{2.5 \text{ kn}}{\phi} = 7.04 \times 10^7 \text{ erg/gm deg}$$

$$q = L/V = 3.45 \times 10^{33} / 3.0 \times 10^{31} = 1.13 \times 10^2 \text{ erg/g s}$$

With $\Delta T/\Delta R = 1.6 \times 10^{-2}$ deg/cm, it follows that the average velocity of vertical rise is:

$$\omega \geq 1 \times 10^{-4} \text{ cm/sec (with k = Boltzman constant)}$$

Then the time to float out of the core is given by

$$R(core)/\omega = 1.2 \times 10^{14} \text{ sec or } 4 \times 10^6 \text{ years}$$

The number 2.9 in the numerator of k is obtained from 2.54 neutrons per fission in U-238 times $(1 = \sigma_f^8 N^8 / \sigma_f^5 N^5) = 2.54 \times 1.14$.

<u>Remnant Magnetism in Returned Moon Rocks</u> (Ref. 4) indicates that the dynamo which made the Moon's juvenile magnetic field (which magnetised the rocks) decayed with a lifetime of 0.5 b.y. indicating that the power came from an iron-soluble superheavy element with that lifetime. In the sun this superheavy would have been decaying to uranium for thoses 0.5 b.y. during which time the solar uranium breeder would slowly have come into existence.

<u>Nota bene:</u>
The first two authors limit their collaboration to the physics of breeders.

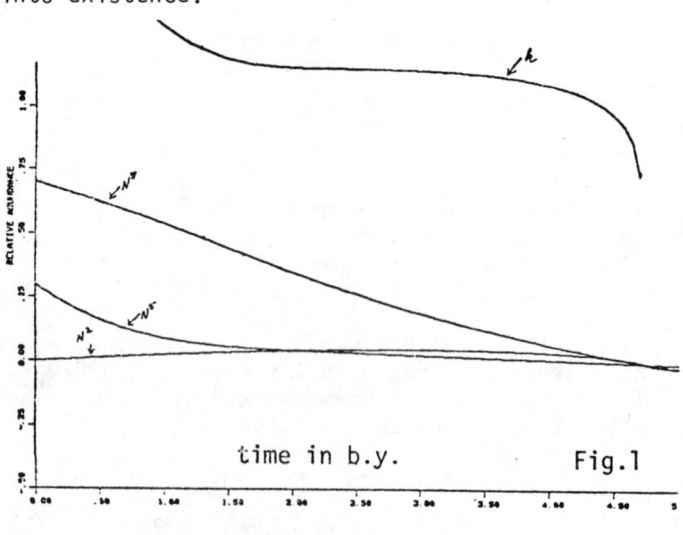

time in b.y.   Fig.1

REFERENCES

1. M. Schwarzschild, <u>Structure and Evolution of the Stars</u>, Princeton University Press, Princeton, N.J. 1958.
2. Private communication from John Marshall, Jr., CTR, LASL, Los Alamos, N.M..
3. G. MacDonald, in <u>Origin and Evolution of the Atmospheres and Oceans</u>, ed. P.J. Brancazio and A. Cameron, John Wiley, NYC. 1964.
4. S.K. Runcorn, New Scientist, October 21, 1982 p 177, Fig. 4.

Chapter V

GEOPHYSICS

Obviously there are no well qualified students of the Earth, and all of us, in different degrees, dig our own small specialised holes and sit in them.

E. C. Bullard, in his response to the award of the Arthur L. Day medal, *Proc. Vol. for 1959, Geol. Soc. Am.*, p. 92 (1960).

SITING, CONSTRUCTING, AND MAINTAINING A DEEP UNDERGROUND SCIENCE LABORATORY

Robert R. Sharp, Jr.
Los Alamos National Laboratory, Los Alamos, NM 87545

ABSTRACT

Experience from a recent site selection study has shown that the important considerations in siting a deep underground science facility are numerous, complicated, and most often based in economics. Even so, there are natural constraints on the maximum depth, size of openings, and construction techniques that can be attained or utilized at any particular site. In general, these are set by the local geothermal gradient, hydrologic regime, and properties of the rock mass or substance that are pertinent to geological, mining, and drilling engineering. At the same time, economics control the type of access, means of dewatering, and methods of sustaining the openings that are most feasible. If major faults can be avoided, sites offering acceptable temperature, a low water table, no outstanding aquifers or badly broken ground above the main openings, and strong, massive, largely impermeable rocks surrounding them at depth are best. Upon completion of a facility, the investment and safety of the occupants demand its rigorous maintenance.

INTRODUCTION

In 1981 the Los Alamos Laboratory took part in drafting a proposal for a National Underground Physics Complex. This involved the location, convenient to its operating environs, of potentially suitable sites for constructing a new facility to provide about 3,500 m water-equivalent (mwe) of shielding for experiments in physics. The required depth was therefore 1,310 m in fairly dense rock, such as granite, or deeper. Other criteria specified: that the facility consist of three large (12 m W x 70 m L x 14 m H and smaller), dry (no dripping water), air-conditioned (to 21°C ambient) research chambers at suitable depth; that use of one or more horizontal entryways (adits), as well as vertical or inclined ones (shafts), be addressed; and that detailed cost and time estimates for exploration, site acquisition, and construction be developed.

The above was done during October, reported that month, and presented later at a physics workshop.[1] Once the initial report was in hand, it was decided that a more detailed one should be provided. It was to set forth all the criteria, discuss the experience with similar structures, give the methods used in evaluation, elaborate on the many geological engineering considerations, and describe all the 14 areas and 25 included sites initially dealt with. The detailed report was to also present the deliberations made in obtaining the cost and time estimates and reaching all conclusions. This second report, too, was completed in 1981. And, while more formally treated therein, the original findings remained unchanged.[2]

In short, the basic conclusions are that: (1) due to problems innate in driving long adits and the length of any that might suffice, a vertical shaft will likely provide the most economical entry if such a facility is to be built anew; (2) in hard, dense, granitic rocks--as exist in four of the areas studied within a 65-km radius of Los Alamos, and elsewhere--large-diameter shafts cannot be economically drilled to the required depths; (3) where lost circulation in drilling softer rocks is a problem--as at Los Alamos, most of the 11 sites studied at the Nevada Test Site (NTS, where the Laboratory also operates), and other places--a 3.66-m ID, conventionally mined and lined shaft will be most economical if otherwise feasible; and (4) while the geothermal gradient, groundwater conditions, and geology in several of the 14 areas treated in the western U. S. would likely prove adequate, a site projected to be in granodiorite, in the southwest corner of Wahmonie Flat, at the Nevada Research and Development Area of NTS, appears most promising from the standpoint of estimated cost and time to completion. Therefore, a 1400-m mined shaft, of the above diameter and type, sited at Wahmonie Southwest--as the prime choice--or 200-300 m deeper in tuffs at one of three locations nearby, is recommended in both reports. Due, however, to the high cost of gaining access to the cover required for the facility, if built anew, a survey of possibly available mine shafts and tunnels in the western U. S. is also recommended.[1,2]

Since 1981, new developments have occurred. Interest in a deep facility has been shown by scientists outside the realm of particle physics. And a search for mined space in New Mexico and adjacent states has presented candidates that meet a good part of the original criteria but are generally less deep than desired. These factors, along with time constraints and escalating costs, have made it expedient to attempt to proceed with a possible phased approach regarding the ultimate depth of the facility. This paper will therefore briefly address siting, constructing, and maintaining a deep underground science facility in general, but will use the criteria and siting of the one above as an example.

SITING A DEEP FACILITY

In siting a deep underground complex, it is important first to critically evaluate the criteria the facility is to meet. These should specify: the acceptable regions for its location; its desired life and interior quality; the general depth, number, size, and disposition of all openings to be required; the type and amount of utilities to be used; the allowable temperature and humidity levels; and any special needs. The criteria should be flexible where possible, account for all anticipated uses, and allow for the maximum size, weight, and safety requirements of all components of equipment to be accommodated. Most importantly, constraints must be set on future modifications or additions, because every critical property of the ground may undergo complex changes in three dimensions. Also, in correctly judging whether or not the criteria can reasonably and economically be met, and in choosing the most advantageous site, a knowledge of the following four aspects is imperative: (1) the past experience with similar structures; (2) the general considerations pertinent to their siting; (3) the geological engineering constraints that can be anticipated; and (4) the engineering requirements and relative costs of the methods available for their construction.

Underground facilities with openings of a depth or size similar to our example include other research installations akin to it, deep mines or tunnels, and underground powerhouse or special-use chambers. In all of these, the loading at depth is--initially, at least--assumed hydrostatic and due to gravity.[3] Laboratories requiring appreciable shielding are being actively pursued in France,[4] Italy[4,5] and the Soviet Union.[6,7] Earlier, such facilities were established in South Africa, India, Switzerland, and the U.S.[2,4] All of these will be, are, or were at depths of 1,400 m or greater (providing about 3,740 mwe or more of shielding in average crustal rocks). Some are or will be far inside tunnels; and others have been or are in the lower levels of mines--though, at present, none at such depth have completed chambers of the size required by our example.[4-7]

Many mines and some tunnels have been carried deeper than 1,400 m. In the three deepest mining districts of the world, the Witwatersrand, South Africa, the Kolar Gold Fields, India, and Morro Velho, Brazil, mines go to 3,800 m,[8] nearly 3,050 m,[9] and 2,590 m,[10] respectively. To reach such depths, all have used conventional shaft sinking. But this, too, can prove troublesome--particularly when going much below 1,500 m.[11] And excess groundwater has sometimes been a major problem.[12] Yet, the critical depth of mining is generally set by where the rock temperature--increasing by an average but variable 1°C per 30 m under continents[13]--reaches about 57°C, rather than by inability to seal or sustain an opening.[2] Even so, notable differences exist between driving deep tunnels or mining deep ore, and constructing large, permanent openings at depth. Deep tunnels, while designed for long life, are kept to the minimum allowable width (or least span), take great distance to attain depth, demand expensive exploratory drilling along their surface alignment, are susceptible to large influxes of water and unstable ground (at near-vertical faults and fracture zones, that may go undetected until encountered), and require massive linings that are costly.[14] And in deep mines, aside from the shafts and haulageways, generally of small span, the economics hardly allow long-term sustainment of the immense, irregular, ore-depleted workings that accrue. Instead, their backfilling or controlled failure is the rule.[15,16]

Thus, where large, permanent-type openings are concerned, the size--and particularly the least span in the horizontal--is also critical. Because, in a given stress field, while the magnitude of stress concentrations around an opening depends solely on its shape (in elastic theory, at least),[17] the volume of ground subjected to the concentrated stress and probability of encountering serious flaws in the wallrock increase with size.[18] So undergound powerhouses and the like provide closer analogies to our science facility in their shape, size, method of construction, and life, albeit few are quite so deep. A survey of such chambers shows more than 240 to have been built; and roughly half are as large as, or larger than, in our comparative case. Their openings usually have least spans in the range of 10-30 m, lengths of 60-180 m, and heights of 15-45 m. And though some are markedly longer or higher, few are any wider. Most have near-vertical walls with arched crowns and were mined in hard, strong, only moderately broken, crystalline rocks, at depths less than 300-900 m.[19]

As for deeper or larger openings of similar type, an 18-m spherical room was mined at 1,750 m in the volcanics of Amchitka Island, Alaska; smaller, cylindrical ones were made at 1,200-1,500 m in the tuffs of NTS; and a study of flat-floored, hemispherical chambers, 45-90 m across and at 600-900 m, found their construction in weaker rocks at NTS would be costly--but within the technical state-of-the-art.[2,20] So, while chambers of the depth and span in our example fall within past experience, there are limits on these parameters. And beyond their dimensions for these criteria, or where groundwater is excessive, the unit costs accelerate.[11,20,21]

The general considerations pertinent to siting an underground laboratory are: climatic suitability for all-season access; locational convenience to quarters, transport, and services; project acceptance by the local populace; land tenure, acquisition, and access outlays; availability of utilities; the time and expense of preparing an environmental impact statement where none exists; the advent of labor problems; and overall cost. Some of these are critical and others not, but all are important. And, referring to our example, while the Wahmonie Southwest site may be seen as lacking some convenience, it scores high in all other general categories.

As for the geological engineering constraints in areas appearing otherwise inviting, these relate to the availability of local geotechnical data and deep drilling and mining experience, to the favorability of local rock types, and to the adequacy of local geothermal, hydrologic, and other geologic conditions. Acquisition of data and experience is costly and any available will provide savings. Strong, massive rocks of low permeability are likely to be most favorable, particularly at depth. The thermal gradient can be above average, as in much of the western U.S.,[22] and may well be critical. Sites with shallow or abundant water should be avoided. Appreciable groundwater can occur in most rocks down to about 300 m, in good aquifers and broken ground to perhaps 1,000 m, but only in the best aquifers and along major faults to depths much greater.[23] And large dikes, folds, and chemically altered zones, as well as other geologic structures, can adversely influence underground openings.[3,9,17,18] Thus, if some local data and experience are available and major faults and adverse structures can be avoided, sites having a likely adequate temperature, a low water table, no outstanding aquifers or badly broken ground above the main target horizons, and strong, massive, largely impermeable rocks surrounding the deep zones of excavation will provide the best selection. These, it happens, are the indicated conditions at the prime and alternate sites recommended in our example.[2]

Because of their often intimate relationships, the combined effects of such engineering factors as the likely rock failure mechanisms, possible in situ stress conditions, probable stress concentrations about the openings (and the optimum shape, spacing, and layout that result for them), various rock properties in addition to strength and permeability, and the advent of noxious natural gases--as well as excess groundwater--are likewise important considerations.[3,9,17,18] So, too, are the risks of natural and manmade hazards, such as from floods, landslides, earthquakes, volcanism, and construction blasting or local activities.[2]

In any event, exploratory drilling, coring, and geophysical and hydrologic testing will be required along the entire alignment of all entryways, as well as around the main openings. For this and other reasons, the shortest route of access below ground is of great benefit. A vertical shaft will provide the shortest route, can be more readily sited to avoid shallow water and high-angle faults, can follow a single exploratory hole to depth along the same alignment, and should require the minimum long-term water removal. Water inflows of 15-25 lps at the bottom can be handled in conventional shaft sinking and lining without extreme delays, but greater amounts require grouting.[21] An adit will be much longer, more susceptible to shallow water, more likely to encounter high-angle faults, and will demand more exploratory drilling--although it can be drained by gravity if sloped slightly up from the portal. An inclined shaft provides the intermediate case in regard to length and encountering faults, but lacks the advantages of both other alternatives and is more troublesome and costly to sink, line, and maintain.

Also, while it may appear that any single type of accessway, or some combination, might be mechanically driven at a cost savings, this is not necessarily so. Tunneling machines are costly, are limited in the range of rock hardness, type of ground, and inclination they can economically handle, and are usually designed and built for lengthy alignments under favorable circumstances.[15,24] Large, vertical drilling rigs will allow a shaft to be drilled, cased, and sealed against water remotely. But they, too, have limitations; and here, most of their attraction fades on recognizing that much of the equipment to conventionally mine a shaft will be required at any rate--to excavate the chambers.

So, in siting a deep laboratory, all of the above considerations must be thoroughly and quantitatively addressed. Available large-scale maps and regional data on the allowable areas are first assembled and evaluated to determine factors that would, a priori, render any unsuitable. More detailed maps and reports on the possibly feasible remaining ones are examined next, in light of the above, to find potential sites. The more promising are then studied in greater detail and examined in the field. Following this, site-specific maps, geologic cross-sections, and preliminary conclusions regarding the likely optimum location, type, depth, shape, size, spacing, and layout of workings for all the best remaining candidates are made, and inputs for their initial cost estimates are formulated. Once the respective costs are estimated, the still-contending sites can reasonably be compared. Finally, after increasingly detailed iterations of the last process, and one or more of the best appearing sites is drilled, cored, tested, and extensively studied in the field, as well as thoroughly assessed on the basis of firm, locally pertinent geological data and engineering studies, a specific site can be chosen. None of the sites recommended in our example has yet been drilled, though many pertinent data concerning each are available. And the Wahmonie site remains the prime choice.

CONSIDERATIONS IMPORTANT IN CONSTRUCTION

Most considerations important in constructing a deep facility should be specified in its design. But much of the exploration, hydrologic, and rock mechanics work needed to economically select and size many of the construction components required must be done along with the excavating.[3,11] For this reason, a contingency is included in the cost estimates.[2] And the choice of all pumping, cooling, ventilating, and mining equipment--along with the headframe and hoist designs, if a shaft is used--should be made with their versatility as well as efficiency and safety in mind.[11,15] Whatever the type of access, increasingly pertinent data on the nature of the ground, the hydrologic conditions, and the effectiveness of different construction methods will become available as the project develops.[14] These data should be continuously evaluated and applied as the work progresses,[17] and be formally recorded for possible future use.[23]

All water encountered must be sealed off or drained and removed. If appreciable inflows under high head are to be sealed off, thick concrete liners must be installed.[5,14] On the other hand, if the water table is deep and the permeability beneath is relatively low, drainage of a slow influx is not a major problem.[21] For example, a circular shaft under the latter conditions can be efficiently lined with prefabricated steel in ring-sections above the water table and sealed by concreting downward to provide a sump below the bottom shaft station.[2] Then, what little water enters the main workings can be drained to the sump and pumped upward to the surface through stations staged along the shaft.[21] Prior to mining off the shaft, however, exploration should be done by pilot holes and drifts.[15]

At depths below about 900 m, a slabbing failure, caused by lateral tension induced by compression near the free-face of openings,[9] generally becomes prominent.[23] The importance of acknowledging this is that only a small lateral force will greatly strengthen the rock in which it occurs (even if already broken),[9] and it must be mainly the rock that supports the openings.[17,18] If allowed to develop, the fracture zone around an opening at depth is generally circular or elliptical--due to the natural distribution of concentrated stress[3]-- and may extend as far above or even below as the opening is wide.[16] However, if sharp corners are avoided and smooth blasting is used,[25] rockbolting over wire mesh,[18] first into the crown and then progressively into the sidewalls, bottom edges or springline, and across the floors, as the openings are expanded downward, can radically curtail development of a fracture envelope.[17] And, if done properly, it will also inhibit movement along pre-existing fractures and prohibit the fall of free-hanging wedges. Thus, with only a substantial concrete floor-slab, and several centimeters of gunite or shotcrete sprayed elsewhere over the wire mesh and around the rockbolt bearing-plates (to inhibit rock deterioration from cyclic moisture changes), well sited chambers of appreciable size can be stabilized.[2,18] Or, if required, more costly cable-tendons, strung through drillholes to galleries surrounding the large openings, can be used.[20]

MAINTAINING A DEEP FACILITY

Once the facility is operational, protection of the investment and personnel safety must be insured. Only certified engineers, operators, or maintenance workers, all with clearly defined responsibilities, should have access to the controls or working components of the mechanical and electrical plants. Likewise, only the scientists or technicians should control the research gear. All reserve power units, cables, hoists, hoist controls, ropes and sheaves, headframes, cages, guides and catches, dewatering pumps, drains, and piping, as well as the lighting, air conditioning, ventilating, rock mechanics monitoring, communications, safety, and other facility equipment and linings should be inspected daily and subjected to periodic maintenance.[14,15] Strict measures should be used against corrosion. Lastly, all wastes must be removed to the surface daily.

REFERENCES

1. A. Mann and R. Sharp, Jr., in Proc. Workshop on Proton Decay, June 1982, Argonne NL, Chicago, D. Ayers, Ed. (in press).
2. R. Sharp, Jr., R. Warren, P. Aamodt, and A. Mann, Rept. LA-UR-82-556, Los Alamos NL, NM (1981).
3. E. Isaacson, Rock Pressure in Mines (Min. Pubs., London, 1960).
4. See appropriate papers by J. Ernwein, A. Zichichi, P. Picci, S. Miyake, R. Davis, or K. Lande, these proceedings.
5. A. Zichichi, Rept. INFN/AE-82/1, Inst. Nuc. Phys., Rome (1982).
6. A. Pomanskiy, Atomnaya Energiya, $\underline{44}$, 4, 376 (1978).
7. Yu. Lapin, Science in The U. S. S. R., 1, 100 (1982).
8. B. Hessian, Mining Survey, Chmbr. Mines S. Afr., 1, 33 (1982).
9. S. Woodruff, Methods of Working Mines (Pergamon, London, 1966).
10. C. Townsend, Papers and Disc., 1964-65, AMM S. Afr., 675 (1966).
11. J. Redpath, in Proc. N. Am. Rapid Excavation and Tunneling Conf., June 1972, Chicago, $\underline{2}$ (AIME, NY, 1972), p. 843.
12. A. Cartwright, Ordeal by Water (J. G. Ince, Johannesburg, 1969).
13. A. Cook, Physics of the Earth (Wiley, NY, 1972), p. 163.
14. K. Szechy, The Art of Tunnelling (Akad. Kiado, Budapest, 1967).
15. W. Hustrulid, Ed., Underground Mining Handbook (AIME, NY, 1982).
16. E. Leeman, Papers and Disc., 1958-59, AMM S. Afr., 357 (1960).
17. L. Obert and W. Duvall, Rock Mechanics (Wiley, NY, 1967).
18. D. Coates, Rock Mechanics Principles (Info. Can., Ottawa, 1967).
19. E. Hoek, Intl. Jrnl. Rock Mechs. Min. Sci., $\underline{12}$, 12, 381 (1975).
20. T. Kipp and R. Kennedy, Rept. 4723T, DNA, Washington, DC (1978).
21. K. Barnhill, in Proc. First Symp. on Mining Techniques, May 1982, NMIMT, Socorro, NM (in press).
22. A. Kron and G. Heiken, Map LA-8467, Los Alamos NL, NM (1981).
23. H. McKinstry, Mining Geology (Prentice-Hall, NY, 1948), p. 523.
24. U. S. NRC, Rept. NRC/CTES/TT-82-1, NTIS, Springfield, VA (1982).
25. U. Langefors and B. Kihlstrom, Rock Blasting (Wiley, NY, 1978).

SUBTERRANEAN GRAVITY AND OTHER DEEP HOLE GEOPHYSICS

Frank D. Stacey
Physics Department, University of Queensland,
Brisbane 4067, Australia

ABSTRACT

The early history of the determination of the Newtonian gravitational constant, G, was closely linked with the developments of geodesy and gravity surveying. The current search for non-Newtonian effects that may provide an experimental guide to unification theories has led to our retracing some of this history. Modern geophysical techniques and facilities, using especially mines and deep ocean probes, permit absolute measurements of G for distance scales up to a few kilometers. Although the accuracy of the very long range determinations cannot equal that of the best laboratory measurements, they are crucial to assessment of the possibility of a scale dependence of G. Preliminary data give values of G on a scale 100–1000 m biased about 1% higher than the laboratory value. Possibilities of systematic error compel us to question this apparently significant bias but it provides the incentive for better controlled large scale experiments. Several are in progress or under development. A particular difficulty concerns the measurement of in situ density. Even for hard rock, release from overburden pressure causes microcracks and pores to open. Natural pore closure is effective only with deep burial and for this reason there are advantages in deep instrument placement for several geophysical studies.

INTRODUCTION : NEWTON'S LAW AND THE GRAVITATIONAL CONSTANT

At the distances of planets and man-made space probes the inverse square law of gravitational attraction remains unassailable. If we write the index of Newton's law as $(2 \pm \delta)$ then $\delta < \sim 10^{-7}$ for ranges of 10^6 m to 10^{11} m. The accuracy with which the inverse square law has been checked on the laboratory scale is, by comparison, very poor indeed. Conversely the value of the constant G in the familiar Newtonian law for the attractive force between point or spherical masses:

$$F(r) = -\frac{G m_1 m_2}{r^2} \qquad (1)$$

is known from several laboratory measurements to a few parts in 10^4, but there are in principle no data that permit determination of its value on a planetary scale. According to a comprehensive 1982 bibliography on G circulated privately (Dr. G.T. Gillies, Bureau International des Poids et Measures, Pavillon de Breteuil, F-92310, Sèvres, France) three recent determinations of G give standard deviations of about 1 part in 10^4 but embrace a total range of 5 parts in 10^4. A recent N.B.S. redetermination[1] gives $G = 6.6726(5) \times 10^{-11}\,m^3\,kg^{-1}\,s^{-2}$. This coincides with the 1973 CODATA value[2] but with an 8 fold smaller uncertainty. Astronomical and geophysical calculations on masses of stars, planets or geological bodies and theories of internal structures all assume this value of G.

The possibility that the laboratory and planetary scale values of G do not coincide has been considered in several papers in the past decade or so. Theoretical reasons for postulating a difference are reviewed by Gibbons and Whiting[3]. Most such theories favour the addition to normal Newtonian gravitational potential of one or more terms of Yukawa form, so that the total potential due to a point mass m at distance r becomes (with a single additional term)

$$V = -\frac{G_\infty m}{r}\left(1 + \alpha e^{-\mu r}\right) \qquad (2)$$

where G_∞ is the Newtonian constant at large distances, α is the magnitude and μ^{-1} is the range of the Yukawa term. Precision of the inverse square law at large distances is ensured if $\mu^{-1} < \sim 10^5\,m$, so that the second term vanishes. If $\mu^{-1} > 10\,m$, then at laboratory ranges, much shorter than this, $\mu r \ll 1$ and gravitational force will be related by a "constant"

$$G = G_\infty\left[1 + \alpha(1 + \mu r)e^{-\mu r}\right] \sim G_\infty(1 + \alpha) \qquad (3)$$

which could indeed be indistinguishable from a constant in spite of a significant value of α because

$$\frac{d \ln G}{d \ln r} \sim -\frac{\alpha(\mu r)^2}{1 + \alpha(1 + \mu r)} \qquad (4)$$

Thus there is a wide range of values of μ^{-1} for which the combination of laboratory and planetary data impose no effective constraint on the possibility of a non-zero value of α. Geophysical measurements, on scales up to 1000 m or so, fill a large part of this gap.

Mikkelsen and Newman[4] examined the constraints to G_∞ imposed by the radius of gyration of the Earth and plausibility arguments concerning its density and by models of the solar structure and concluded that a difference from the laboratory value of G, here referred to as G^*, up to 40% was permitted. Some other authors[5,6] have used theories of stellar structure to infer somewhat tighter constraints, but reliance on such theories appear less than secure. As Gibbons and Whiting[3] assert, the agreement between $G_\infty M_{Earth}$ from satellite data and surface gravity is much better than Mikkelsen and Newman[4] admit and may be as good as 1 part in 10^6. However, this only imposes a mutual constraint on α and μ^{-1}, since any difference is due only to the effect on surface gravity of the mass within a hemisphere of radius μ^{-1}. As McQueen[7] points out no other gross Earth data offer useful evidence. A revised value of G_∞ and consequent inverse revision of M_{Earth} would require similar rescaling

of internal density and elastic constants, leaving seismic
velocities and internal gravity unaffected. No adjustments to free
mode theory or equations of state for the Earth's interior would be
required unless $\mu^{-1} \gg 10^4$ m and even then the theory could
accommodate the anomaly without recognizing it.

The report by Long[8] of experimental evidence for a slight
range dependence of G at laboratory distances (3 cm to 25 cm)
stimulated a flurry of attempts to check his result. With a
recently communicated numerical revision he represents the
dependence as $d \ln G/d \ln r = 1.3 \times 10^{-3}$. Considered as an
indication of a short range force superimposed upon normal Newtonian
gravity, this result implies that the short range force is repulsive,
that is G increases with distance. Several reports[9,10,11,12] claim
to refute this result, but all except one[9] admit experimental errors
far too large to provide any test for the small effect reported by
Long[8]. Spero et al[9] reported a null experiment which sought a
gravitational force on a test mass suspended from a Cavendish
balance so that it hung within a long uniform hollow cylinder that
could be moved about. The absence of an attraction or repulsion by
the wall of the cylinder is evidence for precision of the inverse
square law which is the only force law giving zero gravity inside a
spherical or suitable ellipsoidal shell or an infinite cylinder.
The reported null result imposed a limit to non-Newtonian gravity
more than 10 times smaller than the effect reported by Long, who has,
however, countered that a null experiment cannot in principle test
for non-Newtonian gravity[13]. The argument is that an explanation
for a non-Newtonian effect must be sought in the theory of vacuum
polarization which modifies the Newtonian field but does not cause a
force where no Newtonian force exists. The conventional view[3] is
that vacuum polarization does not have the effect claimed by Long[13].

The controversy has stimulated the interest of experimenters.
There are at least 10 experiments in progress or under development
that are designed to obtain absolute values of G over the range of
scales represented in Fig.1. Most are intended specifically to
examine the possibility of a scale dependence of G. In this
connection we regard experiments over the conventional laboratory
range ($\sim$ 10 cm) as short range experiments. Geophysical measurements
over distances of order 1 km are identified as long range
observations. Oil tank experiments by Yu et al[11] and the Queensland
lake experiment, using distance scales of order 10 m, which is the
geometric mean of the short and long ranges, are referred to as
intermediate range experiments.

GRAVITATIONAL FIELDS OF SHELLS AND EXTENSIVE SLABS

Almost all the G determinations to date have relied upon
measurements of attractive forces between spherical or other simple
but approximately equidimensional masses. In this respect the
larger scale experiments in Fig.1 depart from convention by using
extensive layers as the attracting masses. This geometry has some
important advantages. It greatly simplifies the necessary metrology

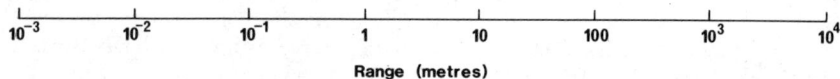

Fig.1: Experiments to determine G in progress or under development. These cover the total range of scales that are accessible by currently contemplated techniques.

and it avoids almost completely corrections for gravitational coupling to parts of the apparatus other than deliberately attracted masses. It happens also that correction terms for the finite extent and curvature (sphericity) of a layer are very simple. Since these features are common to several experiments, the principles are presented here.

As mentioned in connection with the experiment of Spero et al[9], the gravitational field of a uniform spherical shell is zero inside the shell. It follows that if we assume the Earth to be a spherically symmetrical body of radius R and total mass M, then at distance $r < R$ from the centre, i.e., at depth $z = (R-r)$, the gravity is due only to the mass $(M - \Delta M)$ inside r. If G is regarded as unknown then M is also unknown, but the difference in gravity, g, between the surface and a point at depth z can be expressed in terms of ΔM without knowing M:

$$R^2 g(R) - r^2 g(r) = G \Delta M \qquad (5)$$

As Airy[14] appears to have been first to point out, this permits G to be determined from measurements of gravity at depth and of the densities of the overlying layers. Current large scale experiments on G are revivals of Airy's idea, which is clearly the best founded of the geophysical approaches.

The vertical gradient of gravity in the spherical earth model is

$$\frac{dg}{dz} = \frac{2g}{r} - 4\pi G \rho = 4\pi G \left(\frac{2}{3}\bar{\rho} - \rho\right) \tag{6}$$

where ρ is the density of the layer through which the gradient is measured and $\bar{\rho}$ is the mean density of the Earth (strictly the mean inside r). Although the first form of equation (6) is used to determine G, the second form has an interesting practical implication. If we could make measurements within a layer where $\rho = \frac{2}{3}\bar{\rho}$ then the gravity gradient would be zero and a precisely calibrated gravity meter would not be required; it would simply serve as a null instrument to detect the absence of a change in g. This is not very far from being the situation in mines and boreholes, in which the gravity gradient is typically $dg/dz = \sim 0.7 \times 10^{-6}$ s^{-2} (0.07 mgal/m) compared with the free air gradient above the Earth's surface 3.086×10^{-6} s^{-2} (0.3086 mgal/m). This is one respect in which measurements in mines have an advantage over measurements in the sea, although the much better determined density of sea water is an important contrary consideration.

Equation (5) is readily generalized to give the gravity difference between depth z and the surface in the ellipsoidal rotating Earth[15,16]. To a sufficient approximation for the present purpose we write

$$g(z) - g(o) = U(z) - 4\pi G X(z) \tag{7}$$

where $U = 2\frac{g_o}{r_o}z\left[1 + \frac{3}{2}\frac{z}{r_o} - 3J_2\left(\frac{3}{2}\sin^2\phi_o - \frac{1}{2}\right)\right] + 3\omega^2 z(1 - \sin^2\phi_o)$

$$X = \frac{c}{a}\left[1 + 2\frac{z}{r_o} + \frac{1}{2}\left(1 - \frac{c^2}{a^2}\right)\right]\int_o^z \rho \, dz - \frac{2}{r_o}\int_o^z \rho z \, dz$$

$J_2 = 1.08264 \times 10^{-3}$ is the coefficient of ellipticity of the Earth's mass, $c/a = 0.9966472$ is the ratio of polar to equatorial radii, ϕ is geocentric latitude, $\omega = 7.292115 \times 10^{-5}$ rad s^{-1} is the rotation rate and subscript zero indicates a surface value. The original derivation of equation (7) assumed that internal surfaces of equal density all had the same ellipticity but Dahlen[16] derived the same equation without this assumption. Equation (7) has the same form as equation (6) but with corrections for the second order effects of ellipticity and rotation.

It is important to note that equation (7) refers to the densities of the layers in the immediate vicinity of a gravity profile and is unaffected by crustal structure at distances much greater than z. The same principle applies to the smaller scale situation of plane layers or slabs that are extensive but not strictly infinite, or to spherical caps.

The gravity due to an infinite plane layer of thickness t and uniform density ρ is

$$g = 2\pi G \rho t \tag{8}$$

and is independent of position, including distance from the layer. Thus if such a layer can be even approximately realized, its gravitational force on a test mass is unaffected by the shape or homogeneity of the test mass. Further, the only dimension that must be measured precisely to use such an arrangement to determine G is the layer thickness, t. This is the basis of the Splityard Creek gravity experiment[17] which will use layers of water in a lake that will undergo frequent level changes in the course of its use as a hydroelectric pumped-storage reservoir. We can also identify equation (8) with the second term of equation (6) because by (8), the opposite gravity on opposite sides of the layer gives a gravity difference $4\pi G \rho t$.

In generalizing equation (8) to a finite layer of irregular outline, it is convenient to consider first an elementary disc of radius r, thickness dt and density ρ. At a point P at distance $a \ll r$ from the disc along the axis, the gravity of the disc is

$$dg = 2\pi G \rho dt \left[1 - \frac{a}{(r^2 + a^2)^{\frac{1}{2}}} \right]$$

$$= 2\pi G \rho dt \left[1 - \frac{a}{r} + \frac{1}{2} \frac{a^3}{r^3} + \ldots \right] \qquad (9)$$

It is important to the analysis that there is, in this expansion, no term of second order in the small quantity a/r. In the situations envisaged the third order term is negligible. Then we can write the gravity difference between P and another point Q at distance b along the axis on the other side in terms of the total separation $z = (a+b)$ of these two points

$$d(\Delta g) = 4\pi G \rho dt \left[1 - \frac{1}{2} \frac{z}{r} \right] \qquad (10)$$

The fact that this gravity difference does not depend upon a and b individually simplifies the integration over a disc of any thickness t between P and Q. Further, if we restrict our interest to the component of gravity along PQ, the geometry of the attracting mass can be generalized in a simple way to any arbitrary shape between two parallel planes normal to PQ:

$$\Delta g = 4\pi G \rho t \left[1 - \frac{z}{2} <r^{-1}> \right] \qquad (11)$$

where $<r^{-1}> = \frac{1}{2\pi t} \int_0^{2\pi} \int_0^t \frac{d\theta \, dt}{r}$ is an integral over the thickness range t and all azimuths θ in the plane normal to PQ. In the lake experiment $<r^{-1}>$ for any depth range can be determined from the precisely known geometry, but can also be measured gravitationally using different separations z of the test masses. Of course it appears only as a correction term because $z<r^{-1}> \ll 1$.

Replacing the plane layer by a spherical cap of radius of

curvature R >> r, the largest term added to the square bracket in equation (10) is (b - a)/R. If R is the Earth's radius, the effect of curvature, for example of the lake surface, is negligible and it is appropriate to use plane geometry. In fact this is still true for the largest scale geophysical experiments at the accuracy so far achieved.

GRAVITY IN MINES, BOREHOLES AND THE DEEP SEA

The early work on gravity in mines by Airy[14] and R. von Sterneck was directed to the determination of G, but toward the end of the last century it became so obvious that laboratory methods were intrinsically superior that geophysical methods were abandoned. Until our recent revival of the geophysical approach there had been no reports for nearly 100 years. Meanwhile a new interest in subterranean gravity began in the 1930's. It was seen as adding another dimension to gravity surveying, which was, by then, a routine exploration technique. However, it is only since the late 1950's that the accuracy of conventional gravity surveying instruments has been sufficient to provide useful input to the G problem. More recently, reliable borehole gravity meters have become available. There are also instruments for measuring gravity on the sea floor, although in the exploration industry they are considered to have been rendered obsolete by the development of gravity meters for use on surface ships.

The modern literature on subterranean gravity is quite meagre and a careful search[18] yielded only four reports that give sufficient information, including independent density determinations, to permit calculation of G. Similarly, wide enquiries in the geophysical exploration industry have turned up only one data set that is usable for the purpose. Normally gravity is measured only to determine layer densities, where these are not amenable to other techniques, or to explore localized density anomalies and in no case was it used to draw an inference about G. Using all the available data, with one recent measurement made specifically to determine G[15], we have six independent geophysical determinations[18]. The intriguing thing about them is that all six values are higher than the conventional one.

A conventional explanation for the high apparent G must be sought in terms of systematic errors; two possibilities have been considered that cannot at this stage be ruled out: (1) local biasing of the gravity gradient by an extensive gravity anomaly (of the same sign for each set of data), and (2) a systematic underestimation of the in situ density of rock by its removal to a laboratory for measurement, permitting the opening of microcracks and pores held closed by overburden pressure. At least in some cases random observational errors were quite inadequate as an explanation. This is most obvious in the case of measurements by McCulloh[19], whose data yield a value of G with a standard deviation 25 times smaller than the discrepancy with the laboratory value, G^*. Subsets of McCulloh's data covering different depth ranges yield very consistent results[18]. The possibility of an anomalous regional gravity gradient was considered in the case of the Mount Isa mine data[15], by measuring the free air gradient in the head frame above the shaft that was used for measurements. Subsequently S.C. Holding and G.J. Tuck have obtained a 230 m free air profile in the smelter flue stack at the mine. In the upper part of the stack, which is clear of effects of local topography and mine working, the gradient coincides with the theoretical value, within the uncertainty of observation.

A particular reason for seeking data from the deep sea is that sea water density presents no ambiguity or difficulty of measurement and has very much better lateral homogeneity than do the rocks surrounding mines, although the lower density is less favourable. Data held by the Exxon Company's Exploration Department include 1100 km^2 of overlapping sea floor and sea surface gravity surveys from the Gulf of Mexico. 703 pairs of values equally spaced over this area, in water up to 700 m deep, yield one of the high values of G reported by Stacey and Tuck[18]. In this case the possibilities for a conventional explanation are a systematic error in the baseline ties of the two surveys or a very broad gradient anomaly. In any case, caution must be exercised in interpretation of the marine data in terms of a short range gravitational force because the two interfaces involved (air-water and water-sediment) have comparable density contrasts.

The available geophysical data are tantalizing. There is a strong hint of a noticeable non-Newtonian effect, but the data lack the observer control to permit a secure conclusion. Further work, as represented by the last three entries on Fig.1, is in progress. Meanwhile the best estimates of the non-Newtonian parameters in equation (2) from the presently available data are $\alpha = -0.01$, $\mu^{-1} = 20$ m. It is emphasized that this is a very tentative result, obtained from data that do not clearly exclude possibilities of systematic error. However the suggestion[20] that $\alpha = +\frac{1}{3}$ and $\mu^{-1} = 10$ to 1000 m is clearly discounted. It is of interest to note two comments that appear in the literature by other authors concerned about a discrepancy between gravity gradients and densities. Yellin[21] indicated evidence for a systematic discrepancy between sea floor and sea surface data in U.S. Geological Survey records, although the published data appear too limited for a conclusion. Hinze et al[22] comment on a discrepancy between sample density and gravimetrically inferred density in a deep borehole in the Michigan basin.

The immediate need is for better data and Fig.1 gives an indication of what may be expected in the next few years. New mine surveys with very extensive density data from bore cores are in progress and could yield a new result within a few months. The bathyscaphe experiment was carefully planned, but is not presently making progress. We are determined not to let it lapse because it is the largest scale experiment of any and the area of the Gulf of Mexico which it is proposed to use appears to be free of obvious gravity anomalies and is protected from gravitational effects of small crustal irregularities by many kilometers of sediment. On the other hand if μ^{-1} is of order 20 m, as suggested by present data, then intermediate range experiments will be crucial. According to the present schedule for completion of the hydroelectric facility at Splityard Creek[17], the first readings from this experiment should now be expected in late 1983 or early 1984.

EFFECTS OF ROCK POROSITY AND WATER CONTENT

Virtually all rocks are composed of grains of a variety of minerals with different properties and in particular different elasticities and thermal expansion coefficients. It follows that even if the grains of a rock fit together perfectly at a particular temperature and pressure they will not do so under any other conditions. In the case of an igneous rock, the grains are formed with a mutual accommodation of their shapes and sizes at high temperature (and perhaps high pressure) and subsequent differential thermal contraction causes microcracks and pores to open unless the rock is held under a substantial pressure. Even apparently hard and completely solid rocks, such as granites, have measurable porosities which influence their physical properties. One property that is

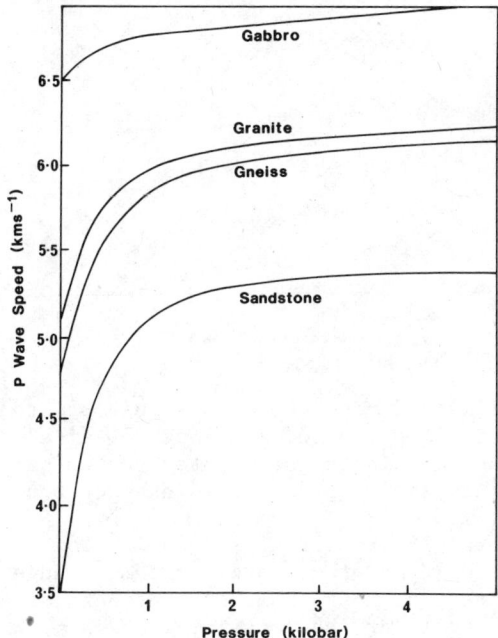

Fig.2: Effect of pressure on speeds of compressional waves in four dry rocks from plots by Gladwin[23], using data by F. Birch, O.L. Anderson and R.C. Liebermann.

very strongly affected and serves as a convenient indicator of open microcracks and pores is acoustic velocity. Fig.2 shows the pressure dependences of compressional wave velocity in four representative rock types. All of them show strong initial increases with pressure, but the initial effect saturates at 1 to 2 kilobars, above which the rock velocities assume the appropriate average values (and very much smaller pressure gradients) for the pure mineral crystals. The initial effect is due to pore and crack closure, bringing the mineral grains into more effective acoustic contact with one another.

For the purpose of the gravity studies the importance of rock porosity is its effect upon density. There are rather few direct measurements, but one data set is represented in Fig.3. This shows

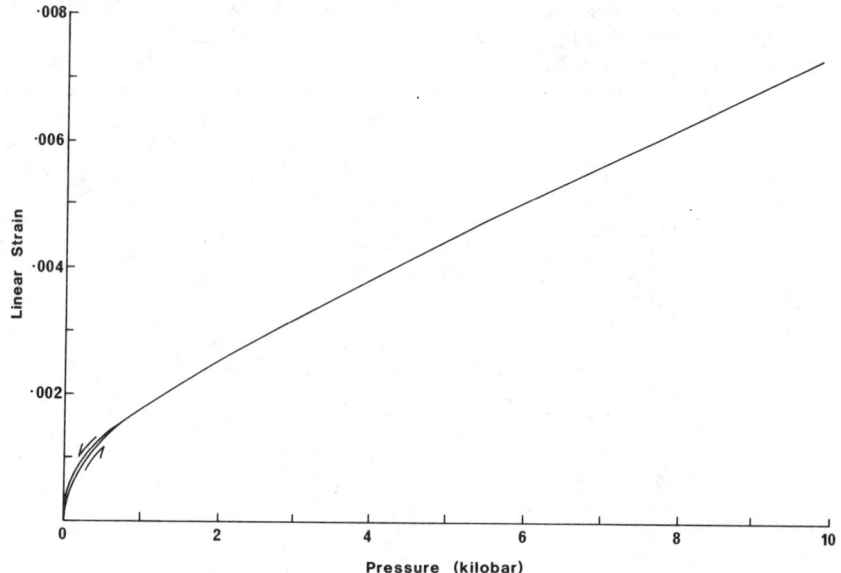

Fig.3: Change in linear dimensions of a sample of Stone Mountain granite with hydrostatic pressure, from data obtained by Brace[24] from electrical resistance strain gauges mounted on the sample. The slight irreversibility at low pressure is exaggerated in this figure.

rather clearly a pore closure effect accounting for a linear dimension change of 0.1% and therefore a volume change of 0.3%. This is typical of granites, although some have higher porosities. A common method of determining the pore volume of a rock is to measure the fluid volume required to saturate it. This is, of course, directly relevant to the pore volume of a wet rock, but the process of wetting a rock itself causes dimensional changes so that the wet and dry pore volumes may not be the same. Thus some care may be required in the interpretation of laboratory densities in terms of in situ densities.

The overburden pressure of 1 km of rock is about 300 bars, sufficient to cause appreciable but not complete pore closure. Thus no more than partial pore closure occurs at any of the depths of the mine gravity measurements. Nevertheless, since a density discrepancy of 1% may suffice to bring the laboratory and mine values of G into agreement, the porosity problem must be treated carefully.

Pumping of a mine to keep it dry creates a permanent sink of groundwater and a conical depression of the water table. Thus the rock in the walls of a mine is, in this respect, not representative of the surrounding rock at the same depth. Also an excavation affects the stress and strain state of a substantial skin of rock surrounding it, more obviously so when blasting methods have been used. Probably the only way to infer the condition and properties of undisturbed rock at the depths of interest is to measure acoustic (seismic) velocities and attenuation for horizontal paths between well separated excavations. Then matching the velocities to laboratory samples, by applying pressure to specimens appropriately water-saturated, permits the condition of undisturbed in situ rock to be inferred. This has not been done systematically for any of the mine measurements so far.

STRESS AND STRAIN MONITORING

The development of sensitive techniques for observing stress and strain in the Earth and particularly their variations with time has been stimulated by interest in earthquake prediction in the past 15 years or so. One of the pitfalls that has come painfully to attention is the ineffectiveness of near surface installations unless the operating instrumental baselines are of kilometer scale. Resort to deep installations is an expensive alternative.

In general it appears that the deeper the installation of an earth strain sensor the better, but the principal considerations are that it should be well below the permanent water table and below the zone of surface weathering of rock. The long term stabilities of modern instruments are good, but the dimensional stability of rock that is not kept both at constant temperature and constant state of wetness or dryness is very much poorer. This situation has led to the development of borehole instruments, of which the Sacks-Evertson strainmeter is probably the most completely de-bugged and best tested, but not every kind of measurement is amenable to this approach. Sensors must be permanently installed and are neither recoverable nor adjustable after installation unless there is a facility for remote control.[25]

The alternative of installation in a mine or tunnel is only a partial solution to the problem because an annual (if not also a diurnal) temperature wave is transmitted by the ventilation system, causing a thermal stress wave to penetrate the mine wall[26] and pumping causes a variation of water flow with weather conditions. In one shallow (40 m) mine installation of a Michelson interferometer[27] a transient shear strain of 2×10^{-6} accompanied one heavy rainfall and the annual peak-to-peak strain amplitude was 10^{-5}. Similarly tilt records from a shallow seismic vault[28] gave equally disturbing transients. Mine and vault installations of very sensitive long period seismic instruments are greatly improved by the environmental isolation provided by sealing against atmospheric changes[29].

The planning of a specially constructed deep underground laboratory invites consideration of the extent to which some of the scientific problems of installations in a conventional mine may be avoided. Sealing the whole excavation against inflow of groundwater is unlikely to be a practical proposition. A much simpler but still very valuable alternative would be to excavate one or more relatively small instrument chambers, well below the level of the lowest laboratory, that would be allowed to flood in normal use but could be pumped dry for access in instrument installation or modification.

CONCLUDING COMMENTS

Underground gravity measurements provide the opportunity for a check on Newton's law of gravity that cannot be achieved in any other way. Measurements above the Earth's surface may permit an examination of the possibility of a short range force, but cannot in principle then also check on the value of G_∞. Without this check even very careful measurements would probably not effectively distinguish non-Newtonian effects from localized anomalies that have been used to explain variability of measured free air gradients[30,31].

The principal difficulties in mine G measurements are in the precise determination of in situ density and its local variability. The existence of a mine almost inevitably implies density heterogeneity, whereas this should be less serious in an underground facility excavated for purely scientific reasons. This is a particular reason for carrying out a mine-type G experiment in such a facility. The density/porosity problem is closely linked with the problems encountered in earth strain and stress monitoring. It appears to be solvable by combining ultrasonic probing of the walls of the excavation with laboratory measurements of the acoustic velocities in rocks under pressure.

REFERENCES

1. G.G. Luther and W.R. Towler, Phys.Rev. Lett. 48, 121 (1982).
2. E.R. Cohen and B.N. Taylor, J. Phys. Chem. Ref. Data 2, 663 (1973).
3. G.W. Gibbons and B.F. Whiting, Nature 291, 636 (1981).
4. D.R. Mikkelsen and M.J. Newman, Phys. Rev. D16, 919 (1977).
5. S.I. Blinnikov, Astrophys. Space Sci. 59, 13 (1978).
6. P. Hut, Phys. Lett. 99B, 174 (1981).
7. H.S.W. McQueen, Phys. Earth Planet. Int. 26, P6 (1981).

8. D.R. Long, Nature 260, 417 (1976).
9. R. Spero, J.K. Hoskins, R. Newman, J. Pellam and J. Schultz, Phys. Rev. Lett. 44, 1645 (1980).
10. V.I. Panov and V.N. Frontov, Sov. Phys. JETP 50, 852 (1979).
11. H.T. Yu, W.T. Ni, C.C. Hu, F.H. Liu, C.H. Yang and W.N. Liu, Phys. Rev. D20, 1813 (1979).
12. H. Hirakawa, K. Tsubono and K. Oide, Nature 283, 184 (1980).
13. D. Long, Nuovo Cimento 55B, 252 (1980).
14. G.B. Airy, Phil. Trans. Roy. Soc. London 146, 297 and 343 (1856).
15. F.D. Stacey, G.J. Tuck, S.C. Holding, A.R. Maher and D. Morris, Phys. Rev. D23, 1683 (1981).
16. F.A. Dahlen, Phys. Rev. D25, 1735 (1982).
17. F.D. Stacey and G.J. Tuck, in N.B.S. Spec. Publ. 617 (in press, 1983).
18. F.D. Stacey and G.J. Tuck, Nature 292, 230 (1981).
19. T.H. McCulloh, Geophysics 30, 1108 (1965).
20. Y. Fujii, Nature Phys. Sci. 234, 5 (1971).
21. M.J. Yellin, U.S. Dept. of Commerce, ESSA Operational Data Report C & GSDR-2 (1968).
22. W.J. Hinze, J.W. Bradley and A.R. Brown, J. Geophys. Res. 83, 5864 (1978).
23. M.T. Gladwin, Int. J. Rock Mech. Mining Sci., in press (1982).
24. W.F. Brace, J. Geophys. Res. 70, 391 (1965).
25. I.S. Sacks, S. Suyehiro, D.W. Evertson and Y. Yamayishi, Papers in Meteorology and Geophysics (Japan) 22, 195 (1971). See also Carnegie Inst. Dept. Terr. Mag. Ann. Repts. 1979-80, p.495 and 1980-81, p.499.
26. M.T. Gladwin and F.D. Stacey, Tectonophysics 21, 39 (1973).
27. S.K. Shamsi and F.D. Stacey, Earth Plan. Sci. Lett. 3, 466 (1967).
28. F.D. Stacey and J.M.W. Rynn, in L. Mansinha et al, Earthquake Displacement Fields and the Rotation of the Earth, 230 (Reidel, Dordrecht, 1970).
29. J.M. Savino, A.J. Murphy, J.M.W. Rynn, R. Tatham, L.R. Sykes, C.L. Choy and K. McCamy, Geophys. J. R. Astron. Soc. 31, 179 (1972).
30. S. Hammer, Trans. Am. Geophys. Un. 19, 72 (1938).
31. J.T. Kuo, M. Ottaviani and S.K. Singh, Geophysics 34, 235 (1969).

Chapter VI

POSSIBLE DIRECTIONS

...theoreticians have always succeeded in providing an understanding for all observed phenomena — even those which later proved to be incorrect.

Anonymous, Observatory **94**, 6P (1974).

MONOPOLES UNDERGROUND*

J.A. Harvey
Princeton University, Princeton, N.J. 08544

ABSTRACT

A new bound on the product of the galactic flux of monopoles and the cross section for monopole catalyzed nucleon decay, $(F_M/cm^{-2}sr^{-1})(\sigma_{\Delta B}/10^{-27}cm^2) \leq 10^{-22}$, and its implications for the detection of monopoles are discussed.

INTRODUCTION

Although we have had no definite evidence of magnetic monopoles in the fifty years since they were proposed by Dirac, both experimental and theoretical interest in monopoles is at a peak. This interest is due in large part to the reported detection by Cabrera[1] of a single magnetic monopole at a flux level of $F_M = 6 \times 10^{-10} cm^{-2}s^{-1}sr^{-1}$. Whether or not this detection is confirmed, the renewed interest in monopoles comes at an opportune time.

As a result of recent theoretical developments we have much better knowledge of how monopoles interact with matter as well as remarkable new limits on their density in the galaxy. Unfortunately, this knowledge comes in the form of good news and bad news for monopole hunters. The good news is that model calculations of monopole-fermion interactions indicate that monopoles should catalyze nucleon decay (e.g., $p + M \to M + e^+ + $ mesons) at a strong interaction rate with $\sigma_{\Delta B} \sim 10^{-27} cm^2$,[2] thus leaving a very distinctive chain of nucleon decays as they pass through, e.g., a proton decay detector. The bad news is that nature has already provided us with detectors which are exceedingly sensitive to this process, namely neutron stars, and they are sending us a very definite negative signal.[3]

Before explaining this last remark I would like to first give a quick review of monopoles. I will then discuss the limit on $F_M \cdot \sigma_{\Delta B}$ that comes from neutron stars and then explain the relevance of this limit to monopole searches.

REVIEW

Magnetic monopoles have fascinated physicists ever since Dirac's seminal paper in 1931. By considering the quantum mechanics of an electron moving in the field of a monopole he derived the celebrated quantization condition

---

*Supported in part by the National Science Foundation under Grant No. PHY80-19754

$$\frac{eg}{\hbar c} = n/2 \quad n = 1, 2, 3\ldots \tag{1}$$

where e is the electric charge of the electron and g is the magnetic charge of the monopole. This result may by derived heuristically by considering the angular momentum stored in the electromagnetic field of an electron and a monopole. The result is

$$L_{em} = \frac{1}{4\pi c} \int d^3r \, \vec{r} \times (\vec{E} \times \vec{B}) = \frac{eg}{c} \hat{r} \tag{2}$$

where $\hat{r}$ is a unit vector pointing from the monopole to the electron. Requiring that L_{em} be quantized in units of $\hbar/2$ then gives Dirac's quantization condition.

The beauty of Dirac's construction is that the existence of a single magnetic monopole implies the quantization of electric charge, something that is a total mystery in the context of ordinary electrodynamics. However, the mass and spin of these Dirac monopoles are completely arbitrary. The monopoles must be introduced as completely new degrees of freedom in the theory.

The modern viewpoint of magnetic monopoles is just the reverse of this. We now believe that the quantization of electric charge follows as a consequence of the imbedding of the U(1) gauge transformations of electrodynamics into a compact unified gauge group (e.g. SU(5), SO(10), etc.). The quantization of electric charge in these theories is analogous to the quantization of the Z component of angular momentum that occurs when the group of rotations about the Z-axis (SO(2) ≈ U(1)) is imbedded into the group of three dimensional rotations (SO(3)). As shown by 't Hooft and Polyakov,[4] the imbedding of electromagnetism into a unified gauge theory results in stable classical solutions with quantized magnetic charge when the unified gauge group breaks down to a subgroup with a U(1) factor. Furthermore, all the properties of these monopoles are determined in terms of the parameters of the unified gauge theory. Thus the modern viewpoint is that charge quantization implies the existence of magnetic monopoles.

A variety of monopole solutions may exist in different grand unified theories. For simplicity we will focus on the monopole with minimal magnetic charge that occurs in the SU(5) grand unified theory of Georgi and Glashow. Recall that SU(5) is presumed to break down to the observed electro-weak-color gauge group $SU(3)_C \times SU(2)_L \times U(1)_Y$ at an energy scale of $\sim 10^{15}$ GeV. One can show that the monopole mass is governed by the scale at which a U(1) factor first appears in the unbroken group so that in SU(5) the monopole mass is $M_M \sim M_X/\alpha \sim 10^{16}$ GeV where M_X is the mass of the X boson in SU(5) which is responsible for proton decay.

The core of the monopole, which contains excitations of the X fields, extends out to a distance of $\sim 1/10^{15}$ GeV $\simeq 10^{-29}$ cm. Between 10^{-29} cm and $\sim 1/M_W \sim 10^{-16}$ cm the effective symmetry group is $SU(3)_C \times SU(2)_L \times U(1)_Y$. The hypercharge Y plays the role of electromagnetism and in this range the monopole has a "hypermagnetic" charge as well as color and $SU(2)_L$ magnetic charge. However the $SU(2)_L$ magnetic charge and the hypermagnetic charge are aligned in just the right way to give a $U(1)_{em}$ magnetic field with strength corresponding to unit Dirac charge, $g = \hbar c/2e$, where e is the charge of the <u>electron</u>. The structure of the long range fields of the monopole are unaffected by the breaking of $SU(2)_L \times U(1)_Y$ to $U(1)_{em}$.

We see that the Dirac quantization condition is not valid when applied naively to quarks. In fact, in the presence of unbroken color interactions the quantization condition is modified. The monopole can have a <u>color</u> magnetic field which has no effect on color singlets like electrons but which restores the single valuedness of the wave function for particles with color electric fields and e/3 electric charge. Since the monopole we are discussing has <u>unit</u> Dirac charge it must also possess a color magnetic field. This field will be screened by the QCD vacuum at a distance of $1/GeV \simeq 10^{-14}$cm. Finally, as shown by Rubahov and Callan,[2] the monopole is surrounded by a fermion condensate which extends out to a distance of $1/M_f$ where M_f is the relevant fermion mass.

The idea that the scattering of fermions in the presence of grand unified monopoles may violate baryon number is not new.[5] Since it is only the monopole core that involves excitation of the B-violating X boson field the naive expectation is that the cross section for this process should be governed by the geometrical size of the monopole core, i.e. $\sigma_{\Delta B} \sim 10^{-58} cm^2$. Rubakov,[2] Wilczek,[6] and Callan[2] have argued that this need not be the case. I don't have time to review their arguments here, but I would like to point out that the process envisioned by Rubakov and Callan is quite independent of that suggested by Wilczek. Wilczek's mechanism relies on a baryon number anomaly which occurs only in the presence of electroweak fields. Since these occur only on distance scales $<1/M_W$ this would give a cross section $\leq 1/M_W^2 \sim 10^{-32} cm^2$. In Rubahov's and Callan's scenario the B violation is caused by the physics at the monopole core and the only anomaly that plays a role allows the fermions to change their helicity but does not violate baryon number. They conclude that the cross section may be as large as a typical strong interaction cross section, $\sigma_{\Delta B} \sim 10^{-27} cm^2$. A simplified description of their analysis is given in Ref. 7.

MONOPOLES AND NEUTRON STARS

There are two simple reasons for considering the effects of monopoles in neutron stars. First, because of their incredibly high density, $\rho_{NS} \sim 3 \times 10^{14} gm/cm^3$, they are able to stop monopoles which strike their surface. Monopoles which strike the earth or sun probably pass through with only a very small energy loss. The relevant quantity to compare is the product of the density and the radius which is $\sim 10^{20} gm/cm^2$ for a neutron star as compared to $\sim 10^9 gm/cm^2$ for the sun. Second, again because of the high density, the rate at which monopoles catalyze nucleon decay will be much larger than it would be in normal matter. The rate in a neutron star will be $\sim n_{NS} <\sigma_{\Delta B} V>$ per monopole where $n_{NS} \sim 2 \times 10^{38} cm^{-3}$ is the nucleon density, $\sigma_{\Delta B}$ is the cross section for monopole catalyzed nucleon decay and V is the relative velocity of the nucleon and monopole which in our case is just the Fermi velocity of the nucleons, $V \sim V_F \sim .3c$. For low energy scattering, as would be relevant for monopoles striking the earth, the calculation of $\sigma_{\Delta B}$ is complicated by the interaction between the monopole and the nucleon magnetic dipole moment as well as by strong interaction effects. However, for incident nucleon energies comparable to the Fermi energy in a neutron star, $E_F \sim 100$ MeV, these effects are small and $\sigma_{\Delta B}$ should be comparable to a strong interaction cross section $\sigma_{\Delta B} \sim 10^{-27}$cm^2. Each monopole then catalyzes $\sim 10^{21}$ $(\sigma_{\Delta B}/10^{-27}\text{cm}^2)$ decays/sec. Since each decay releases ~ 1GeV of energy the total luminosity of the neutron star due to nucleon decay is

$$L \simeq 2.4 \times 10^{18} \, N_M \, (\sigma_{\Delta B}/10^{-27}\text{cm}^2) \text{ erg s}^{-1} \qquad (4)$$

The best limit on the photon luminosity of old neutron stars comes from surveys for serendipitous x-ray sources which are able to see discrete sources with an x-ray luminosity of $L^\gamma_{dis} = 10^{31}$ erg s^{-1} at a distance of 1 kpc.[8] These surveys are sensitive to photon energies between .1 KeV and 4 KeV. A neutron star with an x-ray luminosity of L^γ_{dis} has a surface temperature of $T_s \sim (L^\gamma_{dis}/4\pi R_{NS}^2 a c)^{1/4} \sim .04$ KeV where a is the Stefan-Boltzmann constant and $R_{NS} \sim 10^6$cm is the radius of the neutron star. For a black body spectrum with temperature T_s the peak of the spectrum occurs at an energy of $\sim .11$ KeV and $\gtrsim 75\%$ of the energy is radiated at energies $> .1$ KeV. These surveys should therefore be able to see neutron stars at 1 kpc with a photon luminosity of L^γ_{dis}. Since the average distance between old neutron stars in the solar neighborhood is expected to be much less than 1 kpc,[9] the dearth of sources in these surveys can only be explained if the x-ray luminosity of old neutron stars is less than L^γ_{dis}.

The luminosity given by Eq. (4) will in general be emitted as both photons and neutrinos. For total luminosities less than 10^{33} erg s^{-1} the luminosity is dominated by photons unless pions condense in the interior of neutron stars in which case a photon luminosity of 10^{31} erg s^{-1} corresponds to a <u>total</u> luminosity of 10^{33} erg s^{-1}.[10] By demanding that the luminosity given by Eq. (4) be less than 10^{33} erg s^{-1} we find the bound

$$N_M \, (\sigma_{\Delta B}/10^{-27}\text{cm}^2) \leq 4 \times 10^{14} \qquad (5)$$

We must now relate N_M to the incident galactic flux of monopoles. We will ignore for the moment any monopoles present at the formation of the neutron star. Monopoles which strike the surface of a neutron star will lose energy mainly through electron encounters with[11]

$$\frac{dE}{dX} \sim \frac{4\pi^2 N_e (eg)^2}{P_e c} \frac{V_M}{c} \sim \beta_M \times 10^{11} \text{GeV/cm} \qquad (6)$$

where $N_e \sim 10^{36}$ cm^{-3} is the electron density and $P_e c \sim 100$ MeV is the Fermi momentum of the electrons.

Monopoles in the galaxy will gain an energy of $\sim 10^{10}$ GeV from the galactic magnetic field and may be accelerated to velocities $\sim 10^{-3}$c by the galactic gravitational field if they are sufficiently heavy. Monopoles will strike the surface of a neutron star at a velocity equal to the escape velocity $\sim .5c$. Using Eq. (6) we find that monopoles with $M_M < 10^{17}$ GeV/c^2 will be stopped in the neutron star. Heavier monopoles may pass through the neutron star but since they lose $\sim 10^{16}$ GeV of energy they will be gravitationally bound and will continue to lose energy by passing through the neutron star until they are stopped. Only monopoles with an original kinetic energy $\geq 10^{16}$ GeV can pass through the neutron star and escape. For $\beta_M > 10^{-3}$ this corresponds to $M_M c^2 \geq 10^{22}$ GeV.

The number of monopoles captured by a neutron star is

$$N_M^{cap} = 4\pi F_M \, A_{cap} \, T \qquad (7)$$

where T is the age of the neutron star and the capture area is given by

$$A_{cap} = \pi R_{NS}^2 \left(\frac{1 + 2M_{NS} G/V_M^2 R_{NS}}{1 - R_S/R_{NS}} \right) \sim 4 \times 10^5 \, \pi \, R_{NS}^2 \quad (V_M = 10^{-3} c) \qquad (8)$$

where R_S is the Schwarzschild radius and M_{NS} and R_{NS} are the mass and radius of the neutron star.

The best bound on $F_M \sigma_{\Delta B}$ will come from the oldest neutron stars which should have ages comparable to the age of the galaxy. For $T = 10^{10}$ years Eq. (7) gives

$$N_M \sim 5 \times 10^{36} \, (F_M/cm^{-2} \, s^{-1} \, sr^{-1}) \qquad (9)$$

From (5) we thus find

$$F_M \, (\sigma_{\Delta B}/10^{-27} cm^2) \leq 10^{-22} \, cm^{-2} \, s^{-1} \, sr^{-1} \qquad (10)$$

This is a much more stringent bound than bounds on F_M from the mass density of monopoles or from destruction of the galactic magnetic field which give[12]

$$F_M \leq 10^{-12} - 10^{-16} \, cm^{-2} \, s^{-1} \, sr^{-1} \qquad (11)$$

depending on their mass and distribution.

MONOPOLES UNDERGROUND

I now want to discuss briefly the relevance of all this to underground searches for monopoles. Present and future proton decay decay detectors are also excellent monopoles detectors. There are two ways to use these detectors for monopole searches. The first involves looking for chains of nucleon decays catalyzed by a passing monopole. The effective proton lifetime due to this process is

$$\tau_\rho^{eff} \sim (4\pi F_M \sigma_{\Delta B})^{-1} \tag{12}$$

and since present detectors give a limit $\tau_\rho \leq 10^{30}$ years we can conclude that

$$F_M (\sigma_{\Delta B}/10^{-27} cm^2) \leq 3 \times 10^{-12} \text{ cm}^{-2} \text{ s}^{-1} \text{ sr}^{-1} \tag{13}$$

However the bound from neutron stars is ten orders of magnitude more stringent than this and only allows an effective proton lifetime of $\tau_\rho^{eff} \geq 10^{40}$ years which is unobservably long. I therefore see no hope of using this process to detect monopoles. Either the flux of monopoles is too small or $\sigma_{\Delta B}$ is too small or perhaps both. If we do detect monopoles at a flux level anywhere near that suggested by Cabrera, it must be the case that $\sigma_{\Delta B}$ is much smaller than suggested by the model calculations or that these calculations don't apply. Perhaps because baryon number is in fact conserved in the grand unified theory that gives rise to the monopoles.

The second way to detect monopoles underground and to me the most promising way is by detecting their energy loss. While there is some theoretical controversy over the stopping power of slow monopoles, it appears that even monopoles with $\beta_M \geq 10^{-3}$ should be more than minimum ionizing[11] while for $10^{-4} \leq \beta_M \leq 10^{-3}$ the energy loss is more uncertain and depends on the details of the detector (e.g. scintillator vs. proportional tubes). As we have heard at this workshop, limits for $dE/dX > 1/2 \ I_{min}$ ($\beta_M \geq 4 \times 10^{-3}$) are already several orders of magnitude less than Cabrera's limit. Improvements in the calculation of energy loss in the low β range and new experiments will probably be able to rule out or confirm a monopole flux comparable to Cabrera's for monopole velocities in the expected range of $10^{-4} \leq \beta_M \leq 10^{-2}$.

ACKNOWLEDGEMENTS

I am pleased to acknowledge helpful discussions with my collaborators, S. Colgate and E. Kolb. I would also like to thank J. Ullman for discussions of monopole stopping power and C.G. Callan for discussions of monopole catalysis of nucleon decay.

REFERENCES

1. B. Cabrera, Phys. Rev. Lett. **48**, 1378 (1982).
2. V. Rubakov, JETP Lett. 33, 644 (1981) and Nucl. Phys. **B203**, 311 (1982); C.G. Callan, Phys. Rev. D26, 2058 (1982).
3. E.W. Kolb, S.A. Colgate, and J.A. Harvey, Phys. Rev. Lett. **49**, 1373 (1982); S. Dimopoulos, J. Preskill, and F. Wilczek, Phys. Lett. B to be published.
4. G. 't Hooft, Nucl. Phys. **B79** 276 (1974). A.M. Polyakov, JETP Lett. **20** 194 (1974).
5. See e.g. C.P. Dokos and T.N. Tomaras, Phys. Rev. **D21** 2940 (1982).
6. F. Wilczek, Phys. Rev. Lett. **48**, 1378 (1982).
7. C.G. Callan, "Monopole Catalysis of Baryon Decay," Princeton

preprint, 1982.
8. F.S. Cordova, K.O. Mason and J.E. Nelson, Ap. J. 245, 609 (1981); G.A. Reichert, K.O. Mason, J.R. Thorstenson and S. Bowyer, Ap. J. to be published.
9. D.Q. Lamb, F.H. Lamb and D. Pines, Nature 246, 52 (1973); J.G. Hills, Ap. J. 219 (1978); A.S. Endal, Ap. J. 228, 541 (1979).
10. K.A. van Riper and D.Q. Lamb, Ap. J. 244, L13 (1981).
11. S.P. Ahlen and K. Kinoshita, Phys. Rev. D26, 2347 (1982).
12. E. Parker, Ap. J. 160, 383 (1970); M. Turner, E. Parker, T. Boylan, Enrico Fermi Institute preprint, 1982; J. Preskill, Phys. Rev. Lett. 43, 1365 (1979).

NON-LTE ASTROPHYSICS AND THE ORIGIN OF HIGH ENERGY COSMIC RAYS

Stirling A. Colgate

Los Alamos National Laboratory, Los Alamos, NM 87545

ABSTRACT

The most startling aspect of the universe is the departure from the nonuniform, nonthermodynamic equilibrium state. Mass concentrations and high energy electromagnetic quanta and particles are the primary examples. Of these various departures the extremum of nonexplained phenomena is still the acceleration of cosmic rays. The usual explanation involving multiple mechanisms contributing to various regions of the spectrum is suggested as most unlikely. In particular, the suggestion that the acceleration of the very highest energy cosmic rays is by distant active galactic nuclei is shown to be untenable because of photon particle interactions.

INTRODUCTION

The simplest state of affairs for our universe would be uniform, homogeneous, isotropic, local thermodynamic equilibrium (LTE). The departures from this benign state are the interesting astrophysics. The extremum of these departures are both the mass distributions of the universe in the form of clusters, galaxies, and stars, and the extrema of energy such as cosmic rays, quasars, supernovae, gamma bursts, and masers. The explanation of these phenomena, using the known laws of physics, is the essence of astrophysics. There are very many puzzles. It is surprising how old many of these problems are and how such depressingly small progress has been made in their resolution. On the other hand, it is heartening that astrophysics universally restricts itself to the laws of physics tested here on earth and so far has refrained from attempting to invoke new laws of physics to explain its greatest confrontations. I will select one of these topics, the ultrahigh energy part of the cosmic ray spectrum, and attempt to make the puzzle still more difficult for astrophysics to explain. I note in passing that the radiation mechanisms of quasars, Seyfert galaxies, and active galactic nuclei are as challenging, or as little understood as ever, that supernova consistently either marginally work or marginally don't work, and that gamma bursts are still a high energy phenomenon that await an explanation.

ULTRA-HIGH ENERGY COSMIC RAYS

Cosmic rays are the extremum of nonthermal particles. A quasi power law spectrum extends up to 10^{20} eV, or for protons $\gamma \equiv 1/\sqrt{1-\beta^2} = 10^{11}$. Measurements of variations in this power law as a function of energy have been continuously refined, but nowhere is there allowed the possibility of a departure from a monotonic spectrum of particles from the very lowest energies to the very highest. There is nowhere in the spectrum clear cut evidence for a peak and thus for a contribution from an entirely different mechanism for a given group of particles. The special circumstance that any given mechanism should more or less serendipitously overlap another in just such a fashion as to produce a monotonic spectrum would seem highly unlikely. This is unlikely not only because of the lack of a secondary peak, but also because a second acceleration mechanism should produce more lower energy particles than the presumed primary mechanism.

STOCHASTIC PROPERTY OF ACCELERATION

Acceleration is a progressive departure from thermal and hence low energy particles are nearly always overabundant. The general property of an acceleration mechanism is that the probability of producing a high-energy particle decreases as a function of energy simply because the high-energy particle requires a greater coherence or lifetime within the acceleration region. This is inherently true for all proposed mechanisms. It is not true for a laboratory accelerator because of a perfection of construction whose a priori probability is of measure zero.

DIFFUSION ACCELERATION

It has been suggested that energy diffusion accompanying spacial diffusion within the galaxy should lead to a monotonic distribution from several different energy sources. A diffusion in energy, however, is equivalent to a heating or acceleration of the few particles that drift to high energy. A first order Fermi acceleration across supernova induced shock waves in the galactic medium is now well recognized as a possible primary acceleration mechanism. Particles are scattered back and forth across the expanding shock wave by Alfvén wave turbulence induced in turn by the the particle diffusion. The limit to this process is generally recognized to be roughly 10^{13} eV due to the ratio of particle Larmor orbit size to the maximum (strong) shock diameter induced by the supernova.[1]

HYPOTHETICAL SECONDARY ACCELERATION

If a second mechanism exists to produce high energy cosmic rays, it must become predominant within a factor of 2 in energy and 2 in magnitude of the value where the primary mechanism fails, otherwise there will either be a peak in the spectrum or too large a negative step in flux to be consistent with the near constant power law slope from the measured spectrum $N(> E) \propto E^{-1.7 \pm 0.5}$. Diffusion must spread this secondary source in energy so as not to leave a spectral "bump". Such an energy diffusion will spread the secondary source to higher energies, i.e., a tertiary mechanism. Then why not dispense with the secondary acceleration and just diffuse the upper energy limit of the primary mechanism to still higher energy by the tertiary mechanism? If there were such a diffusion to high energy, this would be the primary mechanism. Instead it is generally believed that galactic diffusion removes energy from the high energy cosmic rays with energy greater than 10^{13} eV and lets high energy particles preferentially escape.[2] Hence the supernova interstellar medium shock mechanism energizes the turbulent diffusion, but this energization leads to a modest upper energy limit.

The second unlikely circumstance is that the proposed high energy mechanism should not produce more low energy particles than the proposed primary mechanism. When one considers the immense scientific effort spent on the problem of the origin of cosmic rays it would seem far less likely to accept the existence of any serendipitous high energy source (E > 10^{13} eV) of just the right source strength, energy, and lack of low energy acceleration than to start from scratch and say I am going to find an entirely new mechanism for the full monotonic spectrum. In other words, a good theory of a low energy source is less likely to be true just because it demands a special high energy one to complete the picture. Whatever may be the high energy source, this source is then more likely to dominate the orignally proposed low energy one.

EXTRAGALACTIC ACCELERATION

The problem of multiple acceleration mechanisms is recognized and taken seriously for energies less than galactic confinement ($E_p < 10^{18}$ eV) but above this energy, it has been assumed that cosmic rays must be accelerated in extra galactic space, more particularly in exotic energetic objects like quasars and active galactic nuclei.[3,4]

Besides the difficulty of serendipitous matching in energy flux and the lack of low energy particles, further difficulties can be found.

The very highest energy cosmic rays were inferred from air shower measurements (10^{19} to 10^{20} eV) and confirmed among various experiments.[5] Greisen[6] pointed out that such high energy particles in the extragalactic environment would lose their energy in a Hubble expansion time due to Doppler shifted collisions with the blackbody radiation. The predicted cutoff, around 10^{20} eV, is definitely not seen and indeed the spectrum becomes flatter above 10^{18} eV, the exact converse of the expectation. Very detailed calculations of spallation of nuclei due to infrared and star light, pair production and pion production were performed by Puget, Stecker, and Bredekamp.[7] Figure 1, reproduced from their article, shows the energy loss time due to the blackbody photons alone. The full radiation spectrum used for the spallation analysis is shown in Fig. 2. Even with the very weak photon flux in the infrared and optical, the Doppler shifted spallation of iron nuclei limits this possible explanation of these ultrahigh energy cosmic rays to either protons or iron nuclei within the local supercluster.

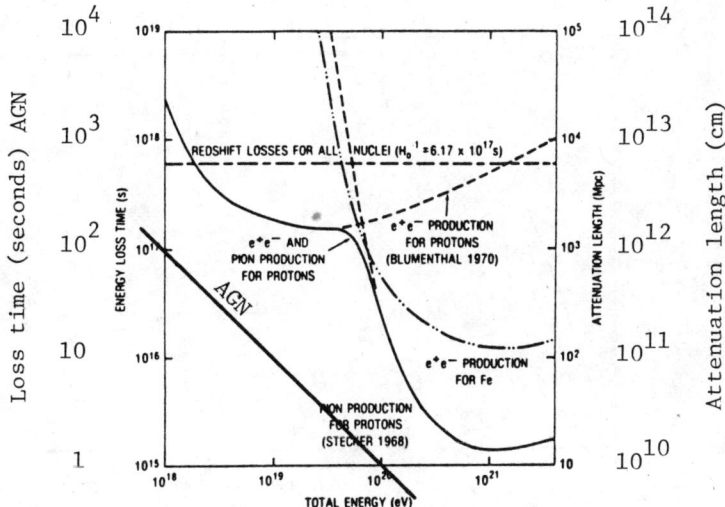

Fig. 1. From Puget et al (1976). The energy loss time and attenuation length for protons from pair production, redshift, and photopion production losses. Also shown is the attenuation length for ^{56}Fe from pair production losses. Separate scales have been added for the equivalent process in the near radiation field of an active galactic nucleus (AGN). The diagonal line added is the approximation appropriate to these added scales.

What we now wish to point out is that the frequent assumption of acceleration in active galactic nuclei in the local supercluster has far worse limitations due to the same radiation damping.

ACCELERATION IN ACTIVE GALACTIC NUCLEI

The assumption of the acceleration of cosmic rays in quasars, BL lac objects, Seyfert galaxies, and AGN (active galactic nuclei) and suffers in the extreme from the same problem of photon damping. The damping or deceleration occurs within the photon energy density of the emission by which we recognize the objects in the first place. A particle can be accelerated inside or outside such an object

"Inside" and "outside" corresponds to an emission surface. In general the most energetic phenomenon should occur inside, but it would be possible for a particle to be accelerated outside in a magnetically confined orbit. Therefore the following assumptions are made as applying to any reasonable mechanism of acceleration in any AGN:
1. The acceleration of a particle must take place within and the particle traverse at least one radius of the object, where the radius corresponds to an emission surface inside of which the photon flux is quasi-isotropic.

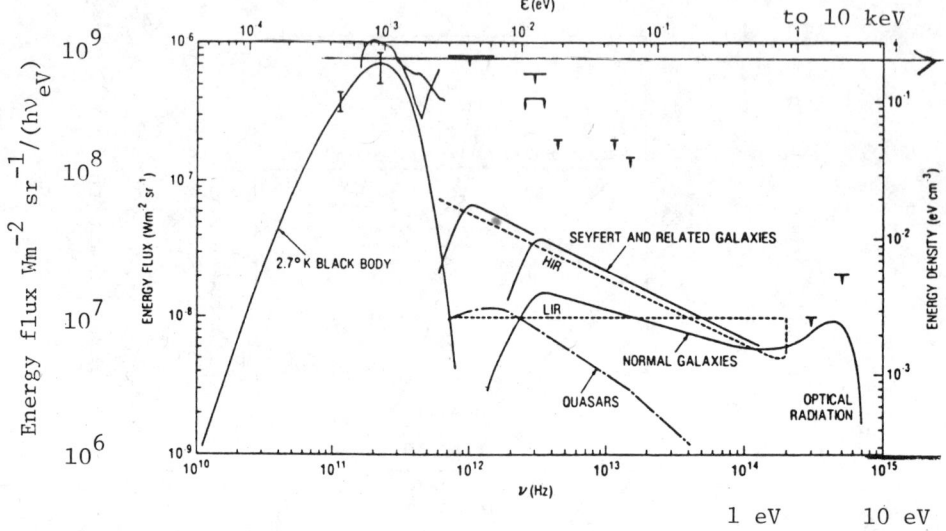

Fig. 2. From Puget et al (1976). Computed background radiation fields from quasars, Seyfert and related galaxies, and normal galaxies. Separate scales and an extension of the graph have been added which are appropriate for the near radiation field of AGN.

2. Acceleration outside an emission surface must require at least one orbit around the object in a presumed magnetic field strong enough to confine the high energy particle.

Both circumstances lead to the impossible loss of the cosmic ray energy due to the two processes of pair production and pion production. In the first case of purely radial traversal of the emission nebula, the photons are quasi-isotropic in the rest frame and in the second case the photon streaming will lead to a primarily orthogonal flux, both of which lead to photon collision energies in the frame of the cosmic ray particle essentially the same as calculated by Puget et al.[7] It should be noted that one cannot use the radial photon flux external to the surface to significantly accelerate the particles further because of red shift and photon drag.[8]

Under these circumstances we can scale the photon damping or cosmic ray energy loss to a standard AGN. Let the standard luminosity be

$$L_{46} = 10^{46} \text{ erg s}^{-1}$$

and radius for 10-day fluctuation period be

$$R_{16} = 10^{16} \text{ cm}$$

so that the photon energy flux at the surface becomes

$$\phi = 10^{13} L_{46} R_{16}^{-2} \text{ ergs cm}^{-2} \text{ s}^{-1}.$$

Some AGN's have a peak in the emission spectrum in the long wavelength IR ($\lambda \cong 100$ microns) about a factor of 10 shorter in wavelength than the cosmic black body radiation. Others, like Seyfert galaxies, have their emission peak in the x-ray region. In general the fluctuation time decreases for shorter wavelength, being months to years in the IR and as short as hours in the x-ray for low luminosity Seyferts. The average energy flux per logarithmic band width, $d\phi/d[\log(h\nu)]$, is conservatively 1/10 the total luminosity so that the average flux at the surface becomes

$$d\phi \cong 10^{12} L_{46}(h\nu_{eV}) \, d(\log h\nu) \text{ ergs cm}^{-2} \text{ s}^{-1}$$

where $h\nu_{eV}$ is the mean photon energy in eV in a band $d(\log h\nu)$ wide and where the fluctuation time has been approximated as being

$$\Delta t \cong 10^6 (h\nu_{eV})^{-\frac{1}{2}} \text{ s so that } R_{16} = (c/3) \Delta t = (h\nu_{eV})^{-\frac{1}{2}}$$

corresponding to a fluctuation time of an hour for 10 keV x-rays and a year at $\lambda = 1$ mm. With this fluctuation time and $L = 10^{46}$ ergs s^{-1}, the photon emission is superluminal below $h\nu = 1$ eV.

This spectral flux is shown as a separate scale and a line on Fig. 2 from Puget et al[7] and also as a time on the cosmic ray photon loss curve, Fig. 1.

Here the rate per particle of electorn pair or pion production is taken as proportional to the photon number density at the relative energy $\Gamma(h\nu)$ necessary to create pairs or pions. Since $n(h\nu) \propto d\phi/(h\nu)$, the loss time becomes independent of photon energy. Since this time is 10^{15} seconds for the black body flux of 10^{-3} ergs cm^{-2} s^{-1}, the AGN deceleration time becomes

$$\Delta t_{loss} = \Delta t_{bb}(\phi_{bb}/\phi_{AGN}) = L_{46}^{-1} \text{ s}.$$

The fluctuation time and hence radius of the surface of an AGN is 10^4 times larger than this energy loss time, even for the hard x-ray upper limit of the spectrum ($\cong 10$ keV for Seyfert galaxies). The cosmic rays accelerated in such an object would therefore likely be truncated for energies above

$$\Gamma(h\nu_{x \text{ ray}}) > 2 \times 10^8 \text{ eV},$$

the pion threshold, or

$$\Gamma > 2 \times 10^4, \text{ for cosmic ray energy } E > 2 \times 10^{13} \text{ eV}.$$

The spallation of heavy nuclei such as Fe would, of course, extend to the energy of the photo-nuclear threshold or a factor of 20 lower in energy.

The large photon damping of the acceleration of high energy cosmic rays is strongest for the highest energy particles in the most energetic quasars, where the emission peaks at 10-100 microns at $L = 10^{47}$ ergs s^{-1} and the luminosity fluctuation time is a year. Here the damping time is 3×10^{-9} s of the particle escape time at an energy of 10^{18} to 10^{19} eV. Hence it seems unreasonable to expect ultra high energy cosmic ray acceleration in AGN.

REMAINING ACCELERATION IN IGM

Currently we recognize the universe external to the Galaxy as being filled with other galaxies, the intergalactic medium (IGM), intercluster medium (ICM), and AGN. The medium between galaxies and between cluster members is similar in the sense that the magnetic field, plasma energy density, and particle density is less than that within galaxies.

The cosmic ray energy limitation to shock wave acceleration in the interstellar medium (ISM) also applies to the IGM, or ICM. The much weaker fields ($\lesssim 10^{-2}$) of the ISM means that the field times the distance (galaxy spacing) or likely maximum Larmor orbit size is not much larger than within the galaxies. The distance varies, at most, by a factor of ten so that it is unlikely that a factor of 10^7 higher acceleration energy (to produce 10^{13} eV to 10^{20} eV energies) can be achieved by any of the presently proposed shock and turbulence acceleration mechanisms.

The only remaining space is between the galaxies themselves. Ours is a good example. I believe we must still look for the one singular mechanism for cosmic ray acceleration within our own galaxy.

I am indebted to Albert Petschek for discussions and review.

This work has been supported by DOE and the Astronomy Section of NSF.

REFERENCES

1. Blandford, R. D. and Ostriker, J. P., Ap. J. Lett. 221, L29 (1978).
2. Cesarsky, C., J. 1980, Ann. Rev. Astron. Astrophys. 18, 289.
3. Ginzburg, V. L. and Syrovatskii, S. I., "The Origin of Cosmic Rays," 1964, Pergamon Press, N. Y.
4. Brecher, K. and Burbidge, G. R. 1972, Ap. J. 174, 253.
5. Bower, A. J., Cunningham, G., Linsley, J. Reid, R.J.O., and Watson, A. A., "Energy of Giant Air Shower Primaries," submitted for publication to J. Phys. G., October 22, 1982.
6. Griesen, K., 1966, Phys. Rev. Lett. 16, 748.
7. Puget, J. L., Stecker, F. W., and Bredekamp, J. H., 1976, Ap. J. 205, 638.
8. Noerdlinger, P. D., 1974, Ap. J. 192, 529.

THOUGHTS ON FAMILY SYMMETRIES*

Frank Wilczek
Institute for Theoretical Physics, University of California
Santa Barbara, California 93106

ABSTRACT

Advantages accruing to theories with spontaneous breakdown of flavor symmetries are reviewed. A possible experimental signature in rare K decays is discussed. One particularly attractive breakdown scheme is pointed out. The history of the U(1) problem is briefly discussed, including recent ideas regarding axions. The relevance of Cavendish-type experiments in testing for axions and familons is mentioned.

§1 ADVANTAGES OF FAMILY SYMMETRIES

Although gauge theories have been remarkably successful in describing many features of the interactions of elementary particles, so far they have been rather disappointing in that little light has been shed on the pattern of fermion masses and mixing angles.

Perhaps the most outstanding feature of the fermion spectrum is the repetition of particles with the same $SU(3) \times SU(2) \times U(1)$ quantum numbers: (e,μ,τ), (d,s,b), ... This pattern on its face suggests a symmetry of which these triplets are the multiplets. Postulating a continuous symmetry of this kind has several attractions:

i) the obvious one, that if the symmetry is broken in a simple pattern we may be able to relate observable fermion masses and mixing angles. This will be illustrated in §3.

ii) in order to understand the smallness of the CP-violating parameter θ in the string interactions, it seems that we need the Peccei-Quinn U(1) quasi-symmetry. This means that there must be field transformations which change the overall phase of the quark mass matrix, while leaving the rest of the Lagrangian unchanged. It is, I think, hard to believe that such a quasi-symmetry can be a fundamental physical law. Fortunately, a much more satisfactory possibility exists. The P-Q. quasi-symmetry can easily arise as an accidental by-product of more general __genuine__ (spontaneously broken) family symmetries, and the usual requirements of renormalizability.

As a toy example, let us consider a family symmetry $SU(3) \times SU(3)$ under which the mass matrix $M \to UMV$ ($U, V \in SU(3)$). Then the potential can contain terms like $\mathrm{tr} MM^+$, $\mathrm{tr}(MM^+)^2$, $(\mathrm{tr} MM^+)^2$ but not for instance $\mathrm{tr} M^2$, $\mathrm{tr} M^4$, ... The allowed terms do not depend on the phase of M ($\det M$ is forbidden by renormalizability), so there is an accidental PQ quasi-symmetry.

Along these lines, remember that the essence of the PQ mechanism consists in making the phase of the determinant of M a dynamical variable. It would then seem attractive a priori that more of M should be dynamical with the low-energy effective form determined by spontaneous symmetry breaking.

iii) The U(1) PQ quasi-symmetry in general contains a discrete subgroup of true symmetries. Spontaneous breakdown of these symmetries leads to the possibility of topologically stable domain walls, which are a grave cosmological embarrassment. If these discrete symmetries lie in a continuous group of family symmetries, the domain walls can relax away harmlessly.

§2 A POSSIBLE CONSEQUENCE: NAMBU-GOLDSTONE BOSONS

If there is a fundamental symmetry between the different families, broken not intrinsically but dynamically, then there will arise characteristic (strictly massless, spin zero, neutral) Nambu-Goldstone bosons. Some of these may be of phenomenological interest. The general form of coupling of these familons f is

$$\mathcal{L}_{int.} = \frac{1}{F} \partial_\mu f \, j_\mu$$

where F is the scale of symmetry breaking and j_μ is the relevant symmetry current, e.g. $j_\mu = \bar{s} \gamma_\mu d$ for $s \leftrightarrow d$ symmetry.

The decay $K^+ \to \pi^+ f$ looks most promising. One finds

$$\frac{K^+ \to \pi^+ f}{K^+ \to \pi^+ \pi^0} \approx 10^{14} \left(\frac{\text{GeV.}}{F}\right)^2$$

The existing limit corresponds to $F \gtrsim 10^{11}$ GeV. I will argue in §4 that $F \approx 10^{12}$ GeV. is a very interesting value; it appears to be accessible in currently planned experiments.

Familons can also destabilize heavy neutrinos on cosmological time scales, circumventing the powerful limits ($m_\nu < 50$eV.) on cosmologically stable neutrinos.

One might worry that truly massless particles might lead to large macroscopic effects, competing successfully with gravity. However the derivative coupling of familons suppresses these effects.

§3 A REPRESENTATIVE MASS MATRIX

I will now present an example of a family symmetry breaking that appears rather attractive.

Recall that in SU(5) fermion masses can be generated through a Higgs multiplet in the $\underline{5}$ representation. The resulting masses are equal in pairs between charge -1/3 quarks and charge -1 leptons e.g. $m_b = m_\tau$. Actually equality only holds for effective masses at unification scales; to compare with laboratory results we must re-normalize these down. This procedure gives us the good relation $m_b \approx 3m_\tau$ but also $m_e/m_\mu = m_d/m_s$ which is way off (m_d/m_s is calculated to be ~1/20 by current algebra methods).

Masses may also be generated by Higgs particles in the 45 representation; this contributes three times as much to lepton as to quark masses (before renormalization).

To first order the observed mass matrix is very simple: large entries for the third family, zeros elsewhere. It is simply accounted for if we group the fermion $\bar{5}$ and 10 representations into two triplets under a family SU(3), and assign the Higgs fields to (5,6) under (SU(5), SU(3)). The six-dimensional representation readily breaks giving a single non-zero component $\langle\phi_{33}\rangle \neq 0$; this yields the zeroth-order mass matrix including the good prediction for m_b/m_τ.

It is now entertaining to consider the hypothesis that the rest of the symmetry breaking is accomplished through a (45,3) Higgs representation. This will lead generically to a mass matrix of the form

$$\begin{pmatrix} 0 & A & 0 \\ -A & 0 & B \\ 0 & -B & C \end{pmatrix}$$

in the charge -1/3 quark sector. For the charged leptons, we get the same thing but with B→3B, A→3A (for a Higgs 45) and an overall 1/3 (for renormalization effects)

The bad relationship $m_e/m_\mu = m_d/m_s$ gets replaced by the much better $m_e/m_\mu = 1/9\, m_d/m_s$. Also, one finds the famous relationship $\tan^2\theta_c = m_d/m_s$ for the Cabibbo angle. Other mixing angles, both in weak and in $\Delta B=1$ decays, are predicted to be small; the precise consequences will be discussed elsewhere. A feature which may be disturbing is that the bare strange-quark mass (renormalized at say 2GeV., before the QCD coupling has become really strong) is quite small, ~35MeV. It is perhaps not excluded that m_s really is this small, although such crude estimates as exist in the literature are typically larger.

The Higgs assignments used are compatible with an extension to SO(10), which might permit us to say something about charge 2/3 quarks.

§4 THE NEW! IMPROVED! AXION

Theoretical expectations for F, the scale at which the PQ quasi-symmetry is spontaneously broken, have varied over the years. The primary physical significance of F is that the mass and coupling strength of the axion are inversely proportional to F.

The original suggestion was F≈300 GeV., i.e. that the PQ quasi-symmetry breaks at the same scale as electroweak SU(2)xU(1). This suggestion was tested in several experiments and refuted, the axion was not found. Astrophysical arguments, based on how axion emission would effect the structure of stars, were used to bound $F \gtrsim 10^8$ GeV.

Later it was pointed out that in many unified models $F \approx 10^{15}$ GeV. i.e. the PQ quasi-symmetry is broken at the same scale as the unification group (e.g. SU(5)). The axion is then exceedingly weakly coupled and therefore inaccessible to laboratory experiment. At this point it seemed we had a consistent, though fruitless, solution to the strong CP problem.

Very recently, however, a number of people realized that the invisible axion scheme has severe cosmological problems. There will typically be energy stored in the axion field, which of course contributes to the overall energy density of the Universe and is observable through its gravitational effects. It is easiest to visualize this process as the creation of a dense gas of axions, utterly cold, when the Universe is at $T \approx 1$ GeV. The mass density of the axion gas is calculated to be (to an adequate approximation) $10^4 \times (F/10^{15}$ GeV.$)$ $\times \rho_B$, where ρ_B is the mass density of baryons. With $F \approx 10^{15}$ GeV., as for the invisible axion, this is too large.

An intriguing possibility is $F \approx 10^{12}$ GeV., which I cannot forbear calling the new! improved! axion. Then axions provide the non-baryonic, non-luminous dark matter for which impressive observational evidence now exists. Such a source of mass is also required for $\Omega=1$, marginal closure, as is demanded by inflationary Universe ideas. Since the axions are produced cold they will behave like ultra-massive (2 GeV.) neutrinos as far as the theory of growth of inhomogeneities into galaxies is concerned. There is no "streaming" as there would be for light (≈ 30 eV.) neutrinos. Thus axions probably lead to a hierarchical clustering picture, as analyzed for instance by Rees and White[1], instead of the Zeldovich fragmentation picture of galaxy formation.

It seems very reasonable, in line with our arguments of §1 that PQ and family symmetries are closely related, that the scale $F \approx 10^{12}$ GeV. should also be characteristic of family symmetry breaking generally. This idea can be tested in $K^+ \to \pi^+ f$, as mentioned in §2.

Finally with $F=10^{12}$ GeV. for family or PQ symmetries we might expect colored scalars with this sort of mass capable of mediating $\Delta B=1$ decays. These tend to give proton decay at rates not incommensurable with existing experimental limits, and mostly into strange channels.

§5 WINDOWS ON FAMILY SYMMETRY

Let me briefly summarize the experimental handles on family symmetries discussed above. Potentially the richest of course is the pattern of fermion masses and mixing angles. A scheme such as that in §3 gives many relationships among these masses and angles as measured in weak and ΔB=1 decays; also possibly CP violation parameters. Of course observation of $K^+ \to \pi^+ f$ would be spectacular confirmation of this whole line of thought. Further cosmological investigations should determine whether we want the dark matter formed cold, as it will be if it is axions.

Another fascinating possibility is that very light axions or familons could be detected by macroscopic experiments. The Compton wavelength of the new! improved! axion is about 1 cm. Unlike familons it is not derivatively coupled and could contribute to Cavendish-type experiments at short distances. Since $F^{-1} >> M_{p\ell}^{-1}$ the axions are in some sense more strongly coupled than gravitons, however their static coherent exchange violates CP and is thereby suppressed, though perhaps not hopelessly so. Both axions and familons could mediate spin-dependent $1/r^3$ forces, with a range of 1 cm. or ∞ in the two cases.

Experimental input from direct observation of Higgs particles and their couplings would of course be most helpful. We must also be alive to the possibility, suggested in §4, that there can be substantial contributions to proton decay due to exchange of scalars.

Note: Much more detailed accounts of the material in §3-§5 are being prepared, which will include adequate references. For §1-§2, see Phys. Rev. Lett. 49, 1549 (1982).

REFERENCE

1. S. D. M. White and M. J. Rees, Mon. Nat. Roy. Astr. Soc. 183, 341 (1978).

UNDERGROUND NEUTRINO ASTRONOMY

David N. Schramm
The University of Chicago and Fermilab

ABSTRACT

A review is made of possible astronomical neutrino sources detectable with underground facilities. Comments are made about solar neutrinos and gravitational collapse neutrinos, and particular emphasis is placed on ultra-high energy astronomical neutrino sources. An appendix mentions the exotic possibility of monopolonium.

INTRODUCTION

Neutrino astronomy divides itself into four areas: 1) Solar neutrinos, 2) Neutrinos from gravitational collapse, 3) High energy background neutrinos, and 4) High energy point source neutrinos, all of which require underground detectors. Since 1 and 2 are covered in other papers in this volume I will concentrate on the High Energy Neutrinos. At the end of the paper I will also briefly mention a possible exotic source of high energy neutrinos and other high energy particles, namely, monopolonium.

SOLAR NEUTRINOS

Before going into the high energy neutrino situation, I feel it is my duty to mention a viewpoint on the theoretical solar neutrino situation which is slightly different from that presented by the others in this volume (Fowler[1], Ulrich[2]). As was pointed out by Filippone and Schramm[3], the difference centers on the formal estimate of the uncertainty in the standard model calculation. Given a set of selected input parameter values we all agree on the best estimate of the number of SNU's predicted. The question comes as to how one should treat the uncertainties on those input parameters and how those uncertainties propagate into an uncertainty in the predicted number of SNU's. On this latter point it should be noted that <u>even if the same estimate of errors</u> is used for the input parameters, a Monte Carlo analysis of the type used by Filippone and Schramm[3] gives a significantly larger (± 2 SNU's vs. ± 1 SNU) estimated uncertainty to the standard model than does the linear least square technique used by Bahcall et al.[4]. In addition, the Monte Carlo uncertainty is not symmetric about the standard value. Since solar models are extremely non-linear, it seems reasonable to assume that the Monte Carlo treatment is a more accurate estimate of the uncertainties.

In addition to the Monte-Carlo versus least square analysis, there is also the statistical versus systematic error estimate question. It has been shown that frequently, the input parameters for the calculation as measured by different groups are outside of each others statistical errors. Examples are the ^{3}He$(\alpha,\gamma)^7$Be experiment, the ^{7}Be$(p,\gamma)^8$B experiments and the opacity and abundance estimates. When several experiments with different techniques converge on a value then one can perhaps ignore the problem value, but when no such convergence occurs it seems to me that one can be mislead as to the confidence one has in the SNU prediction if one ignores data which is outside of the statistical errors of a selected value. To treat that systematic error one can either use input errors which include both statistical and systematic errors as was done by Filippone and Schramm[3] or one can do calculations with all the systematically different inputs and present all results together (see Filippone, 1982)[5]. New ^{7}Be$(p,\gamma)^8$B values[5] are beginning to converge on a value almost 40% less than the "standard". The ^{3}He(α,γ) rate is converging near the standard but the low value of Rolfs' group has still not been explained. In addition, the two leading opacity calculations alone yield differences of about 1 SNU. It seems to me that to ignore these possible systematic differences and to just quote an overall error of ± 1 SNU on the standard model is to give a false sense of confidence. When we include some conservatively estimated systematic errors based on existing differences, with the statistical we increase our Monte Carlo error estimate from ~ 2 to ~ 3 SNU's even without including the disputed Rolfs value.

It is interesting to note that Bahcall's "best estimates" from the past decade scatter with a standard deviation of ~ 3 rather than his stated formal error of ± 1 SNU. Although I agree with Willy Fowler that some of these systematic errors can in principal eventually be eliminated, they have not yet been so eliminated on all pieces of input and as one gets minimized, new ones seem to crop up. Thus it appears that the solar neutrino "problem" may be less than a "2σ" problem.

The problem as I see it is not the magnitude of the descrepancy but whether or not there is a real descrepancy. As Filippone and Schramm demonstrated, the ^{37}Cl experiment is probably incapable of resolving this problem and the best solution is one upon which all sides agree, namely we need the Gallium experiment.

GRAVITATIONAL COLLAPSE

Gravitational collapse events produce $\sim 10^{53}$ ergs of ~ 10 MeV neutrinos. The Homestake Gold Mine Detector of Ken Lande, and the Mount Blanc Tunnel Detector of the Torino Group should be able to see a gravitational collapse event any place in our galaxy. From looking at the statistics of supernovae in other galaxies, it seems that the rate of such galactic collapse events should be ~ 1 every thirty years; even though the rate of visual supernovae within our part of the galaxy is only one every two hundred years. This higher number comes from the fact that a large fraction of our Galaxy is obscured visually from us. It would be nice if future proton decay detectors are designed so that as a by-product they can detect these gravitational collapse neutrinos. Unfortunately, the $1/r^2$ factor makes it prohibitively expensive with present technology to have a detector large enough to see supernovae in the Virgo cluster of galaxies where a supernovae goes off every few weeks.

HIGH ENERGY BACKGROUND

The High Energy background neutrinos come from proton-proton collisions producing π's and K's which then yield neutrinos. The major source of background (see Margolis, Schramm and Silberberg,[6] and Stecker and Learned,[7] DUMAND) will come from cosmic rays hitting the earth's atmosphere. This neutrino flux has been detected by Reines and his collaborators in a South African gold mine, and by experiments in the Kolar gold fields.

It is also conceivable that there could be high background fluxes due to bright early phases in the formation of galaxies when there may have been significantly higher cosmic ray fluxes, (such models have been proposed by Berezinsky and Zatsepin[8]). However, it is unlikely that a proton decay detector with only 10,000 tons would be able to detect such fluxes, even if they did exist.

There would also be background fluxes from cosmic rays hitting the galactic center. As evidenced by the observed π^0 γ-ray background, those fluxes would be even lower still, and tend to require detectors of at least $\sim 10^9$ tons. Very high energy neutrino backgrounds coming from cosmological distances may have their associated γ-rays degraded by γ-γ collisions with the 3^0 background[10]. However, energy conservation leads to the scattered γ's coming out at lower energy with higher multiplicity. It would take an exotic special model to get the observed low energy photons to be associated with a high energy ν flux.

However, there is an interesting limit that can be put on a very important cosmological problem; namely, deuterium production. Standard cosmological models have deuterium produced during Big Bang nucleosynthesis[11]. However, it is conceivable that deuterium might be able to be made by high energy spallation reactions in the early universe[12]. As Dave Eichler[13] has emphasized, such spallation reactions, in addition to producing deuterium, will also produce π's and K's and thus neutrinos. Since the amount of deuterium needed to be of cosmological significance is ~ 1 part in 10^5 of the mass of the universe, tremendous amounts of spallation reactions would have to take place, and thus there would be a very high neutrino background from any such process. Reines' experiment already had put severe limits on such models for deuterium production. New, more sensitive proton decay detectors would be able to improve these limits significantly, and should effectively rule out these spallation models for deuterium production.

Another background that may be interesting could be due to the decay of long-lived exotic particles produced in the Big Bang. Appendix I presents an example of one such object, monopolonium which is a monopole anti-monopole bound state.

POINT SOURCES

Neutrinos from point sources[14] could be exceedingly interesting for large neutrino detectors of $\gtrsim 10^9$ tons. This is true if the angular resolution is sufficiently fine so as to enable ν fluxes from certain precise directions to stand out above the diffuse background. However, we do know that there are severe limits on fluxes of such neutrinos, assuming the neutrinos are accompanied by π^0 gamma rays, as would occur if the proton-proton collisions in the sources were occurring without significant shrouding, obscuring the gamma rays. There are high energy γ-ray sources observed in Cygnus X-3, Centaurus A and the crab nebula[15]. From these γ-fluxes, it is clear that the neutrino fluxes would be so low, that they would not be detectable, with a 10^4 ton detector, and even their detectability with 10^9 tons is very model dependent.

The one possibility of getting neutrinos out without having these gamma ray limits imposed would be if there is suitable shrouding in matter (or radiation) to thermalize the γ's, and convert them into infrared radiation. Even then, from the limits on the infrared radiation from sources, it is clear that one would need a detector significantly larger than 10^4 tons before detecting such point sources. Thus, the only way around the limits from electromagnetic backgrounds of various sources would be if there is an object like a gravitational collapse event, which produces more energy in neutrinos than any form of electromagnetic radiation. At present, it seems very difficult to envision such a source, since to produce high energy particles, tends to require relatively low densities from which some form of radiation would get out.

One final point source which may be interesting, would be an extraordinarily energetic solar flare of the type which went off in 1956. Such flares may produce detectable neutrino bursts[13].

Another intriguing possibility concerns the Soudan I multi-muon events which seem to have a directionality towards the North Galactic Pole (the Virgo Cluster). Since the energy of these events is $\sim 10^{15}$ eV it is clear that to have such directionality requires a neutral primary. (The cyclotron radius for a 10^{15} eV proton in the galactic magnetic field is ~ 1 light year and yet the disk is <u>at least</u> several hundred light years thick.) Since neutrons (and other massive neutrals) are ruled out by time-of-flight considerations we are left with protons and/or neutrinos. As mentioned above any standard high energy neutrino source would have π^0 gamma's associated with it, so one would expect photons to be present. However, photons preferentially produce electrons, not muons so the multiplicity of the events is curious. Also, if the photons were removed by thick blanketing at the source or by scatterings of the 3° background if the energy were in error, then there should be about as many upward as downward moving events. If we are dealing with a photon source, then the question arises as to why high energy gamma-ray detectors have not seen such an intense source. Hopefully new γ-ray searches will help clarify this. At present this Soudan observation is still a mystery and needs further study. The answers may be encouraging to neutrino detectors.

An important point I wish to reiterate is that there is no "sure" high energy neutrino source, either diffuse or point source. Even for a 10^9 ton detector much less a 10^4 ton one, all predictions are model dependent. Even for those few sources where high energy gamma rays are seen, we always have to be aware that purely electromagnetic processes may be responsible for those γ's. (Of course, the discovery that a sensitive neutrino detector did not see ν's from those sources would be interesting since it would confirm the pure electromagnetic option.) However, the fact that there are no "sure" detectable sources should not stop people from looking and from developing new technology to look. (In this regard neutrino astronomy is almost in the identical position as gravitational wave astronomy.) The new technology may be beneficial for a variety of reasons and breakthroughs in sensitivity are certainly possible and of course, the most optimistic neutrino source models may be correct. But, perhaps the most compelling point is the fact that in the past the universe has always shown itself to be more inventive than theorists and probably for neutrino astronomy as with radio, infrared, X-ray and γ-ray astronomy before, the most exciting discoveries will not be model predicted ones.

APPENDIX I

Monopolonium

Chris Hill[16] has shown that monopolonium "molecules" will live the age of the universe if they have radii of $\gtrsim 0.1$ Å. Hill, in collaboration with J. B. Bjorken and I have looked into the astrophysical consequences of such objects. It can be shown that they will be currently radiating in the radio due to spin down radiation. In particular, thermally produced cosmological distribution of monopolonium will lead to a background radiation with a spectral peak at ~ 1 GHz for GUT monopoles of 10^{16} GeV mass.

The final annihilation due to the decay of the monopolonium will produce $\sim 10^{16}$ GeV events which will yield $\sim 10^6$ gluon jets. These jets will have angular spread of ~ 0.1 deg for standard GUT monopoles. Such jets may lead to time correlated high energy cosmic ray events. A detailed paper on the subject is currently being prepared.

ACKNOWLEDGEMENTS

I would like to acknowledge DOE grant DE AC02 80ER10773 and NSF grant AST 81 16750 at the University of Chicago, as well as the Hospitality of Fermilab, where the paper was prepared. I thank Chris Hill and James (BJ) Bjorken for interesting discussions, as well as my former collaborators Brad Filippone, Dave Eichler, Steve Margolis and Rein Silberberg.

REFERENCES

1. W. Fowler, in Proceedings of the Los Alamos Conference of Underground Physics, 1982.
2. R. Ulrich, in Proceedings of the Los Alamos Conference of Underground Physics, 1982.
3. B. Filippone and D. N. Schramm, Ap. J. 253, 1 (1982).
4. J. N. Bahcall, H. F. Huebner, S. H. Lubow, P. D. Parker, and R. K. Ulrich, Rev. Mod. Phy. (1982).
5. B. Filippone, "Nuclear Physics and the Calculation of the Solar Neutrino Flux", Ph.D. Thesis, December, 1982.
6. S. H. Margolis, D. N. Schramm, and R. Silberberg, Ap. J. 221, 990 (1978).
7. J. G. Learned and F. W. Stecker, in Proceedings of Neutrino 79, (Bergen, 1979), p. 461.
8. V. S. Berezinsky, and V. Zatsepin, in Proceedings of DUMAND 76, (Fermilab, Batavia, IL, 1977).
9. C. Fichtel, Phys. Rep. (to be published, 1983).
10. M. M. Shapiro, and R. Silberberg, in Proceedings of DUMAND 80, (Hawaii DUMAND Center, 1980) p. 262.
11. D. N. Schramm, and R. V. Wagoner, Phys. Today, 27, 40 (1974).
12. R. Epstein, J. Lattimer, and D. N. Schramm, Nature 263, 198 (1976).
13. D. Eichler, in Proceedings of DUMAND Ocean 77, (Naval Ocean Systems Center, San Diego, 1978).
14. D. Eichler and D. N. Schramm, Nature 225, 704 (1978).
15. T. Weekes, in Proceedings of DUMAND Ocean 77, (Naval Ocean Systems Center, San Diego, 1978).
16. C. Hill, Nuc. Phys. (in press, 1983).

Chapter VII

GRAVITY WAVES

Two beads slide almost freely on a smooth stick; only slight friction impedes their sliding. ... Plane gravitational waves impinging on the stick, push the beads back and forth. The resultant friction of beads on stick heats the stick; and the passage of the waves is detected by measuring the rise in stick temperature. (Of course, this is not the best of all conceivable designs.)

From an idea of H. Bondi in C. W. Misner, K. S. Thorne, and J. A. Wheeler, *Gravitation* (Freeman, San Francisco, 1970), p. 444.

AN HEURISTIC INTRODUCTION TO GRAVITATIONAL WAVES

Vernon D. Sandberg
Los Alamos National Laboratory
Los Alamos, New Mexico 87545

ABSTRACT

We describe in physical terms the phenomenon of gravitational waves. The philosophy of William Gilbert is used.[1]

"Since in the discovery of secret things and in the investigation of hidden causes, stronger reasons are obtained from sure experiments and demonstrated arguments than from probable conjectures and the opinions of philosophical speculators of the common sort; therefore to the end that the noble substance of that great loadstone, our common mother (the earth), still quite unknown, and also the forces extraordinary and exalted of this globe may the better be understood ..."

INTRODUCTION

The gravitational interaction is the weakest of all known elementary physical interactions and on the size scale of particle physics it has been generally ignored. In the macroscopic world it is the force we deal with on a daily basis and it defines to a great extent our physiological and psychological conceptions of the world. On a cosmic scale it is the medium of the Universe itself. Gravity was the first physical interaction to be given an analytic treatment and Newton's synthesis of the world has remained valid over a very impressive range of phenomena. Einstein's general theory of relativity has extended this range to include the universe itself as a dynamic object and has brought forth a series of exotic objects, such as black holes, into our view of the world.

General relativity was constructed by Einstein through an enormous intellectual feat, based on carefully thought out comparisons of different areas of physics and by demanding a consistent and unified view. General relativity was produced theoretically. The experimental consequences of this theory of gravity only became known afterwards and then as a list of tests: The bending of light by the sun, the advance of the peribelion of the orbit of Mercury, and the gravitational redshift of light. The prediction of a dynamic universe was not accepted until Hubble's observations forced the idea.

The technological advances of experimental physics are affording us the possibilities of observing non Newtonian gravitational fields. In addition to being interesting as tests of general relativity, detection of these fields will give rise to a new astronomy to be compared with the present electromagnetic and weak interaction (e.g. neutrino) astronomies. Neutrino astronomy is exemplified by R. Davis's solar neutrino flux observations, discussed previously in this proceedings. The experimental searches for gravitational waves and inductive gravitational fields are described in W. Oelfke's, R. Spero's, and R. W. P. Drever's following articles. Experimental tests of terrestrial Newtonian gravity have been described by F. Stacey in the proceeding session. The purpose of this present article is to provide a rough and somewhat heuristic theoretical background and introduction to gravitational radiation, its generation, and its detection based on Einstein's general theory of relativity. For details and further analysis the reader is invited to consult references 2, 3, or 4.

GENERAL RELATIVITY

General relativity theory describes the "gravitational field" as the "geometry" of a four dimensional mathematical manifold and the remainder of all physics are the fields (scalar, spinor, vector, tensor) that may be assigned to this manifold. The notion of a "geometry" is given an analytic realization in the form of the metric tensor field $g_{\mu\nu}(x)$ and its physical interpretation comes from the infinitesimal line element

$$ds^2 = g_{\mu\nu}(x) \, dx^\mu \, dx^\nu \,.$$

This determines either the temporal or spatial distance between two points with coordinates x^μ and $x^\mu + dx^\mu$, respectively separated by a timelike or spacelike interval. The metric field describes everything of physical interest. For example, test particle trajectories are given by the paths of extremal interval as determined by $g_{\mu\nu}(x)$. In flat spacetime, in what are usually referred to as Galilean coordinates, $ds^2 = c^2 dt^2 - dx^2 - dy^2 - dz^2$, $g_{\mu\nu}(x) = \eta_{\mu\nu} \equiv \text{diag}(1,-1,-1,-1)$ and the extremal paths are "straight lines". This is Newton's law of inertia, for in a flat spacetime geometry we would say there is no gravitational field acting on the particle.

Due to the freedom of choice of coordinates to label the points of the manifold, the unambiguous determination of equivalent gravitational fields is difficult. Metric fields with different functional forms may be transformable into one another by a coordinate transformation. The coordinates are only labels and in general carry no physical significance beyond being labels. The Galilean coordinates of flat spacetime possess additional geometric properties that makes inertial motion look simple in terms of them. This is reflected in the simple form the flat spacetime metric field takes when written in terms of Galilean coordinates (c.f. writting the spatial terms in spherical-polar coordinates). An invariant description of a geometry is given in terms of the curvature, a fourth rank tensor field derived from $g_{\mu\nu}(x)$ and it first and second derivatives. Geometrically this tensor measures the "curvatures" of two dimensional surface elements and its vanishing provides the necessary and sufficient conditions for the spacetime to be flat. Physically the curvature tensor is the manifestation of a "real" gravitational field. (It is the appropriate analog of the electromagnetic field). Consider two nearby observers following their metrically defined "straight" paths (i.e. two geodesics infinitesimally close). Physically they correspond to two freely falling observers. The curvature tensor describes how the distance they measure between them varies as they move through the spacetime. (In fact this equation of geodesic deviation takes on the form of Newton's second law with one observer measuring the local acceleration of the neighboring observer and putting this equal to a term involving the curvature tensor and the observers' four-velocities.) This distance is coordinate independent and provides the physical manifestation of the gravitational field through the curvature.

Einstein's field equations describe how the density of stress and energy in spacetime curves and bends the geometric properties of the spacetime. They are a set of second order nonlinear partial differential equations for the $g_{\mu\nu}(x)$ with source terms constructed from the stress-energy tensors of the "matter" fields. An interesting consequence of Einstein's equations is that they comprise an overdetermined set. They are consistent, however; because they satisfy a number of constraints. These constraints impose conditions upon the matter field sources that become the equations of motion or field equations for the sources themselves.

A general procedure for solving the field equations does not exist due to their enormous complexity. The exact solutions that exist are based on exploiting some symmetry to simplify the field equations. There are exact solutions known for the astrophysically important cases such as the Robertson-Walker metrics that describe open and closed homogeneous and isotropic cosmologies and the Kerr-Newman metrics that describe electrically charged rotating black hole solutions. (The Schwarzschild black hole is the neutral, nonrotating case of the later.)

In spite of their complexity, it is possible to make predictions about the physical processes that go on. The approach is to use a perturbation analysis about one of the exact solutions. The dominant strong field nonlinear features are then taken care of at lowest order and smaller deviations, that can be calculated, may be used to describe the variations. This way, for example, the dynamical evolution of a star undergoing gravitational collapse can be followed and estimates of the radiation it generates (electromagnetic, gravitational, neutrino) can be made.

GRAVITATIONAL WAVES

If the field equations are expanded about the flat spacetime solution a set of equations describing a spin-2 massless field that propagates as a wave traveling at the speed of light results. These waves are transverse and have two polarization states. They manifest themselves as shears in the geometry. Consider a set of freely falling observers on which a gravitational wave is impinging. If one observer is selected as the reference, then as the wave passes the distance she measures to the other observers will oscillate in time with the frequency of the wave and the magnitude of the change in distance will be directly proportional to the original baseline distance, i.e., the wave's amplitude is described by a strain. If she measures the distance to a ring of observers surrounding her, she will find that at any given instant two diametrically opposite observers will be moving towards her and at a right angle to them two other similarly situated observers will be moving away from her. The motion of the ring of observers is that of a pulsating ellipse oscillating from a condition of maximum eccentricity, through to a circle, to minimum lateral extent, and then back out. The area of the ring will remain constant, only its eccentricity will change. The observers will find that their displacement is transverse to the direction of propagation of the wave. For any such wave the observers' motion can be described as a linear superposition of two elliptical figures, one being rotated 45° relative to the other. These represent the two states of linear polarization of a gravitational wave. As with an electromagnetic wave states of circular polarizations can be defined by retarding one state of a linearly polarized wave a quarter wavelength and adding it to the other linearly polarized state. The result is an ellipse that appears to rotate either clockwise or counterclockwise.

GRAVITATIONAL WAVE DETECTORS

In practice the strain induced in the spacetime geometry by a wave is measured by the displacement it induces in either an elastic bar or the change in optical path length of an interferometer.[5] In either case the length to be measured is infinitesimal. For the expected gravity wave strains of 10^{-17} from astrophysical sources the displacement of a one meter long aluminum bar is one hundredth the diameter of an atomic nucleus! This measurement has meaning only is the statistical sense of making position measurements averaged over the $\sim 10^{23}$ atoms on the face of the bar. Gravitational waves of kilohertz frequencies with strains on the order of 10^{-21} are still highly classical fields, however; the antenna systems being built to detect them are highly quantum mechanical systems. In an aluminum bar at a few millidegrees above absolute zero the first level of oscillation of the lowest normal mode of the bar has an energy of the same size as that associated with the displacement induced by a gravitational wave of strain amplitude 10^{-21}. This is referred to as the (apparent) "quantum limit of detectability". It can be circumvented by a more careful choice of measurement strategies than a simple position measurement.

The weakness of the interaction of a gravitational wave with an antenna and the subsequent attempts to make measurements on the antenna force us into treating the antenna as a quantum mechanical object. To an excellent approximation the antenna can be regarded as a simple harmonic oscillator that is driven by a classical force (the gravity wave) and by Nyquist forces (i.e. thermal noise). The

Nyquist forces may be quantified by the "Q" of the particular normal mode we are observing. This measures the coupling of this particular mode to all the other modes in terms of how quickly this mode looses energy. By using bars of crystaline material (e.g. saphire) of good quality and by operating at cryogenic temperatures the Nyquist forces can (at least in principle) be made negligible. (This is a major technological feat and should not be underestimated!) The idealized model to be examined then consists of a quantum mechanical harmonic oscillator interacting with a classical force and a measuring apparatus. The interaction of the oscillator with the measuring apparatus will necessarily disturb the oscillator (for example a position measurement will impart an uncertainty in the oscillators' momentum) and depending upon the disturbance generally contaminate the result of any subsequent measurement. This appears as an irreducible "noise" produced by the quantum nature of the oscillator-measurement apparatus interaction. The key to its elimination is in the measurement interaction. Instead of measuring the position which disturbs the momentum and the subsequent evolution of the oscillator's motion, we enlarge the class of possible measurements by allowing time dependent coupling of the measurement apparatus to the oscillator. By using a position transduce and a velocity transduce and modulating their couplings to the oscillator, we can arrange to couple to a dynamical variable of the oscillator that is not distrubed by the back reaction of the measurement interaction. It is in principle possible to measure in detail the excitation of the oscillator by the gravitational wave and the quantum mechanical nature of the oscillator is not a limiting principle. The study of these ideas has collected the name "quantum nondemolition measurements" and is elaborated in detail in reference 6.

FINAL COMMENTS

In addition to radiative solutions, the flat spacetime background expansion also has solutions that correspond to magnetic like gravitational effects and weak corrections to Newton's equation for the gravitational field. These "non-Newtonian" fields provide additional interesting tests of general relativity. Experiments to detect gravitational waves and to measure departures from Newtonian gravity require a reasonable amount of isolation as would be offered by an underground laboratory. The specific type of disturbance and amount of isolation required are very experiment dependent. These are discussed in detail in the following papers.

REFERENCES

1. William Gilbert, <u>On the Loadstone and Magnetic Bodies</u>, translation into English contained in <u>The Great Books</u> (Univ. of Chicago Press, Chicago, 1952).

2. L. D. Landau and E. M. Lifshitz, <u>The Classical Theory of Fields</u>, 4th Ed. (Pergamon Press, N.Y. 1975).

3. C. W. Misner, K. S. Thorne, and J. A. Wheeler, <u>Gravitation</u>, (W. H. Freeman and Co, San Francisco, 1973).

4. S. Weinberg, <u>Gravitation and Cosmology Principles and Applications of the General Theory of Relativity</u> (John Wiley and Sons, N.Y., 1972).

5. L. Smarr, Editor, <u>Sources of Gravitational Radiation</u>, (Cambridge University Press, Cambridge, 1979).

6. C. M. Caves et al., "On the Measurement of a Weak Classical Force Coupled to a Quantum - Mechanical Oscillator I," Rev. Mod. Phys. <u>52</u>, 341 (1980).

REVIEW OF RESONANT BAR GRAVITY WAVE EXPERIMENTS

W. C. Oelfke*
Louisiana State University, Baton Rouge, LA - 70803

ABSTRACT

A review is given of the experimental efforts since 1958 to detect gravitational waves using Weber-type resonant bar antennas. Special attention is paid to the continued funding of and collaboration between research groups and the importance of this kind of support in experimental searches for rare events. Evidence for an atmosphere of mutual support between gravitational wave groups on an international level is seen in the rapid progress toward greater antenna sensitivity and improved measurement techniques.

INTRODUCTION

Experimental efforts to detect gravitational radiation were begun in 1958 by Josph Weber. After an extensive study of antenna-transducer designs[1] Professor Weber assembled a room temperature gravitational wave detector consisting of quartz piezoelectric strain sensors bonded to a 1400 kg aluminum cylinder, and in 1969 he reported the first tentative evidence of gravitational radiation of cosmic origin[2]. Immediately, efforts to detect gravitational radiation at room temperature were begun at Bell Labs, Rochester, Munich, Frascati, Moscow and Glasgow while more sensitive low temperature systems were initiated at L.S.U., Rome, and Stanford. By 1975 it had become clear that the room temperature resonant bar antennas would not provide sufficient sensitivity to detect any but the strongest bursts produced by nearby collapses which are no more frequent than one per century. A great cooperative effort was begun to find and develop a second generation of antennas with greatly improved sensitivities to gravitational wave bursts. The cryogenic systems had provided not only a low antenna temperature but had also opened the door to a family of low temperature transducers based on SQUID and superconducting devices. These transducers offered a means of approaching quantum limited antenna sensitivity

$$\Delta x \simeq (\hbar/m\omega)^{1/2},$$

for an antenna of effective mass m and resonant frequency ω. Also it had been realized that the effective noise temperature of the system not only depended on the antenna temperature, but also varied inversely with the mechanical antenna Q factor. Efforts were therefore made to develop second generation antenna systems with low-antenna temperatures and high mechanical Q factors.

---

*Department of Physics, University of Central Florida, Orlando, FL - 32816

Special materials such as single crystal sapphire and silicon were used in place of aluminum for antennas at Rochester[3] and Maryland[4]. Niobium was found to have high mechanical Q. Since it is a superconducting material it can be supported by magnetic levitation[5].

The four groups in the United States that have continued to develop new resonant bar gravitational wave antenna systems are located at L.S.U.[5], Maryland[1], Rochester[3], Stanford[6]. Members of this group have shared data, equipment and ideas and have constructed some large pieces of apparatus on small budgets. Nevertheless, "levitation" is still used to refer to a low temperature magnetic effect rather than to the "no visible means of support" mentioned by A. K. Mann[7]. In fact the gravitational wave search in the United States has been supported with enthusiasm by the Air Force, N.A.S.A. and other agencies and has been carried forward by the continued support of the National Science Foundation.

SOURCES OF GRAVITATIONAL RADIATION

This experimental work of the last decade has been accompanied by a large effort in theoretical physics to expand the body of knowledge about sources of gravitational radiation[8,9] behavior of graviatational wave antennas[10,11] and the nature of quantum-limited measurements of gravitational wave-induced strains[12,13]. Currently it is believed that the most energetic gravitational wave sources in the frequency range, 500 to 2000 Hz, of the resonant bar antennas are stellar collapse events associated with supernovae[14]. According to our best knowledge of the astrophysics in our galaxy, these events will produce one event every 10 to 50 years with a local strain amplitude $\Delta \ell / \ell$ of 10^{-17} to 10^{-19}. In order to see collapse events in the Virgo Cluster of Galaxies at a rate of one per month it will be necessary to achieve antenna strain sensitivities of 10^{-20} to 10^{-22}. Steve Detweiler[15], in his presentation on gravitational wave sources, pointed out that large mass collapsed objects could give rise to gravitational waves of greater amplitudes and lower characteristic frequencies. Ron Drever[16] gave an in-depth description of the laser interferometer systems currently being constructed to detect these lower frequency waves.

TRANSDUCERS

The most widely used transducer for detecting resonant bar antenna vibrations has been the piezoelectric strain gauge. This transducer, coupled to the bar near the point of maximum strain provides strain sensitivities to below 10^{-17} for both room temperature and cryogenic antennas. These transducers have been used by gravitational wave groups at Maryland, Bell Labs, Rochester, Stanford, L.S.U., Munich, Rome, and China. A DC biased capacitive transducer has been developed at Maryland[17] to be used as an accelerometer whose coupling to the antenna, β, is increased through resonant matching. Although simple in design and adaptable to a wide variety of antenna systems, both of these transducers are bilateral and therefore couple amplifier current noise back into the antenna. This back action noise then sets the limit of sensitivity of such systems at a level somewhat higher than that needed to observe the collapse events in the local galaxy.

The low temperature environment enables the use of accelerometer transducers with gain and reduced internal noise levels. A low temperature inductive accelerometer has been developed at Stanford[18] and is also being used at Maryland. The proof mass of this accelerometer, resonantly coupled to the antenna, consists of a niobium diaphram whose movement induces magnetic flux changes in nearby superconducting coils. The resulting magnetic field changes are sensed by a DC SQUID. The Stanford antenna has been operated at an effective temperature of 30 mK with such a transducer and this group is presently attempting to push T_{eff} to 0.1 mK, the SQUID noise limit for their system.

Accelerometers based on the high phase sensitivity of superconducting rf cavities have been developed at L.S.U., Western Australia, and Rochester. These transducers exhibit high parametric gain and lend themselves to special coupling schemes that offer a means of controling the back action noise so as to reduce its effect on the antenna signal[19]. This year David Blair[20] of the Western Australia group reported operating such a transducer at an effective temperature of 3 mK.

ANTENNA SENSITIVITY

Strain sensitivities of resonant bar antennas have been calculated for a variety of noise conditions, however, the condition most commonly encountered at the present level of antenna development is the case where Nyquist forces dominate and the amplifier is matched to the transducer. Under these conditions the limit to sensitivities is given by[21]

$$\Delta \ell/\ell \geq \pi (15/16)^{1/2} (mc_s^2)^{-1/2} (32kT_{amp} kT_{ant}/\beta Q)^{1/4},$$

where m, c_s and Q are the effective mass, sound velocity and mechanical Q factor of the antenna respectively, β is the transducer-to-antenna coupling coefficient and T_{amp} and T_{ant} are the amplifier noise and antenna temperatures respectively.

Since antenna materials and sizes very somewhat from research group to research group, although we all share a common state-of-the-art in transducer and amplifier development, system sensitivities tend to be described in terms of an overall effective noise temperature

$$T_{eff} = (2T_{ant} T_{amp}/\beta Q)^{1/2}.$$

This definition points directly to the four system properties upon which each group focuses its research efforts. Recently the gravitational wave group in Tokyo[22] reported that the aluminum alloy 5056 possessed much lower acoustical losses than the softer alloys presently being used by the resonant bar groups. With the prospect of increasing antenna Q factors from 10^6 to 10^8 all of these groups are now preparing 5056 antennas. Likewise new developments in low-noise amplification, ultra low-temperature refrigeration and methods of increasing transducer coupling are met with equal enthusiasm.

Perhaps the most striking thing about the gravitational wave effort is that although it is a rare event search that has yet to produce dramatic evidence of confirmed gravitational wave pulses, the development has continued at an unflagging pace over the past decade. We are now operating atennas at sensitivities sufficient to see not only the strongest events allowed by physics but the strongest events most likely to be occurring in our galaxy. The history of this antenna development, represented in Fig. 1, is a tribute to the continued support of this research and the excellent cooperation between research groups. As in many rare event searches, we are each aware of the importance of the largest possible number of observations of the event when it occurs. The development of a national cooperative effort or special facility such as the proposed National Underground Science Facility can greatly enhance our ability to study rare events.

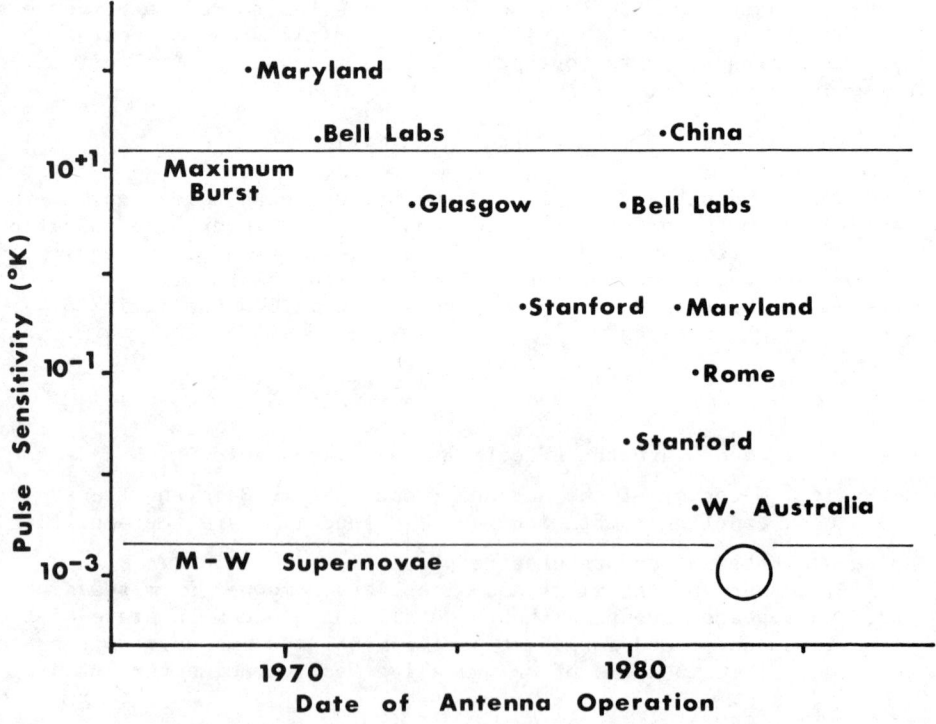

Fig. 1. A developmental chart of the pulse sensitivities of operating gravitational wave antennas expressed in terms of effective antenna temperature. Also shown are the levels at which the strongest possible events would appear and the most likely Milky Way galaxy collapse events would appear.

REFERENCES

1. J. Weber in General Relativity and Gravitation, Vol. 2, edited by A. Held, Plenum Publishing Corporation (1980).
2. J. Weber, Phys. Rev. Lett. 22, 1302 (1969).
3. D. H. Douglass in Gravitazione Sperimentale, Roma Accademia Nazionale dei Lincei, Roma (1977).
4. J. Weber in Gravitazione Sperimentale, Roma Accademia Nazionale dei Lincei, Roma (1977).
5. W. O. Hamilton, T. P. Bernat, D. G. Blair and W. C. Oelfke in Topics in Theoretical and Experimental Gravitation Physics (Erice), V. de Sabbata and J. Weber (eds.), Plenum Publishing Company, New York (1977).
6. S. P. Bougn, W. M. Fairbank, R. P. Giffard, J. N. Hollenhorst, M. S. McAshan, H. J. Paik and R. C. Taber, in Gravitazione Sperimentale, Roma Accademia Nazionale dei Lincei, Roma (1977).
7. A. K. Mann, Proposal for a National Underground Science Facility, these proceedings.
8. L. Smarr, ed., Sources of Gravitational Radiation: Proceedings of the Battelle Seattle Workshop July 24 to August 4, 1978, Cambridge University, Cambridge, England (1979).
9. K. S. Thorne in Theoretical Principles in Astrophysics and Relativity, N. R. Lebovitz, W. H. Reid and P. O. Vandervoort (eds.), University of Chicago, Chicago (1978).
10. C. W. Misner, K. S. Thorne and J. A. Wheeler, Gravitation, Freeman, San Francisco (1973).
11. R. P. Giffard, Phys. Rev. D 14, 2478 (1976).
12. V. B. Braginsky and Y. I. Vorontsov, Usp. Fiz. Nauk. 114, 41 (1974).
13. C. M. Caves, K. S. Thorne, R. W. P. Drever, V. D. Sandberg and M. Zimmerman, Rev. Mod. Phys. 52, 341 (1980).
14. K. S. Thorne, Rev. Mod. Phys. 52, 285 (1980).
15. S. Detweiler, Theory of Gravity Waves and Sources, invited talk presented at this workshop.
16. R. W. P. Drever, Laser Interferometers, these proceedings.
17. J. P. Richard, Rev. Sci. Instrum. 47, 4 (1976).
18. H. J. Paik, J. Appl. Phys. 47, 1168 (1976).
19. W. C. Oelfke in Quantum Optics, Experimental Gravitation and Measurement Theory, P. Meystre and M. O. Scully (eds.) Plenum, London, to be published.
20. D. G. Blair, Reported at the Third Marcel Grossmann Meeting, Shanghai (August 30 to September 3, 1982).
21. R. Weiss in Sources of Gravitational Radiation, L. Smarr (ed.), Cambridge University, Cambridge (1979).
22. T. Suzuki, K. Tsubond and H. Hirakava, Phys. Lett. A 67. 2 (1978).

LASER INTERFEROMETER GRAVITATIONAL RADIATION DETECTORS

R.W.P. Drever
California Institute of Technology, Pasadena, California, 91125
and University of Glasgow, Glasgow G12 8QQ, Scotland

ABSTRACT

Some techniques proposed or currently under development for detection of gravitational radiation by laser interferometers are reviewed, with particular emphasis on experiments covering the lower frequencies potentially accessible to ground based instruments.

INTRODUCTION

Surveys of expected sources of gravitational radiation[1] suggest that gravitational waves are arriving at the earth over a wide frequency spectrum and in a variety of forms - including pulses and bursts from stellar or black hole collapse or collision processes, periodic signals from pulsars or binary systems, and a general stochastic background. Most of the experimental work done in this field so far has used resonant bar gravitational wave detectors in searches for millisecond pulses which might be produced in collapse of stars to give supernova or black holes. An alternative detection technique, in which laser interferometers are used to sense relative motions of "free" test masses, is showing promise for future experiments of this type which may be significantly more sensitive; and also for wideband searches over an extensive range of frequencies for signals from various different kinds of source. We will review here some of these experimental developments, and possibilities which look likely to open up.

The experimental difficulties of gravitational wave detection arise almost entirely from the extremely small magnitude of the effects to be observed, and it may be useful to give an indication of these magnitudes. It is convenient to think here in terms of the amplitude of the gravitational radiation - which corresponds approximately to the maximum fractional change in apparent distance between two free test masses which can be produced by the passing wave. In these terms, a gravitational wave pulse from a star in our galaxy collapsing to give a supernova or a black hole might have an amplitude of order 10^{-18} to 10^{-20}, and a duration of order 1 millisecond. This is a small amplitude, and to bring it into perspective it might be noted that the most sensitive gravity wave detector to date - the cryogenic bar detector at Stanford University - has only recently approached this sensitivity. Moreover the number of such collapse events in our galaxy is expected to be small - perhaps only one per 10 to 30 years. It would clearly be desirable to observe signals more frequently than this; and one way of doing this would be to improve sensitivity sufficiently to detect events in a large number of galaxies. To detect signals of this type at the rate of once per month it is estimated that a detection sensitivity in the range 10^{-20} to 10^{-22} may be required. Considerations of this type make a sensitivity of about 10^{-21} seem a good target for future experiments, and indeed a number of other possible radiation mechanisms look likely to give signals of about this amplitude also. This target is not an easy one, however. It corresponds to achieving a gravitational wave flux sensitivity better by a factor of 10^6 than that of current instruments; and in test masses 1 kilometer apart it would correspond to relative motions of only 10^{-18} m.

Signals of larger amplitude may be expected at lower frequencies, from collapse or collision processes involving supermassive black holes. It has been suggested[2], for example, that black holes of mass 10^5 times the solar mass, at the Hubble distance, could lead to bursts of duration about 10 seconds and amplitude of order 10^{-18}. For this and other reasons there is considerable incentive for extending gravitational wave searches to lower frequencies.

Detection of periodic gravitational waves is also of considerable interest. It has been shown[3] that pulsars may in general be expected to radiate gravitational waves at frequencies near the radio pulse repetition frequency and at twice this frequency, and measurements of signal amplitudes would give information on the pulsar shape, orientation and structure. Estimates of amplitudes are uncertain at present, but for the Crab pulsar, for example, a gravitational wave amplitude of order 10^{-26} might be expected at a frequency of 60 Hz. In searches for a signal of known frequency, such as this, sensitivity can be significantly improved by suitable data analysis and integration over an extended observing time, and detection of this kind of signal may become quite practicable, although outside the range of current experiments.

The orders of magnitude of expected gravitational wave signals just quoted give some idea of the problems to be faced, and give good but difficult targets for planning experiments. It is not at all impossible that there are stronger signals than these, produced by mechanisms not yet considered, and the amplitudes could be much larger than those indicated without violating any basic ideas. Experiments at lower levels of sensitivity are well worth carrying out but the targets mentioned are interesting ones to aim at in the long term. We will consider now what might be done using laser interferometer techniques.

BASIC ARRANGEMENT

In this approach changes in separation between test masses induced by the gravitational wave are sensed optically; and to avoid need for extreme absolute stability a differential measurement is made along two baselines perpendicular to one another, which may be oppositely affected by a gravitational wave. A possible arrangement, in principle, is indicated in Fig. 1. Here three test masses are suspended like pendulums, with periods long compared with the period of the gravitational waves of interest, and are placed to

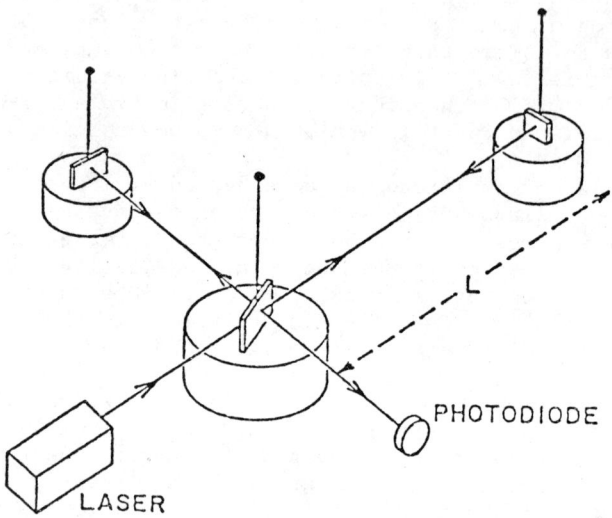

Figure 1

form a pair of perpendicular baselines, each of length L. Gravitational waves travelling in a vertical direction may give a differential motion which could be detected by a Michelson interferometer which monitors distances between two mirrors and a beamsplitter attached to the masses as indicated.

Several factors could limit the sensitivity of this simple gravity wave detector. A fundamental one is the quantum limit to detectable change in position of the test masses corresponding to the uncertainty principle, which sets a limit to displacement detectable over time τ of $(2\hbar\tau/m)^{\frac{1}{2}}$ where m is the mass and $2\pi\hbar$ is Planck's constant. With a baseline of 40 m, and 100 kg masses, the corresponding gravity wave amplitude for a 1 millisecond pulse is about 10^{-21}. This quantum limit is near our target sensitivity but it can be reduced by increasing the baseline or the size of the masses, and is unlikely to be the most serious difficulty.

A more important practical limit may come from photon counting error. Statistical fluctuations in photon counting rate in the output light from the interferometer set a limit to the change in optical path which is detectable. The corresponding limit to detectable gravitational wave amplitude is given approximately by $(\lambda\hbar c/8\pi L^2 I\tau)^{\frac{1}{2}}$, where I is the power of the laser illuminating the interferometer, λ is the wavelength of the light, c is the velocity of light, and the photodiode is assumed to have unity quantum efficiency. With a 40 m baseline and a laser giving 1 watt of light at 500 nm the photon counting limit to gravitational wave amplitude sensitivity is of order 2×10^{-17} for a 1 millisecond pulse duration - which is far from our target sensitivity. Sensitivity might be improved by increasing baseline length or laser power, but it would be difficult to reach our target by doing this alone, and other methods are necessary. It may be noted here, however, that pioneering experiments with a laser interferometer gravity wave detector were carried out with a configuration essentially similar to this one by Forward[4], who achieved performance near the photon noise limit with a low power laser.

The displacement sensitivity of a Michelson interferometer may be improved considerably by arranging that the light within each arm is reflected back and forwards many times between mirrors attached to the test masses before recombining at the beamsplitter, so that the optical path difference resulting from a given mirror motion is increased. This multireflection technique was suggested in this context by R. Weiss[5], who proposed use of Herriott delay lines to achieve the many reflections.

MULTIREFLECTION MICHELSON INTERFEROMETERS

A schematic diagram of one arrangement for a multireflection Michelson interferometer gravity wave detector is shown in Fig. 2.

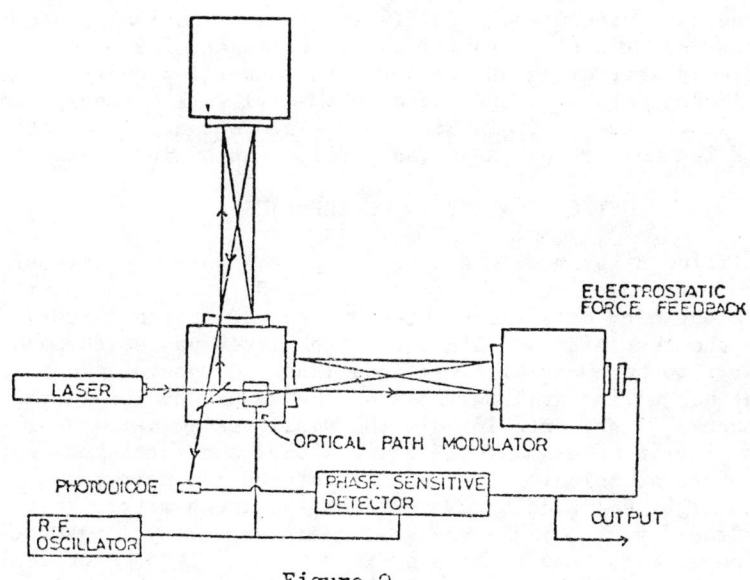

Figure 2

Here one of the optical paths is phase modulated at a high radio-frequency and an electrostatic force feedback system is used to control the position of one of the masses, and lock the interferometer onto a dark fringe of the interference pattern, where in this case the optimum photon-noise-limited sensitivity and useful rejection of low-frequency laser intensity noise is obtained.

If reflection losses in the mirrors are negligible, then the improvement in gravity wave sensitivity obtained by this type of multireflection system is proportional to the number of round-trip traversals between the test masses made by the light. In a large instrument with low-loss mirrors a limit to the number of useful reflections may be reached when the light spends a time within the system comparable to the time-scale of the gravitational wave, and the photon noise limit to sensitivity then becomes independent of the arm length L, and is better than that of a simple Michelson interferometer by the factor $2L/c\tau$. For a 1 millisecond pulse and a laser power of 1 w, this limiting sensitivity is about 2×10^{-21} (at unity signal to noise ratio).

Early experimental work on multireflection Michelson interferometers at the Max Planck Institute in Munich and at the University of Glasgow showed that incoherent scattering at the mirrors can give serious noise in these systems, since light which reaches the photodetector by a path different from that of the main beam can have its phase modulated significantly by quite small fluctuations in laser frequency. The Munich group suggested[6] avoiding this problem by carefully controlled frequency modulation of the laser, designed to cause the phase of scattered light to average to zero; while the Glasgow group suggested use of a different type of optical system[7] based on Fabry-Perot optical cavities, which looked likely to be less affected by scattering, and to have other useful properties in addition. A considerable amount of experimental work has now been carried out at Glasgow, and also in a newer project at the California Institute of Technology, on these Fabry-Perot cavity interferometers.

OPTICAL CAVITY INTERFEROMETERS

A simplified diagram of one type of optical cavity gravitational wave detector is shown in Fig. 3. Here Fabry-Perot optical cavities are formed along each baseline between highly-reflecting mirrors attached to the test masses. Light from the laser passes through a beamsplitter into both cavities. If the length of one of the cavities matches an integral multiple of the laser wavelength then resonance occurs, light entering via the small transmissivity of the input cavity mirror makes many reflections back and forth between the cavity mirrors, building up to a high stored intensity, and the phase of the light emerging backwards from the input mirror then varies very rapidly with small changes in either cavity length or laser wavelength. The cavity behaves in some ways like an optical delay line with all the beams folded on top of one another. The phase difference between the light emerging from within the cavity

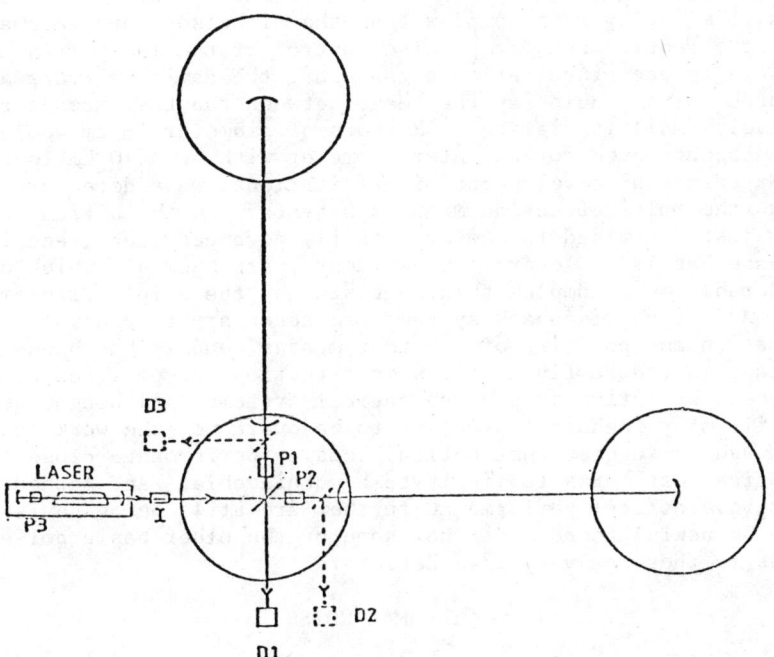

Figure 3

and the light from the laser can be monitored by phase modulating the input light at a high radiofrequency using electro-optical modulators P1 and P2, and synchronously demodulating the signal from photodiodes D2 and D3 which detect part of the light coming back from the cavity mirror. For each of the cavities, the optical phase difference obtained is a measure of the amount by which the cavity resonance deviates from the wavelength of the light. If one of the phase signals is used to stabilise the laser wavelength, and lock the laser to that cavity, and the other is used to adjust the average length of the second cavity to match the laser, then it would be possible to look for gravity wave signals by taking a difference between the two phase error signals. Slightly higher sensitivity may in principle be achieved by directly measuring the difference in phase between light from the two cavities by means of photodetector D1, if the modulators P1 and P2 are driven in antiphase with one another.

This complete interferometer behaves in some respects like a multiple reflection Michelson interferometer; and it can be shown that the sensitivity is effectively similar to that of a multi-

reflection Michelson with the same light storage time, provided
absorption losses in the input mirrors are not important. The
system is slightly more complex than the Michelson interferometer
due to the requirement for precise control of the laser frequency,
but there is the practical advantage that the cavity mirrors and
the vacuum pipes enclosing the beams between the test masses can be
relatively small in diameter. Mirrors of diameter 18 cm would be
quite adequate even for an interferometer with arms 10 km long.

Experimental development of gravitational wave detectors based
on both the multireflection Michelson system and the optical cavity
system just described is now at a fairly advanced stage, and in
each case has led to overall experimental arrangements which are
considerably more complex than suggested by the simple diagrams
here. Additional feedback systems are necessary to control
orientation and position of the test masses, and it has been found
important to reduce fluctuations in direction and position of the
laser beam by active or passive optical systems. Although there
are still many technical problems to be overcome, the work has gone
far enough to suggest that optical sensing performance close to
theoretical estimates is likely to be achievable. And methods which
may improve optical performance further are still being devised[8].
It may be useful to consider now some of the other basic noise
sources in these gravity wave detectors.

THERMAL NOISE

Thermal noise is much less important in laser interferometer
gravity wave detectors than in resonant bar detectors since there is
no direct connection between the test masses at the gravitational
wave frequency. Some thermal noise may enter the system through
losses in the pendulum suspension, however, along with Brownian
noise from collisions of residual gas molecules. A separate source
of thermal noise arises in the internal thermal motions of the test
masses themselves and of associated structures supporting mirrors
and other components. We will not take space to discuss these noise
sources in detail here, since they are covered in the companion paper
by Spero[9] and in other accounts[5,8], but we note that with careful
mechanical design it would seem practicable to keep this noise small
in the kiloherz region of the spectrum. At lower gravity wave
frequencies - approaching 1 Hz or lower, suspension and gas noise
may become a major problem.

ISOLATION FROM SEISMIC DISTURBANCES

Seismic noise is a well recognised problem for all gravitational
wave detection experiments, and very effective isolation techniques
have been developed for frequencies in the neighbourhood of 1
kiloherz for work with resonant bar detectors. Stacks of alternate
layers of rubber and lead or steel give good attenuation at these
frequencies, and even the simple pendulum suspension of the test
masses in a free mass detector can give useful high frequency

isolation. As we consider looking for gravitational waves at lower frequencies, however, the problems become rapidly more difficult. Seismic noise amplitudes increase, and as the attenuation of a simple spring-mass isolator at a frequency f, well above its resonance frequency f_o, is proportional to $(f_o/f)^2$ it becomes harder to achieve good passive isolation. There do exist several special suspension systems with very low natural frequencies, and we have considered, for example, a differential torsion balance arrangement as shown in Fig. 4. as a low frequency gravitational wave detector[10], with an interferometer to sense changes in

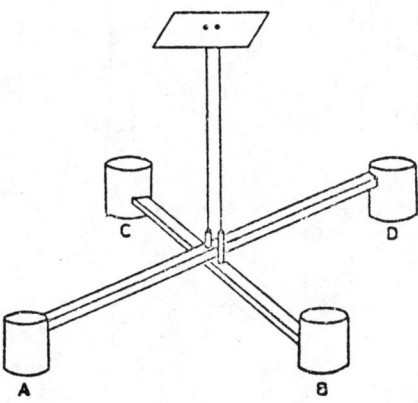

Figure 4

distances between the four test masses. However this would be an awkward arrangement to make on a large scale. For experiments in the frequency range from 1 kHz down to about 10 Hz there are several ways of improving the effective isolation achievable with a pendulum suspension. One relatively simple possibility is to monitor changes in inclination of the wire suspending the test mass, and from this deduce, and correct for, accelerations due to ground motion. To make such horizontal acceleration monitoring independent of tilts of the ground we introduced the idea of a "reference arm" - a bar freely pivoted close to its center of gravity so that it retains its orientation unchanged over the time scale of interest - and the measurements are made relative to this as indicated in Fig. 5. This reference arm technique has been experimentally developed by N. Robertson et al. at Glasgow[11]. Such an arrangement is effectively a horizontal accelerometer or displacement monitor using the whole test mass, and if the sensing to the reference arm is done with an interferometer comparable to that used between the test masses high sensitivity can be achieved, and accurate compensation for seismic noise can in principle be made. Systems of this type can be extended

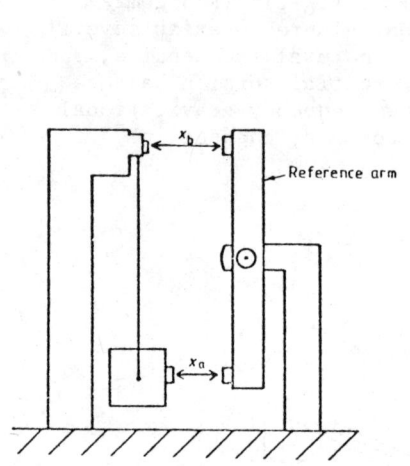
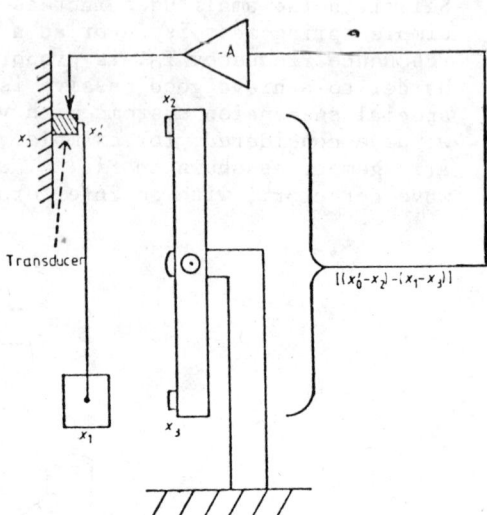

Figure 5   Figure 6

to directly reduce the seismically induced motion of the test mass by using feedback to move the point from which the test mass is suspended, as indicated in Fig. 6., and experimental work on several arrangements of this type has been done at Glasgow[11], while some preliminary tests have also been made at Caltech. It may be noted also that a feedback antiseismic system equivalent in principle to these but operating in a vertical direction - where tilts are unimportant - has been developed by Faller and Rinker[12] at JILA for measurements of acceleration due to gravity.

We have discussed seismic isolation at some length here because it seems to be generally considered that seismic noise will be the main factor limiting low frequency gravitational radiation experiments, and we think it more likely that changing gravity gradients from moving objects will be a more serious problem. It is possible that some of these gravity gradient fluctuations may be reduced by performing the experiments in a suitably quiet underground location, and an underground laboratory may well be a suitable site for a range of gravitational wave experiments.

We may conclude this review by noting that the laser interferometer gravitational radiation detectors described probably still require a considerable amount of development before detection of predicted gravitational waves can be expected. The prospects look good, however, and possibilities for real development of gravitational wave astronomy look interesting and exciting.

ACKNOWLEDGEMENTS

The author would like to acknowledge the essential contributions of his colleagues at the California Institute of Technology and at the University of Glasgow in relation to discussions of principles and ideas and to underlying experimental work. The research at Caltech is supported by the National Science Foundation (Grant NSF PHY82 04056) and at the University of Glasgow by the Science and Engineering Research Council.

REFERENCES

1. See, for example: V.D. Sandberg, these proceedings;
K.S. Thorne, Rev. Mod. Phys. $\underline{52}$, 285 (1980);
R. Epstein and J.P.A. Clark in L. Smarr (ed.) <u>Sources of Gravitational Radiation</u> (Cambridge University Press, Cambridge, 1979), and other papers in this volume;
K.S. Thorne in R.N. Lebovitz, W.H. Reid and P.O. Vandervoort (eds.) <u>Theoretical Principles in Astrophysics and Relativity</u> (University of Chicago Press, 1978) 149;
D.H. Douglass and V.B. Braginsky in: S.W. Hawking and W. Israel (eds.) <u>General Relativity: An Einstein Centenary Survey</u> (Cambridge University Press, 1979);
J.A. Tyson and R.P. Giffard, Ann Rev. of Astro. Astrophys., $\underline{16}$, 521 (1978);
W.H. Press and K.S. Thorne, Ann. Rev. of Astro. Astrophys. $\underline{10}$, 335 (1972).

2. K.S. Thorne and V.B. Braginsky, Astrophys. J. Lett. $\underline{204}$, L1 (1976).

3. M. Zimmerman and E. Szedenits, Jr., Phys. Rev. $\underline{D20}$ 351 (1979).

4. R.L. Forward, Phys. Rev. $\underline{D17}$ 379 (1978).

5. R. Weiss, Progress Report 105, Res. Lab Electronics, MIT, 54 (1972).

6. R. Schilling, L. Schnupp, W. Winkler, H. Billing, K. Maischberger and A Rudiger, J. Phys. E. Sci. Instrum. $\underline{14}$, 65 (1981).

7. R.W.P. Drever, J. Hough, A.J. Munley, S-A Lee, R. Spero, S.E. Whitcomb, H. Ward, G.M. Ford, M. Hereld, N.A. Robertson, I. Kerr, J.R. Pugh, G.P. Newton, B. Meers, E.D. Brooks III and Y. Gursel, Proc. of the NATO Advanced Study Institute, Bad Windsheim, West Germany 1981, in press.

7. Cont....

   R.W.P. Drever, G.M. Ford, J. Hough, I. Kerr, A.J. Munley, J.R. Pugh, N.A. Robertson and H. Ward, 9th International Conference on General Relativity and Gravitation, GR9, Jena (1980), in press.

   R.W.P. Drever, J. Hough, A.J. Munley, S-A Lee, R. Spero, S.E. Whitcomb, H. Ward, G.M. Ford, M. Hereld, N.A. Robertson, I. Kerr, J.R. Pugh, G.P. Newton, B. Meers, E.D. Brooks III and Y. Gursel, Proc. of the 5th International Conference on Laser Spectroscopy (Springer-Verlag, 1981), p. 33.

8. R.W.P. Drever, Interferometric Detectors for Gravitational Radiation in: Gravitational Radiation (eds. T. Piran and N. Derouelle) Proceedings of the Les Houches Summer Institute, June 1982 (Caltech Orange Aid Preprint OAP-648, 1982).

9. R. Spero, Prospects for Ground Based Detectors of Low Frequency Gravitational Radiation - these proceedings.

10. R.W.P. Drever, J. Hough, W.A. Edelstein, J.R. Pugh, W. Martin, Proc. of the Intern. Sympos. on Experimental Gravitation, Pavia 1976, B. Bertotti (ed.) (Accad. Nazionale dei Lincei, 1977).

11. N.A. Robertson, R.W.P. Drever, I. Kerr and J. Hough, J. Phys. E: Sci. Instrum. 15, 1101 (1982).

12. J.E. Faller and R.L. Rinker, "Super-Spring", Dimensions, 25 (September, 1979).

PROSPECTS FOR GROUND BASED DETECTORS OF LOW FREQUENCY GRAVITATIONAL RADIATION

Robert Spero
California Institute of Technology, Pasadena, Ca. 91125

ABSTRACT

Substantial effort will be required to minimize seismic and thermal noise in low frequency (below 100 Hz) gravitational wave detectors. The ultimate limit to the sensitivity of earthbound detectors is time-varying gravitational gradients due to atmospheric activity; a practical lowest frequency for sensitive operation may be about 0.1 Hz.

INTRODUCTION

It may be possible to build ground based gravitational wave detectors which work at frequencies much lower than 100 Hz, the cutoff conventionally assumed for free mass detectors. The cost of low frequency detectors and the effort required to build them would be similar to that of higher frequency detectors currently under development. As long as the detectors are confined to the ground, unavoidable practical considerations coupled with the noisy environment of the earth's surface dictate a lowest frequency for sensitive operation. I will estimate this minimum frequency in light of predicted source strength, noise internal to the detector, and the backgrounds associated with any gravitational wave detector situated on (or slightly below) the surface of the earth. Apropos of the theme of this conference, I will mention some arguments for and against siting such a detector underground.

SOURCES OF LOW FREQUENCY GRAVITATIONAL RADIATION

Astrophysicists and relativity theorists in recent years have refined their estimates of the flux of gravitational radiation that bathes the earth. There are large uncertainties in the predicted amplitude and event rate, but the estimates serve as a guide in determining the required sensitivity and frequency response of detectors. The expected sources include periodic signals from binary star systems and rotating neutron stars, burst signals from violent events such as supernovae and black hole formation, and a stochastic background remnant of an early epoch.

As a concrete example and a benchmark to evaluate detectors, consider the burst radiation expected from supermassive black holes. Thorne and Braginsky[1] find the duration, τ, and the strength of the wave, measured by the dimensionless strain, h, of a burst signal from the formation of or collision between black holes:

$$\tau \sim 1 \text{ sec } (M/10^4 M_\odot)$$
$$h \sim 2 \cdot 10^{-19} (M/10^4 M_\odot) \qquad (1)$$

Here $M/M_\odot$ is the mass of the black hole in solar masses, ranging from ~3 for "ordinary" black holes to 10^9 for the supermassive black holes thought to occupy the centers of some galaxies. The black hole is taken to be at a distance corresponding to a redshift of $z = 2.5$; that is, effectively at the Hubble distance.

The estimates (Equation 1) are probably accurate to within an order of magnitude. There is very large uncertainty, however, in estimates of how frequently such events occur (assuming they ever occur!). Thorne and Braginsky place the repetition rate within the range once per week to once per 300 years.

SENSITIVITY LIMITS OF DETECTORS

Consider a three-mass laser gravitational wave detector of the type first used by Forward[2] and now under intensive development.[3,4,5] These detectors are "free" (in the sense of being composed of masses defining an inertial reference frame) only for frequencies well above the lowest resonance of the mass suspension, typically 1 Hz. Below about 100 Hz, seismic motion penetrates the isolation provided by the suspension, accelerating the masses more than a strong gravity wave. A technique proposed by Drever, the antiseismic reference arm, may extend the useful frequency much lower, perhaps to 10^{-3}Hz. Alternatively, a detector composed of crossed torsion pendula hanging from a common support and read out by laser interferometers monitoring the ends of the balance arms[6] would also be isolated from low frequency seismic noise. The two designs differ in practical details, but they are subject to the same noise sources additional to seismic noise, viz. (1) photon counting statistics, (2) the standard quantum limit to position measurements, (3) thermal noise, and (4) noise from fluctuating gravitational gradients due to moving masses in the environment.

The photon counting limit for burst events depends on the time τ_s that the light is stored in the optical system, be it in a Fabry-Perot or a multi-reflection Michelson interferometer. If τ_s is equal to the gravitational wave period τ, then the minimum detectable signal is limited by photon counting statistics to

$$h_o = \frac{1}{2\pi} \frac{\lambda_{op}}{\lambda_{gw}} \langle n \rangle^{-\frac{1}{2}}, \qquad \tau_s = \tau \qquad (2)$$

where λ_{op} is the wavelength of light in the interferometer, λ_{gw} is the wavelength of the gravitational radiation, and $\langle n \rangle$ is the number of photons between the interferometer mirrors (for input laser power I, $\langle n \rangle = I \tau_s \lambda_{op}/2\pi \hbar c$. If the optical storage time is less than the period of the gravitational wave (the most likely circumstance for low frequency detectors), the sensitivity is reduced by a factor of τ/τ_s : $h = h_0 \tau/\tau_s$ or

$$h = 5 \cdot 10^{-23} \left(\frac{10W}{I}\right)^{\frac{1}{2}} \left(\frac{\lambda_{op}}{500nm}\right)^{\frac{1}{2}} \left(\frac{msec}{\tau_s}\right) \left(\frac{sec}{\tau}\right)^{\frac{1}{2}} \qquad \tau_s < \tau \qquad (3)$$

The photon counting noise increases with frequency by $f^{\frac{1}{2}}$, and limits the sensitivity at 1 Hz to $5 \cdot 10^{-23}$ if the optical power, wavelength, and storage time indicated in (3) are used.

There is a quantum uncertainty in the position of the free masses, equal to $(2\hbar\tau/m)^{\frac{1}{2}}$, where m is the mass and τ is approximately equal to the gravitational wave pulse duration. For a one second pulse sensed by one ton masses separated by 100 meters, the minimum detectable h in the presence of this quantum fluctuation is $5 \cdot 10^{-21}$.

Thermal noise or kT fluctuation is the most serious internal noise source at low frequencies. With a low loss suspension and a resonant frequency less than $1/\tau$, the position noise due to thermal fluctuations is $\pi^{-2} (kT/m\tau^*)^{\frac{1}{2}} \tau^{3/2}$ where τ^* is the damping time of the suspension. Losses due to friction in the suspension and collisions of the residual gas molecules with the masses must be minimized to keep τ^* large; even with a damping time of 10^{10} sec, room temperature thermal fluctuations exceed quantum fluctuations at frequencies below 0.2 Hz. Damping times this long have not yet been achieved, although it may be possible with a long period torsion pendulum in an ultra-high vacuum environment; alternatively, the detector could be cooled, perhaps to millidegree temperatures.

Gravitational gradients from moving masses in the vicinity of the detector will produce a background signal indistinguishable from gravitational radiation. One might hope to discriminate against the gravitational gradient background by constructing an array of detectors or by rejecting signals with the wrong pulse shape, but the time varying gradients manifest all the variety of life and all the unpredictability of the weather. A mass M moving with velocity v and located a distance r from the detector will simulate a gravity wave of amplitude $h \sim GMv \tau^3/r^4$. Here the mass is assumed to change its position during a time τ equal to the duration of a gravity wave burst. For example, a 1 kg rabbit hopping around 200 meters from the detector would produce signals stronger than the black hole burst (Equation 1). At closer distances, smaller animals (even insects) become noise sources.

Finally, for frequencies much below 1 Hz, the changing gravitational gradient due to atmospheric fluctuations becomes significant. The spectrum of air pressure fluctuations has a broad peak near 60 seconds corresponding to the motion of several tons of air contained within one cubic kilometer[7]; the gravitational gradient from this fluctuation might swamp a 60-second black hole burst. Weather fronts carry along much greater masses of air, but primarily at frequencies corresponding to tens of hours; the high frequency tail of the gradient spectrum due to moving regions of high or low air pressure is probably small. A heavy rainfall has higher frequency components, but less mass is involved.

SUMMARY AND CONCLUSIONS

A low frequency gravitational wave detector sensitive enough to see predicted burst events will require several orders of magnitude improvement over the current technology of seismic isolation and thermal noise reduction. The effort required to achieve these advances would be similar to that required for large-scale high frequency laser-based detectors now under development (time of ~10 years and cost of ~$3\cdot 10^7$ dollars). The result of this effort will be to extend the lowest frequency for sensitive detection of gravity waves downward from 100 Hz, the limit for detectors now under development, to perhaps 0.1 Hz.

To operate successfully at low frequency the detector must be isolated from moving masses, such as human and animal traffic. Siting the experiment underground may be the best way to achieve isolation from these gradient effects, providing that gradients associated with the underground facility such as geological settling, water flow, ventilation, or operating machinery are small. Even so, an underground laboratory would be subject to gravitational disturbances due to the atmosphere. Atmospheric effects probably preclude earth based detectors below 0.1 Hz; spacecraft detectors will be needed for these very low frequencies.

REFERENCES

1. K. Thorne and V. Braginsky, Astrophys. J. Lett. 204, L1 (1976).
2. R.L. Forward, Phys. Rev. D 17, 379.
3. R. Weiss, Quarterly Prog. Report, Research Lab. Electronics, MIT, 105, 546, (1972).
4. R.W.P. Drever, lectures delivered at the Les Houches Summer Institute in Gravitational Radiation, 1982 (in press), and references cited therein.
5. H. Billing et al., J. Phys. E12 (1979) 1043.
6. R.W.P. Drever, proceedings of this conference.
7. A.S. Monin, Weather Forecasting as a Problem in Physics (MIT Press, Cambridge, 1972), p. 7 ff.

Chapter VIII

DOUBLE BETA DECAY

Pray you, sir, stay.
Rather than I'll be bray'd sir, I'll believe
That alchemy is a pretty kind of game,
Somewhat like tricks o' the cards, to cheat a man
With charming.

Ben Johnson,
The Alchemist,
Act II, Scene *iii*.

DOUBLE BETA DECAY, MASSIVE NEUTRINOS, AND LEPTON CONSERVATION

S. P. Rosen
National Science Foundation, Washington, D.C. 20550
and
Purdue University, West Lafayette, Indiana 47907

ABSTRACT

The phenomenon of double beta decay is reviewed and its implications for neutrino properties such as mass, right-handed currents and Majorana self-conjugacy are described. At present the only indication that the lepton number violating no-neutrino mode occurs is an indirect one based upon a comparison of ^{128}Te and ^{130}Te lifetimes. The history and latest status of this argument are discussed. Should the no-neutrino mode be observed, a whole program of experiments will be needed to determine the mechanism responsible for it.

INTRODUCTION

Although it is the rarest phenomenon presently known in physics, double beta decay has become an important testing ground for the grand unification of strong and electro-weak forces. Grand unified theories have created a general expectation that many simple conservation laws are likely to be violated, albeit at very low levels. For example, proton decay is expected to violate baryon number conservation at a level of order 1 in 10^{30} years. Double beta decay is known to have a much shorter lifetime, of order $10^{21\pm}$ years, and the principle question is whether or not it violates the law of lepton number conservation. The discovery of a violation would be interpreted as a signal for grand unification, especially if it could be associated with the existence of massive Majorana neutrinos; the absence of lepton number violating decays would not be fatal to theories of grand unification, but it would impose serious constraints upon them.

In my talk today I shall describe the two modes of double beta decay for which searches have been made, how these searches have been carried out, and the data they yield. I shall then show how information about the properties and interactions of neutrinos might be extracted from the data and review the prospects for entirely new interactions. In the course of the talk, it will become apparent that there are important questions of nuclear physics associated with the calculation of double beta decay rates. To close, I shall talk about prospects for a program of experiments for the future.

THE PHENOMENON OF DOUBLE BETA DECAY

The systematics of ground-state energies in even-even nuclei are such that the following situation occcurs not infrequently: the even-even nucleus (A,Z) is lighter than its odd-odd neighbour (A,Z+1), but heavier than (A,Z+2). Transitions from (A,Z) to (A,Z+1) are therefore forbidden by energy conservation, but transitions from (A,Z) to (A,Z+2) are allowed. In contrast to ordinary single beta decay, such transitions increase the charge on the nucleus by two units instead of one, and they must therefore be accompanied by the simultaneous emission of two electrons instead of one. Are the two electrons always accompanied by neutrinos?

Two-neutrino double beta decay,

$$(A,Z) \rightarrow (A,Z+2) + 2e^- + 2\bar{\nu}_e \qquad (1)$$

is expected to occur as a second order effect of the same Hamiltonian H_W as gives rise in first order to single β-decay. This process obviously conserves lepton number and its lifetime has been estimated to be $10^{21 \pm 2}$ years. No-neutrino double beta decay,

$$(A,Z) \rightarrow (A,Z+2) + 2e^- \qquad (2)$$

may also occur if neutrinos are Majorana particles, or if some new, hitherto undetected interaction exists. Lepton number changes by two units ($\Delta L=2$) in this process, and its detection would establish that lepton number is not an exactly conserved quantity.

As illustrated in Fig. (i), ground-state to ground-state transitions between even-even nuclei involve energy releases of order 2 - 3 Mev and are associated with spin and parity changes of the type:

$$0^+ \rightarrow 0^+$$

In many cases, there is also a $J = 2$ excited state about 500 kev above the ground-state of the daughter nucleus; as will be discussed below, the detection of double beta decay transitions to these excited states may have important implications for the precise mechanism of lepton nonconservation.

Experimental searches for double beta decay fall into two main classes: (i) geochemical detection of daughter nuclei occluded in ores of known age, typically several billion years, which are rich in the parent nucleus;[3] and (ii) laboratory detection of the electrons in tracking chambers or scintillating crystals.[4] The daughter nuclei in geochemical experiments are chosen to be noble gases because, being chemically inert, they accumulate over long periods of time without losses induced by chemical processes, and because volatile gases are expelled at the time of formation of the ore. Thus any anomaly in the relative abundance of an isotope, say Xe^{130}, as compared with the atmospheric gas, must be radiogenic in origin. The problem then is to rule out all possible background processes and to show that double beta decay is the only source of the anomaly: From the magnitude of the anomaly and independent measurements of the age of the ore, one can then deduce the half-life for the decay.

Laboratory experiments have been carried out on many parent nuclei, ranging from ^{48}Ca to ^{150}Nd, with a variety of different techniques. In principle, they can be used to distinguish the two-neutrino mode from the no-neutrino mode by measurements of the sum of the electron energies: for the two-neutrino mode, the energy sum will follow a continuous spectrum, broadly peaked at about one-half the total energy release; while for the no-neutrino mode, the energy sum will yield a spike at the total energy release. So far no one has detected the spike at the end of the spectrum (or, if you prefer, the pot of gold at the end of the rainbow). Professor Wu and her colleagues did come very close in a ^{48}Ca experiment:[5] they had one event with an energy sum approximately equal to the total energy release, but could not say for sure that it was the no-neutrino mode because of a contamination by a radium isotope of 1 part in 10^{13}! This example indicates how carefully one must guard against low-level activities in laboratory experiments.

Because there is presently no direct evidence for no-neutrino decay from laboratory experiments, one must turn to the geochemical experiments. They cannot give direct evidence because they do not detect the electrons, but they can give indirect evidence for the existence of no-neutrino decay based upon comparisons of different isotopes. We shall discuss such a case in tellurium later in this talk.

A selection of the presently available data is shown in Table I. It begins with geochemical measurements of the half-life for $^{130}Te \rightarrow {}^{130}Xe$; the first such measurement was carried out by Inghram and Reynolds[6] in 1951 and the most convincing demonstration of double beta decay was performed by Professor Kirsten and his colleagues[7] in 1968. The best value of the half-life is about 2×10^{21} years.

Next comes a series of measurements of the ratio of half-lives for $^{128}Te \rightarrow {}^{128}Xe$ and $^{130}Te \rightarrow {}^{130}Xe$. The authors of the first measurement[8] found that their sample could be seriously contaminated by backgrounds. Ten years later the Missouri group of Manuel and co-workers[9] re-measured the ratio and found a much larger value of about 1500 in an ore containing excess Xenon from several sources. Very recently Professor Kirsten[10] has measured the ratio and, as we shall hear in his talk, he finds a value greater than 3100 at the 2σ limit in an ore with a very clean signal for the $^{130}Te \to {}^{130}Xe$ double beta decay. The implications of this measurement will be discussed below.

Besides the tellurium isotopes, the most widely studied case has been the transition $^{82}Se \to {}^{82}Kr$. Two geochemical measurements[11,12] yield half-lives in the range $(1 - 3) \times 10^{20}$ years, while a recent cloud chamber experiment by Moe and Lowenthal[13] gives a much shorter value of 1×10^{19} years. In general, the geochemical measurements are all roughly consistent with one another, and consistent with other methods of radio-active dating;[3] therefore, it there is anything wrong with them, there would have to be something wrong with the entire field of radio-active dating and cosmo-chemistry—a not very likely possibility[3]. On the other hand, Professor Moe[13] has approximately 15 examples of what appear to be pairs of electrons from double beta decay in his cloud chamber. Are they the real thing, or could they result from some unrecognized contaminant in his source?

Various laboratory searches for the 2ν and 0ν modes have been made in ^{48}Ca and ^{76}Ge, and the limits obtained are shown in the lower part of Table 1.

THEORY

The two-neutrino mode of double beta decay is a second-order effect of the standard Hamiltonian for single beta decay, H_w^β, and the decay amplitude is:[2]

$$M((A,Z) \to (A,Z+2) + 2e^- + 2\bar{\nu}_e)$$

$$= \mathcal{A}^e_{12} \, \mathcal{A}^\nu_{12} \, m, int.$$

$$\sum \left\{ \frac{<f; e_1^- e_2^-; \bar{\nu}_1, \bar{\nu}_2 | H_w^\beta | m; e_1^-, \nu_1 > \times <m; e_1^-, \bar{\nu}_1 | H_w^\beta | i>}{(E_m - E_i + E_{e_1^-} + E_{\bar{\nu}_1})} \right\} \quad (3)$$

where the $\mathcal{A}$ are anti-symmetrization operators for the electrons and neutrinos. To estimate the lifetime we restrict ourselves to $0^+ \rightarrow 0^+$ transitions, and make a series of "reasonable" approximations.[2]

Chief amongst these is the replacement of the energy difference between initial and intermediate nuclear states, i and m, by an average value $<W_{mi}>$ and the summation over m by means of closure; this is tantamount to assuming that only a limited number of low lying intermediate nuclear states makes an important contribution to the sum. In addition, position operators are averaged over all directions, and Fermi-type matrix elements, $<f|\sum_{m,n} \tau_m^+ \tau_n^+|i>$, are neglected on grounds of isopin.

The half-life can then be estimated in terms of the Gamow-Teller type matrix elements $M_{GT} = <f|\sum_{m,n} \tau_m^+ \tau_n^+ \vec{\sigma}_m \cdot \vec{\sigma}_n|i>$:

$$T_{1/2}(2\nu) = \left(\frac{3 \times 10^{20} \text{yrs}}{f_{2\nu}(E)}\right) [F_{coul.}(Z)]^{-2} \left[\frac{g_A^2}{g_V^2} \frac{M_{GT}}{<W_{mi} + \frac{1}{2}E_0 + 1>}\right]^{-2}$$

$$f_{2\nu}(E) = \left(\frac{E^7}{8!1980}\right)(E^4 + 22E^3 + 220E^2 + 990E + 1980) \quad (4)$$

where E is the energy release in units of the electron rest energy. The phase space factor $f_{2\nu}(E)$ depends on relatively high powers of E because there are four fermions in the final state; consequently its value ($\approx 2 \times 10^{-3}$) for $Te^{128} \rightarrow Xe^{128}$, for which $E \approx 1.7$, is very much smaller than the value (≈ 15) for $Te^{130} \rightarrow Xe^{130}$ with $E \approx 5$. The Coulomb correction factor $[F_{coul}(Z)]^{-2}$ is reasonably well represented by the standard nonrelativistic approximation $(1-\exp(-2\pi\alpha Z))/2\pi\alpha Z^2$ for low Z, but for high Z, relativistic corrections reduce it by a factor between 3 and 5.[14]

By far the largest uncertainty in the calculation occurs in the nuclear matrix element M_{GT}. In our original estimate,[2] Primakoff and I treated it as a product of two single beta decay matrix elements, each of which had a value between 1 (super allowed) and 0.1 (allowed). The range for M_{GT} was therefore between 1 and 0.01; we chose a value in between, but allowed ourselves errors of a factor of ten in either direction:

$$(M_{GT})_{PR} \approx 0.1 \times 10^{\pm 1} \quad (5)$$

Haxton, Stephenson, and Strottman[14,15] (HSS) have recently calculated these matrix elements using the shell model and they find them to be roughly an order of magnitude larger than our (conservative) central value; Zamick and Auerbach[16] have used a simpler model, but with similar results.

Taking the energy denominator $\langle W_{mi} + 1/2E_o + 1\rangle$ to be 25 in units of $m_e c^2$, and using the relativistic corrections for the Coulomb factor, we find from Eq-(5) that the 2ν lifetime for $^{130}\text{Te} \to ^{130}\text{Xe}$ is $10^{22\pm 2}$ years. The larger matrix elements of HSS[14] reduce this estimate by a factor of $\simeq 100$. If we assume that the matrix elements for the ^{130}Te and ^{128}Te decays are equal to one another, then the ratio of two-neutrino half-lives is just the ratio of phase space factors:

$$\frac{T_{1/2}(2\nu)\,(^{128}\text{Te})}{T_{1/2}(2\nu)\,(^{130}\text{Te})} \simeq \left(\frac{f_{2\nu}(5)}{f_{2\nu}(1.7)}\right) \simeq 7 \times 10^3 \qquad (6)$$

The assumption of equal matrix elements is both reasonable and borne out by calculations of Vergados[17] and HSS;[14] however, as we shall see later, there may be subtle arguments against it.

Let us now turn to the theory of no-neutrino double beta decay. First, we emphasize that, if the no neutrino process of Eq. (2) is definitely detected, then we will have to conclude that lepton number is not absolutely conserved. What then is the nonconserving interaction or mechanism?

One logical possibility is that some entirely new interaction is at work. But what is its strength? It is far more illuminating to predict the strength ahead of time than to fix it after the fact. The idea of a new interaction was originally discussed by Winter[18] and by Pontecorvo,[19] and it has recently been revived by several authors.[20]

A second approach, and one that we follow here, is to continue to regard the process as a second-order effect, but to modify the Hamiltonian and the properties of the neutrino so as to allow for lepton nonconservation.[2,21]

The most general such Hamiltonian can be written as

$$H_M = \frac{G_F}{\sqrt{2}} \sum_{\lambda=V,A,S,T,P} (\bar{\psi}_p \Gamma^{(\lambda)} \psi_n) \mathcal{L}^{(\lambda)} + \text{h.c.} ; \quad \mathcal{L}^{(\lambda)}$$

$$= C_\lambda (\bar{\psi}_e \Gamma^{(\lambda)} (1+\delta_\lambda \gamma_5) \psi_\nu) + D_\lambda (\bar{\psi}_e \Gamma^{(\lambda)} (1+\eta_\lambda \gamma_5) \psi_{\nu c}) \quad (7)$$

and it is the simultaneous appearance of the neutrino field ψ_ν and its charge conjugate $\psi_{\nu c}$ ($\equiv C\tilde{\bar{\psi}}_\nu$ where $C\gamma_\mu C^{-1} = -\gamma_\mu$) that leads to a breakdown of lepton number conservation. Two special cases correspond to Majorana[22] neutrinos: in one, $C_\lambda = D_\lambda$, $\delta_\lambda = \eta_\lambda$ and the neutrino enters (λ) in the C-even combination $\phi(+) = \frac{1}{\sqrt{2}}(\psi_\nu + \psi_{\nu c})$; in the other, $C_\lambda = -D_\lambda$, $\delta_\lambda = \eta_\lambda$ and the neutrino appears in the C-odd combination $\phi(-) = \frac{1}{\sqrt{2}}(\psi_\nu - \psi_{\nu c})$. Also of interest is an "almost Dirac" neutrino for which $D_\lambda \ll C_\lambda$; here the Dirac neutrino carries with it a small admixture of its anti-particle.

The basic mechanism for no-neutrino double beta decay is shown in Fig. 3. One neutron (or down quark) inside the nucleus undergoes a β-decay through the C_λ part of the leptonic current, and a second neutron (or down quark) then undergoes a neutrino capture through the D_λ part of the current to become a proton (or up quark) plus an electron. There is also a second diagram in which the roles of C_λ and D_λ are interchanged.

Now one of the lessons learned a long time ago was that neutrons always emit right-handed neutral leptons in beta decay, and they can only absorb left-handed ones. Therefore the neutrino exchanged by the neutrons in Fig. 2 must flip its helicity as it travels from the C_λ vertex to the D_λ one. This requires either a small admixture of right-handed current in the dominantly left-handed leptonic current, or a non-zero neutrino mass, or both.

It is this necessity, especially the non-zero mass, which makes double beta decay of such relevance to grand unified theories. Although these theories do not predict a definite value or range of values for the neutrino mass, they operate on the principle that if there is no good reason for something to vanish then it most likely does not do so. Moreover these theories tend to make the neutrino into a Majorana particle when they endow it with mass.[23]

In general, the amplitude for the no-neutrino mode depends upon combinations of coupling constants like

$$\left[\bar{C}_\lambda D_\mu (1 - \delta_\lambda \eta_\mu) + (\lambda \leftrightarrow \mu)\right] \text{ and } m_\nu \{C_\lambda D_\mu (1 + \delta_\lambda \eta_\mu) + (\lambda \leftrightarrow \mu)\}$$

which vanish when either: (i) all $D_\mu = 0$ (pure Dirac neutrino); or (ii) $m_\nu = 0$, $\delta_\lambda = \eta_\mu = \pm 1$ for all λ, μ (pure helicity suppression). For the purposes of this discussion, we shall make use of the C-even Majorana case and allow for small admixtures of right-handed currents by taking:[2]

$$C_\lambda = D_\lambda = \frac{1+\eta}{\sqrt{2}} \; ; \; \delta_\lambda = \eta_\lambda = \frac{1-\eta}{1+\eta} \; ; \; m_\nu \neq 0 \qquad (8)$$

The no-neutrino amplitude will consist of one set of terms proportional to η, and another set proportional to m_ν. Later in the discussion we shall consider how to determine whether η or m_ν is the dominant parameter; we shall also examine the possibility raised by Wolfenstein[24] that two or more neutrinos with opposite charge conjugation properties contribute to the amplitude, as well as the intriguing possibility of an "almost Dirac" neutrino which has recently been emphasized by Valle.[25] For the moment, however, we limit ourselves to the exchange of one Majorana neutrino for which and $m = (m_\nu/m_e)$ are small parameters.

Using the same types of approximation in the nuclear matrix element as for the two-neutrino mode, we estimate the half-life for the no-neutrino mode to be:[2]

$$T_{1/2}(0\nu) = \left(\frac{2.1 \times 10^{15} \text{yrs}}{F_{0\nu}(\hat{m}, \eta)}\right) \left[\mathcal{F}(Z)_{\text{coul.}}\right]^{-2} \left[\frac{g_A^2}{g_V^2} \frac{M'_{GT}}{m_p}\right]^{-2}$$

$$M'_{GT} = \langle f | \sum_{m,n \neq m} \zeta_m^+ \zeta_n^+ \frac{\vec{\sigma}_m \cdot \vec{\sigma}_n}{r_{mn}} | i \rangle$$

$$F_{0\nu}(\hat{m}, \eta) = \frac{1}{15} \{(\hat{m})^2 f_m(E) + \hat{m}\eta f_{m\eta}(E) + \eta^2 f_\eta(E)\}$$

$$f_m = E\left[E^4 + 10E^3 + 40E^2 + 60E + 30\right]$$

$$f_{m\eta} = \frac{E^2}{9}\left[6E^3 + 60E^2 + 10E - 60\right]$$

$$f_\eta = \frac{2E^2}{567}\left[16E^5 + 224E^4 + 1183E^3 + 630E^2 - 315E + 210\right]$$

where the Coulomb correction factor is the same as before, and the nuclear matrix element M_{GT} differs from the one for the two-mode by the factor of the distance ν_{mn} between the mth and n^{th} neutrons. The overall scale for the no-neutrino half-life is much smaller than that for the two-neutrino half-life (2.1×10^{15} yrs as compared with 3×10^{20} yrs) because the average energy of the virtual neutrino $\langle E_\nu \rangle$ is much greater ($30 - 100$ Mev) than that of the real ones $\langle E_r \rangle$ ($1 - 3$ Mev) and the ratio of lifetimes goes roughly as $[\langle E_r \rangle / \langle E_\nu \rangle]^4$. This difference between real and virtual neutrinos also shows up in the phase space factor $F_{o\nu}(\hat{m},n)$ which depends upon lower powers of the energy release E than does $f_{2\nu}(E)$. There is also a difference in the power of the pure mass-mechanism phase space f_m as compared with that for the right-handed current; it comes about because the mass mechanism is an S-wave process and the right-handed current is a P-wave one.

If, as before, we compare the half-lives for ^{128}Te and ^{130}Te assuming that they have the same nuclear matrix elements, then we find that:

$$\frac{T_{1/2}(o\nu)\,(^{128}Te)}{T_{1/2}(o\nu)\,(^{130}Te)} = \begin{cases} 320 & \hat{m} = o \\ 80 & \hat{m} = \eta \\ 30 & \eta = o \end{cases} \quad (10)$$

Even though the predicted value is sensitive to the admixture of $\hat{m}$ and η, the entire range in Eq. (10) is very much smaller than the ratio predicted for two-neutrino decay in Eq. (6).

THE TELLURIUM RATIO ARGUMENT

As has been emphasized in §2, we have no direct evidence for the occurrence of no-neutrino decay, and so we must look for whatever indirect evidence might be available. The Tellurium ratio argument is the best we can find.

The argument was originally invented by Pontecorvo[19] to explain an anomalously "short" lifetime for ^{128}Te found by Takaoka and Ogata.[8] In 1968 Kirsten, Schaeffer, Norton, and Stoenner[7] made the first definitive measurement of double beta decay and found a lifetime of approximately 2×10^{21} years for ^{130}Te. By itself this number is perfectly consistent with the predictions for two-neutrino decay; however, two years earlier, Takaoka and Ogata had measured the ratio of ^{128}Te and ^{130}Te lifetimes and found a value of ($10^{1.2\pm0.6}$) which, despite the large errors, was much closer to the expectations for no-neutrino decay.

To resolve this apparent paradox, Pontecorvo proposed a new interaction, a Δ(lepton number)=2 companion to the $\Delta S=2$ superweak interaction introduced by Wolfenstein[26] to describe CP violation. Such an interaction would automatically explain the small lifetime ratio, and at the same time it would have just the right strength ($\simeq 10^{-3} G_w^2$) to account for the ^{130}Te lifetime.

Primakoff and I, in 1969,[27] took a more conventional view and attributed the no-neutrino double beta decay to second-order effects of an H_w^β in which the neutrino is a Majorana particle. We introduced the small admixture η of right-handed currents and estimated its magnitude to be $\eta \simeq 10^{-3} - 10^{-4}$ from the measured half-life of ^{130}Te. This is a much more sensitive measure of η than either the original Davis experiment,[28] or electron helicity measurements in single beta decay.

Unfortunately for Pontecorvo and for us, Hennecke, Manuel, and Sabu[9] re-measured the half-live ratio in 1975 and found a much larger value than that of Takaoka and Ogata. This value of approximately 1600 is much closer to the expectation for the two-neutrino mode, and so we were inclined to regard the matter as settled.

Bryman and Picciotto,[29] however, observed that the ratio of 1600 is still a significant factor below the value of roughly 7000 expected for two-neutrino decay. They therefore suggested in 1978 that both modes are actually present in the ^{130}Te and ^{128}Te decays. From the half-life ratio, they calculated the lepton nonconserving parameter to be:

$$\eta = (4.3 + 0.1) \times 10^{-5} \quad (11)$$

Two years later Doi, Kotani, Nishiura, Okuda and Takasugi[30] adapted the Bryman-Picciotto argument to the case of the neutrino mass as the lepton-nonconserving mechanism, and they estimated the mass to be:

$$m_\nu \simeq 34 \text{ ev} \quad (12)$$

This value falls well within the range $14 \leq m_\nu \leq 45$ ev resulting from the ITEP tritium β-decay experiment,[31] and in fact it is very close to the "best" value.

Very recently Kirsten, Richter, and Jersberger[10] (1982) have remeasured the Tellurium ratio, and they find it consistent with a pure two-neutrino decay. At the 2σ limit, their ratio is ≥ 3100 and it yields limits on the mass and η-parameter of:

$$m_\nu \leq 5.6 \text{ ev} \quad \text{or} \quad \eta \leq 2.4 \times 10^{-5} \qquad (13)$$

I would now like to analyse the argument more carefully to see the extent to which the latest results rule out the existence of the no-neutrino mode.

If we assume that the nuclear matrix elements for the ^{128}Te and ^{130}Te decays are equal to one another, then the ratio, R, of their half-lives will be a function of $\hat{m}, \eta$, and the ratio, ρ, of nuclear matrix elements for the two-neutrino and no-neutrino modes respectively. Defining ρ to be

$$\rho \simeq \left[\frac{\langle f | \sum_{m,n} \zeta_m^+ \zeta_n^+ \sigma_m \cdot \sigma_n | i \rangle}{\langle f | \sum_{m,n} \zeta_m^+ \zeta_n^+ \frac{\sigma_m \cdot \sigma_n}{r_{mn}} | i \rangle} \cdot \frac{m_p}{\langle W_{mi} + \frac{1}{2}E_o + 1 \rangle} \right] \qquad (14)$$

we obtain the general expression:

$$(6600 - R)\frac{\rho^2}{10^8} = 21.5 \, (R-31)\hat{m}^2 + 4 \, (R-324)\eta^2$$
$$+ 2.1(R-121)\hat{m}\eta \qquad (15)$$

For very large values of R, this yields the approximate limits:

$$\hat{m} \simeq \frac{\rho}{10^4} \sqrt{\frac{6600-R}{21.5R}} \quad ; \quad m_\nu \simeq 11\rho \sqrt{\frac{6600}{R} - 1} \text{ (ev)}$$

or

$$\eta \simeq \frac{\rho}{10^4} \sqrt{\frac{6600-R}{4R}} \qquad (16)$$

The interference term between $\hat{m}$ and η does not disturb these limits too much.

To estimate ρ, Primakoff and I[2] replaced r_{mn} in the no-neutrino matrix element by an average which we took to be the nuclear radius, and with $\langle W_{mi} + 1/2 \, E + 1 \rangle \simeq 25$, we found

$$\rho_{PR} \simeq 1.2 \, (A/130)^{1/3} \qquad (17a)$$

Haxton, Stephenson, and Strottman$^{(14,15)}$ evaluated both nuclear matrix elements using a large shell model code, and this had the effect of replacing r_{mn} by roughly 0.4 times the nuclear radius; Minkowski,$^{(32)}$ on the other hand, has suggested a much smaller average value. The estimates of ρ by these authors are

and
$$\rho_{HSS} \simeq 0.4 \rho_{PR} \qquad (17b)$$

$$\rho_M \simeq 0.05 \rho_{PR} \qquad (17c)$$

For a ratio R=3100, these estimates of ρ yield mass and η parameters:

$$m_{PR} \simeq 14 \text{ ev}, \quad \eta_{PR} \simeq 6.4 \times 10^{-5} \qquad (18a)$$

$$m_{HSS} \simeq 5.6 \text{ ev}, \quad \eta_{HSS} \simeq 2.6 \times 10^{-5} \qquad (18b)$$

$$m_M \simeq 0.7 \text{ ev}, \quad \eta_M \simeq 3 \times 10^{-6} \qquad (18c)$$

The estimates of Doi, Kotani, Nishiura, and Takasugi$^{(33)}$ are essentially the same as those of HSS.$^{(14,15)}$

Let me bring this discussion to a close by considering three basic questions concerning the Tellurium ratio argument:

(1) How good is the experimental data?

It is obvious that there are no order of magnitude discrepancies between the various geochemical measurements of the ^{130}Te lifetime; indeed they are all consistent with one another within a factor of two or three. The same holds true for the geochemical measurements of ^{82}Se decay. Furthermore these lifetimes also fit in consistently with a general pattern of age determinations obtained from other radio-active decays, for example K-Ar dating.$^{(3)}$ In addition, as Professor Kirsten has emphasized, they are also consistent with age-independent comparisons involving other long-lived processes such as the spontaneous fission of U^{238}.

With regard to the ^{128}Te/^{130}Te ratio, it is likely that the very first measurement$^{(8)}$ was distorted by large background sources of ^{128}Xe. In choosing between the 1975 measurements$^{(9)}$ and the most recent ones,$^{(10)}$ I think that, given the possibility of backgrounds, a theorist would be well advised to err on the side of pessimism and work with the larger value of R.

In short, there is a very strong case in favor of the data.

(2) Are the matrix elements for ^{128}Te and ^{130}Te really equal to one another?

Equality between the matrix elements for the two isotopes is the basis of the theoretical argument, and where it has been tested, it seems to work well. Vergados[17] finds that equality holds to within 10-15%, while Haxton, Stephenson, and Strottman (HSS)[14] find that it is valid to better than 1%. However these latter authors raise a subtle argument against the notion of equality.

They observe[14] that while the calculated ratios of matrix elements are very close to one, the absolute values yield a much shorter lifetime than has been measured for ^{130}Te. Therefore, they argue, there must be very large cancellations in the individual matrix elements. Now if this is indeed the case, there is no guarantee that the difference of two large numbers, A(130) and B(130), for ^{130}Te will equal the difference of two large numbers, A(128) and B(128), for ^{128}Te even though A(130) $\simeq$ A(128) and B(130) $\simeq$ B(128).

In other words, even though the assumption of equality is a very "reasonable" one, the actual matrix elements may behave in a most "unreasonable" manner.
(3) Does the apparent conflict between the latest Tellurium ratio mass limit ($m_\nu \leq 5.6$ ev, Eq. 18b) and the ITEP tritium β-decay range ($14 \leq m_\nu \leq 45$ ev) imply that the neutrino is a perfect Dirac particle?

Here we treat the equality of nuclear matrix elements as a valid assumption, and the ITEP result[31] as a firm bound on the neutrino mass. It is then certainly possible to conclude from the conflicting limits that the neutrino is a Dirac particle and that lepton number must be conserved in double beta decay. There are, however, other interpretations possible.

Wolfenstein[24] has recently observed that the exchange of two neutrinos with opposite C-eigenvalues ($C_\lambda^{(1)} = +D_\lambda^{(1)}$ for one, and $C_\lambda^{(2)} = -D_\lambda^{(2)}$ for the other in eq. (7)) in the no-neutrino diagram (Fig.(3)) can lead to large cancellations in the amplitude. In the limit of equal coupling strengths and equal neutrino masses ($C_\lambda^{(1)} = C_\lambda^{(2)}$), this cancellation would be complete simply because the two Majorana neutrinos would be equivalent to one Dirac neutrino and its non-identical anti-particle. For unequal masses, the amplitude depends on the difference in mass between the Majorana mass eigenstates.

When both mass eigenvalues are light ($\leq$ 100 ev), the cancellation remains complete for the right-handed current mechanism, and only the mass-mechanism itself can give rise to no-neutrino decay. In this case, the 5.6 ev limit is a bound on the mass <u>difference</u>, and it does not conflict with the ITEP experiment which measures the average mass of the two neutrinos.

If one neutrino is light, and the other heavy ($\geq$ 100Mev), a possibility realized in some models, then the analysis is more complicated.(34) The 5.6 ev becomes a far from simple function of the masses. The ITEP experiment, however, gives bounds only on the light neutrino, the heavy one being too massive to be emitted in single beta decay. Again there is no conflict.

Valle(25) has emphasized another approach in which, rather than considering the Majorana type possibilities with $|C_\lambda| = |D_\lambda|$, one takes $D_\lambda = \xi C_\lambda$ with $\xi \ll 1$. This corresponds to an "almost Dirac" neutrino represented dominantly by the particle operator ψ_ν, but carrying a small admixture of its anti-particle $\psi_{\nu c}$. The 5.6 ev now becomes a limit not on the m alone, but on the product $m_\nu \xi$; thus m_ν could fall well within the ITEP range as long as the parameter ξ is sufficiently small.

Cancellations between the mass and right-handed current mechanisms are possible in principle, but unlikely in practice. The $\hat{m} - \eta$ interference terms (Eq. (9)) tend to be rather small for "typical" values of the energy release, and they only become significant in the few cases of a large energy release, for example ^{48}Ca where E = 8.4. Nevertheless, we see from the above discussion that there are ways to resolve the discrepancy between the geochemical and tritium β-decay results without giving up the Majorana neutrino.

DIRECT DETERMINATIONS OF $\hat{m}, \eta$.

Where we do not have the half-lives of different isotopes to compare, we can try to extract limits on $\hat{m}$ and η from the data on a single isotope. If, for example $T_{1/2}(2\nu)$ is known for a particular isotope, we can determine the value of the nuclear matrix element M_{GT} from Eq. (3). Using this M_{GT} and the theoretical value of ρ (Eqs. 14 and 17), we can then place bounds on $\hat{m}$ and η from any known bound on $T_{1/2}(o\nu)$ and Eq. (9). If the only information we have is an experimental limit on $T_{1/2}(o\nu)$, then we must calculate M_{GT} instead of evaluating it from a measured (2ν) - half-life.

For ^{82}Se $\to ^{82}$Kr, we find that $m_\nu \leq 52$ ev when we use the geochemical lifetime(11,12) for $T_{1/2}(2\nu)$, and $m_\nu \leq 9$ ev when we use laboratory measurement of Moe and Lowenthal.(13) Calculating M_{GT}, and using only the bound on the oν half-life, Haxton, Stephenson, and Strottman(15) have found $m_\nu \leq 13$ ev.

For ^{76}Ge $\to ^{76}$Se there are no experimental limits on $T_{1/2}(2\nu)$ and so one must rely entirely on theoretical estimates of M_{GT}. Using the latest experimental limit on $T_{1/2}(o\nu)$, HSS(15) find that $m_\nu \leq 11$ ev, while Doi et.(33)

al. find that $m_\nu \leq 40$ ev. A similar situation holds for $^{48}Ca \rightarrow {}^{48}Ti$, and the limits on m_ν have been estimated to by Haxton, Rosen, and Stephenson[35] to be in the range 38 - 87 ev depending on various nuclear physics assumptions. (Making an extreme assumption about $\langle 1/r_{mn} \rangle$, Minkowski[32] has estimated a limit $m_\nu \leq 3$ ev in this case.)

All of those limits are still within the ITEP range, but they show signs of moving below it. The limits on η corresponding to these masses are all in the ball-park of a few x 10^{-5}.

DISTINGUISHING BETWEEN LEPTON NONCONSERVING MECHANISMS

In the event that the existence of no-neutrino double beta decay is definitely established, it will be important to determine the mechanism by which lepton number is not conserved. There are at least three ways in which this can be done: (i) the energy dependence of lifetimes; (ii) the angular distribution of the electrons in $0^+ \rightarrow 0^+$ transitions; and (iii) the detection of $0^+ \rightarrow 2^+$ transitions. It is apparent from Eq. (9) that the phase space for the mass mechanism, $f_m(E)$, is a polynomial of the fifth degree in the energy release E, while the right-handed current (RHC) phase space, $f_\eta(E)$, is a seventh degree polynomial. Therefore a careful study of the lifetimes of several isotopes with different energy releases could be used to determine whether the mass mechanism alone is responsible for the decay, or whether it is the RHC mechanism alone, or some combination of RHC and neutrino mass. Besides sufficient data, this method requires an accurate knowledge of the nuclear matrix elements.

Not dependent on the nuclear matrix elements is the angular distribution test. In the case of mass-induced no-neutrino decay, the electrons are created with exactly the same helicity, $(-v/c)$, and consequently the conservation of angular momentum in $0^+ \rightarrow 0^+$ transitions requires that the electrons tend to come out back-to-back, with an angular distribution:

$$A_m(\theta_{12}) \simeq (1 - \frac{v_1 v_2}{c^2} \cos\theta_{12}) \qquad (19a)$$

where θ_{12} is the angle between the electrons and v_1, v_2 are their velocities. In the case of the RHC-induced decay, the electrons are created with opposite helicities, and therefore they must emerge in a parallel configuration to conserve angular momentum in the direction of motion.

The RHC angular distribution is therefore

$$A_\eta(\theta_{12}) \simeq (1 + \frac{v_1 v_2}{c^2} \cos\theta_{12}) \qquad (19b)$$

For two-neutrino decay, the electrons again have the same negative helicity, and the Pauli Principle requires that, in $0^+ \to 0^+$ transitions, they be emitted with essentially the same distribution as in the mass-induced, no-neutrino mode. Therefore, significant deviations from $A_m(\theta_{12})$ in the angular distribution are a signal for RHC-induced no-neutrino decay. A careful measurement of the angular distribution would be required to determine whether the mass-induced decay is also present.

Another way of detecting the presence of right-handed currents is to look for transitions to the low-lying 2^+ excited states which are present in most of the daughter nuclei of interest. Because the electrons from the mass-mechanism are emitted in a relative S-wave, the transitions induced by this mechanism must obey the selection rule $\Delta J = 0^+$ [30,36]. The RHC mechanism produces electrons in a relative P-wave, and it induces transitions of the type $\Delta J = 0^+, 1^+$, and 2^+ [37]. Therefore the occurrence of $0^+ \to 2^+$ no-neutrino decay modes is an unambiguous signal for the presence of the RHC mechanism. Moreover, if the appropriate nuclear matrix elements are known, one can determine the magnitude of η from the $0^+ \to 2^+$ lifetime and use it to extract the neutrino mass from the $0^+ \to 0^+$ data.

HEAVY NEUTRINOS AND NON-NEUTRINO MECHANISMS

Up to now, we have examined the exchange of one light neutrino as the basic mechanism for no-neutrino double beta decay. There are, however, other possibilities. For example, neutrino exchange as in Fig. (3) gives rise to a propagator factor

$$P \equiv \left(\frac{m_\nu \langle p_\nu \rangle}{m_\nu^2 + \langle p_\nu \rangle^2} \right) \qquad (20)$$

in the amplitude, where $\langle p_\nu \rangle \simeq 1/\langle r_{ij} \rangle \approx 30 - 100$ Mev. The low mass case arises when $M_\nu \ll \langle p_\nu \rangle$, and it yields a propagator factor $P \simeq (m_\nu/\langle p\nu\rangle)$; an alternative, high mass limit exists in which $m_\nu \gg \langle p\nu\rangle$ and $P \simeq (\langle p\nu\rangle/m\nu)$. Limits on no-neutrino lifetimes can therefore tell us either that m_{light} is less than so many ev, or that m_{heavy} is greater than so many Gev.

A rough equivalence between these mass bounds is given by:[2,34]

$$m_{Heavy} \simeq 0.42 \ln\left[\frac{3.3\times 10^7}{(A)^{2/3} m_{light}(ev)}\right] \text{Gev} \qquad (21)$$

According to this formula, m_{light} in the range 1 - 100 ev corresponds to m_{heavy} in range 6 - 4 Gev. Much larger limits on m can be obtained from the quark model as long as N* resonances are present in the nucleus at a level of order 1% in amplitude.

For neutrino masses which are comparable with $\langle p \rangle$, we must work with the full Yukawa factor $\langle \frac{e^{-m_\nu r_{ij}}}{r_{ij}}\rangle$ in the matrix element. The resulting expressions are then much more complicated.[34]

Another alternative to the exchange of light neutrinos is some entirely new mechanism or interaction. Winter,[18] in 1955, constructed a six-fermion effective interaction while Pontecorvo,[19] in 1968, invented the $\Delta L = 2$ companion to superweak interactions which discussed in §4. More recently, Gelmini and Roncadelli, and Georgi and Glashhow[20] have invented a Goldstone boson ϕ associated with a global lepton number; they then anticipate processes of the type $(A,Z) \simeq (A,Z+2) + 2e^- + \phi$ which would show some of the characteristics of the two-neutrino mode. Mohapatra and Vergados[20] have discussed a model based on the exchange of Higgs scalars; however it is unlikely to make a dominant contribution to no-neutrino decay.[35]

CONCLUSIONS

To summarise this discussion, I would make the following points:

1) The indirect geochemical evidence for no-neutrino decay is much weaker then it was one year ago; however the nuclear matrix elements may have "unreasonable" properties, and/or there may be destructive interference between two Majorana neutrinos.

2) Top priority should be given to the detection of real $\beta\beta$ decay events in the laboratory for several

reasons. Measurement of the two-neutrino rate is necessary to resolve the disagreement between the geochemical and laboratory experiments on $^{82}Se \rightarrow {}^{82}Kr$, and to show whether we know how to calculate nuclear matrix elements; in this connection I would mention Resonance Ionization Spectroscopy[38] as a possible laboratory technique. Detection of the electron sum spectrum is also essential to determine whether that peak at the end of the spectrum is there or not; it is, perhaps, unfortunate in this regard that to maximize the absolute rate one needs the largest possible energy release, whereas to maximize the no-neutrino branching ratio one needs a small energy release.

3) If the no-neutrino mode is actually detected, then we will have to undertake a major experimental program to unravel the mechanism of lepton nonconservation. This will require studies of angular distributions and $0^+ \rightarrow 2^+$ transitions as well as careful measurements of the lifetimes as a function of energy release.

In conclusion let me say that positive evidence for lepton nonconservation and Majorana neutrinos would have far-reaching consequences for our understanding of fundamental interactions.

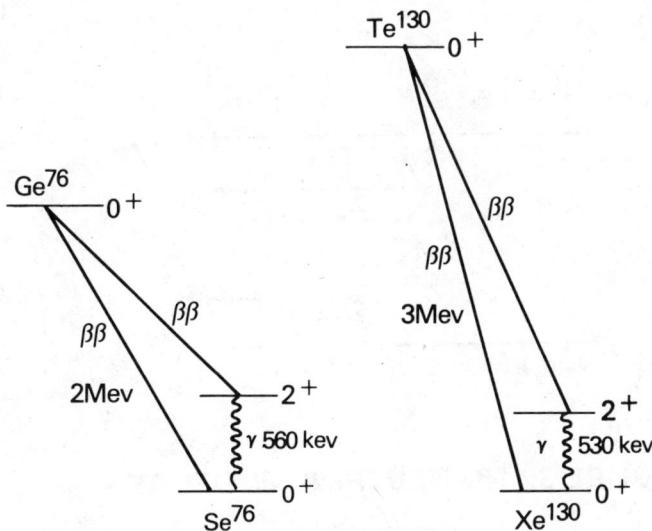

FIGURE 1: TYPICAL $\beta\beta$ LEVEL SCHEMES

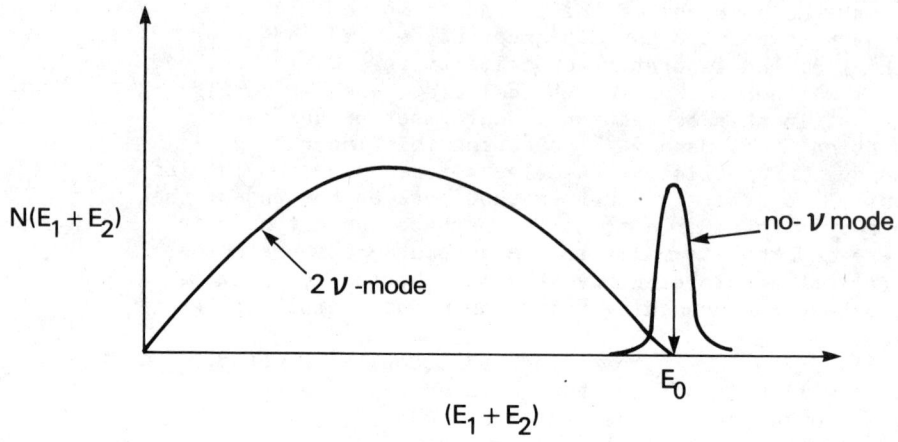

FIGURE 2: ELECTRON ENERGY SUM SPECTRUM

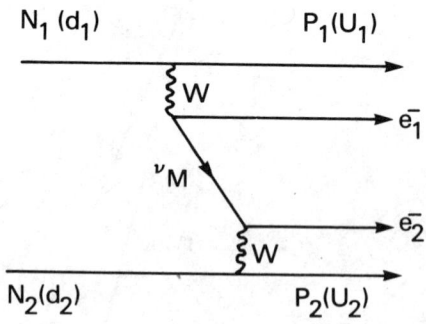

FIGURE 3: NO-NEUTRINO $\beta\beta$ DECAY

REFERENCES

1) For a review see P. Langacker, Phys. Rep. C72, 185 (1981).
2) For general reviews see H. Primakoff and S. P. Rosen, Ann. Rev. Nucl. Sci. 31, 145 (1981); S. P. Rosen Proceedings of Neutrino '81, Vol. II p. 76 (1981); W. Haxton (Comments on Nuclear and Particle Physics, to be published); and papers cited therein.
3) For a review of geochemical experiments see the talk by T. Kirsten in this volume.
4) For a review of laboratory experiments see the talk by C. S. Wu in this volume.
5) R. Bardin, P. Gollon, J. Ullman and C. S. Wu, Nucl. Phys. A158, 337 (1970).
6) M. Inghram and J. Reynolds, Phys. Rev. 78, 822 (1950).
7) T. Kirsten, O. Schaeffer, E. Norton and R. Stoenner, Phys. Rev. Lett. 20, 1300 (1968).
8) N. Takaoka and K. Ogata, Z. Naturforsch. 21a, 84 (1966).
9) E. Hennecke, O. Manuel, and D. Sabu, Phys. Rev. C11, 1378 (1975).
10) T. Kirsten, H. Richter, and E. Jessberger, Phys. Rev. Lett. (to be published); MPI H - 1982 - v 19.
11) T. Kirsten and H. Muller, Earth and Plan. Sci. 6, 271 (1969).
12) B. Srinivasan, E. Alexander, R. Beaty, D. Sinclair, and O. Manuel, Econ. Geol. 68, 252 (1973).
13) M. Moe and D. Lowenthal, Phys. Rev. C22, 2168 (1980).
14) W. Haxton, G. J. Stephenson Jr., and D. Strottman, Phys. Rev. D25, 2360 (1982).
15) W. Haxton, G. J. Stephenson Jr., and D. Strottman, Phys. Rev. Lett. 47, 153 (1981).
16) L. Zamick and N. Auerbach, Phys. Rev. C (in press).
17) J. D. Vergados, Phys. Rev. C12, 865 (1976); ibid C24, 640 (1981).
18) R. Winter, Phys. Rev. 99, 88 (1955), and 100, 142 (1955).
19) B. Pontecorvo, Phys. Lett. 26B, 630 (1968).
20) G. B. Gelmini, and M. Roncadelli, Phys. Lett. 99B, 411 (1981); H. M. Georgi, S. L. Glashow and S. Nussinov, Harvard University preprint HUTP-81/A026 (1981); R. Mohapatra and J. D. Vergados, Phys. Rev. Lett. 47, 1713 (1981).
21) W. Pauli, Nuovo Cim. 6, 204 (1957); C. D. Enz, ibid 6, 250 (1957); D. L. Pursey, ibid 6, 266 (1957). G. L. Luders, ibid 7 171 (1958).
22) E. Majorana, Nuovo Cim. 14, 171 (1937); G. Racah, ibid 7, 322 (1937).

23) P. H. Frampton and P. Vogel, Phys. Repts. C82, 339 (1982).
24) L. Wolfenstein, Phys. Lett. 107B, 77 (1981).
25) J. Valle, Syracuse & Lewes preprints (1982).
26) L. Wolfenstein, Phys. Rev. Lett. 13, 562 (1964).
27) H. Primakoff and S. P. Rosen, Phys. Rev. 184, 1925 (1969).
28) R. Davis, Phys. Rev. 97, 766 (1955).
29) D. Bryman and C. Picciotto, Rev. Mod. Phys. 50, 11 (1978).
30) M. Doi, T. Kotani, H. Nishimura, K. Okuda and E. Takasugi, Phys. Lett. 103B, 219 (1981).
31) V. Lubimov, E. Novikov, V. Nozik, E. Tretyakov, and V. Kozik, Phys. Lett. 94B, 266 (1980).
32) P. Minkowski, Nucl. Phys. B201, 269 (1982).
33) M. Doi, T. Kotani, H. Nishiura, and E. Takasugi, Osaka University preprint OS-GE 82-43 (1982).
34) A. Halprin, P. Minkowski, H. Primakoff and S. P. Rosen, Phys. Rev. D13, 2567 (1976).
35) W. Haxton, S. P. Rosen, and G. Stephenson, Phys. Rev. D26, 1805 (1982).
36) S. P. Rosen in Gauge Theories, Massive Neutrinos and Proton Decay, Edited by A. Perlmutter (Plenum Press, New York, 1981) p. 333.
37) H. Primakoff, Phys. Rev. 85, 888 (1952); S. P. Rosen, Can. J. Phys. 37, 780 (1959).
38) C. S. Hurst, this volume.
39) Y. G. Zdesenko, Pisma Zh. Eks. Teor. Fiz. 32, 62 (5 July 1980).
40) B. Cleveland, W. Leo, C. S. Wu, L. Kasday, A. Rushton, P. Gollon, and J. Ullman, Phys. Rev. Lett. 35, 737 (1975).
41) E. Bellotti, E. Fiorini, C. Liguori, A. Pullia, A. Sarracino and L. Zanotti, Lett. Nuovo Cim. 33, 273 (1982); F. T. Avignone III, R. Brodzinski, D. Brown, J. Evans, W. Hensley, J. Reeves, and N. Wogman (to be published).

FIGURE CAPTIONS

Figure 1: Energy Levels of Typical Double Beta Decaying Nuclei

Figure 2: Energy Sum Spectrum Showing Broadly Peaked Distribution for 2 -Decay and the Spike for 0 -Decay.

Figure 3: Feynman Diagram For 0 -Decay.

TABLE I

Selected Data

DECAY	$T_{1/2}(10^{21}$ yrs)	GEO/Lab	REF.
$Te^{130} \to Xe^{130}$	1.4 + 0.3	Sweden	(6)
	2.2 + 0.6	Colorado	(7)
	0.97 + 0.1	Australia	(9)
	>1.2 + 10	(0_ν)	(39)
$(Te^{128} \to Xe^{128})$			
$(Te^{130} \to Xe^{13})$	(3.2 + 3.2)X10	Japan	(8)
	(1.5 + 0.2)X10	Australia	(9)
	>3.1 X 10 (2_ν)	Colorado	(10)
$Se^{82} \to Kr^{82}$	0.14 + 0.03	Western Moravia	(11)
	0.28 + 0.09	Sweden	(12)
	>3.2	(0_ν)	(40)
	0.01 + 0.004	(2_ν)	(13)
$Ca^{48} \to Ti^{48}$	>2 X 10^{21}	(0_ν)	(5)
	>3.6 X 10	(2_ν)	
$Ge^{76} \to Se^{76}$	>1 X 10^{22}	(0_ν)	(41)

DOUBLE BETA DECAY - AN EXPERIMENTAL REVIEW

Chien-Shiung Wu
Columbia University, New York, NY 10027

ABSTRACT

In this review, a brief discussion on the experimental objectives and theoretical conceptions of the neutrinos was introduced. Two different types of experimental methods and their results were presented and discussed.

INTRODUCTION

We have just enjoyed listening to Prof. S. P. Rosen's[1] eloquent and comprehensive review of the theory of "Double Beta Decay, Massive Neutrinos, and Lepton Number Conservation". Some of the experimental results and theoretical predictions are not in good agreement at the present moment. This is due to the difficulties and treacheries of the measurements of these rare events, the difficulty one experiences when trying to detect them, and the fact that they are also easily confused because of the many kinds of 'look-alike' double beta decay backgrounds. From the most recent geochemical determination of the ratio "S" of double beta decay lifetime of $T_{\frac{1}{2}}(^{128}Te)$ to $T_{\frac{1}{2}}(^{130}Te)$, a much larger ratio was obtained. Much reduced limits for the mass of neutrino $m_\nu \leq 5.6ev$ and of the lepton number violation $\eta \leq 2.4 \times 10^{-5}$ have been concluded with a 95% C.L.[2]. The experimental physicists are now further motivated to improve the sensitivity of the Direct Detection Method by one or two orders of magnitude, hoping to eventually observe the unique picture of the single sum peak of the $(no\nu)\beta\beta$ process, or at least to confirm or to extend the lower limits which had been set on the neurtino mass and lepton number violation by the recent geochemical determination.

EXPERIMENTAL OBJECTIVES AND THEORETICAL CONCEPTIONS OF THE NEUTRINOS:

The theoretical study of double beta decay was originally initiated in 1935 by M. Mayer[3]. It has been continuously developed[4] alongside our changing conceptions of the neutrino fields in the charged weak current. There are, in general, three very marked periods in this development:
(I) The Pre-1957 Era,
(II) After the Overthrow of parity in Beta Decay, and
(III) The interest stirred up by the proposed Grand Unified Theories:

(I) The Pre-1957 Era:
In the period parity non-conservation had not yet been discovered in beta decay. The relationship between double beta decay and the neutrino was simply related to the identity of the neutrino: whether it was a Majorana or a Dirac neutrino.

(A) Majorana Neutrino[5]:

A particle is said to be a Majorana particle if it is identical with its anti-particle (charge conjugate partner) operationally

$$\nu \equiv \nu^c \equiv \nu_M$$

The neutrino is unique among the fermions in that it possibly possesses such an identical anti-particle. All other fermions, having either a charge or a measurable magnetic moment, are ruled out as having an identical anti-particle.

A Majorana neutrino, emitted into a virtual state with the emission of the first electron, might be reabsorbed in the subsequent emission of the second electron. Thus a (noν)$\beta\beta$-decay process results:

$$n \rightarrow p + e^- + \bar{\nu}_e \quad \text{or} \quad n + \nu_e \rightarrow p + e^-$$

$$\text{If} \quad \nu_e \equiv \bar{\nu}_e \quad \text{then}$$

$$2n \rightarrow 2p + 2e^-$$

$$\text{or} \quad (A,Z) \rightarrow (A,Z+2) + 2e^- \qquad (1)$$

Fig. 1 sketches the second-order (noν) double beta decay by emission and reabsorption of a neutrino.

(B) Dirac Neutrino[6]:

A Dirac neutrino is defined as distinct from its charge conjugate partner (anti-particle)

$$\nu \neq \nu^c$$

Therefore only the two-neutrino double beta decay, (2ν)$\beta\beta$-decay can occur:

$$(A,Z) \rightarrow (A,Z+2) + 2e^- + 2\bar{\nu}_e \qquad (2)$$

The ($\beta\beta$) Sum-Energy Distribution.

The two-neutrino double beta spectrum is shown by the continuous energy distribution as in Fig. 2. The (noν) double beta decay is indicated by the vertical sum-peak at the maximum energy of the two-neutrino spectrum. It is very distinctive if (noν)$\beta\beta$-decay exists.

The Phase Space Advantage in (noν)$\beta\beta$-decay:

The (noν)$\beta\beta$-decay also enjoys a considerable phase space advantage over the (2ν)$\beta\beta$ decay as the neutrino in the intermediate state of the (noν)$\beta\beta$-process is virtual. Because of this, its energy need not be conserved and may be assumed up to approximately to 35 Mev (i.e. its de Broglie wave length is approximately equal to the radius of the nucleus). The ratio of the decay life times is essentially equal to the fourth power of energy ratio

$$T_{\frac{1}{2}}(0\nu)/T_{\frac{1}{2}}(2\nu) \sim (E_{2\nu}/E_{0\nu})^4$$

$$\simeq \left(\frac{1 \text{ Mev}}{35 \text{ Mev}}\right)^4 \simeq 10^{-6}$$

So in the Pre-1957 era, the life time of the $(0\nu)\beta\beta$-process by Majorana neutrino was expected to be a factor $10^{-5} - 10^{-6}$ times that of the $(2\nu)\beta\beta$-decays by Dirac neutrinos. The sensitivity which this enhencement factor imparted to the search for $(0\nu)\beta\beta$-decay played an important factor in the Pre-1957 era.

(II) <u>After the Overthrow of Parity in Beta Decay</u>[7]:

With the discovery of maximal parity non-conservation in beta-decay, the two component theory of the neutrino was generally accepted. The amplitude of the $(0\nu)\beta\beta$-decay would always be strongly inhibited by either the lepton conservation $\Sigma \ell_e \equiv 0$ or the perfect or near perfect Chirality of the neutrino. Under these conditions, all double beta decay would more likely proceed through the slower $(2\nu)\beta\beta$-process. Even at the present moment, with greatly reduced radiation backgrounds by both passive and active shieldings and with much improved detector sensitivities, the lower limit of the life time of $(0\nu)\beta\beta$-decay has been pushed up to 2×10^{22} years, one order of magnitude higher than the previous limits set at a few 10^{21} years. However, it can be said that even a slight indication of $(0\nu)\beta\beta$-decay has yet to be reported in direct detection.

(III) <u>The Interest stirred up by the proposed Grand Unified Theories (GUTS)</u>[8].

Recently, great interest has been generated in the study of double beta decay by the proposed theories of Grand Unification of Strong and Electroweak Interactions, using modern Gauge theories. Only a few years ago, with the introduction of the Unification of Weak and electromagnetic interactions by Weinberg, Salam and Glashow[9], the properties of neutrinos used in the models were still massless and the conservation laws of lepton and baryon numbers were perfectly intact. But in the formulation of the GUTS, where the mass scale governing the strong-electroweak unification is enormous, it seemed people were reluctant to come out and postulate apriori global conservation laws. Instead, they suggest that many of the conservation laws are now assumed <u>only approximate</u>. It is quite natural to suggest that the neutrino is a <u>massive Majorana particle</u>. As this mass will break the γ_5-invariance of the weak current, it could also implicitly introduce wrong helicity mixing of the neutrino.

<u>Explicit and Implicit Effects on the Probability of $(0\nu)\beta\beta$-decay by Neutrino Masses and the Lepton Number violation</u>.

Soon after the overthrow of parity in beta decay, Pauli and Pursey[10] pointed out that the most general form of the weak current for β-decay can be written as

$$L_\lambda \equiv (\bar{\psi}_\lambda \gamma_5 \{C(1+\delta_\lambda \gamma_5)\psi_\nu + D(1+\eta_\lambda \gamma_5)\psi_{\nu c}\} \qquad (3)$$

↑ neutrino field   ↑ Its charge conjugate

In connection with the basic tests for distinguishing between Dirac and Majorana neutrinos, it was also shown by EnZ[11] in 1957 that the so called "Racah sequences"

$$n \to p + e^- + \nu_M$$
$$\nu + n \to p + e^-$$

which are allowed for Majorana but not Dirac neutrinos, could be altered by varying certain combinations of coupling constants. Their methods are:

(1) Explicitly add small right-hand components in the weak leptonic currents such as

$$I = CD(1 - \delta_\mu \eta_\lambda)$$
$$J = CD(\delta_\mu - \eta_\lambda) \qquad (4)$$

It requires $D \neq 0$ and $(1-\delta_\mu \eta_\lambda) \neq 0$. It <u>is not a pure Dirac</u> neutrino and it is also <u>mass-independent</u>.

(2) Implicitly let m_ν serve as the helicity breaking parameter, in which case its amplitude then depends on two combinations:

$$I' = m_\nu CD(1+\delta_\mu \eta_\lambda) \quad \text{and}$$
$$J' = m_\nu CD(\delta_\mu + \eta_\lambda) \qquad (5)$$

It definitely requires $D \neq 0$ and $m_\nu \neq 0$; the neutrino is a massive one.

It should be noted that the neutrino mass mechanism can not give rise to $0^+ \to 2^+$ but only to $\Delta J^P \equiv 0^+$. For example, in ^{76}Ge($\beta\beta$) decay, the $0^+ \to 2^+$ transition can be produced only by the chirality mixing, and not the mass term. An attempt to detect the low energy (noν)$\beta\beta$ transitions ($0^+ \to 2^+$) as shown in ^{76}Ge (<u>Fig. 3</u>) would be very informative.

It is hard to imagine that 25 years ago, physicists already earnestly speculated on all these possibilities which are now directly applicable to the sensitive tests being carried out on the properties of neutrinos and symmetry laws in double beta decays.

EXPERIMENTAL METHODS

There are two quite different types of methods of detecting the double beta decays. One is the <u>Direct Detection Method</u> in which the sum spectrum of the two electron energies along with other properties are used for identification. In neutrinoless decay, $(\beta\beta)_{0\nu}$, they must add up to the available decay energy; in two-neutrino decay $(\beta\beta)_{2\nu}$, the electron sum energy spectrum will be a broad distributuion as shown in (Fig. 2). So far, no direct observation of $(\beta\beta)_{0\nu}$ decay has been reported indicating the smallness of the lepton-violation parameter "η". The direct observation of the evidence of the $(\beta\beta)_{2\nu}$ decay from ^{82}Se has been reported but its life-time determination may be questionable because of the background uncertainties in the source and in the environment.

In opposition to <u>the Direct Detection Method</u> is <u>the Geochemical Method</u>, in which, the daughter nucleus of the $(\beta\beta)$ decay must be a rare gas and occluded in a rock of known old age ($\geq 10^9$ years). The rare gas can be thermally extracted from the rock and its isotopic contents can be accurately determined by a highly sensitive mass spectrometer. Although leptons are not detected in these experiments, much insight concerning lepton conservation can be gained by comparing the measured ratio of the double beta decay half lives such as $T_{\frac{1}{2}}(^{128}Te)/T_{\frac{1}{2}}(^{130}Te)$ to the theoretically predicted ratios based on $(\beta\beta)_{0\nu}$ or $(\beta\beta)_{2\nu}$ decays, or comparing the double beta decay life of ^{82}Se from the Geochemical Method to that of $(\beta\beta)_{0\nu}$ from Direct Method. The most recent results by the Max Planck group [12] on the ratio $T_{\frac{1}{2}}(^{128}Te)/T_{\frac{1}{2}}(^{130}Te)$ illustrated clearly the unique and delicate power of the Geochemical Method.

(I) Direct Detection Method

At first glance, the coincidence technique of simultaneous detection of the two e^- (or $2e^+$) in double beta decay may be sufficient to observe this rare process. Many experiments had been carried out with coincidence techniques prior to 1967 in the search for $(\beta\beta)$ decay, but the sensitivity of those experimental searches were limited by <u>background counting rates</u> much larger than the expected $(\beta\beta)_{2\nu}$ rate.

In the usual coincidence counting experiments (Fig. 4) the background results principally from local gamma rays either Compton scattering from counter to counter (4B) or producing recoil electrons in one counter which strike another counter (4C). The other mechanisms from (D) to (J) are more rare than (B) and (C) but all these events can be reduced or eliminated by surrounding the double beta decay source with some device showing the tracks of particles leaving the source and placing the apparatus in a magnetic field. Such a device, by the Columbia group[13], is shown schematically in <u>Fig. 5</u>. A source, unstable to double beta decay, is mounted inside a track visualizing device (such as a streamer chamber, a cloud chamber or a TPC). Outside the chamber are mounted a number of counters. A coincidence between any two counters makes the chamber operate and produces a picture of all the particle tracks in it. A magnetic field is directed normal to the plane of the source. The background was further reduced by running the experiment in the low radiation environment of a salt mine 2000 ft. below ground level. <u>Fig. 6</u> is a comparison of gamma spectra taken with a 12.7 and x 12.7 cm NaI crystal under various conditions on the surface and in the mine. The "intrinsic" background in the mine was found by placing the NaI crystal in a freshly dug pit in the salt, well-shielded from all other man-made objects. The 1.46 MeV gamma ray from the decay of ^{40}K in the salt was the only significant radiation naturally present in the salt. The 2.62 MeV line of the Th C" (^{208}Th) was visible throughout the experimental area and appeared to be associated with objects brought from the surface of the earth.

A large component of the background came from airborne radon gas (^{222}Em, with a 3.8d half-life) coming from the decay of ^{226}Ra. The level of this activity increased noticable <u>in rainy weather</u> even in the deep mine.

The data film was scanned for possible $(\beta\beta)_{0\nu}$ events by first selecting all events with a total energy above 3 MeV containing a single pair of tracks meeting at the source. In 1103 hrs. of running with the ^{48}Ca source, 191 such events were found. Of the remaining 119 events, all but 23 were rejected for track curvature clearly of the wrong sign or particle momentum clearly inconsistent with recorded scintillation information. These were mostly high-energy Compton recoils or cosmic-ray muons crossing the chamber. Only one of the surviving ^{48}Ca events was within two standard deviations of 4.24 MeV, the expected $(\beta\beta)_{0\nu}$ sum energy. A half-life limit for the $(\beta\beta)_{0\nu}$ decay in ^{48}Ca was arrived at

$$T_{\frac{1}{2}}(0\nu) \geq 1.6 \times 10^{21} \text{ yr.} \quad \text{at the 80\% C.L.}$$

The scanning of the two-neutrino decay $(\beta\beta)_{0\nu}$ events obtained requires changing the criteria above. A lower limit of the two-neutrino decay obtained at an 80% confidence level is

$$T_{\frac{1}{2}}(2\nu) \geq 10^{19.56} \text{ yr.}$$

The apparatus and measurement technique similar to that used for the study of ^{48}Ca by Bardin et. al. were also applied to the study of the $(\beta\beta)$ decay of ^{82}Se by Cleveland et. al.$^{(14)}$ in 1975. The selenium source contained approximately 46g of metallic powder enriched to 56.5% of ^{82}Se formed the center plate (Diam. = 50 cm.; t = 58 mg/cm^2) of a helium-filled double gap streamer chamber. The selenium was purified by precipitation of radium with separated ^{138}Ba and by multiple passes through ion-exchange columns. In order to reduce background, the experiment was also conducted in the Morton salt mine (600m below ground level) at Cleveland. The total number of events recorded was 65,500. The vast majority of these were background induced, primarily Compton scattering of γ-rays in a scintillator, the recoil electron then passing through the chamber and hitting a second counter. Only 201 events are two-track events with the signature of $\beta\beta$ decay. Many of these were caused by double scattering within the source and it was difficult to distinguish them from true $\beta\beta$ events. By restricting the energy range to between 2.4 and 3.2 MeV and imposing appropriate acceptance criteria on the track curvature, the overall selection efficiency became 19% and no events were found in this energy region. At a confidence level of 68%, this null result implies the following lower limit on the half-life for no-neutrino $(\beta\beta)$ decay of ^{82}Se:

$$T_{\frac{1}{2}}^{0\nu} \geq 3.1 \times 10^{21} \text{ year}$$

Combining this with the measurement of the overall half-life by Srinivasan et. al.[15] of $(2.76 \pm 0.88) \times 10^{20}$ year, we find the branching ratio, R, to be

$$R = \frac{\text{non-neutrino rate}}{\text{total } (\beta\beta) \text{ rate}} \leq 9\%$$

This is the first experimentally determined branching-ratio limit in $(\beta\beta)$ decay.

Moe and Lowenthal's new approach

In the Direct Detection used by Bardin et. al. in 1967 and Cleveland et al. in 1975 for the study of the $(\beta\beta)$ decay of ^{48}Ca and ^{82}Se, respectively, among the "2e⁻" backgrounds they noticed unusually large numbers of double-beta-like events indistinguishable from the true double beta decays. After thorough investigations, it was concluded that the principal radioactive contaminant responsible for these mechanisms was ^{214}Bi, a member of the naturally occuring uranium series; its multiple $(\beta-\gamma)$ cascades cause "two electron" events to emerge from the source resembling the $(\beta\beta)$ decay. (See Fig. 4I and 4J). The second electron was produced by Møller or Compton scattering or by internal conversion. Although the ^{214}Bi nucleus is short lived, its presence after a few days is maintained by its long-lived progenitor ^{226}Ra. It had been hoped that by chemical purification substantially lower levels of radiation in isotopically enriched double beta sources could be achieved. However, success had not been adequately demonstrated. The level of ^{226}Ra contamination that would prove troublesome in this case only needs to be on the order of one part in 10^{15}.

Moe and Lowenthal[16] have made special efforts to modify the direct detection method in order to identify and eliminate a large fraction of double-beta-like events from ^{214}Bi. The general principle is follows: ^{214}Bi beta decay is followed in 164 μs by emission of a 7.7 MeV alpha particle. The ^{214}Bi induced "two-electron" events thus can be identified if the α-particle can be detected (see Fig. 7). For the 164 μs delayed α-particle from ^{214}Bi to be detected, <u>a track visualization chamber of longer sensitive duration of post trigger period as compared to 164 μs must be used.</u> A large cloud chamber was chosen by Moe and Lowenthal for this purpose and its duration of post trigger sensitivity was found to easily cover the delay of alpha-particles attending ^{214}Bi decay.

Observed Events

At the end of 37 live days, 36 negatron pairs from the ^{82}Se source (13.34 g) strips had been observed. Of these, 20 were clean events (2e⁻) and 16 were accompanied by α-particles (2e⁻ + α) track pictures.

The observed total number of 2e⁻ events was 20 and the calculated (2e⁻) from ^{214}Bi with α-particles trapped in the source was 4.8 ± 1.2. Thus, (2e⁻) from ^{82}Se was only (15.2 ± 4.6).

Using the calc. overall detection efficiency as 0.022 ± 0.007, the half life for two neutrino (2ν) decays can be calculated to be

$$T_{1/2}^{2\nu} = 1.0 \pm 0.4 \times 10^{19} \text{year},\quad \text{with sum energy below 3.3 MeV.}$$

Because of its extremely low source mass and trigger efficiency, it is not efficient for detecting neutrinoless $(\beta\beta)_{0\nu}$ decays. No events at high energies were seen.

Energy Sum Method in Ge-Detector[17]

Another ingenious direct detection of double beta decay was applied to ^{76}Ge by Fiorini et. al.[17] using a high resolution Ge(Li) detector in 1973. A similar principle was also applied to detection of $(\beta\beta)_{0\nu}$ in ^{48}Ca by Goldhaber and des Mastesian.[18] However, the background was much too high. Ge(Li) detectors are fabricated out of high purity Ge metal with a natural abundance of ^{76}Ge of 7.67%. The Ge(Li) detector used by Fiorini et. al. had an active volume of 68.5 cc and an energy resolution of 6 kev at 2.615 MeV. A diagram of the experimental apparatus is shown in (Fig. 8). The experiment was located in the Mont Blanc tunnel connecting Italy to France and the crystal was shielded from local radioactivity by layers of paraffin, cadmium, low-activity lead, bi-distilled mercury, nylon and high purity electrolytic copper. After 4400 hours of running, no peak was found in the 2.045 MeV region of expected neutrinoless $(\beta\beta)$ decay (Fig. 9). Tests indicated that the residual backgrounds observed in this energy region were likely due to ^{40}K, ^{235}U, ^{238}U and ^{232}U contaminants of less than 10^{-5}ppm which originated inside the local shield, probably in the Ge(li) crystals crystat structure. Possible ^{222}Rn contaminants in the liquid nitrogen coolant were also suspected. Fig. 9 shows the spectrum obtained in the final run of the experiment lasting 2300 hours. The background counting rate in the 2.045 MeV region was $(2 \pm 0.2) \times 10^{-4}$ counts keV^{-1}hr^{-1}, allowing a limit to be set on the half-life for the neutrinoless $(\beta\beta)$ decay of ^{76}Ge, $T_{1/2}^{0\nu} \geq 5 \times 10^{21}$ year at a 68% confidence level. With improved background, the limit has improved to $T_{1/2} \geq 2 \times 10^{22}$ yr. in the summer of 1982 (See Table I).

(II) Geochemical Methods

By utilizing the high sensitivity of noble gas mass spectrometry, as early as in 1950, Inghram and Reynolds[29] determined the half-life of double beta decay of ^{130}Te $\rightarrow$ ^{130}Xe as 3.3×10^{21} yrs. The tellurium ores generally contain sizable amount of Uranium contaminant which greatly complicates the analysis as α-particles from U or Th may induce $(\alpha,2n)$ or (α,n) reactions and Uranium fission may create some excesses of Xe-isotopes. In general, to accept the results of geochemical methods one must also rule out all possible productions of rare gas excesses in ore by $(n-\gamma)$ reaction.

In the middle of the sixties, the activities of geochemical method in the study of double beta decay were greatly rejuvenated. Kirsten, et. al.[12] obtained unequivocal proof of double beta decays in Te and

Se ores of known ages and yielded $T_{\frac{1}{2}}(^{130}Te) = 10^{21.34 \pm 0.12}$ yr. A similar result was obtained by Srinivasan et. al.$^{(15)}$ in 1972 with $T_{\frac{1}{2}}(^{130}Te) = 10^{21.38 \pm 0.10}$ yr. The results of Kirsten et. al. on native tellurium ore from the Good Hope Mine in Colorado were particularly significant as the uranium concentration in this ore was four orders of magnitude less than in samples used earlier (See Fig. 10). On the otherhand, a large excess of ^{130}Xe was found unaccompanied by any other anomalies (as might be caused by nuclear fission and brought about by neutrons from uranium fission). This could only be from the double beta decay of ^{130}Te in the Ore. Even in that early stage, the double beta decay rate calculated from the age of the ore and the concentration of ^{130}Te and ^{130}Xe in it was consistant with lepton conservation. An upper limit could be set on the lepton number violating fraction of the beta interaction amplitude of $\alpha \leq 3 \times 10^{-3}$.

Life-Time Ratio "S" $\equiv T_{\frac{1}{2}}(^{128}Te)/T_{\frac{1}{2}}(^{130}Te)$.

In an earlier experiment, by Takaoda & Ogata$^{(19)}$ (1966), using different tellurium ores, an excess of ^{128}Xe was found; but the authors cautioned that it was difficult to assign the excess of ^{128}Xe entirely to ^{128}Te double beta decay, because a small, persistent background in the mass spectrometer disturbed exact measurements at ^{128}Xe. If one takes the total amount of excess ^{128}Xe observed as due to ^{128}Te double beta decay then the ^{128}Te half-life is $10^{22.5 \pm 0.5}$ yr. Furthermore, it is not unreasonable to assume that nuclear matrix elements for $^{128}Te \rightarrow {^{128}Xe}$ and $^{130}Te \rightarrow {^{130}Xe}$ are approximately equal. Under this assumption, the ratio of double beta decay rates should be proportional to the ratio of the available phase spaces, which is the 7th through the 11th power of energy release for $(\beta\beta)_{2\nu}$ decay and the 1st through 7th power of energy release for $(\beta\beta)_{0\nu}$ decay. Since the energy release for ^{130}Te is three times that for ^{128}Te, one would expect $S \equiv T_{\frac{1}{2}}$ $(^{128}Te)/T_{\frac{1}{2}}(^{130}Te) \simeq 3^{8.4} = 10^{4.0}$ for $(\beta\beta)_{2\nu}$

" " $\simeq 3^{4.6} = 10^{2.2}$ for $(\beta\beta)_{0\nu}$.

When the specific warning given by the authors about a small persistent background on the ^{128}Xe excess was not only ignored, instead, used to calculate the ratio "S", then we have

$$"S" = T_{\frac{1}{2}}(^{128}Te)/T_{\frac{1}{2}}(^{130}Te) = 10^{22.5 \pm 0.5}/10^{21.34 \pm 0.12}$$
$$= 10^{1.2 \pm 0.6}$$

This ratio seems to be in better agreement with that predicted for no - neutrino double β-decay.

Pontecorvo proposal$^{(20)}$:

On the basis of this rather questionable experimental indication, Pontecorvo proposed that the decays of ^{130}Te and ^{128}Te are, predominantly, the first order effect of a new super-weak ($\Delta Q = \pm 2$, $\Delta S = 0$) interaction which mediates non-neutrino double beta decay, rather than the second-order effect of the usual weak ($\Delta Q = \pm 1$, $\Delta S = 0$) interaction. This super-weak interaction also causes the observed slight CP violation in K^0 decay.

In 1975, Hennecke, Manual and Sabu[21] carried out another investigation on the ratio "S" of $T_{1/2}(^{128}Te)/T_{1/2}(^{130}Te)$ by extracting Xenon by stepwise heating of the sample, and its isotopic abundances were measured in the mass region A = 122 to A = 136 in a mass spectrometer. For ^{130}Xe the excess over atmospheric abundance was $(^{130}Xe$ excess$)/(^{130}Xe$ atmospheric$) = 712\pm2$, which was more than an order of magnitude greater than in the previous experiments. Many sources of systematic errors in measuring the individual ββ decay half-lives, such as errors in ore age T, in the tellurium determination $N(^{130}Te)$ and in the Xenon content $N^{excess}(^{130}Xe)$, cancel out in the determination of the ratio "S" of the half-lives. The experimental results found by Hennecke, Manuel and Sabu were:

$$T_{1/2}(^{130}Te) = 10^{21.34\pm0.10} \text{ yr.}$$

$$T_{1/2}(^{128}Te) = 10^{24.54\pm0.12} \text{ yr.}$$

and $"S" = 10^{24.54\pm0.12}/10^{21.34\pm0.10} = 10^{3.20\pm0.01}$

Since $"S" = 10^{3.20}$ which lies between the predictions for (2ν) and (0ν)(ββ) processes, it was natural to conclude that the final state must be a combination of these two decays. In 1978, Bryman and Picciotto[22] employed the lepton number non-conserving parameter "η" introduced by Primakoff and Rosen[6] in the lepton current

$$(j_L^{lep} + \eta\, j_R^{lep})$$

and obtained a value for $\eta = (4.3 \pm 0.1) \times 10^{-5}$ from the ratio $"S" = 10^{3.20}$.

By early 1980, the theoretical and experimental interests[8] in searching for the solutions of the mass of neutrino by using double beta decay results were greatly aroused. Several theoretical attempts[1,4], using implicit helicity breaking by neutrino mass m_ν, to interpret the observed "S" value (S=3.20) were reported. The mass values obtained vary from ~10ev by Haxton[23] to ~34ev by Doi[24]. These values can be compared with the mass value

$$14 \leq m_\nu \leq 45 \text{ev}$$

from the ^3H β-spectrum of ITEP.[25]

Heidelberg Recent Results[12]

Because of the fundamental importance of studying the mass of the electron neutrino and lepton conservation, Kirsten et al.[12] have remeasured the half-lives of ^{128}Te and ^{130}Te and the ratio "S". The exceptionally high purity of the Te ore which was in the possession was <u>particularly suitable</u> for the determination of the relatively small ^{128}Xe excess volume. Any contamination in the ore would effect sensitively this small fraction of rare gas excess from DBD.

The results obtained by Kirsten et al.[12] in 1982 were

$$T_{\frac{1}{2}}(^{128}Te) \geq 8 \times 10^{24} \text{ yr.} \quad (2\sigma)$$

$$T_{\frac{1}{2}}(^{130}Te) = 2.60 \pm 0.28 \times 10^{21} \text{ yr.}$$

$$\rho = \frac{^{128}\lambda}{^{130}\lambda} = \frac{1}{"S"} = (1.03 ^{+1.13}_{-1.03}) \times 10^{-4}; \quad (\text{where } \lambda = \frac{\ln 2}{T_{\frac{1}{2}}})$$
$$\uparrow \text{ decay const.}$$

Actually, the half-lives of ^{130}Te determined from various attempts are rather in good accord such as $(2.27\pm0.60)\times10^{21}$yr.$^{(12)}$ and $(2.72\pm0.27)\times10^{21}$yr.$^{(26)}$ and the present one $(2.60\pm0.28)\times10^{21}$yr.$^{(2)}$. The suggested "best value" is $^{130}T_{\frac{1}{2}} = (2.55\pm0.20) \times 10^{21}$ yrs..

Continuous Reduction of Value of "ρ".

While the $^{130}T_{\frac{1}{2}}$ determined were in rather good accord; however, the value of "ρ" has varied from 295×10^{-4} from Takaoka's experiment ('66) to 194×10^{-4} by Srinivasan's study('72). In 1972 Kirsten et. al. obtained a much lower "ρ" value equal to 5×10^{-4}, but it was kept from publication in order to improve on it. In 1975 and 1978, Hennecke et. al. published their results on $\rho = (6.29\pm0.20)\times10^{-4}$ and $(6.37\pm0.41)\times10^{-4}$. It is indeed a long way for ρ to come down to the present value$^{(2)}$

$$\rho = (1.03 ^{+1.13}_{-1.03}) \times 10^{-4}$$

The theoretically expected ratio for pure Dirac decay is ($\rho_{2\nu} \sim 2 \times 10^{-4}$). Since the measured value is $\rho = (1.03 ^{+1.13}_{-1.03}) \times 10^{-4}$, it implies that there is very little room for an admixture of Majorana decay or no-neutrino decay in DBD.

Fig. 11 is shown the relationship between the value of "ρ" and "η" with various values of m_ν for ^{128}Xe and ^{130}Xe $^{(2)}$. It is quite obvious, from the Heidelberg results$^{(2)}$, limits on $m_{\nu_e} \leq 5.6$ev (95% confidence) and on the lepton violation parameter "η" $\leq 2.4\times10^{-5}(2\sigma)$ can be concluded from the curves. However, it is also consistent with zero neutrino mass $m_\nu = 0$ and for the lepton conservation $\Sigma \ell_e \equiv 0$.

^{82}Se Double Beta Decay

The double beta decay of ^{82}Se was the only one which has been studied by both the Direct Detection method and the Geo-chemical method. The half-life of the $(2\nu)\beta\beta$ decay determined by Moe and Lowenthal$^{(16)}$ using direct detection yielded a value $^{82}T_{\frac{1}{2}}(\text{Irvine})=(1\pm0.4)\cdot \times10^{19}$ yrs. This is more than one order of magnitude shorter than that obtained by geochemical method$^{(15)}$ which gave $^{82}T_{\frac{1}{2}}(\text{Missouri})=2.76\times 10^{20}$ yrs. This life time was again measured recently by Heidelberg group$^{(2)}$ in a systematic study containing both ^{82}Se and ^{130}Te. Their value $^{82}T_{\frac{1}{2}}(\text{Heidelberg}) = 1.4\pm0.7 \times 10^{20}$ yr. in the same order of magnitude 10^{20} yr. as Missouri's value. In addition, the theoretical predictions$^{(27)}$ have given a shorter life time $^{82}T_{\frac{1}{2}}$(Theoretical) = 2.35 × 10^{19} yr. However, in order to improve the statistics and uncertain background of the Irvine result, the Irvine Group has completely replaced their Cloud Chamber detection by a Time Projection Chamber. It will be very interesting to see the final outcome of this three-sided contest. Should Irvine's new ^{82}Se lifetime determination reaffirm the previous short lifetime ($\sim 10^{19}$yr.), one may have to investigate carefully the reliability of the geochemical method used as an <u>absolute</u> lifetime determination. If the $^{82}T_{\frac{1}{2}}(2\nu)$ is of $\sim 10^{20}$ yrs., then the theoretical calculation of the nuclear matrix may need further improvement.

OUTLOOK

The study of the very rare natural processes such as the double beta decay has now assumed paramount importance in shedding crucial information on the properties of electron neutrino and the conservation laws of leptons. The determination, by the Geochemical method, of the ratio of lifetime of ^{128}Te to ^{130}Te has given us much lower limits on both "m_ν" and "η". To improve the power of the Direct Detection method, we must greatly reduce the background radiation either in the visualizing chamber (such as in the Time Projection Chamber) or in the detector system which also serves as DBD source (such as ^{76}Ge in the Ge-detector). Great efforts have been spent on selecting construction materials with low natural radioactivities. Underground laboratories have been sought and established below 2000 MWE. Large area anti-coincidence counter shieldings have been provided wherever they are necessary. At the present moment, the lower limit of the (noν)$\beta\beta$-decay lifetime may be capable of reaching $T_{\frac{1}{2}} \geq 5 \times 10^{22}$ yrs. This limit may increase another order of magnitude in the near future.

An ambitious contemplation of utilizing single <u>atom counting</u> in ^{82}Se → ^{82}Kr has been proposed. But to count such rare events even by the newly developed Resonance-Ionization-Spectroscopy Method may also encounter the interference problem from the overwhelming background.

In Table (I) and (II), we summarized the DBD Results from the Direct Detection methods and the Geochemical methods respectively. In Table (III) we list existing and planned double beta decay experiments which were mentioned or discussed during this Work Shop. Time did not permit me to consult with every one of you before making up the list. My very best to your valliant endeavor.

REFERENCES

1) S.P. Rosen; "Double Beta Decay, Massive Neutrinos and Lepton Number Conservation." Talk given at the Underground Science Workshop, LASL,N.M. 9/27-10/01, (1982). Also, talk given at Neutrino Masses Workshop at Cable, Wis., 2-4 Oct., (1980)

2) T.Kirsten et. al. "Rejection of Evidence for Non-zero Neutrino Restmass from Double Beta Decay." Talk given at the Underground Science Workshop LASL, N.M. 9/27-10/01, (1982). Submitted to PRL, Sept., (1982)

3) M. Goeppert-Mayer, Phys. Rev. $\underline{48}$ 512 (1935)

4) W.H. Furry, Phys. Rev. $\underline{56}$, 1184 (1939); E.J. Konopinski, USAEC LAMS (1949); H. Primakoff and S.P. Rosen, Rept. Prog. Phys. $\underline{22}$, 121 (1959); E. Greuling and R.C. Whitten, Ann. Phys. (N.Y.) $\underline{11}$, 510 (1960); W.C. Haxton, G.J. Stephenson,Jr., and D. Strottman, Phys. Rev. Lett. 47, 153, (1951) also Phys. Rev. D. $\underline{25}$, 2360, (1982); M. Doi, T. Kotani, H. Nishiuro, K. Okuda and E. Takasugi, "Double Beta Decay", Preprints of Osaka Univ., also Phys. Lett. $\underline{103B}$, 219 (1981); $\underline{11313}$, 513 (1982); V.D. Vergados, CERN Reports or Ioannina preprints.; E. Fiorini, ReVista del Nuovo Cimento $\underline{2}$, 1 (1971), Also, D. Bryman and C. Picciotto, Rev. Mod. Phys. $\underline{50}$, 11 (1978).

5) E. Majorana, Nuovo Cimento $\underline{14}$, 171 (1937), or R.E. Marchak, Riazuddin and C.P. Ryan, "Theory of Weak Interaction in Particle Physics (1969)

6) H. Primakoff and S.P. Rosen, Rept. Prog. Phys. $\underline{22}$, 121 (1959); Also, G. Racah, Nuovo Cimento $\underline{14}$, 322 (1937)

7) C.S. Wu, E. Ambler et al., Phys. Rev. $\underline{105}$, 1413 (1957); H. Frauenfelder et al., Phys. Rev. $\underline{106}$, 386 (1957); R. Garwin, L. Lederman and M. Weinrich, Phys. Rev. $\underline{105}$, 1415 (1957); J.I. Friedman and V.L. Telegdi, Phys. Rev. $\underline{105}$, 1081, (1957)

8) For an introduction to GUTS see e.g. H. Georgi and S.L. Glashow, Phys. Rev. Lett. $\underline{32}$, 438 (1974); H. Georgi and D. Nanopoulos, Nucl. Phys. B155, 52 (1979)

9) Nobel addresses on the subject delivered by S. Weinberg A. Salam and S.L. Glashow in Rev. Mod. Phys. $\underline{52}$ 515; 525; 539 (1980)

10) W. Pauli, Nuovo Cim. $\underline{6}$, 204 (1957); D.L. Pursey, Nuovo Cim. $\underline{6}$ $\underline{266}$, (1957); G. Luders, ibid. $\underline{7}$, 171 (1958)

11) C.D. EnZ. Nuovo Cim. $\underline{6}$, 250 (1957)

12) T. Kirsten, O. Schaeffer, E. Norton, and R.W. Stoenner, Phys. Rev. Lett. $\underline{20}$, 1300 (1968); T. Kirsten, W. Gentner, and O.A. Schaeffer, A. Physik, $\underline{202}$, 273 (1967); T. Kirsten, H. Richter and E.K. Jessberger will appear in Phys. Rev. Lett.

13) R.K. Bardin, P.J. Gollen, J.D. Ullman, C.S. Wu, Nucl. Phys. A158 337, (1970); Phys. Lett. 26B, 112 (1967)

14) B.T. Cleveland, W.R. Leo, C.S. Wu, L.R. Kasday, A.M. Rushton, P.J. Gollen and J.D. Ullman, Phys. Rev. Lett. 35, 737 (1975)

15) B. Srinivasan, E.C. Alexander, Jr., B.D. Beaty, D. Sinclair and O.K. Manual, Economy Geology 68, 252 (1973)

16) M.K. Moe and D.D. Lowenthal, Phys. Rev. C22, 2186 (1980)

17) E. Fiorini, A. Publia, G. Bestolini, F. Capellani and G. Restelli, Nuovo Cimento A13, 747 (1973)

18) E. der Mateosian and M. Goldhaber, Phys. Rev. 146, 810 (1966)

19) N. Takaoka and G. Ogata, Z. Naturforsch 21A, 84 (1966)

20) B. Pontecorvo, Phys. Lett. 26B, 630 (1968)

21) E.W. Hennecke, O.K. Manual and D.D. Sabu, Phys. Rev. C11, 1378 (1975)

22) D. Bryman and C. Picciotto, Rev. Mod. Phys. 50, 11 (1978)

23) W.C. Haxton, G.J. Stephenson, Jr., and D. Strottman, Phys. Rev. D25, 2360 (1982). Also, Doi et. al. 'Preprint Osaka University, OS-Ge 82-43 (1982)

24) M. Doi, T. Kotani, H. Nishiura, K. Okuda and E. Takasugi, Phys. Lett. 103B, 219 (1981); 11313, 513 (1982)

25) V. Lubimov, E. Novikow, V. Nozik, E. Tretyakov and V. Kozik, Phys. Lett. 94B, 266 (1980)

26) B. Srinivasan, E. Alexander and O. Manual, J. Inorg. Nucl. Chem. 34, 2381 (1972)

27) W.C. Haxton, G.J. Stephenson, Jr., and D. Strottman, Phys. Rev. Lett. 47, 153 (1981)

28) G.S. Hurst, M.G. Payne, S.D. Kramer and J.P. Young, Rev. Mod. Phys. 51, 767 (1979). Also, G.S. Hurst, M.G. Payne, S.D. Kramer and C.H. Chen, Phys. Today 33, 24, Sept. (1980)

29) M.G. Inghram and J.K. Reynolds, Phys. Rev., 76, 1265 (1949); 78, 822 (1950)

Table I. Direct Detection Results on ($\beta\beta$) Decays.

($\beta\beta$) Decay	Authors	Year	$T_{1/2}$	
			(0ν)	(2ν)
$^{48}_{20}$Ca → $^{48}_{22}$Te 4.27 Mev	Bardin, Gallon, Ullman & Wu$^{(13)}$	'67	$\geq 1.6 \times 10^{21}$ at 80% C.L.	$\geq (3.5 \pm 0.9)$ $\times 10^{19}$ tentatively
^{76}Ge → ^{76}Se 2.04 Mev	Fiorini, Pullia, Bestolini, Capellani and Restelli$^{(17)}$	'73 '82	$\geq 5 \times 10^{21}$ at 68% C.L. $\geq 10^{22}$	
^{82}Se → ^{82}Kr 3.00 Mev	Cleveland, Leo, Wu, Kasday, Rashton, Gallon & Ullman$^{(14)}$	'75	$\geq 3.1 \times 10^{21}$ at 68% C.L.	
^{82}Se → ^{82}Kr	Moe & Lowenthal$^{(16)}$	'79	see no events	$= (1 \pm 0.4) \times 10^{19}$ yr.
^{82}Se → ^{82}Kr	Haxton, Stephenson, Jr. and Strottman$^{(27)}$	'81		$= 2.35 \times 10^{19}$ yr. calc.

Table II. Geochemical Results on (ββ) Decays

(ββ) Decay	Authors	Year	$T_{1/2}$(yr.) or $S=T_{1/2}(^{128}Te)/T_{1/2}(^{130}Te)$
$^{130}Te \rightarrow ^{130}Xe$ 2.54 Mev	Inghram & Reynolds[29]	'50	$T_{1/2} = 3.3 \times 10^{21}$
$^{130}Te \rightarrow ^{130}Xe$	Kirsten, Gentner, Schaeffner[12] Kirsten Schaeffer, Norton, and Stoner[12] (Max-Planck Inst.)	'67 '68	$T_{1/2} = 21.34 \pm 0.12$ $= (2.27 \pm 0.60) \times 10^{21}$
$^{130}Te \rightarrow ^{130}Xe$	Srinivasan, Alexander, Manuel[15] (Missouri)	'72	$T_{1/2} = 10^{21.38 \pm 0.10}$ $(2.72 \pm 0.27) \times 10^{21}$
$^{128}Te \rightarrow ^{128}Xe$ 872 Kev	Takaoka & Ogata[19] (Osaka)	'66	$S = 10^{1.2 \pm 0.6}$
$^{130}Te \rightarrow ^{130}Xe$ $^{128}Te \rightarrow ^{128}Xe$	Hennecke, Manuel, & Sabu[21] (Missouri)	'75	$T_{1/2} = (1.54 \pm 0.17) \times 10^{24}$ $S = 10^{3.20 \pm 0.01}$
$^{130}Te \rightarrow ^{130}Xe$ $^{128}Te \rightarrow ^{128}Xe$	Kirsten, Richter, Jessberger[2] (Max-Planck)	'82	$^{130}T_{1/2} = (2.60 \pm 0.28) \times 10^{21}$ $^{128}T_{1/2} \geq 8 \times 10^{24}$ yrs. (2σ) $S = 9.71 \times 10^3 \simeq 10^4$
$^{82}Se \rightarrow ^{82}Kr$ 3.00 Mev	Kirsten, et. al.[2]	'71	$^{82}T_{1/2} = (1.4 \pm 0.7) \times 10^{20}$
$^{82}Se \rightarrow ^{82}Kr$	Srinivasan, Alexander, Beaty, Sinclair, Manuel[15]	'73	$^{82}T_{1/2} = \begin{array}{l}(1.5 \pm 0.45)\\(2.6 \pm 0.8)\end{array} \times 10^{20}$
Lifetime ratio of ^{130}Te & ^{82}Se	Kirsten et al.[2] SABSM[15]	'82	$T_{1/2}(^{130}Te)/T_{1/2}(^{82}Se) = 12.4 \pm 2$ $^{130}T_{1/2} = (2.55 \pm 0.20) \times 10^{21}$ $^{82}T_{1/2} = (2.06 \pm 0.15) \times 10^{20}$

Table III. Existing and Planned (ββ) Experiments

Direct Detection Methods		
(ββ)-decays	Authors	Comments
^{76}Ge	E. Fiorini et al. (Milano) (Mt. Blanc Tunnel)	135 c.c. Ge(Li) $0^+ \to 0^+ \geq 3\times10^{21}$ yr. $0^+ \to 2^+ \geq 2\times10^{21}$ yr. Summer of '82, $T_{\frac{1}{2}} \geq 10^{22}$ yr. Intrinsic Ge(late '82)
^{136}Xe	"	Developing a Xenon Time-Projection-Chamber
^{136}Xe	Chen, Mahler and Doe (Irvine)	Developed Liquid Xe T.P.C. (1000 c.c.) Feasibility study for one year.
^{76}Ge	Avignone, Brodzinski and Wogman (Battelle Pacific Northwest, and Univ. of South Carolina)	250 c.c. Ge(li), NaI shield $T_{\frac{1}{2}} \geq 1.3\times10^{22}$ yr.(95%CL) $T_{\frac{1}{2}} \geq 2.0\times10^{22}$ yr.(68%CL)
^{76}Ge	U.C. Santa Barbara Caldwell, Eisberg & Witheral	1000 c.c. Intrinsic Ge, NaI shielding Goal 2×10^{23} yr.
^{76}Ge	F. Boehm (Cal. Tech)	100c.c. Ge. No molecular sieve. Goal $>5\times10^{22}$ yr.
^{76}Ge	J.J. Simpson et al. (Guelph)	200-500 c.c. Ge Goal $T_{\frac{1}{2}} \geq 3\times10^{22}$ yr.
^{82}Se	Cleveland, Ullman, Hurst, and Chen (BNL, Oakridge, Lehman)	single atom counting 10 kg. of Se.
^{82}Se	Moe, Hahn, Reines (U.C. at Irvine)	40 gm. enriched ^{82}Se; TPC; Goal $(ov)T_{\frac{1}{2}} \geq 3\times10^{22}$ yr.
^{100}Mo	Tripp, Kenney, Muller Garnjost & Nicholson (UC at Berkley)	200 gms of ^{100}Mo will be sandwiched between several Ge hexagon detectors. Feasibility study underway.

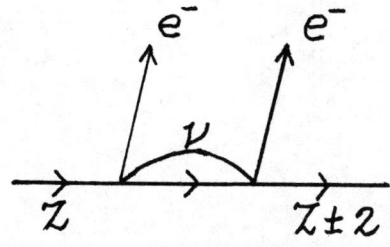

Fig. 1. Second-order neutrino-less double beta decay by emission and reabsorption of a neutrino

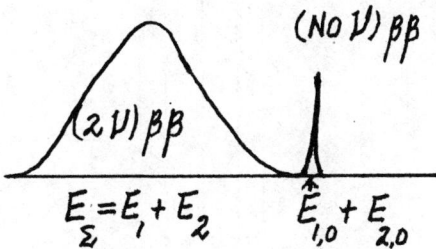

Fig. 2. The sum energy of $E_{e_1} + E_{e_2}$) from two-neutrino double beta decay is shown by the continous energy distribution. The no-neutrino double beta decay is indicated by the vertical peak at the maximum energy of the two-neutrino($\beta\beta$) decay.

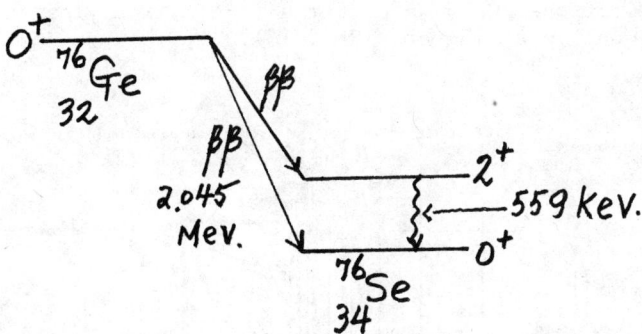

Fig. 3. Decay scheme for $^{76}\text{Ge} \xrightarrow{\beta\beta} {}^{76}\text{Se}$ Double Beta Decay.

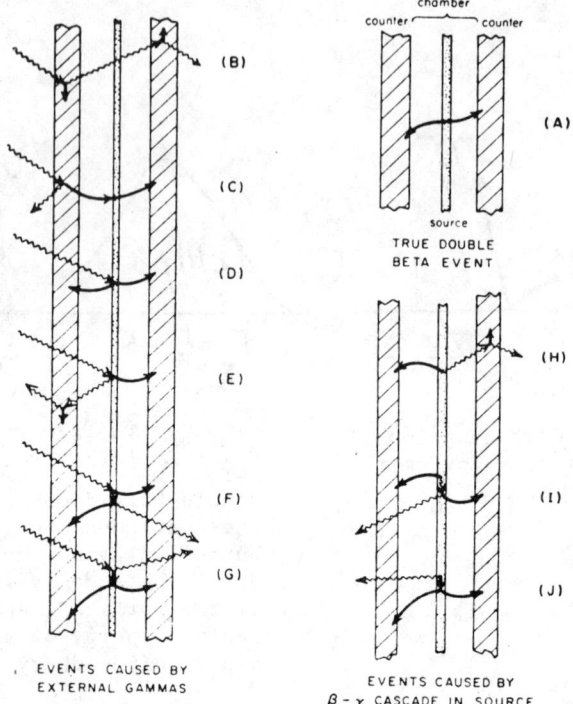

Fig. 4. Appearance of various events in an apparatus with a source mounted inside a track-visualizing chamber, triggered by coincidences between two counters outside the chamber. A magnetic field is directed out of the paper. Wavy lines indicate gamma rays; solid lines indicate electrons. Only an electron (solid line) leaves a track or makes a signal in a counter. (Ref. 13).

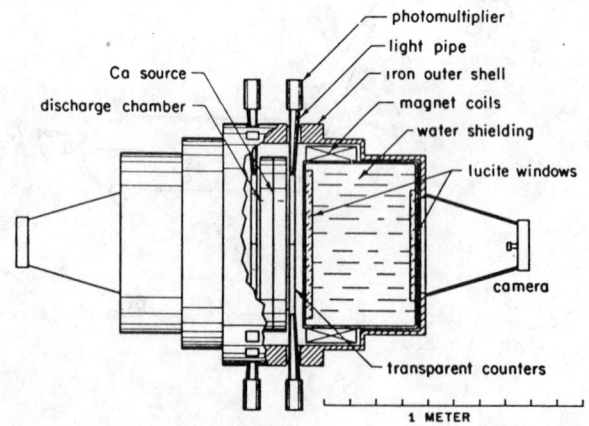

Fig. 5. The cutaway drawing of double beta decay apparatus used by Bardin et al. in the ^{48}Ca[13] and by Cleveland et al. in the ^{82}Se[14] investigation.

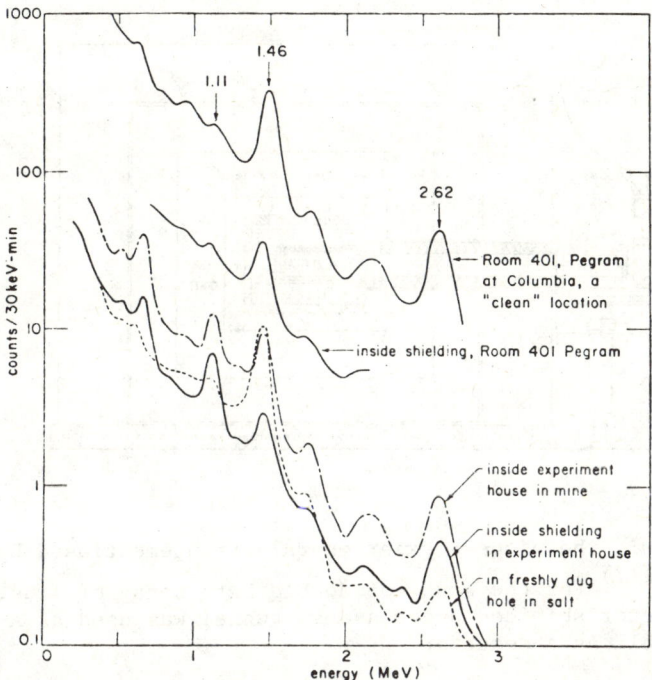

Fig. 6. Gamma-ray background at various experimental sites measured with 12.7cm x 12.7 cm NaI scintillator. Fresh salt pit gives the lowest background. (Ref. 13).

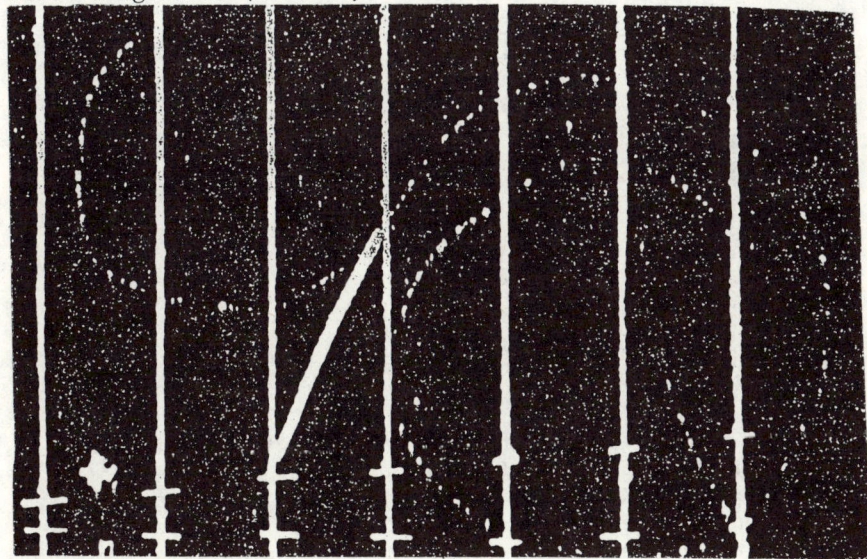

Fig. 7. An electron pair from ^{214}Be contamination as revealed by an accompanying alpha particle ($2e^- + \alpha$). (Ref. 16).

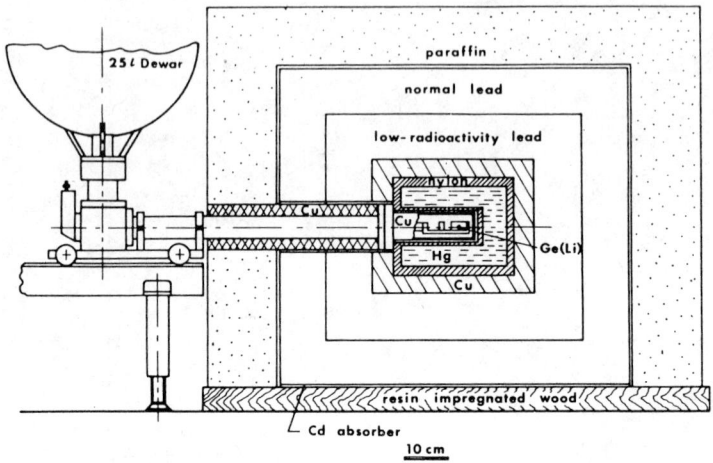

Fig. 8. This is the pioneer experimental arrangement used by Fiorini et al.[17] for the study of double beta decay of ^{76}Ge. The Ge(Li) detector shielded under a deep tunnel was used as a DBD source as well as a detector.

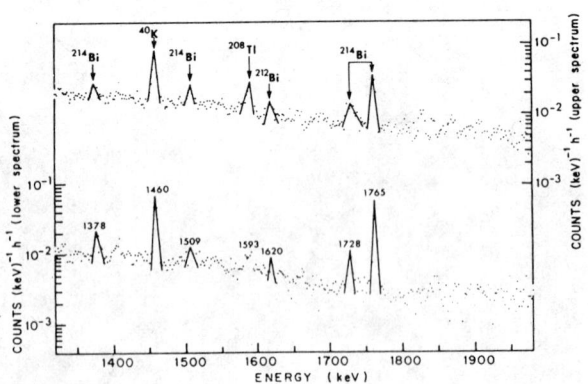

Fig. 9. Showing background radiations in the underground Ge(Li) detector (68.5 c.c.) of Fiorini's pioneer experiment[17]. Noticably they were ^{40}K, ^{212}Bi, ^{214}Bi, ^{208}Te, U, Th and etc. No peaks were found in the energy region (2.045 MeV) expected of ^{76}Ge neutrinoless ββ decays after 4400 h. of running time (Ref. 17).

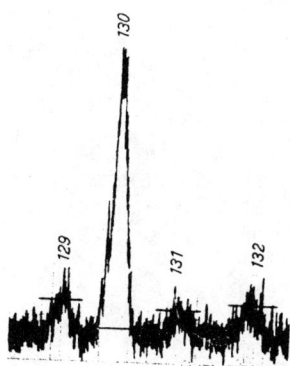

Fig. 10. Mass spectrum of Xe-isotopes extracted from the pure tellurium ore from Goodhope Mine, Colorado. Each short horizontal bar marks the expected maximum amounts of atmospheric Xenons in each isotopes. There were no noticable anomalies observed in the other three isotopes except the ^{130}Xe which had a large excess as shown, thus unequivocally gave the first evidence of double beta decay in ^{130}Te $\xrightarrow{\beta\beta}$ ^{130}Xe by Geochemical Method. (Ref. 12).

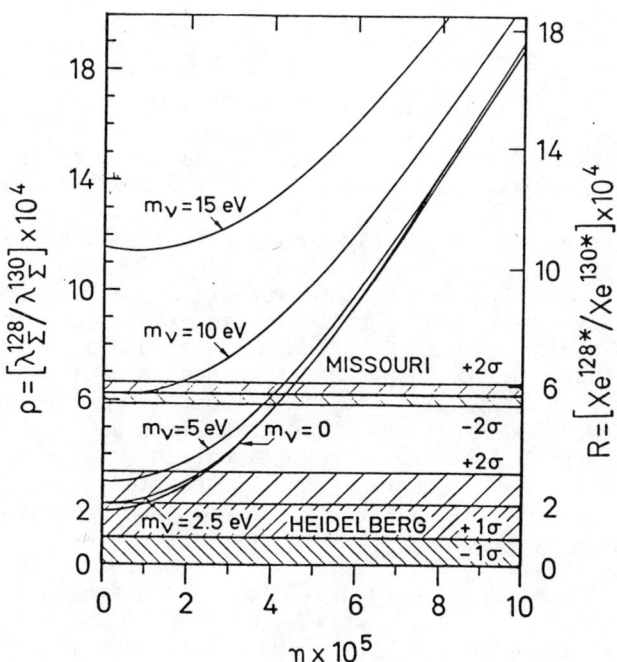

Fig. 11. This figure was given in one of Kirsten et. al.'s reports on ^{128}Te and ^{130}Te in 1982. The left hand scale gives $\rho = {}^{128}\lambda/{}^{130}\lambda$ where λ is the decay const.. The right hand scale gives $R = {}^{128}$Xe*/^{130}Xe* where Xe* is the Xenon excess. η is the lepton non-conservation parameter. The relationship between ρ or R with η and m_ν is taken from Doi et al.'s paper (Osaka preprint OS-Ge 82-43, 1982). (Ref. 2).

GEOCHEMICAL DOUBLE BETA DECAY EXPERIMENTS

T. Kirsten
Max-Planck-Institut für Kernphysik, Heidelberg, Germany

ABSTRACT

Double Beta Decay provides a very sensitive test of lepton-number conservation. However, decay rates are so small that it has not yet been possible to unequivocally detect a Double Beta Decay event directly; so much the more important is the data obtained by the "geochemical" method where one detects the Double Beta Decay products which have accumulated in natural minerals during long geological time periods. In that order, I discuss
- the principles of the geochemical method, its strength and its limitations
- the available data
- new results on the ^{128}Te-^{130}Te system
- theoretical implications of the data concerning the electron neutrino restmass and lepton-number conservation.

A limit of $m_\nu \leq 5.6$ eV (95% confidence) has been obtained from our recent ^{128}Te-^{130}Te-measurements.

THE "GEOCHEMICAL" METHOD

The basic concept of what is somewhat inappropriately called the "geochemical" method for Double Beta Decay (DBD) detection is actually not restricted to DBD but is a useful approach for various kinds of "rare" processes, often too rare to be directly detectable. It consists in the application of the "Accumulation Principle" over long geological time periods, whereby the products of nuclear transmutations are collected in a closed system during the accumulation time t and detected by their integral effect. The accumulation principle is commonly applied to stable product isotopes in radiometric age dating of minerals [1], but it also finds application for radioactive products of stimulated reactions which accumulate up to their saturation level in case of continuous production (e.g., Solar Neutrino detection [2] or Cosmic Ray exposure ages of meteorites [1]).

For the extremely rare, second order weak DBD-process, expected decay rates are at best of order 10^{-21} per year, hence for a medium weight element, the event rate is certainly less than 5 per gram and year. On the other hand, if a natural mineral specimen having a geological age of, say, 10^9 yrs, contains 1 gram of a DBD-active nuclide, some 10^9 stable product atoms would have accumulated in our example:

$$N_d = \frac{dN_p}{dt} \cdot t = N_p \cdot \lambda \cdot t$$

($t \ll \lambda^{-1}$ = decay constant, d=daughter, p=parent). If the accumulation time t is known from an independent age determination, the DBD-rate follows from

$$\lambda_{\beta\beta} = \frac{N_d}{N_p} \cdot \frac{1}{t}$$

N_p is normally determined from a bulk chemical analysis and from the natural isotopic abundance of the prospective DBD-active nuclide while the stable decay product N_d is usually detected by mass spectrometry.

The major experimental difficulty is not the detection of, say, 5×10^9 (or even 10^5) product nuclei, but the fact that these nuclei are admixed to whatever the natural concentration of that species in the natural sample is. In principle, trace quantities of all elements of the whole periodic table occur in every natural sample, and a concentration even as low as 1 part in 10^9 (ppb) would exceed said quantity of 5×10^9 atoms in our (favourable) example by a factor of $\sim 10^3$. The difficult task then would be to significantly distinguish, relative to a suitable reference isotope j of the sample element, x, a ratio

$$r' = \frac{i_x(1+10^{-3})}{j_x} \quad \text{from} \quad r = \frac{i_x}{j_x}.$$

For even smaller decay rates as in our example (say, $10^{-25} y^{-1}$ instead of $10^{-21} y^{-1}$) it is immediately clear that geochemical DBD-detection is only possible if an extremely large natural depletion of the daughter element D has occurred relative to the parent element P before or during mineral formation. The change of the isotope ratio $i_r(D) = i_D/j_D$ due to DBD of the nuclide i_P is

$$\frac{i_{r'}}{i_r} - 1 = \ln 2 \cdot \frac{t}{i_{T_{1/2}}} \cdot \frac{[P] \cdot I(i_P)}{[j_D] \cdot \frac{I(i_D)}{I(j_D)}} \tag{1}$$

with I : natural isotopic abundances
t = mineral age
$i_{T_{1/2}}$ = halflife for DBD of isotope i_P

and brackets indicate absolute concentrations per analyzed specimen.

The present state of art in mass spectrometry allows to determine the left-hand quantity of eq. 1 at best down to $\sim 10^{-5}$. A most favourable value for $t/i_{T_{1/2}}$ would at best be 10^{-12}, hence it follows that the abundance ratio requirement when the sample forms is at least $|i_P|/|i_D| > 10^7$. Such strong fractionations among chemical elements with $\Delta Z = \pm 2$ occur in nature exclusively for rare gases, for principal geo- and cosmochemical reasons. The cosmochemical cause is the primary depletion of chemically inactive volatiles during formation of the terrestrial planets by many orders of magnitude relative to the condensable elements. The geochemcial reason is the additional nearly complete expulsion of inert volatiles during mineralization and cooling of the ores. This is why only Double Beta Decays leading to rare gas isotopes are accessible to the "geochemical" method. They are:

$^{130}Te \rightarrow {}^{130}Xe$; E=2.53 MeV; $I(^{130}Te)$=34.48 %; $I(^{130}Xe)$=4.08 %
$^{128}Te \rightarrow {}^{128}Xe$; E=0.87 MeV; $I(^{128}Te)$=31.79 %; $I(^{128}Xe)$=1.92 %
$^{82}Se \rightarrow {}^{82}Kr$; E=3.0 MeV; $I(^{82}Se)$= 9.19 %; $I(^{82}Kr)$=11.56%.

These decays have been studied by various authors through analysis of natural Tellurium- and Selenium-ores. The requirements to be

obeyed are:
- availability of sufficient quantities (since these ores are rare, this is a non-trivial condition)
- high Te(Se)-concentration
- low concentration of adsorbed, dissolved, or occluded Xe(Kr) of normal terrestrial ("atmospheric") composition
- low concentrations of U, Th to minimize side reactions due to fission, neutrons, or α-particles which could mask or simulate DBD. In this context, it is very important to measure <u>all</u> isotopes of Xe(Kr), since in many cases the presence or absence of such interfering reactions can be judged on the basis of the effects which they have or would have on the abundances of certain other Xe(Kr)-isotopes. For a more thorough discussion of this aspect, see [3,4].
- Sufficiently high and known gas accumulation time ("gas retention age"). The importance of the last condition deserves its own paragraph.

AGES AND ABSOLUTE DECAY RATES

The very fact that natural minerals are in principle capable of retaining radiogenic rare gas isotopes over billions of years is a well established observation in geochronology. 4.5 b.y. old meteorites as well as terrestrial and lunar rocks up to 4 b.y. old have all been successfully dated by one of the most widely used dating technique, the K-Ar method, which is based on the retention of ^{40}Ar in the mineral lattices in which it is formed by ^{40}K decay [1].

On the other hand, it is true that partial gas losses may occur depending on the type and the actual geological history of the mineral. The gas retention properties of the more common rock-forming minerals have been intensively studied, but this is admittedly not the case for such uncommon minerals as Tellurium or Selenium ores. However, for the few of them which have been dated by the K-Ar or U-Xe-fission method (another dating scheme based on rare gas retention), the results were in general context with the geological and geochronological situation of the ore occurrences as deduced from dating of adjacent rocks by other dating schemes such as e.g. Rb-Sr. We caution, however, that it is not safe to entirely rely on ages determined on adjacent rocks or the known age of certain geological formations in which the ore veins occur since it is well known that in some situations ore metamorphism or late intrusions of ore veins have occurred, hence the ages of ores may be younger than the rock formations in which they are imbedded.

All these uncertainties can be circumvented if one determines a gas retention age of the ore mineral itself. Then, even if the gas retention age should deviate from the geological age of mineralization, it would affect the retention of the DBD-produced rare gas isotopes in the same way and the effective accumulation time is properly obtained.

In this way, DBD-decay rates can be directly tied to well established other decay constants and become independent of the actual age of the ore. As an example, ^{82}Kr from DBD and ^{86}Kr$_{fiss}$ from

^{238}U-fission was measured in a Selenium ore which also contained some Uranium [5]. The DBD-rate of ^{82}Se is then directly inferred from the known decay rate of spontaneous fission of ^{238}U, from the ^{82}Kr/^{86}Kr isotopic ratio and from the Se/U ratio easily obtained by a chemical analysis. No assumptions about ages or gas retention properties had to be made in this case. Similarly, ores containing both, Te and Se [15,8] tie the DBD-rate of ^{130}Te to that of ^{82}Se, hence these two decay constants are tied together and cannot be evaluated independently.

AVAILABLE DATA

The available data from the literature are summarized in Tables I, II and Figures 1,2. Two major conclusions can be drawn from the Figures:

a) The halflives for ^{130}Te(^{82}Se) deduced by various laboratories agree at least within a factor of 5 for many ores from widely varying proveniences with ages ranging from some ten million to 2 1/2 billion years and Te(Se) concentrations ranging from about half a percent to 100 %. There is, with other words, a general proportionality between the amount of excess ^{130}Xe(^{82}Kr) and the product of age and parent isotope concentration. Based on this evidence and on the large magnitude of the radiogenic anomalies in some samples, it is undisputed that these isotope anomalies are in fact due to DBD. In addition, a consideration of all possible side reactions has failed to give a reasonable production mechanism to explain the general excess of ^{130}Xe(^{82}Kr) by any other means than DBD.

b) Good numerical agreement results from selecting those data which are based on the determination of internal gas retention ages (K-Ar, U-Xe$_{fiss}$) and on an appreciably large measured ratio: ^{130}Xe/^{132}Xe$_{fiss}$>1.

For ^{130}T$_{1/2}$: $(2.27+0.60) \cdot 10^{21}$yrs [9]
$(2.72+0.27) \cdot 10^{21}$yrs [7]
$(2.60+0.28) \cdot 10^{21}$yrs [4].

We infer from these data a "best value" or recommended mean of $(2.55+0.20) \cdot 10^{21}$yrs for ^{130}T$_{1/2}$(=T$_{1/2}$(^{130}Te)).

For Selenium two different approaches are possible. First, we have determined the weighed mean value of all absolute determinations which are based on gas retention ages (Table II). The result is

$$T_{1/2}(^{82}Se)=(1.45+0.15) \cdot 10^{20} \text{yrs}.$$

Another approach relies on the ratio of $^{82}\lambda/^{130}\lambda$ for samples which contain both, Se and Te. In this way, one arrives at a halflife of $(2.05+0.30) \cdot 10^{20}$yrs, taking ^{130}T$_{1/2}$=$(2.55+0.20) \cdot 10^{21}$yrs as deduced above. From this it becomes evident that the uncertainty of the ^{82}Se DBD halflife is ∼25 %; but certainly not a factor of 10-20 as would be needed to agree with the counting experiment by Moe and Lowenthal[16] (see Figure 2).

The results for DBD of ^{128}Te will be covered in a later section. Here, we summarize that "geochemical" DBD decay rates are considered to be reliable within some 20-30 %. Order of magnitude deviations

Table I. Survey of geochemical determinations for the DBD-halflife of ^{130}Te

Te-Mineral	Origin	Te (%)	Age (10^9yrs)	Age type	$\frac{^{130}Xe^*}{^{130}Te} \times 10^{13}$	$T_{1/2}(10^{21}$yrs)	Evaluation[x] (a)	(b)	Ref.
Bi_2Te_3,Telluro-bismuthite	Boliden, Sweden	11.5	1.56 +0.10	U-Xe	3.16+0.3	3.42+0.40	+	–	6
		8.1	1.56 +0.10	U-Xe	5.74	2.11+0.27	+	–	7
		23.4	1.56 +0.10	U-Xe	4.96	2.18+0.27	+	–	7
		32.6	1.56 +0.10	U-Xe	3.54	3.06+0.35	+	–	7
		46.7	1.56 +0.10	U-Xe	3.97	2.72+0.27	+	+	7
$Pb_2(BiSb)_2(S,Se)_5$ Selenokobellite	Boliden, Sweden	0.46	2.4 +0.7	(Se-Kr)	7.5 +2	~2.45	(+)	–	8
Te, Native Tellurium	Goodhope, Colorado	99.4	1.31 +0.14	K-Ar	3.99+0.67	2.27+0.6	+	+	9
		100	1.31 +0.14	K-Ar	3.49+0.17[c]	2.60+0.28[c]	+	+	4,10[c]
PbTe,Altaite	Kirkland, Ontario	53.1	1.83 +0.04	Rb-Sr (adjac.rock)	5.64+0.95	2.25+0.35	–	–	11
(Au,Pb,Bi...Te) X-Telluride	Kalgoorlie, Australia	51.7	2.46 +0.08	Rb-Sr (adjacent rocks)	6.02+0.6	2.83+0.30	–	–	12
		58.6	2.46 +0.08		4.18+0.4	4.08+0.43	–	+	12
		47.5	2.46 +0.08		17.7+1 [c]	0.96+0.10[c]	–	+	13[c]
Bi_2Te_2S Tetradymite	Oya, Japan	22.4	0.091+0.003	K-Ar (adjacent rock)	0.85+0.03	0.74+0.04	–	–	14
		22.4	0.091+0.003		0.76+0.05	0.83+0.07	–	–	14
		29.8	0.091+0.003		0.67+0.03	0.94+0.06	–	–	14
X-Tellurides	Fiji	33.7	< 0.013	Geol.estimate	~ 0.018	≲ 5	–	–	12
	Facebaja,Rom.	17.3	< 0.06	Geol.estimate	~ 0.083	≲ 4.8	–	–	12
	Moctezuma,Mex.	69.9	< 0.06	Geol.estimate	~ 0.14	≲ 3	–	–	12

x) To infer the best numerical mean value of $T_{1/2}(^{130}Te)$ data should pass two criteria (marked: +):
(a) Age based on gas-retention method applied to the mineral itself (see text).
(b) Appreciably large isotope anomaly of ^{130}Xe (that is, favorable mixing ratio Xe_{rad}/Xe_{atmos}). We assign "+" if the measured ratio $^{130}Xe/^{132}Xe>1$, that is 6.6 times the atmospheric ratio of 0.15136. The respective mean is
$^{130}T_{1/2}=(2.55\pm0.20)\cdot 10^{21}$yrs.
(c) These are the samples which were also used for the determination of $\rho=^{130}T_{1/2}/^{128}T_{1/2}$ as discussed in the second part of this paper. Note that the Kalgoorlie sample on which $\rho_{Missouri}$ is based gave a halflife for ^{130}Te which deviates strongly from most others (see also Figure 1).

Table II. Survey of geochemical determinations for the DBD-halflife of ^{82}Se

Se-Mineral	Origin	Se (%)	Age (10^9yrs)	Age type	$\frac{^{82}Kr^*}{^{82}Se} \times 10^{11}$	$T_{1/2}(10^{20}yrs)$ [a]	Ref.
Bi$_2$(Te,Se)$_3$ Tellurobismuthite	Boliden, Sweden	1.15 1.55	1.3 ±0.2 1.46 ±0.22	Te-Xe[b] Te-Xe[b]	0.60 ±0.15 0.393±0.10	1.5 ±0.45[b] 2.6 ±0.8[b]	15 15
Cu$_{4-x}$Se$_2$ Umangite	Habri, Moravia	42.4	0.285±0.015	U-Xe	0.147±0.018	1.37±0.17	5
PbSe;(Cu$_{2-x}$Se$_x$) Clausthalite	Harz, Germany	13.45	0.325±0.025	K-Ar	0.134±0.020	1.69±0.29	5,8
PbSe;(Cu$_{2-x}$Se$_x$) Clausthalite	Cacheuta, Argentina	21.6	0.171±0.020	K-Ar	0.089±0.019	1.33±0.31	5,8
Pb$_2$(BiSb)$_2$(S,Se)$_5$ Selenokobellite	Boliden, Sweden	4.56	2.4 ±1	Te-Xe[b]	1.204±0.15	1.4 ±0.7[b]	5,8
NiSe$_2$,PbSe,Ag$_2$Se, n(Cu$_{2-x}$Se$_x$)HgSe Blockite	Pakajake Bolivia	57.65	0.080±0.020	Geol.estim.	0.043±0.007	(1.28±0.4)	5,8

a) The weighed mean halflife from all data based on gas retention ages is $(1.45\pm 0.15)\times 10^{20}$yrs.

b) These samples contain both, Se and Te, hence the ratio $^{130}T_{1/2}/^{82}T_{1/2}$ can be inferred without explicit knowledge of the age. The mean value for this ratio is 12.4 ± 2, corresponding to a halflife for ^{82}Se of $^{82}T_{1/2} = (2.05\pm 0.30)\times 10^{20}$yrs.

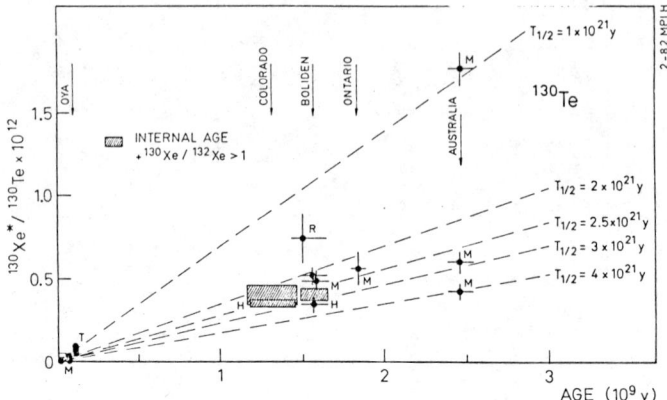

Fig. 1. ^{130}Xe*/^{130}Te vs. ore age for tellurium ores described in Table 1. M=Missouri; H=Heidelberg.

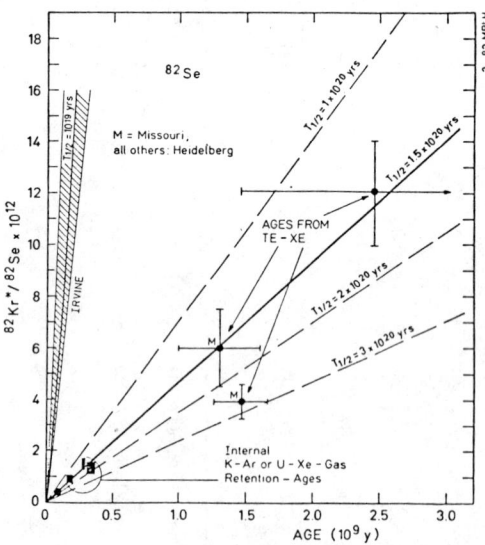

Fig. 2. ^{82}Kr*/^{82}Se vs. ore age for selenium ores described in Table 2.

would lead to unacceptable consequences with respect to the geological context of the ores.

THEORETICAL ASPECTS

The general theoretical aspects of DBD are treated in excellent review articles [17,18,19]. Here it may suffice to note that the two most likely mechanisms which could give rise to neutrinoless (Majorana)-DBD are
(1) An admixture of a right-handed leptonic current to the regular V-A weak interaction (violation of lepton number conservation, amplitude η).
(2) Implicit helicity breaking due to a non-vanishing neutrino rest-mass $m_\nu \neq 0$.

Quantitative theoretical treatments for these mechanisms as well as for Dirac decay (DBD with emission of 2 neutrinos) have been worked out by various authors [20,21,22,23,24,25], leading to decay rate predictions of type

$$\lambda_{2\nu}(A) = \text{(kinematical factor)} \ |M_{GT}^{2\nu}(A)|^2 \qquad (2)$$

and

$$\lambda_{o\nu}(A) = \text{(kinematical factor)} \ |M_{GT}^{o\nu}(A)|^2 \cdot f(m_\nu,\eta) \qquad (3)$$

from which, in principle, m_ν and/or η can be deduced if $\lambda_{o\nu}$ is measured. (M_{GT} stands for Gamov-Teller Matrix element; the function f is quadratic in m_ν and η). However, the geochemical method of DBD-detection measures effective decay rates $\lambda_\Sigma = \lambda_{o\nu} + \lambda_{2\nu}$ without possibility to distinguish between Majorana- and Dirac-decay. Nevertheless, useful limits on η, m_ν can be deduced from $\lambda_{o\nu} < \lambda_\Sigma$. Since the predicted ratio $\lambda_{o\nu}/\lambda_{2\nu}$ increases strongly with decreasing decay energy, there are limiting cases where (a) $\lambda_\Sigma \approx \lambda_{2\nu}$ (decay energy E large) or (b) $\lambda_\Sigma \approx \lambda_{o\nu}$ (E small). ^{130}Te and ^{82}Se certainly belong to category (a), geochemical results alone can therefore not lead to very stringent limits on η, m_ν for these isotopes. However, they are very useful to deduce $|M_{GT}^{2\nu}|$, the absolute magnitude of the Dirac matrix element and to use this result to improve the limits on η, m_ν deduced from experimental limits on $\lambda_{o\nu}$ from advanced direct counting experiments by changing eq. (3) into

$$\lambda_{o\nu}(A) = \text{(kinematical factor)} \cdot \xi^2(A) \cdot |M_{GT}^{2\nu}(A)|^2 \cdot f(m_\nu,\eta) \qquad (3a)$$

where $\xi(A) = |M_{GT}^{o\nu}|/|M_{GT}^{2\nu}|$.

The ratio $\xi(A)$ can be theoretically predicted [24] with much more confidence than the absolute magnitude of nuclear matrix elements, which is the largest uncertainty in the theoretical prediction. In Table III we compile theoretical decay rate predictions and matrix elements as well as experimental results and the values for m_ν, η and $|M_{GT}^{2\nu}|$ deduced from the combined geochemical and counting data.

We note that there is a discrepancy between the "geochemically

Table III. Experimental and theoretical DBD-halflives and inferred limits on m_ν and η

		^{82}Se	^{130}Te	^{128}Te	^{48}Ca	^{76}Ge	$\rho=^{130}T_{1/2}/^{128}T_{1/2}$
Experimental halflives [yrs]	$T^{0\nu}_{1/2}$	$>3.1\cdot10^{21}(26)$	$>1.2\cdot10^{21}(27)$	$>8\cdot10^{24}(a)$	$>2\cdot10^{21}(28)$	$>1.7\cdot10^{22}(29)$	
	$T^{2\nu}_{1/2}$	$\sim10^{19}$ (16)					$<3.3\cdot10^{-4}(2\sigma)$
	$T^\gamma_{1/2}$	$1.5\cdot10^{20}(a)$	$2.55\cdot10^{21}(a)$	$>8\cdot10^{24}(a)$	$>3.6\cdot10^{19}(28)$	$>2.8\cdot10^{19}(29)$	
Theoretical halflives (f) [yrs]	$T^{2\nu}_{1/2}$ (g)	$1.5\cdot10^{20}(b)$	$2.5\cdot10^{21}(b)$	$1.3\cdot10^{25}(c)$	$3.6\cdot10^{19}(d)$	$2.3\cdot10^{21}(e)$	$\sim 2\cdot10^{-4}$
	$T^{0\nu}_{1/2}, m_\nu=0$	$6\cdot10^{12}/\eta^2$	$6\cdot10^{13}/\eta^2$	$10^{16}\eta^2$	$3.5\cdot10^{12}/\eta^2$	$3.5\cdot10^{13}/\eta^2$	$62\cdot10^{-4}$
	$T^{0\nu}_{1/2}, \eta=0$	$1.3\cdot10^{13}\left(\frac{m_\nu}{m_e}\right)^2$	$10^{14}\left(\frac{m_\nu}{m_e}\right)^2$	$2.3\cdot10^{15}\left(\frac{m_\nu}{m_e}\right)^2$	$1.5\cdot10^{13}\left(\frac{m_\nu}{m_e}\right)^2$	$3.6\cdot10^{13}\left(\frac{m_\nu}{m_e}\right)^2$	$420\cdot10^{-4}$
Matrix elements (i) $\|M^{2\nu}_{GT}\|$	HSS (24)	1.88	2.95	2.95	0.44	2.56	~ 1 (ratio)
	DKNT (25)	0.79	0.24	0.24	0.41	1.07	~ 1 (ratio)
Limits on m_ν and η (k)	m_ν[eV]; ($\eta=0$)	≤ 33	≤ 100 (h)	≤ 8.7	≤ 44	≤ 24	≤ 5.6 (2σ)
	$\eta[10^{-5}]$; ($m_\nu=0$)	≤ 4.6	≤ 15 (h)	≤ 3.5	≤ 4.2	≤ 4.5	≤ 2.4

(a) "Geochemical" data; others are from counting experiments. For source of data, see text. (b) Assuming $\lambda_{2\nu}(\text{exper.})=\lambda_{2\nu}(\text{"theory"})$, to deduce the applicable matrix elements $|M^{2\nu}_{GT}|$("theory"). (c) From ^{130}Te as described in (b) and the theoretical ratio $^{128}\lambda_{2\nu}/^{130}\lambda_{2\nu}$. (d) Assuming $\lambda_{2\nu}(\exp)=\lambda_{2\nu}(\text{"theory"})$. (e) Theoretical value of HSS scaled according to the relation $|M^{2\nu}_{GT}|_{HSS}-|M^{2\nu}_{GT}|_{DKNT}$ in case of ^{82}Se. (f) following Doi et al. 25. (g) Entirely theoretical predictions for 2ν-decay, partially in contradiction with experimental results, can be calculated by multiplying the halflives in this line with the factor $(|M^{2\nu}_{GT}|_{DKNT}/|M^{2\nu}_{GT}|_{HSS})^2$. (h) deduced from the geochemical data which yield more stringent limits than the counting limit on $^{130}T_{0\nu}$.
(i) HSS=Haxton, Stephenson, and Strottman, she'l model calculations; DKNT= Doi, Kotani, Nishiura, Takasugi, applying "geochemical" calibration (see text). (k) Limits would become more stringent if the larger matrix elements of HSS are used. Stated limits of the two kinds should not be confused.

calibrated" matrix elements and those obtained from detailed shell model calculations by Haxton et al. [24]. For ^{82}Se, the discrepancy is not too large, but the calculations have a relatively high degree of confidence. For ^{130}Te, the discrepancy is much larger, but for this heavy nucleus, the shell model calculations are extremely difficult and hence relatively uncertain. In this context, we note also a discrepancy between the geochemical and the counting data for $\lambda_{2\nu}(^{82}\text{Se})$. As discussed above, it is very unlikely that all the geochemical experiments are in error by more than an order of magnitude. The limits deduced for m_ν,η are not too stringent. A considerable improvement of these limits is discussed in the next paragraph (see also last line of Table III).

THE ^{128}Te-^{130}Te-SYSTEM

As was first pointed out by Pontecorvo[30], uncertainties caused by insufficient knowledge of the nuclear matrix elements can be overcome by considering DBD rate ratios of pairs of similar nuclei such as ^{128}Te-^{130}Te rather than absolute decay rates.

The underlying assumption that in this case the ratio of the respective nuclear matrix elements should be near unity may be questioned in view of the disagreement of theoretical and experimental matrix elements for ^{82}Se and ^{130}Te. If this is due to cancellations, there is a possibility that they could also cause differences for "similar" nuclei. This is, however, an unsettled question and for the following we assume the correctness of Pontecorvo's original argument.

In the case of ^{128}Te-^{130}Te (isotopic abundances are 31.79 % and 34.48 %, respectively) the decay energies are very different (869 vs. 2533 KeV). Consequently, the ratio

$$\rho_{0\nu} = {}^{128}\lambda_{0\nu}/{}^{130}\lambda_{0\nu} \qquad \text{becomes much larger than}$$

$$\rho_{2\nu} = {}^{128}\lambda_{2\nu}/{}^{130}\lambda_{2\nu}.$$

Hence even very small Majorana contributions will drastically increase the measurable ratio $\rho = {}^{128}\lambda_\Sigma/{}^{130}\lambda_\Sigma$ above the base level of $\rho_{2\nu}$ ($\rho_{exp} > \rho_{2\nu}$). (As it turns out, $^{130}\lambda_\Sigma$ is in fact entirely dominated by $^{130}\lambda_{2\nu}$.) Only kinematical factors and the ratio $\xi = |M^{0\nu}_{GT}/M^{2\nu}_{GT}|$ need to be considered in the theoretical treatment of the ^{128}Te-^{130}Te system as performed by various authors [21,22,23,24,25,31]. Defining $\rho = \rho_\Sigma = {}^{128}\lambda_\Sigma/{}^{130}\lambda_\Sigma$, one obtains, following Doi et al.[25]

$$\rho = 1.972 \left(\frac{1+0.5566(m_\nu^2 - 0.1091 m_\nu \eta + 0.2276\eta^2)}{1+0.00272(m_\nu^2 - 0.3193 m_\nu \eta + 1.522\eta^2)} \right) \qquad (4)$$

where η is given in units of 10^{-5}, m_ν in units of $10^{-5} m_e$ (=5.11 eV) and ρ in units of 10^{-4}. In good approximation, for $m_\nu \times \eta < 10$, this yields

$$\rho \approx 1.97 + m_\nu^2 + \left(\frac{\eta}{2}\right)^2, \text{ and, for } \eta=0,$$

$$m_\nu \sim 5.1 \sqrt{\rho - 1.97} \text{ eV} . \tag{5}$$

The expectation value for pure Dirac decay is $\rho_{2\nu}=1.97\cdot 10^{-4}$ only, but for Majorana-decay (alone) much higher ratios result. Depending on the mechanism, we have

$$\rho_{o\nu} = 62\cdot 10^{-4} \text{ for } m_\nu = 0; \eta \neq 0 \text{ and}$$

$$\rho_{o\nu} = 420\cdot 10^{-4} \text{ for } m_\nu \neq 0; \eta = 0 \text{ (see Figure 3).}$$

It is this difference which allows one to deduce values for m_ν, η from eq. (4) if ρ_Σ is experimentally determined.

EXPERIMENTAL DATA ON ρ_Σ

Apart from theoretical advantages, there are also important experimental advantages in determining a decay rate ratio rather than an absolute decay rate. The ratio $\rho = {}^{128}\lambda_\Sigma / {}^{130}\lambda_\Sigma$ is directly deduced from isotopic ratio measurements, therefore errors in the effective accumulation time (absolute age error, or incomplete gas retention over geological time) or in the absolute Xe-calibration cancel. For the important experimental details of mass spectrometric Xe-analysis, we refer to the original[10] and to a more technical[4] paper. Here we only note that the crucial quantity ρ is directly obtained from a measured isotopic ratio

$$R_{exp} = \frac{{}^{128}Xe^{*\,x)}}{{}^{130}Xe^*} = \frac{{}^{128}r - 0.07136}{{}^{130}r - 0.15136} , \text{ where} \tag{6}$$

0.07136 and 0.15136 are the atmospheric ${}^{128}Xe/{}^{132}Xe$ and ${}^{130}Xe/{}^{132}Xe$ ratios and ${}^{i}r = {}^{i}Xe/{}^{132}Xe$. In this way, ${}^{132}Xe$ serves to subtract the ever present admixture of non-radiogenic ${}^{128}Xe$ and ${}^{130}Xe$ which comes from adsorbed or dissolved Xe having atmospheric composition. Then,

$$\rho = 1.085 \cdot R_{exp} \tag{7}$$

accounts for the different isotopic abundances of ${}^{128}Te$ and ${}^{130}Te$. The results of previous determinations of ρ (Table IV) are illustrated in Figure 3.

The positive result of the Missouri group[13,33] is far above the theoretically expected ratio for Dirac decay ($\rho_{2\nu} \sim 2\times 10^{-4}$), hence it seemed to indicate the occurrence of Majorana decay and was interpreted by the authors in terms of $\eta \neq 0$.

With the advent of the quantitative theoretical formalism concerning the implicit helicity breaking by $m_\nu \neq 0$, values of $m_\nu \sim 34$ eV[22] or more recently, of $m_\nu \sim 10$ eV[25] have been inferred from $\rho_{Missouri}$ (for $\eta=0$). The importance of this result, also in view of reported evidence for a non-vanishing neutrino-restmass of that same order from the shape of the tritium Beta-spectrum[34] has led us to redeter-

---
x)* denotes excess, possibly due to DBD.

mine ρ in a sample from the same Precambrian tellurium ore which we had used 14 years ago when it was first unambiguously demonstrated for ^{130}Te that DBD occurs in nature[9]. Here we summarize the results which are described in much more detail in Ref. 4 and 10. The excess-quantities of DBD-Xenon found are

$$^{130}Xe^* = (2.11 \pm 0.08) \cdot 10^{-11} ccSTP/g \text{ Te and}$$
$$^{128}Xe^* = (0.20 (\pm) 0.22) \cdot 10^{-14} ccSTP/g \text{ Te.}$$

Expressed as ratio, the directly measured value

$$(\frac{^{128}Xe}{^{132}Xe})_{sample} = (7.175 \pm 0.043) \cdot 10^{-2} \text{ compares with}$$
$$(\frac{^{128}Xe}{^{132}Xe})_{atmos.} = 7.136 \cdot 10^{-2}.$$

This means that any possible excess ^{128}Xe from DBD is within 1 standard deviation of its detectability. Consequently, these results (Table IV, Figure 3) contradict the Missouri data and are fully compatible with Dirac decay alone. As for the cause of the discrepancy, we note that before assigning an excess ^{128}Xe to DBD, in principle it would be necessary to exclude all other possible sources of excess ^{128}Xe, such as neutron capture on traces of ^{127}I in the ore, or experimental artifices such as the "memory"-effect. These potential sources of producing larger apparent ^{128}Te-decay constants must be definitely excluded before a positive result on ^{128}Xe* is assigned to DBD. However, since all these potential sources of error can only increase ^{128}Xe*, and since in our case the measured quantity leads only to an upper limit of the ^{128}Te decay constant, these effects could only reduce this upper limit even further. Therefore, we have assumed $^{128}Xe^* = ^{128}Xe_{DBD}$, the most conservative assumption with respect to the inferred halflives and theoretical consequences. This leads to the final result $\rho = ^{128}\lambda_\Sigma / ^{130}\lambda_\Sigma = (1.03 \pm 1.13) \cdot 10^{-4}$ and a 2σ (95% confidence) limit for the halflife of ^{128}Te of $^{128}T_{1/2} > 8 \cdot 10^{24}$ yrs, in disagreement with $\rho = (6.29 \pm 0.20) \cdot 10^{-4}$ and $^{128}T_{1/2} = (1.54 \pm 0.17) \cdot 10^{24}$ yrs reported in [13].

THEORETICAL IMPLICATIONS

Our result leaves very little room for an admixture of Majorana decay to the total decay rate of ^{128}Te. At the 95 % confidence level (2σ), at most 40% of the ^{128}Xe* could come from Majorana decay. On the other hand, the expectation value for Dirac decay alone is reproduced within 1σ of our measurement, even though the nominal value is below the expectation value. The latter fact simply implies that even the ^{128}Xe* from Dirac decay is still within 1σ of its detectability. (Halflife of order 10^{25} yrs!).

From the experimental data for ρ we can now deduce the theoretical implications for η and m_ν from eq. (4). The allowed region for combinations of $\bar{m} \neq 0$, $\eta \neq 0$ is shown in Figure 4, both for the Missouri result and for ours. Clearly, they are incompatible. Also shown in

Table IV: Experimental data on $\rho = {}^{128}\lambda_\Sigma / {}^{130}\lambda_\Sigma$.

Year of publication	ρ	Reference
1966	$\leq 295 \times 10^{-4}$ (1σ)	14
1969	$\leq 194 \times 10^{-4}$ (1σ)	11
1972	$\leq 111 \times 10^{-4}$ (1σ)	7
1972	$\leq 5 \times 10^{-4}$ (1σ)	32
1975	$(6.29 \pm 0.20) \times 10^{-4}$ (1σ)	13
1978	$(6.37 \pm 0.41) \times 10^{-4}$ (1σ)	33
1982	$\leq 3.29 \times 10^{-4}$ (2σ)	10, 4

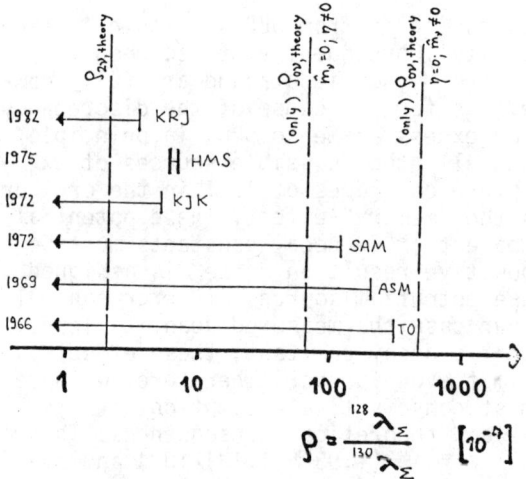

Fig. 3. Available data on the decay rate ratio ρ for DBD of ^{128}Te and ^{130}Te.

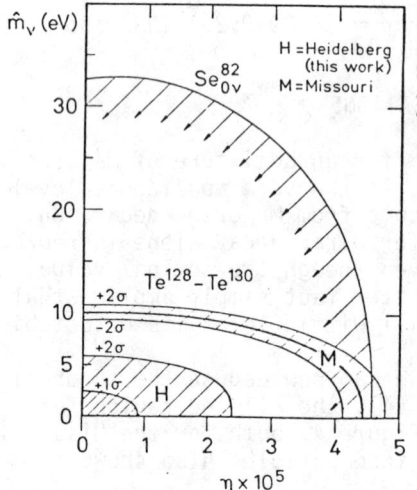

Fig. 4. Allowed regions for m_ν, η-combinations deduced from determinations of ρ by (H): Ref. 4,10; (M): Ref. 13, 33. The ^{82}Se-curve is calculated from the counting data of Ref. 26. Theoretical treatment according to Ref. 25. The 2σ-limit on m_ν is $m_\nu \leq 5.6$ eV.

Figure 4 is the best x) respective limit obtained so far from the (negative) search for neutrinoless DBD in coincidence experiments (^{82}Se, $\bar{m}_\nu \leq 33$ eV, $\eta \leq 4.6 \cdot 10^{-5}$, matrix element "calibration" based on geochemical ^{82}Se decay rate, similar as described for ^{130}Te).

As final result, we note that the Majorana mass of the electron neutrino is ≤ 5.6 eV at the 95 % confidence level. In detail, we have
1σ: $m_\nu \leq 2.12$ eV; $\eta \leq 0.88 \cdot 10^{-5}$
2σ: $m_\nu \leq 5.62$ eV; $\eta \leq 2.36 \cdot 10^{-5}$,
all compatible with $\bar{m}_\nu = \eta = 0$.

OUTLOOK

A further substantial improvement of limits on $\bar{m}_\nu$ from ^{128}Te-^{130}Te experiments is not to be expected since, below a few eV, the Dirac contribution to ^{128}Xe becomes dominant and prevents conclusions at levels below $\bar{m}_\nu \sim 1-2$ eV. With respect to other potential DBD-systems, including the pair ^{80}Se-^{82}Se - ^{80}Kr-^{82}Kr the advantage of increasing Majorana dominance in going to lower decay energies is counterbalanced by the disadvantage of dealing with lower and lower absolute rates. One would need measurement of halflives on the order of 10^{27}yrs, which is made impossible by the ever present admixture of "normal isotopes" in natural samples. Thus, it seems advisable to direct every effort to the improvement of direct counting experiments.

We acknowledge stimulating discussions with W. Hampel, W. Haxton, T. Kotani, O. Manuel, P. Minkowski, S. Rosen, and E. Takasugi as well as the efforts and hospitality of the Workshop organizers.

REFERENCES

1. T. Kirsten in "The Origin of the Solar System", S. Dermott (ed.), 267 (1978); J. Wiley, publishers.
2. W. Hampel, these Proccedings.
3. T. Kirsten, Fortschritte Physik 18, 449 (1970).
4. T. Kirsten, H. Richter, E. Jessberger, submitted to Z. Physik C (1982).
5. T. Kirsten, H.W. Müller, Earth Planet.Sci.Lett. 6, 271 (1969).
6. T. Kirsten, W. Gentner, O. Müller, Z. Naturforsch. 22a, 1783 (1967).
7. B. Srinivasan, E.C. Alexander, O. Manuel, J.Inorg.Nucl.Chem. 34, 2381 (1972).
8. H.W. Müller, Dissertation Heidelberg University (1971).
9. T. Kirsten, O.A. Schaeffer, E. Norton, R.W. Stoenner, Phys.Rev.Lett. 20, 1300 (1968).
10. T. Kirsten, H. Richter, E. Jessberger, submitted to Phys.Rev.Lett. (1982).
11. E.C. Alexander, B. Srinivasan, O.K. Manuel, Earth Planet.Sci.Lett. 5, 478 (1969).
12. B. Srinivasan, E.C. Alexander, O.K. Manuel, Econ. Geology 67, 592 (1972).

---

x) The new ^{76}Ge-data of Ref. 29 are now more stringent with respect to m_ν (see Table III).

13. E. Hennecke, O. Manuel, D. Sabu, Phys.Rev. C11, 1378 (1975).
14. N. Takaoka, K. Ogata, Z. Naturforsch. 21a, 84 (1966).
15. B. Srinivasan, E.C. Alexander, R. Beaty, D. Sinclair, O. Manuel, Econ. Geology 68, 252 (1973).
16. M. Moe and D. Lowenthal, Phys.Rev. C22, 2186 (1980).
17. H. Primakoff, S. Rosen, Ann.Rev.Nucl. Part.Sci. 31, 145 (1981).
18. D. Bryman, C. Picciotto, Rev.Mod.Phys. 50, 11 (1978).
19. H. Frampton, P. Vogel, Phys.Rept. 82, 339 (1982).
20. A. Halprin, P. Minkowski, H. Primakoff, S. Rosen, Phys.Rev. D13, 2567 (1976).
21. W. Haxton, G. Stephenson, D. Strottman, Phys.Rev.Lett. 47, 153 (1981).
22. M. Doi, T. Kotani, H. Nishiura, K.Okuda, E. Takasugi, Phys.Lett. 103B, 219 (1981).
23. S. Rosen, Proc. 1981 Internat.Conf. Neutrino Physics and Astrophysics, Univ. Hawaii, 2, 76 (1981).
24. W. Haxton, G. Stephenson, D. Strottman, Phys.Rev. D25, 2360(1982).
25. M. Doi, T. Kotani, H. Nishimura, E. Takasugi, Preprint Osaka Univ. Os-Ge 82-43 (1982).
26. B. Cleveland, W. Leo, C. Wu, L. Kasday, A. Rushton, P. Gollon, J. Ullman, Phys.Rev.Lett. 35, 757 (1975).
27. Y. Zdesenko, Pis'ma Zh.Eksp.Teor.Fiz. 32, 62 (1980).
28. R. Bardin, P. Gollon, J. Ullman, C. Wu, Nucl.Phys. A158, 337 (1970).
29. F. Avignone, R. Brodzinski, D. Brown, J. Evans, W. Hensley, J. Reeves, N. Wogman, Preprint Univ. S.Carolina (1982).
30. B. Pontecorvo, Phys. Lett. 26B, 630 (1968).
31. P. Minkowski, Nucl.Phys. B201, 269 (1982).
32. T. Kirsten, E. Jessberger, J. Kiko, Ann.Rep. Max-Planck-Inst. Kernphysik, Heidelberg, 199 (1972); see also Ref.13 for citation.
33. E. Hennecke, Phys.Rev. C17, 1168 (1978).
34. V. Lubimov, E. Novikov, V. Nozik, E. Tretyakov, V. Kozik, Phys. Lett. 94B, 266 (1980).

THE IRVINE ^{82}Se EXPERIMENTS

M.K. Moe and A.A. Hahn
University of California, Irvine, CA 92717

ABSTRACT

The large difference in half lives implied for ^{82}Se by a cloud chamber experiment and by geochemical methods remains unresolved. The need for a definitive direct counting measurement of the 2ν double beta decay half-life, and the great interest in searching for 0ν decay, have motivated the construction of a new TPC experiment sensitive to both modes. The cloud chamber and TPC experiments are briefly described.

INTRODUCTION

There is a great disparity between the very careful geochemical measurements reported for ^{82}Se by Prof. Kirsten, and the half-life implied by the cloud chamber observations reported by the Irvine group.[1]

$T_{1/2} = 1.45 \pm 0.15 \times 10^{20}$ yr   Geochemical

$T_{1/2} = 1.0 \pm 0.4 \times 10^{19}$ yr   Cloud Chamber

First, a brief review of the cloud chamber experiment is given, since what went on there is suggestive of a double beta decay rate much faster than geochemical results, and because the cloud chamber tracks and backgrounds are a good indication of what we will have to deal with in the new TPC detector, to be described later in the talk.

THE CLOUD CHAMBER EXPERIMENT

In the early work of the Columbia group[2] with ^{48}Ca, the troublesome background was idenified as decay of ^{214}Bi within the calcium source. This uranium series contaminant can mimic double beta decay in several ways, but the one operative in the cloud chamber was internal conversion. ^{214}Bi, with a Q_β of 3.2 MeV, usually beta decays to an excited level of Po, which then normally de-excites by gamma emission. About 1% of the time the excitation energy is given to a conversion electron. The β particle paired with a conversion electron results in two electrons from a single atom, and gives the appearance of a double beta decay.

In a ^{82}Se experiment the bismuth is particularly threatening because of strong similarity of the bismuth and selenium sum spectra. However, ^{214}Bi beta decay is followed by a 164 µs delayed alpha particle. A cloud chamber has sufficiently long post-trigger sensitivity to include the alpha delay, and was employed

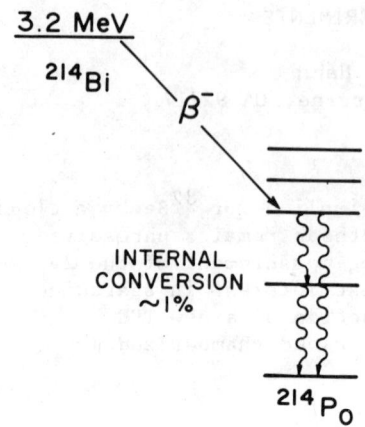

Fig. 1. Simplified ^{214}Bi decay scheme.

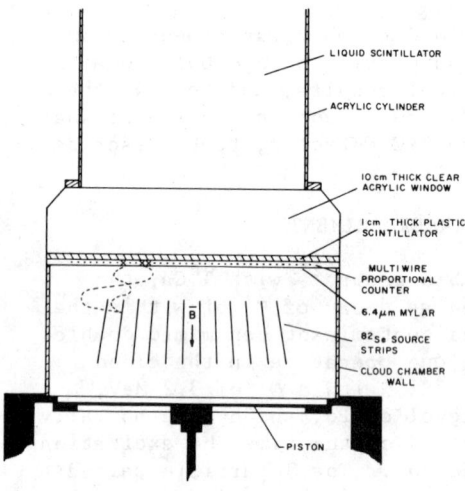

Fig. 2. Cloud chamber cross section.

specifically to exploit the appearance of the alpha particle to discriminate against ^{214}Bi.

^{214}Bi follows within minutes of 3.8 day radon gas in the decay chain. Radon, in turn, follows radium, whose long half life makes it the steady source of bismuth within the sealed cloud chamber.

A cross section of the chamber is shown in Fig. 2. The selenium is distributed in 12 rectangular strips in thin aluminized mylar envelopes. Both beta particles must travel upward to cause a trigger.

The source strips were prepared by evaporation of selenium in an atmosphere of argon, where the selenium vapor promptly condensed to a particulate suspension. In the course of 24 hours the suspension settled to the bottom of the rectangular glass container and formed a uniform deposit on a mylar substrate. There is some possibility that the evaporation process contributed to the purification of the selenium by leaving less volatile radium compounds behind.[3] The TPC selenium deposit will be made in the same way but instead of twelve rectangular strips, the source will be on one large octagon.

Fig. 3 is a cloud chamber photograph of a beta decay ^{214}Bi. The bright alpha track is unmistakable at the origin on the selenium strip. Note how the alpha particle penetrates the neighboring source strip.

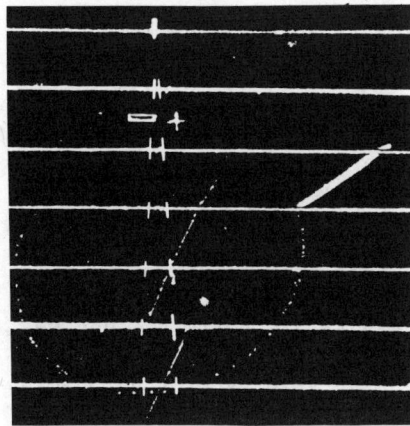

Fig. 3. Beta decay of ^{214}Bi

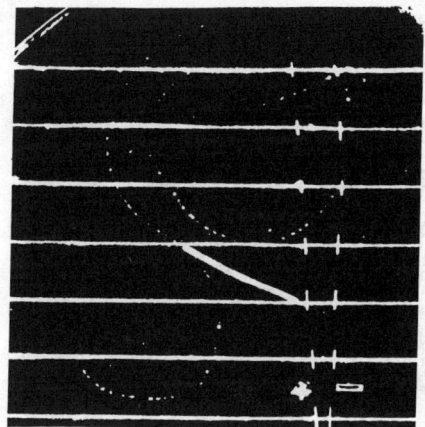

Fig. 4. Beta decay of ^{214}Bi with internal conversion.

The sources were made very thin to allow the alphas to escape. Seventy two such beta-plus-alpha events were recorded.

The dark field in these photographs is black velvet on the cloud chamber piston, a material that assumes some significance, as will be seen in a moment.

In Fig. 4 we see the beta-internal-conversion sequence, which would have been a serious background had the alpha particle not given it away. Sixteen such events were seen. The probability of the alpha being trapped in the source is 23%. The source strips were maintained at alternating positive and negative voltages to provide a clearing field to rid old ions from the chamber. The bismuth events showed a two-to-one preference for the negative strips -- a ratio to be expected if the bismuth descended from radon in the chamber gas. Radon decay products are positively charged two-thirds of the time, and therefore are preferentially attracted to the negative strips. Since the activity came from the chamber gas, the selenium itself was relatively free of bismuth and its radium progenitor.

Fig. 5 is an example of the 20 clean events recorded having the characteristics of double beta decay. Energy was measured from track curvature, in this example about 2 MeV, well below the 3.0 MeV of 0ν decay. Note the large-angle scatter in the argon gas. We plan to use helium in the TPC. Since this kind of scattering goes as Z^2 we expect it to be much reduced in the lighter gas.

The ability to actually see the tracks is a powerful discrimination against background. There were 18,000 cloud chamber triggers for these 20 cases of <u>two negative</u> electrons emerging from <u>one</u> point on the ^{Se} source.

It was possible to demonstrate that the only mechanism capable of producing false double beta events with significant probability

Fig. 5. A double beta decay candidate.

in the thin source is beta plus internal conversion. With ^{214}Bi, essentially eliminated by the alpha particle, we could find no other naturally occurring contaminant capable of escaping our detection, and yet able to mimic the ^{82}Se double beta spectrum.

If the geochemical results are correct, then we will have to find some explanation other than double beta decay for these clean two-electron events.

The experiment was terminated at 20 events because we replaced the black velvet on the piston with new material that turned out to harbor so much radium that the bismuth events increased by a factor of 10. The old velvet had been thrown away, and we were unsuccessful at finding any sufficiently black substance as clean as the original material.

In Fig. 6 the 20 clean two-electron events (circles) are plotted along with two hundred Monte Carlo events (dots) generated from the theory of Primakoff and Rosen[4] for 2ν double beta decay of ^{82}Se, and including the thresholds of the apparatus. The distributions appear to be consistent. No events are seen with a sum energy near the 3.0 MeV, 0ν line.

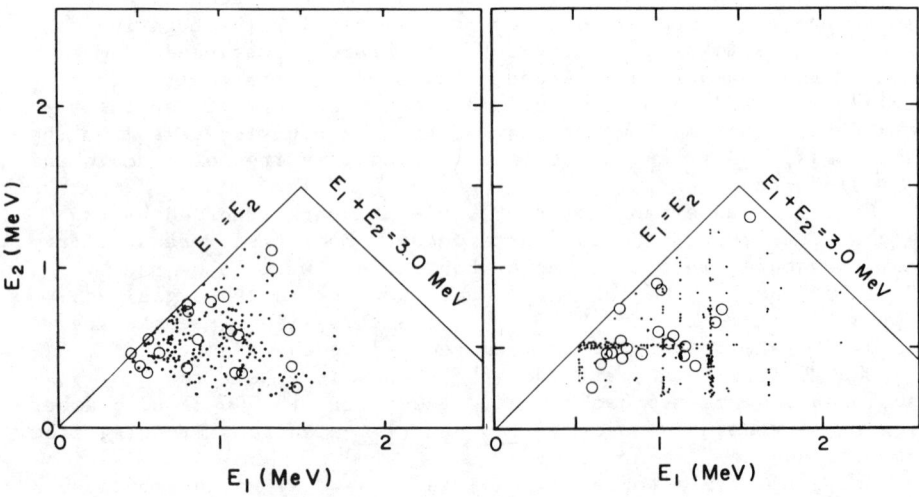

Fig. 6. The 20 clean events.   Fig. 7. ^{214}Bi events.

The distribution is quite different for the bismuth events (Fig. 7). Again, the dots are Monte Carlo points, this time from the beta-internal-conversion sequence in ^{214}Bi. Note how they cluster along lines. One member of each pair is a conversion electron, and the conversion energies are discrete. The circles are observed bismuth beta-plus-internal-conversion events. With improved energy resolution they would fall more closely on the conversion lines, providing another means of rejecting background in experiments that measure the two electron energies separately. Some conversion line rejection should be possible in the TPC.

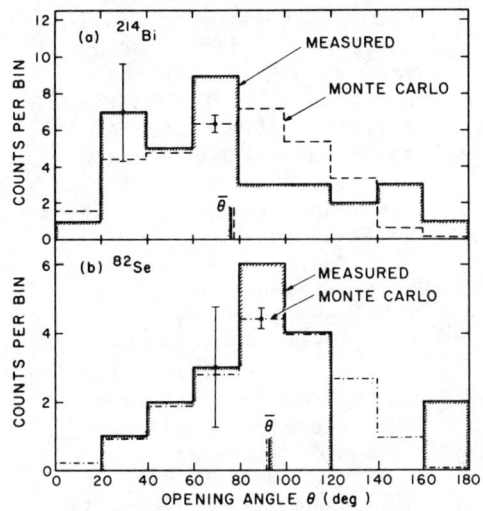

Fig. 8. Opening angle distribution for bismuth events and for clean two-electron events. Bismuth events beyond the original 16 came from radon subsequently added to the chamber.

A proper opening angle distribution would be the best evidence that the clean events are actually double beta decay. The 1-cosθ factor suppresses small angles for double beta decay as opposed to background processes. The distributions are compared in Fig. 8 for the clean events and the bismuth events. The clean events do show relative small-angle suppression, and the mean angles of the distributions differ by two standard deviations. However, this crucial test suffers from two few counts to be definitive.

The need for more counts and improved statistics on 2ν decay, and for greater sensitivity to 0ν decay, provides the incentive for a better experiment.

THE TPC

A cross section of the TPC is shown in Fig. 9. A 700 gauss magnetic field, uniform to 1% over the active volume, is supplied by Helmholtz coils. Beta particles follow helical paths from which their charge, momentum, and opening angle can be determined. Ionization electrons along the helices in the He-CH$_4$ chamber gas drift away from the selenium source under the influence

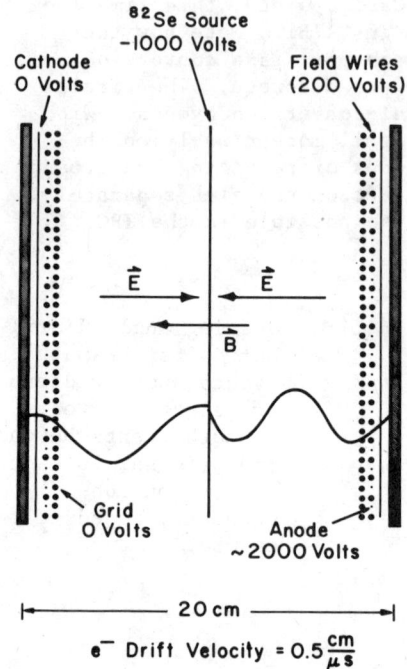

Fig. 9. A cross section of the TPC. (not to scale)

of an electric field. The shape of a drifting track is preserved up to the grid, where the electrons pass through and avalanche at the anode. Pulses produced on the anode and neighboring orthogonal cathode wires are amplified, passed through discriminators, and clocked into shift registers at 1 MHz. At a drift velocity of 5mm/µs, a "time bucket" corresponds to 5mm in the drift (Z) direction. As the anode and cathode are spaced at 5 mm, this resolution is also characteristic of the x and y directions.

There is no hardware trigger beyond the requirement that ionization be distributed over the drift distance on at least one side of the source. All such events are recorded for software analysis.

Very satisfactory tracks have been observed with 16 channels of electronics. Six hundred channels are required to complete the detector.

A schematic axial view appears in Fig. 10. The wire-support frame is Lexan polycarbonate. The usual G-10 glass-epoxy laminate is much too radioactive. The apparatus is nearly complete, with the exception of electronics, which are still under construction. The experiment is located in the Irvine Physical Sciences basement laboratory in which the cloud chamber experiment was done.

Replacement of the cloud chamber with a TPC is expected to yield numerous advantages. Any troublesome radon can be flushed from the TPC by continous changing of the gas, whereas the cloud chamber was difficult to operate with gas flowing. The long recovery time of the cloud chamber meant a very restrictive trigger was needed to avoid excessive dead time. The trigger operated with an efficiency of 2.5% at 50% dead time. The software trigger for the TPC should be nearly 50% efficient with close to zero dead time, for an overall sensitivity increase of a factor of 40.

Reduced scattering in the TPC should result in greatly improved resolution for energy and opening angle. Whereas one had to scan $\sim 10^3$ cloud chamber photographs per event, the TPC software is expected to weed out most of the uninteresting activity automatically.

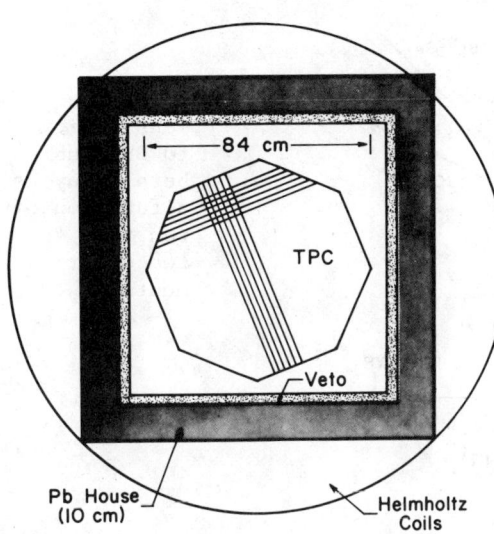

Fig. 10. Axial view of TPC.

Previous searches for 0ν double beta decay of ^{82}Se in track visualizing devices have had <u>no background</u> in the neighborhood of the 3.0 MeV peak. We expect this situation to continue well above a half life sensitivity of 10^{22} years because of the ability to reject bismuth decays. (The TPC will also be sensitive to the delayed alpha particle.) In the absence of background, the measurable half life limit is proportional to the first power of the run time

$$T_{1/2} = 0.693 n \varepsilon t \quad \text{(no background)}$$

where n is the number of double-beta-decay-unstable atoms, ε is the detector efficiency, and t is the run time. Detectors without track visibility have a background rate R, and the half life limit is

$$T_{1/2} = 0.693 n \varepsilon \sqrt{t/R} \quad \text{(background rate R)}$$

These equations are plotted in Fig.11 for the 38g of enriched ^{82}Se we have on hand, (ε= 0.5, no background), and for a typical ^{76}Ge detector (with ε= 1 and the lowest background rate yet reported for a germanium diode.[5]) Time favors the no-background detector. At some point the selenium experiment may encounter background, and move parallel to germanium as shown by the dashed line. The substantially larger germanium experiments now being prepared will have to fight the square root on "n" as well as on "t", since to first order, R $\propto$ n.

If the γ_5 invariance is broken by neutrino mass, then 0ν double beta decay should show up at a shorter half life for ^{82}Se than for ^{76}Ge by about a factor of 3. [6]

These considerations indicate that the TPC should be capable of making a significant contribution to the search for 0ν double beta decay.

CONCLUSION

The cloud chamber hinted strongly at a 1 x 10^{19} year half life for 2ν double beta decay of ^{82}Se. Because of the importance of the

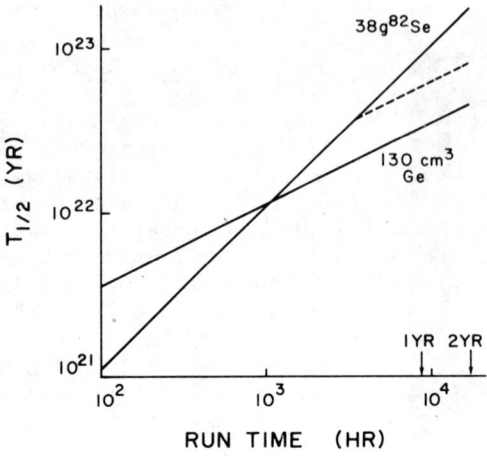

Fig. 11. Half-life sensitivity to 0ν double beta decay as a function of run time for two experiments.

2ν rate in checking the nuclear physics needed to deduce neutrino mass and/or right handed currents from 0ν data, we are attempting a definitive 2ν experiment in a TPC. We expect to have it running by Spring. If the cloud chamber half life is correct, we will see several hundred 2ν events per month. A sensivity to 0ν decay for ^{82}Se in the neighborhood of $T_{1/2} = 5 \times 10^{22}$ years will be achieved in a year of running.

REFERENCES

1. M.K. Moe and D.D. Lowenthal, Phys. Rev. C22, 2186 (1980).
2. R.K. Bardin, et.al., Nucl. Phys. A158, 337 (1970).
3. R. Davis, Jr., private communication.
4. H. Primakoff and S.P. Rosen, Rept. Progr. Phys. 22, 121 (1959).
5. E. Bellotti, Proc. 1982 Int. Conference on Neutrino Physics and Astrophysics, Balatonfüred, (1982).
6. W.C. Haxton, et. al., Phys. Rev. Lett. 47, 153 (1981).

EARLY RESULTS FROM THE BATTELLE-CAROLINA ^{76}GE DOUBLE BETA DECAY PROJECT

F. T. Avignone, III
Department of Physics, University of South Carolina,
Columbia, South Carolina 29208

R. L. Brodzinski, D. P. Brown, J. C. Evans, Jr.,
W. K. Hensley, J. H. Reeves, and N. A. Wogman
Battelle Pacific Northwest Laboratories,
Richland, Washington 99352

ABSTRACT

A search for no-neutrino double beta decay of ^{76}Ge using an anticoincidence shielded Ge spectrometer is reported. A new lower limit of $T_{1/2} \geq 1.7 \times 10^{22}$ y at a 90% CL was determined using a maximum likelihood analysis on a 5 keV wide energy bin centered at 2041 keV. Combining this result with the shell model calculations of Haxton, Stephenson and Strottman, we obtain $\bar{m}_\nu \leq 10$ eV and $|\eta| \leq 2.4 \times 10^{-5}$.

INTRODUCTION

Experimental values or limits on the level of lepton conservation and neutrino masses play important roles in testing Grand Unified Theories (GUT). A recent review by Primakoff and Rosen[1] places the subject of double beta decay in a proper modern context, giving references to earlier reviews as well as important theoretical and experimental work. Strong motivation for the recent activity in ^{76}Ge $\beta^-\beta^-$-decay is in part due to the report by Lubimov et al.[2] that 14 eV $\leq m_\nu \leq$ 46 eV. If one uses this result with recent shell model calculations of Haxton, Stephenson and Strottman,[3] it is seen that $T_{1/2}^{0\nu}(^{76}\text{Ge}) \leq 7.7 \times 10^{21}$ y, which is greater than the detection limit of the earlier experiment by the Milano group.[4] The most recent result given in ref. 4 is $T_{1/2}^{0\nu} \geq 3.2 \times 10^{21}$ y (90% CL); however, a result closer to our present limit is being reported at this conference.[5] The limit of ref. 4

corresponds to $\bar{m}_\nu \leq 22$ eV and $|\eta| \leq 5.4 \times 10^{-5}$, where the bar over m_ν allows for the fact that Majorana ν_e may be a mixture of mass eigenstates which could have CP eigenvalues of opposite sign. Wolfenstein[6] argues that this can lead to different values of m_ν deduced from single-β and from no-neutrino double-β experiments.

The mass $\bar{m}_\nu$ enters the rate calculation directly in the four-component Majorana spinor in the leptonic current j_λ^L, while one can also allow for an explicit breaking of γ_5 symmetry by writing the current as follows:

$$j_\lambda^L = \psi_e^+ \gamma_4 \gamma_\lambda \{(1+\gamma_5) + \eta(1-\gamma_5)\}[\psi_\nu^- + \psi_\nu]. \qquad (1)$$

Recent results of the Heidelberg group,[7] presented at this conference, set limits of $\bar{m}_\nu \leq 5.6$ eV and $|\eta| \leq 2.4 \times 10^{-5}$ (95% CL), based on the ratio of the total $\beta^-\beta^-$-decay half lives of ^{128}Te and ^{130}Te. Confirmation of these results, particularly in a direct search for the no-neutrino mode, would, when combined with the Lubimov result, suggest that the electron neutrino is not a Majorana mass eigenstate.

THE EXPERIMENT

The apparatus is an intrinsic Ge detector in a commercially available low-background cryostat, with a NaI(Tl) anticoincidence shield as shown in Fig. 1. The effective volume is 125 cm³ after accounting for the surface escape of β^- particles. Data were accumulated for a total of 4054 hrs. in periods of several days each. The effective energy resolution from the sum of all of the data was 4 keV at 2615 keV and 3.4 keV in the region of the decay Q-value which is 2040.9 ± 2.5 keV.[5] The data from the different counting periods were combined after minor energy renormalization, correcting for slight gain shifts, with a code which determines the number of channels between the centroids of the 2614 keV

^{208}Tl peak, from the decay of $^{228, 232}$Th, and the 1460 keV peak due to ^{40}K contamination. The data sets are then each normalized to 1 keV per channel. This procedure ensures that only events of the same energy are added.

The experimental limit on the half life, the number of ^{76}Ge atoms present N, (7.76% abundance), the total running time t and the number of counts, C, which with some probability can be attributed to no-neutrino $\beta^-\beta^-$ decay, are connected by,

$$T_{\frac{1}{2}} \geq \frac{(0.693)Nt\varepsilon}{C} . \qquad (2)$$

The efficiency ε contains the probability that electrons escape from the active volume as well as accounting for the fraction of real events which fall outside of the energy interval used to obtain C. No statistically significant peaks were evident and the data were analyzed from 2038 to 2052 keV.

We have invested a significant effort in reproducing the background γ rays, with the correct relative intensities, using radioactive sources of ^{228,232}Th, ^{226}Ra, ^{234m}Pa, ^{60}Co, ^{137}Cs, ^{40}K and a PuBe neutron source. The sources were located at a variety of points with a variety of energy degraders until the proper line shapes and relative intensities were achieved. The reference spectra collected in this manner were normalized to the background peaks in the data and subtracted leaving a smooth continuum due to cosmic rays. This was fit to a function $y = ax^b$ where x is proportional to energy and a and b are constants. This procedure was used to obtain the mean background near the Q-value and also to estimate a lower limit $T_{\frac{1}{2}}^{2\nu} \geq 2.8 \times 10^{19}$y, which can be compared to the theoretical[3] value 3.7×10^{20}y. We must then reduce the background by more than an order of magnitude to hope to observe 2-ν double beta decay using the same apparatus.

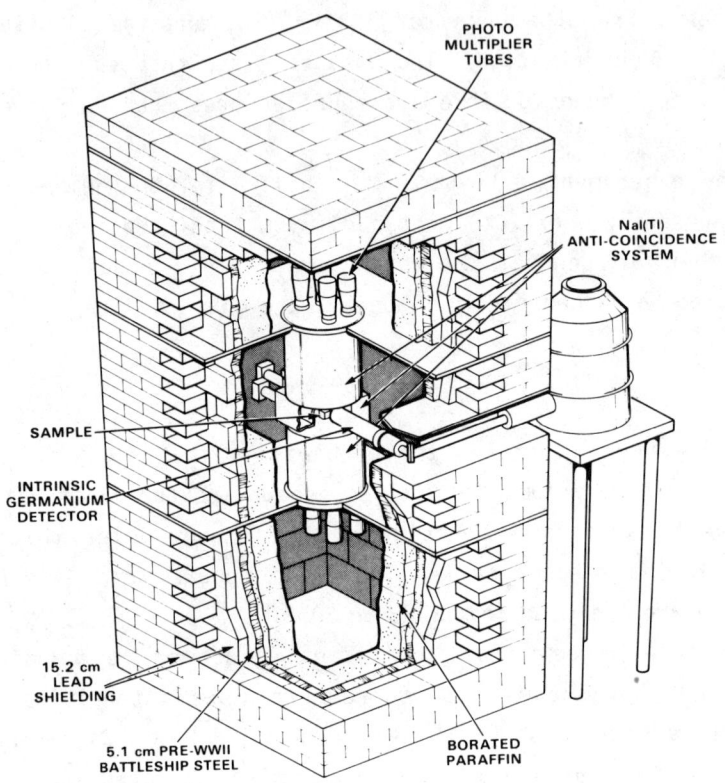

Fig. 1 Bulk-and Anticoincidence-Shielded Spectrometer

Table I. Experimentally determined background levels in no-neutrino and two-neutrino energy regions.

Source	% total (2041 keV)	% total (1000-1100 keV)
^{226}Ra	44%	24%
^{228,232}Th	22%	13%
neutrons	13%	4%
^{234m}Pa	-	24%
^{40}K	-	16%
^{60}Co	-	2%
Cosmic Ray Continuum	21%	17%

ANALYSIS AND RESULTS

The data were examined using a maximum likelihood analysis. Determination of the minimum value of the mean background in the region of the Q-value simplifies this procedure. The result is a set of probability curves dependent on the number of hypothesized double beta decay counts in 5 keV energy bins. The probability calculated is that for observing more counts than actually observed. Values for 90% CL appear in Table II below. The value at 2040.9 keV at the 90% CL is $T_{1/2}^{0\nu} \geq 1.7 \times 10^{22}$ y. These limiting values for the half life can be interpreted on the ξ-η plane, where $\xi \equiv m_{\bar{\nu}}/m_e$, using the following equation in which r is dimensionless, $r = \sqrt{n^2 + \xi^2}$

$$r^{-2}(\theta) = 1.12 \times 10^{-24} \{12.7 - 2.8\cos2\theta + 2.9 \sin2\theta\} T_{1/2}(y). \qquad (3)$$

Table II. Limiting values of $T_{1/2}^{0\nu}(^{76}Ge)$ and $\bar{m}_\nu$ (90% CL)

Energy bin	$T_{1/2} \geq$ (y)	$\bar{m}_\nu \leq$ (eV)
2038	1.47×10^{22}	10.7
2040	1.65×10^{22}	10.1
2042	1.85×10^{22}	9.6
2044	1.72×10^{22}	9.9
2046	1.26×10^{22}	11.6
2048	1.18×10^{22}	12.0
2050	1.03×10^{22}	12.8
2052	1.11×10^{22}	12.3
2055	1.31×10^{22}	11.4

This equation was derived from new results given by Haxton and Stephenson[3] and reflect the η-ξ interference. The corresponding (90% CL) values at 2041 keV are $\bar{m}_\nu \leq 10$ keV and $|\eta| \leq 2.4 \times 10^{-5}$.

CONCLUSIONS AND FUTURE DIRECTIONS

Under the assumption that the electron neutrino is a Majorana mass eigenstate, the present result contradicts the neutrino mass measurement of the ITEP group[2]; however, this conclusion is nuclear model dependent. The calculated value[3] of $T_{1/2}^{\beta\beta}(^{82}Se) = 3.07 \times 10^{19}$ years is in conflict with the mean value from geological determinations[8] which is $(1.45\pm0.15) \times 10^{20}$ y. The theoretical value is, however, greater than the value $(1.0 \pm 0.4) \times 10^{19}$ y measured by Moe and Lowenthal[9] in a cloud chamber experiment. If we assume, with no real justification, that the shell model overestimates the nuclear matrix elements of ^{76}Ge to the degree implied by the ^{82}Se geological measurements, our limit becomes $\bar{m}_\nu \leq 21$ ev. This result may be less sensitive to nuclear theory uncertainties because it depends only on the theorist's ability to calculate ratios of matrix elements. If on the other hand, the measurement of ref. 9 is correct, then our experiment implies $\bar{m}_\nu \leq 6.4$ eV by the same argument.

An improvement over the present limit can be made by increasing t or N and by decreasing the background relative to the present rate (see Fig. 2). Alpha is a multiplier of N or t. If, for example, we count for 1 year with the same volume of Ge but reduce the background by a factor of 10, we gain as much sensitivity as having 14 times the volume without improving the radiopurity. Careful consideration of Table I and Fig. 2 strongly implies that "cleaning up" the materials in the system is the most important step, but that going underground will be necessary to reduce the background by more than a factor of 5. In addition the brute force technique of simply adding more Ge is not justified unless a significant background reduction is also been achieved. This is the highest priority of our program.

We wish to thank Y. Aharonov, T. Ahrens, W. C. Haxton, T. Kirsten, C. Liguori, S. P. Rosen, M. Rushton and G. J. Stephenson for stimulating and helpful discussions. This work was supported by the Department of Energy under contract DE-AC06-76RLO 1830 and The National Science Foundation under grant PHY-8209562.

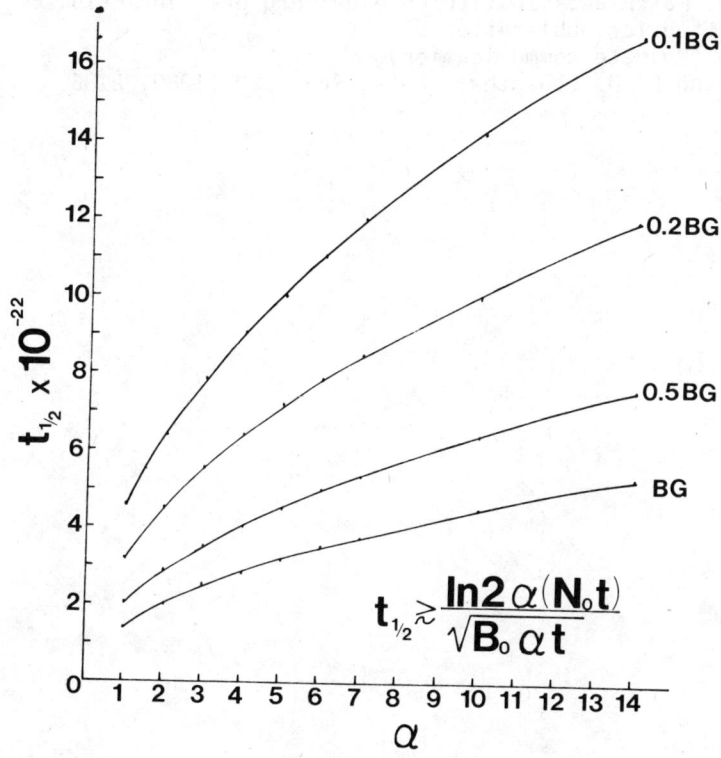

Fig. 2 Experimental sensitivy as a function of background reduction, Ge volume and time.

REFERENCES

1. Henry Primakoff and S. Peter Rosen, Ann. Rev. Nucl. Sci. 31 (1981) 145.
2. V. A. Lubimov, E. G. Novikov, V. Z. Nozik, E. F. Tretyakov, and V. S. Kosik, Phys. Lett. 94B (1980) 266.
3. W. C. Haxton, G. J. Stephenson and D. Strottman, Phys. Rev. Lett. 47, (1981) 153 and W. C. Haxton (Private communication).
4. E. Fiorini, A. Pullia, G. Bertolini, F. Cappallani and G. Restelli, Nuovo Cim. 13 (1973) 747; E. Bellotti, E. Fiorini, C. Liguori, A. Pullia, A. Sarracino and L. Zanotti, Lett. Nuovo Cim. 33 (1982) 273; C. Liguori (Presentation at this conference).

5. C. Ligouri (Private communication).
6. Lincoln Wolfenstein, Phys. Lett. 107B (1981) 77.
7. T. Kirsten, Max-Planck-Institute-Heidelberg preprint H-1982-V19 (Submitted for publication).
8. T. Kirsten (Private communication).
9. M. K. Moe and D. D. Lowenthal, Phys. Rev. C22 (1980) 2186.

THE STATUS OF THE UCSB-LBL ^{76}GE DOUBLE BETA DECAY EXPERIMENT

Michael S. Witherell
University of California
Santa Barbara, California 93106

ABSTRACT

This paper describes an experiment to look for neutrinoless double beta decay in ^{76}Ge which is now under construction. Eight large high-purity germanium detectors are surrounded by an active shield of sodium iodide. The problem of reducing background sources is discussed in some detail.

THE EXPERIMENT

We have already heard at this session comprehensive talks on the physics motivation of double beta decay and on the experimental results and problems. I will therefore limit my discussion to specific characteristics of the experiment we are building to look for the neutrinoless double beta decay ($\beta\beta_0$) of ^{76}Ge. The group includes D. Caldwell, R. Eisberg, and myself at the University of California, Santa Barbara, and F. Goulding, N. Madden, R. Pehl, and A. Smith at the Lawrence Berkeley Laboratory. The goal of the experiment is to substantially improve the sensitivity for the neutrinoless decay mode compared to the original ^{76}Ge experiment.[1] To achieve this we are using a) a much larger Ge source/detector, b) active shielding around the germanium, c) a radioactively cleaner environment, and d) a multi-detector array.

The major constraints on the design are set by the nature of the most serious background. The signal which is being sought in this experiment is a peak in the energy spectrum at 2.041 GeV, corresponding to the decay ^{76}Ge $\rightarrow$ ^{76}Se + 2e$^-$, with no associated neutrinos. In the Fiorini experiment[1] the background continuum of about $2 \cdot 10^{-3}$ counts/hour/GeV can be explained as being predominantly from Compton scatters of the 2.6 MeV gamma from ^{208}Tl in the thorium decay chain. To reduce this background we must use materials with very low levels of natural radioactivity and we need to detect the 0.5 MeV photon escaping from the Compton scatter. We will have NaI counters surrounding the germanium, but to get good efficiency we must also reduce the amount of absorption in the inert materials. We therefore will use as little mass as possible, and it will be of material with low atomic number and which has been selected for radioactive cleanliness.

We chose to use a multi-detector germanium array, for three reasons. To assemble a very large Ge source with detectors of a reasonable size, many detectors are needed. In addition, it will be of some assistance in analyzing the background sources that do remain by using the high resolution of the germanium array to see any peaks in the background spectrum. There is also a physics benefit to having many separate high-resolution detectors. Besides the transition to the 0^+ groundstate, ^{76}Ge can also decay to a 2^+ excited state of ^{76}Se, with an energy release of 1.48 MeV, followed by a 0.56 MeV gamma ray. A closely packed multi-detector array is the best way to see the coincidence of the 0.56 MeV signal in one detector with the 1.48 MeV in another. Although the ground state transition can occur because of either a finite neutrino mass or right-handed currents, only the right-handed current mechanism also allows the $0^+ \to 2^+$ transition.

With these constraints in mind I can describe the experiment, which is shown in Figure 1. The core of the experiment is a closely packed array of eight coaxial Ge detectors, each of volume 170 cc. This gives a total of about 1.3 liters of germanium, or around 7 kg, compared to the original experiment's 65 cc. The germanium is completely surrounded by a 6" thick NaI shield, constructed of ten separate counters, surrounded by a passive shield. The only penetration of the active shield is by two cold plates, each connected to a liquid nitrogen dewar outside the lead. All materials inside the sodium iodide are chosen to be extremely low Z and low in radioactivity. Some of the materials which we have

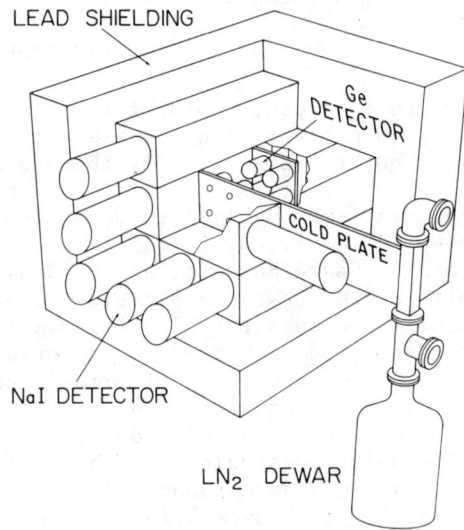

Fig. 1 - Schematic lay-out of the apparatus

found to be acceptable for this purpose are quartz, magnesium, and beryllium oxide. Almost anything one normally uses in a standard experiment is unacceptable. The largest inert mass inside the active shield is the cold plate on which the germanium is mounted, for which we presently plan to use a large piece of a single-crystal silicon. We have obtained a large crystal of silicon and measured it to be extremely low in radioactive contamination, although it is not of a quality sufficient for use in electronics. Silicon is a good thermal conductor at liquid nitrogen temperatures. Our only uncertainty is about the mechanical characteristics of the silicon, since it is not often used for a structural member. The actual piece of material used for each component must be counted in the low-level counting facility at Berkeley, because different pieces of materials such as magnesium can vary wildly in thorium content. Because of the unorthodox nature of the germanium mounting and cryostat, the packaging of the germanium detectors will be done by the LBL group.

The major components for the experiment are on order: the germanium, sodium iodide, etc. The cryostat is now under construction, to be ready for the first Ge detectors when they are delivered. The experiment will first be assembled with one array of four counters, both to take the first data and to study the backgrounds. Unfortunately no facility for counting backgrounds is as sensitive as the one represented by the experiment itself.

Finally, in this workshop on science underground, I should say something about the environment needed for a $\beta\beta$ decay experiment of this type. It is instructive to compare proton decay, where the energy release is 1 GeV and the lifetime range is 10^{31} years, to germanium $\beta\beta$ decay, in which the energy is 2 MeV and the lifetime being investigated is 10^{23} years. The difference in energy scale means that the $\beta\beta$ decay experiment is much more sensitive to natural radioactivity. The difference in lifetime sensitivity that can be reached means that proton decay is much more sensitive to cosmic rays. We will first construct an experiment above ground with the best possible environment and good active shielding. If we succeed at suppressing the major background, natural radioactivity, to the point where cosmic rays are the dominant background, then we will have earned the right to take the experiment to an underground site. Of course we will also be trying to veto the cosmic ray induced backgrounds with the active shield. Although Liguori presented data showing the high level of background in the Milano experiment above ground, Frank Avignone's data in the previous talk shows that most of this is vetoed by the active shield. One also needs to be careful in choosing underground sites, because radon can be a very bad problem. A shallow site with "clean" walls would be ideal.

REFERENCE

1. E. Fiorini, et al., Nuovo Cimento <u>13A</u>, 747 (1973).

RECENT RESULTS OBTAINED IN THE MT. BLANC EXPERIMENT ON ββ DECAY OF ^{76}Ge.

E. Bellotti, E. Fiorini, C. Liguori, A. Pullia, A. Sarracino and L. Zanotti

Dipartimento di Fisica dell'Università - Milano
I.N.F.N. - Sezione di Milano, Via Celoria 16 - 20133 Milano

ABSTRACT

Results are reported of a search for double beta decay of ^{76}Ge. The experimental set-up has been running under Mt. Blanc ($\sim$5000 m w.e. depth) for 3000 hours. No evidence appears for either $0^+ \to 0^+$ or $0^+ \to 2^+$ neutrinoless transitions with half lifetime limits of $1.1 \cdot 10^{22}$ and $4.2 \cdot 10^{21}$ years, respectively, at 90% c.l.

INTRODUCTION

Other people's talks[1-4] in this Conference dealt with the topic of double beta decay. We will only remind you that in neutrinoless double beta decay the two emitted electrons share the total available energy: the spectrum of the sum of their kinetic energies therefore should be a line while in the two-neutrino mode it should be distributed over a broad energy interval (see fig. 1). In this experiment[5,6] we looked for

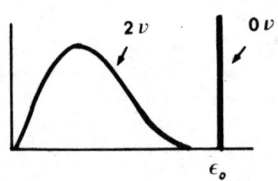

Fig.1. Sum of the kinetic energies of the two emitted electrons.

the double beta decay of ^{76}Ge (i.a.: = 7.76%). The more likely transitions should occur from ground state (0^+) of the father nucleus to ground (0^+) or first excited state (2^+) of the daughter ^{76}Se. The energy release ε_0 is, respectively, 2040.9±2.5 keV and 1481.9±2.5 keV[7].

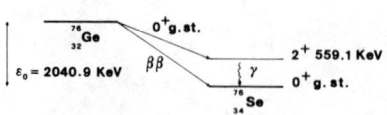

Fig. 2. Diagram of ββ transitions in ^{76}Ge.

EXPERIMENTAL DETAILS

We use as source and detector a Ge(Li) crystal (P.G.T. DL 422) with an energy resolution of less than 2 keV at 1 MeV. The fiducial volume of our detector is 116 cm^3 containing $4 \cdot 10^{23}$ nuclei of ^{76}Ge. Shielding against cosmic radiation and natural radioactivity background is achieved by operating the detector in the road tunnel under Mt. Blanc (at a depth of ∼5000 m w.e.) surrounded by consecutive layers of 4.5 cm of high purity Hg and 25÷30 cm of low activity Pb.

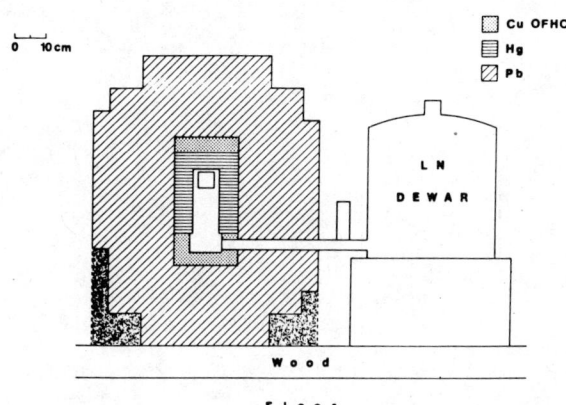

Fig. 3. Cross-section of the experimental set-up.

The total counting rate is shown in fig.4. Residual background is due to small radioactive contaminations contained in the materials constituting the detector assembly[8]. The electronic chain is a conventional one with 8K channel ADC and MCA. A spectrum stabilizer keeps resolution at the nominal value over long lasting runs.

RESULTS

Results discussed here refer to a total run time of 3061 hours to date. The experiment is still in progress.

No evidence has been obtained so far for double beta decay. Using a maximum likelihood method, we can set upper limits of .25±5.60 counts and 0±7.65 counts for the two transitions at 2040.9 and 1481.8 keV, respectively (see fig.5). These limits account for energy resolution and nuclear mass uncertainty and correspond to

$$T_{\frac{1}{2}}[(0\nu), 0^+ \rightarrow 0^+] \geq 1.8 \cdot 10^{22} \text{ yrs at } 68\% \text{ c.l.} \qquad (1)$$
$$1.1 \cdot 10^{22} \text{ yrs at } 90\% \text{ c.l.}$$

$$T_{\frac{1}{2}}[(0\nu), 0^+ \rightarrow 2^+] \geq 7 \cdot 10^{21} \text{ yrs at } 68\% \text{ c.l.} \qquad (2)$$
$$4.2 \cdot 10^{21} \text{ yrs at } 90\% \text{ c.l.}$$

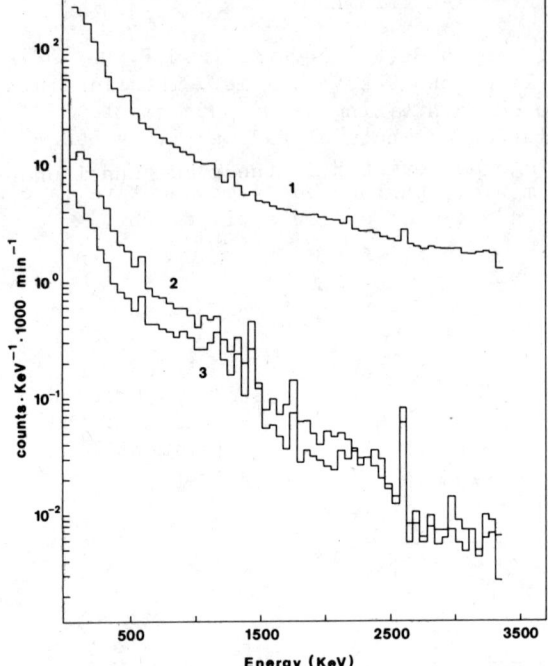

Fig.4. Total counting rate Curve 1 refers to a ground level measure; curves 2 and 3 refer to Mt. Blanc measures with Cu and Hg inner shieldings respectively.

where in (2) we have corrected for the probability of detecting the de-excitation γ-ray emitted by ^{76}Se. Analyzing the total energy spectrum shape we can also quote a half-lifetime limit for the two neutrino mode:

$$T_{\frac{1}{2}}[(2\nu), 0^+ \to 0^+] \geq \begin{array}{l} 4 \cdot 10^{18} \text{ yrs at } 68\% \text{ c.l.} \\ 2.4 \cdot 10^{18} \text{ yrs at } 90\% \text{ c.l.} \end{array} \quad (3)$$

Let us now compare the result given in (1) with theory. Following S.P. Rosen[1] we obtain limits on neutrino mass $<<m_\nu>>$ and on right handed current admixture parameter η:

$$<<m_\nu>> \leq 11 \text{ eV}; \quad |\eta| \leq 2 \cdot 10^{-5} \quad (4)$$

both at 90% c.l.. Following more recent calculation by M. Doi et al.[9] we obtain:

$$<<m_\nu>> \leq 29 \text{ eV}; \quad |\eta| \leq 5.4 \cdot 10^{-5} \quad (5)$$

A theoretical prediction for the $0^+ \to 2^+$ transition rate would be interesting since this transition can occur only in the presence of right handed currents. Unfortunately a precise one is not available at the moment and a rough estimate allows us to set a limit of $|\eta| \leq 5 \cdot 10^{-5}$

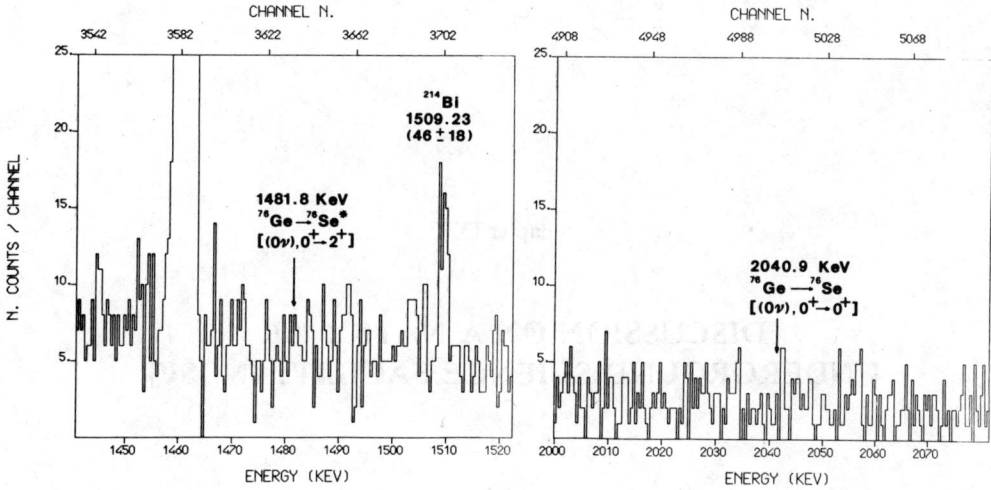

Fig. 5. Regions of interest for $0^+ \to 0^+$ and $0^+ \to 2^+$ transitions.

REFERENCES

[1]. S.P. Rosen, Report to this Conference.
[2]. C.S. Wu, Report to this Conference.
[3]. F.T. Avignone, Report to this Conference.
[4]. M. Witherell, Report to this Conference.
[5]. E. Fiorini et al., Nuovo Cimento A13, 747 (1974).
[6]. E. Bellotti et al., Lett. al Nuovo Cimento 33, 273 (1982).
[7]. A.H. Wapstra, private communication.
[8]. C. Liguori et al., Nucl. Instr. and Methods, in press.
[9]. M. Doi et al., preprint OS-GE 82-43, Osaka (1982).

Chapter IX

DISCUSSION ON A NATIONAL UNDERGROUND SCIENCE FACILITY (NUSF)

The earth has roots, and the roots belong to the soil.
If you cut a hole in the soil you have damaged the earth.
You must therefore be certain it is necessary.

From the thoughts of the Taos Indians (Taos Pueblo, New Mexico) contained in Nancy Wood, *Hollering Sun* (Simon and Schuster, New York, 1972).

PANEL

William A. Fowler, Caltech, Chairman
Ray Davis Jr., Brookhaven National Laboratory
Tom Gaisser, Bartol Research Foundation
Alfred K. Mann, University of Pennsylvania
S. Peter Rosen, Purdue University and the
                  National Science Foundation
Frank D. Stacey, University of Queensland
William Wenzel, Lawrence Berkeley Laboratory

WILLIAM FOWLER

This week has demonstrated the remarkable vitality, exciting promise, and great breadth and range of modern day science. We have heard about plans and prospects in solar, atmospheric and cosmic neutrinos, nucleon decay, cosmic rays, gravity and gravitational waves, and double beta decay. It's indeed an incredible range of proposed activity in science for the future and it is clear, at least to me, that much of that science must be done underground. How much and when and where are decisions to be made in the future. We have done our job this week in bringing to light the exciting possibilities and promise in science underground. Of course, we also learned, in the delightful talk by Bob Sharp on Thursday morning, about plans here at Los Alamos to make that science possible.

In this concluding session we will not discuss specific proposals, but rather the requirements placed on underground experimentation by the various areas of science that have been discussed this week. There will be six brief statements by the gentlemen surrounding me, followed in each case by two or three questions or comments from the audience. In view of the brief time we have, if you wish to make a comment please think about it a little beforehand and make it as brief as possible.

So, with that introduction let me turn things over to Ray Davis, our expert on solar neutrino experiments.

RAY DAVIS

In the conference we heard about the chlorine solar neutrino experimental results and the theoretical interpretation of these results. This is the only experiment now in operation. Many of us have worked a long time on a second experiment, and we have heard various ideas that could lead to a second experiment. We heard about the gallium experiment of Wolfgang Hampel. It appears to be developed, and is essentially ready to go, all that is needed is the gallium. The bromine experiment was described by Sam Hurst, it is a promising approach but needs further development. Work on this detector is moving along and the laser detection of ^{81}Kr looks feasible. The time projection chamber approach using liquid argon would test if the low flux that's observed in the chlorine experiment is due to the solar model, to the basic ideas of how the sun's producing energy, or to the particle physics.

As I mentioned, the gallium experiment is apparently ready to go. The next step is to do a source experiment. That source experiment needs space somewhere. We hope that the source experiment will be done in the very near future and at some site other than the Homestake mine. At least that's the plan.

For the solar neutrino experiments that people have discussed so far, there are certain requirements imposed on a national underground science facility. I think a very important one is to have it as deep as possible. The gallium experiment is the one that could be put at the most shallow depth and that needs at least 3,500 meters of water equivalent. I think the kind of experiment that would go into a national science underground facility would be like the one Herb Chen proposed. That needs to be as deep as possible. There are serious background effects with all of the other solar neutrino experiments. All of these solar neutrino experiments need a rather large space so I think they would need a separate room. Many of them would. Particularly the one proposed by Herb Chen.

ROCKY KOLB: You stressed during your talk the importance in your operation of the personal relationship with the people running the mine and how much is on a personal basis. Have you ever awakened in a sweat in the middle of the night from a dream that the mine has been sold? Is that a real possibility, that the management could change and everything turn around?

RAY DAVIS: Well certainly we worry about that. Remember the Homestake goldmine been operating for one hundred years, it really has. When we first went there they said that if the price of gold remained fixed at President Roosevelt's value they would be there for seven years.

The sort of future experiment that we are talking about doing, if we were to build the experiment, we'd be happy if we could run for three to five years. But a large mine, even if it does cease mining, would probably keep open for quite a few years just to maintain the facility.

BOB SHARP: Would you venture to guess how deep you would like to go ultimately? I know this is futuristic, but you say as deep as you can go.

RAY DAVIS: Well, say we do the lithium solar neutrino experiment. That's one that we people have worked on. We'd like to go deeper than 5,000 meters of water equivalent. As to the experiment of Herb Chen, I don't know whether he has determined his backgrounds, but what you worry about is muon going through and producing something like lithium-8 that sits there for awhile and then decays. That's the background that is rather serious so 5,000 feet is a reasonable depth.

WILLIAM FOWLER: I want to make just one comment for the record. The gallium experiment and any other one that measures the proton-proton neutrinos is not only independent of solar models, but it also is independent of all this argument about nuclear reaction rates that I talked about on Monday. It's really independent, and that's why it is so appealing to me.

Next I'd like to turn to Bill Wenzel who will talk to us about the exciting prospects and requirements from proton decay-nucleon decay, I suppose I should call it.

WILLIAM WENZEL

The standard electroweak model has been so successful that unifying gauge models has acquired very strong asthetic and intellectual appeal. As a result hundreds of physicists and more than a dozen experiments have shown up to test these models in the best way that we could think of, which is to look for nucleon decay. Ironically, this effort has proceeded so rapidly that nucleon decay may be detected before the vector bosons which are the cornerstone of the standard electroweak model.

We have heard from a few of the running experiments, and many more are waiting in the wings or trying to get on the air. There are a few candidate events, surrounded by lots of background. They are coming from the smaller and, in most respects, less sophisticated of the detectors. If any of these candidates survive, then we are talking about a lifetime range of 10^{30}-10^{31} years. This would make possible some very interesting experiments in the second round. In the next year the data-taking of the new experiments plus the continuation of the old ones should increase our information by two orders of magnitude.

The experiments should study very carefully what the backgrounds are. The second round depends very much on how well we can handle them. I compare the measurement of nucleon branching ratios to the K studies of the 1950's: understanding parity, strangeness, isospin rules, as well as seeing hints of charm and confronting the still mysterious neutral kaon system. The study of nucleon decay, if it can be done well, will have parallel implications for grand unified theories.

There are some important differences between the sources of kaon and nucleon decay have strong implications for the second-round detectors. Even before the kaon was identified as a single particle decaying by various modes, the available fluxes of decaying kaons from the accelerators were enormous compared to anything that will ever be available in nucleon decay studies. Another thing is that accelerators were available to go well beyond the kaon mass in energy. For the grand unified theories this may not happen. The accelerators may be running out.

Relatively sophisticated detectors would be needed to study the branching ratios. Of course, I am assuming all along that the decay actually occurs with a manageable lifetime. We have to think about the size and material, including the possibility of hydrogen, in this detector, and I'll throw in homogeneity and isotropy.

We must understand background sensitivities, tracking and particle identification.

The liquid argon TPC studied by Chen and Doe has already done a very nice job of showing that this is feasible. They've shown that purification works and that there's reason to believe cost scaling is favorable. I mean the cost of a big system goes less than linearly with its size. There are variations on this design. At LBL we're studying high pressure gas which has advantages in signal to noise. The event reconstruction you can get with such systems may be absolutely necessary for the kind of problems we have to look at. To match the five major decay modes of the kaon, it would be nice to get several decay modes well measured for the nucleon.

As far as the implications for NUSF, I guess I'm taking a little longer, Willy, but I had to say this.

WILLIAM FOWLER: Be our guest.

BILL WENZEL: The Indians say, "Don't build it unless it's needed." I think it's very important that Los Alamos is studying this: looking at possible sites. Bob Sharp told me yesterday that he wants all the help he can get. I think that labs like LBL, Hanford, Argonne, and I'm sure others, I just don't know who have special expertise, could help. It is too early to give definite specs on what the lab should be, and that's why it's important to keep looking.

The kind of symbiosis that can be developed with the military for the Nevada Test Site might also be possible with Waste Management, the Coast Guard, and Navy or the railroads. Highways are very useful in Europe, but not here. However, railroads and other activities offer possibilities for getting together on a site to lower its cost. It's important to avoid exact duplication of European capabilities. The sizes and depths of the European labs might be ideal, so we should be prepared to diversify.

One aspect of diversification is the continuation of the Davis style symbiosis with the miners. This probably will not go on to a great extent because I don't think there will be as many second generation facilities as first generation. (I'm talking about nucleon decay, now.) At least I hope not, I don't think we can afford it.

Underwater deployment represents a diversity of the program to the extent to which DUMOND will overlap. Other detectors such as a high pressure gas system could be put underwater. We've heard about surface experiments too. These are all examples of the diversity of the program.

In trying to decide the importance of NUSF in all of this, we should know what the dollar-unitarity rule is, or make some guess about it. This lab won't do everything for non-accelerator science. We must have direct funding for the universities continuing. as Fowler pointed out earlier in this program, and funding for experiments that have nothing to do with NUSF must continue.

Finally I've heard from some physicists a concern that the present approach to NUSF is more provision-of-service-oriented than it is science-oriented. I think that criticism would go away in time because Los Alamos and Al Mann have taken an approach which they think is the fastest one to get something done. It may not be the best one. It's undoubtedly too soon to say.

WILLIAM FOWLER: Comments?

RON BRODZINSKI: We have been asked all week long to come up with some cost figures for what the excavation at the Hanford site would cost. So, I called back to find out. I found out that my Bentley has been recalled. It has been replaced with a Chevrolet. Due to the whims of politics and economics, whichever comes first, they have delayed the twelve foot shaft for some undefined period of time, and they're putting in a six foot diameter shaft. The cost for drilling the twelve foot shaft to a depth of 3,800 feet, not 3,200, is $16,500,000. Drifting a fourteen foot diameter shaft would cost $1,140 per linear foot. Those are the cost figures in plateau basalt.

LARRY PRICE: Bill made a point rather implicitly that I would like to make very briefly, explicitly. He talked about considering NUSF and its capabilities in connection with these detectors capable of complete continuous activity and complete reconstruction of the events. I think that those two ideas may be well matched. But certainly in time scale the NUSF idea is probably not well matched to what should be the next generation of nucleon decay. There should be something that gets going essentially immediately. That is, a thousand, three thousand, five thousand ton fine grain calormeter should be operating within three years. I don't see any way that NUSF can be ready in time.

AL MANN: Let me just answer that quickly. I believe that that's a mistake. I believe that the present schedule for NUSF, as we have proposed, would have it ready in time for any next generation experiment. Having just built experiments of ten percent-twenty percent of the size of anything you're talking about, I have a fairly secure idea of how much time it will take. I believe that if we were to move fast enough somewhere, perhaps not the NUSF that we're talking about here, but somewhere, we can have a laboratory by 1986.

WILLIAM FOWLER: I won't permit a rebuttal. That debate could go on for some time.
I'd like to turn to an item which I have just called leptons in general. Tom Gaisser's going to tell us what he thinks about that.

Well, one certainly thinks of the cosmic ray experiments as being secondary to the solar neutrino and proton decay experiments. This is specially true with regard to proton decay experiments. However, given the opportunity to use these new, large, deep detectors of high resolution, it may turn out that the cosmic ray experiments will be the least speculative and most certain of producing some physics of some interest. It's important to think about how to make the most of that opportunity.

Of course there are muons and neutrinos. The most straightforward thing to do with the multiple coincident muons is to try to make some inferences about the primary cosmic ray composition. That's a promising direction because of the coincidence capabilities of the new detectors. Direct measurements of both primary composition and high energy interactions at around 10^{14} eV can be used to calibrate the experiments and to probe composition out to 10^{15} or 10^{16} eV.

Somebody asked me whether it would be useful to have a variety of depths and I hadn't thought about it much. But on reflection it seems that it might be useful. One could sample different places on that average-number-of-muons-per-primary curve at the same time by having underground detectors at a variety of depths.

Another thing that came up during the week, both in the description of the Mont Blanc detector and in Professor Miyake's talk, is the question of muon bundles. Some of these events are hard to understand in terms of ordinary multiple muons so there might be something. It's important to find out what's going on there.

On the subject of neutrinos, it looked to me from David Schramm's talk as if it's going to be hard to do high energy neutrino astronomy even with a full sized DUMAND, unless there's something completely unknown. But his talk showed a possiblity that there might be some astrophysical interest in the low energy neutrinos. So, one needs to do a calculation. In any case, I think it's become clear during this meeting, and it was probably clear to many people before, that we need to do a new calculation of the low energy, cosmic ray neutrino background. Then we would be in a position both to check the neutrino interaction rates with the proton decay experiments and to look for excess sources of low energy neutrinos that might be of astrophysical interest.

Finally, if you have a large ten kiloton or so detector underground, you might be able to probe an interesting region of neutrino oscillations by looking for the difference between upward going and downward going neutrino interactions contained in the volume of the detector.

WILLIAM FOWLER: Thank you.
I was interested in your first statement - maybe I'm stating it incorrectly - that you think the cosmic ray experiments are more important then proton decay?

TOM GAISSER: No, No, No! Absolutely, of course not! But the physics is less speculative. The proton decay lifetime may turn out to be 10^{34} years, but the cosmic ray muons will surely be there.

WILLIAM FOWLER: Any other comments? You can see they don't have to be at a very high level, just as long as they're reasonable.

If there are no further comments for Tom then the next item is double beta decay. None other then Peter Rosen will talk about double beta decay.

PETER ROSEN

Double beta decay has always been associated with the question of lepton non-conservation and as such it deals with a fundamental question in particle physics. I think it is also correct to say that it deals with some extremely important questions in nuclear physics. Namely, can we calculate the nuclear matrix elements associated with the double beta decay? The two are, of course, connected; especially now because there is such interest in the existence of massive neutrinos of the Majorana variety, which many of the grand unified theories tell us is likely to be the case. So, there is no doubt at all in my mind that the study of double beta decay is one of the fundamental programs that we should pursue.

As Professor Wu emphasized so strongly in her talk and as has come out in some of the other talks, from an experimental point of view perhaps the most important question you have to deal with in the first place is getting a clean source, understanding how pure your source is, and learning what possible contaminants there might be in the source. After that, there is the possibility of contaminants in your detector to worry about. I agree with the statements that were made this afternoon that you really have to solve those problems before you begin to go underground. Nevertheless, I look forward to the day when one can indeed take one's double beta decay experiments underground to study all the properties.

However, there could indeed be one ultimate limit, particularly for no-neutrino double beta decay. As we know, there is something of an ultimate limit for proton decay in that no matter how deep down you go you can never escape the neutrino background. That ultimately limits the lifetime that you can see with those experiments. There may indeed be such a limit as well with the double beta decay. Namely, if there are some low energy neutrinos around they might at some level simulate a double beta decay.

I believe Ken Lande estimated a few years ago that the kinds of lifetimes were around 10^{25} years. If that's the case, and you take the tellurium-130 lifetime as on the order 10^{21} years, the present sorts of limits on the branching ratio for the no-neutrino mode in tellurium-130 runs in the ball park of 1% to 1/10 of 1%. This means that you are conceiveably approaching the 10^{24} year lifetime. So, at some level you might come dangerously close to the neutrino background. That is clearly a question that should be investigated again.

Well, I'll stop at that point.

HERB CHEN: The neutrino background that you were mentioning, does that apply only for the geological experiments?

PETER ROSEN: Ken? You made the estimate.

KEN LANDE: They were geological.

PETER ROSEN: Those were geological limits, so then perhaps it's not as pessimistic.

KEN LANDE: I was struck in Dr. Wu's talk about the great sensitivity to radioactivity. This was brought out further in Wendel's talk. In fact, the problem is not a question of great depth. The reduction scales only a factor of an hundred in cosmic ray intensity, even if you went down to Mont Blanc. You can do that in a fairly shallow mine. What's really important is that the radioactivity in the surroundings must be comparable to the radioactivity in the material of the detector. That is a very sensitive point. It occurred to me that if I were thinking of this I would get myself a large bowl of water, maybe like IMB, and I would put the detector in the center of it. That would be a very pure wall to anything.

PETER ROSEN: Yes, I meant to mention that. Thank you for reminding me. You do have to worry down a mine about the local radioactivity. You may be worse off if it's too radioactive a mine.

WILLIAM FOWLER: Any other questions?
Well, I might just say on this question of the radioactivity that the committee that's advising Los Alamos has suggested that a careful study be made of the radioactive levels at each possible site. I think Bob Sharp agrees that that will be done.

BOB SHARP: Definitely.

WILLIAM FOWLER: Right.
Going on, I came away from the sessions on gravity and gravitational waves just shaking. I hope that Frank Stacey can say something that'll calm me down.

FRANK STACEY

Well, thank you. I hope it wasn't because you were mad with us.

WILLIAM FOWLER: No. I was shaking from excitement.

FRANK STACEY: Perhaps gravity has a place here by virtue of the long-term possibilities that eventually it can be unified with the other forces of physics. But you particle boys have now got a "grand" unified field theory. So what are we going to call the thing that will incorporate us? Total unified theory?

WILLIAM FOWLER: We've already got a name, SuperGUTS.

FRANK STACEY: SuperGUTS. That'll do very well.
Perhaps for this purpose I should count gravity to be incorporated in a broader geophysical scene, and answer for both. Gravity itself has three aspects as far as the discussions of this meeting are concerned.

If I deal with each of them in turn, the first concerns the radial dependence of the gravitational constant or, if you like, the question of the inverse square law. In that sense a nicely controlled deep hole would be a site of opportunity and would not lead to any permanent recording or any permanent installation. The experiments for that part of the project, which we have generally called gravity, would be complete before any of the other instruments had been installed and no permanent site would be required. It would be a useful opportunity but doesn't impose any requirement at all on specific space.

With regard to gravity waves, I was talking at lunch time with Ron Drever who certainly would like to keep his options open on the possibility of a deep instrument. The point is that he would seek a kind of isolation which would permit extending the measurements to frequencies much lower than would be possible on the surface. It should be mentioned that there is also planned an experiment in space and the technology there is, I suppose, more difficult.

So, a deep hole is of interest, but there are some requirements as well as some simplifications on requirements that have to be borne in mind. The site would have to be very free of mechanical disturbances or moving masses. It is possible the conventional ventilation might present a problem. You would need some arm or limb of the excavation which would be removed from the main stream of ventilation. However, you would not require a massive chamber. One would be quite happy with relatively long thin tunnels set at right angles, which would in fact be much simpler to excavate than something very large.

The third category of gravity experiment we really haven't considered very adequately. I think none of us here was really competent to consider it. That is the G experiments. Some G experiments certainly are astronomical and of no relevance to our considerations. But the possiblity of a very precise rotating system, magnetically levitated, I know is under consideration. This is relevant to the deep hole because it would be influenced by gravitational gradients or variations in gradient of the same kind that influence gravity wave detectors. But I really can't say more than that. I don't know what the space requirements would be. I suspect not very great.

Let me just extend slightly to other kinds of geophysics. Having such a nicely controlled hole with well known geology (especially in an area where the hole was not excavated because the geology was inhomogeneous and contained ores), it would be a pity not to install one or two or three of the new generation of strain experiments which have been developed with the earthquake prediction program in mind. For this purpose what I would suggest is not to use any of the main chambers with a normal level but to drill below the main chambers, below the sump for pumping. One would install something which would be non-recoverable in a limited array of bore holes which might extend 50 meters or so below the general level of the laboratory space.
Thank you.

WILLIAM FOWLER: Thanks Frank. Questions or comments?

BOB SPERO: A couple of classes of experiments we didn't get a chance to talk about might even be easier than gravity wave experiments. One class is tests of relativistic gravity in the lab; magnetic gravity and so-called kinetic gravity, the interaction of gravity with the spinning of gravity. That can be done on a fairly small scale with a room-sized torsion pendulum. You have to know the gravitational gradients to very high precision. Another class of experiments are frame-dragging experiments.

WILLIAM FOWLER: I must say that the requirement of quiet in a nuclear underground facility gives me great concern. I would hope that those of you in the gravitation wave business would do some considerable thinking about just how far that has to go. At the same time I would like to suggest to those who are thinking about an underground facility to consider how far you can go because there are going to be people moving around from time to time and maybe by some scheduling or something it could be done. We don't want to discuss it now but I hope both groups will try to get something that might match because it looks like a big mismatch to me at the moment. But clearly I'm an amateur.
Any other comments? Ron.

RON DREVER: Yes, one does not require quiet all the time. It just needs to be quiet, say three-fourths of the time. That's just fine.

WILLIAM FOWLER: Well, that's just the sort of thing, Ron, that I think has to be brought out. Maybe you don't even need three quarters, you see, and so forth.

RON DREVER: Maybe half would do.

WILLIAM FOWLER: All right.

Well, the man who had the first word deserves to have the last word and so I have pinned him down. Al Mann is to paint the broad picture.

AL MANN

That's a big job.

I hope all of you have found the science that's been discussed here this week as stimulating as I have. I must confess there are certain areas with which I was completely unfamiliar, particularly the gravity discussions and some others. And just as Willie said, I found myself jumping with real joy to hear this.

On the other hand, having spent so many years as a physicist working at accelerators and having been involving in many accelerator workshops before one had the accelerator, I think, if you go back to those studies, you will find that in the main one did not do many of the experiments that were originally planned in those studies and the ones that were done weren't done the way people had thought they were going to be done. All of that gives me reason to believe that that's why one wants to have some kind of a more general facility - some kind of a more national facility, because the science is alive under our hands. It keeps changing from year to year and we don't do things next year the way we thought we were going to do them this year. The interest changes, our capacity changes, and our techniques change.

The way to exploit this vitality and flexibility of our science is to have a facility which is a more general one than any one of us can afford to have by himself.

Now it's certainly true that this limits our independence to a certain extent. We are all very private individuals with respect to our science, in a certain sense, but we've learned that a certain measure of cooperation buys us as individuals a great deal more then we can get by maintaining complete independence. I think that's another reason why one wants to think about a science facility. I don't completely like to use the word "national" because it sounds so grandiose. But that's the general idea, something that involves some consortium.

Let me close by saying the following. I have had no formal connection with Los Alamos National Laboratory and I still don't have any. In the year in which we've been looking at the possibility of a national facility I have had informal conversations with people at other national laboratories and at mines. Wherever I thought there was a possibility for a competitive place I have tried to talk informally, either with the administration or with someone involved in that laboratory or that mine, to encourage them to consider submitting a proposal or being competitive in some way.

So far, the only institution which has stepped out front and made a serious effort in the direction of a national facility (something that would accomodate people in general and not serve the immediate use of some possessers of that particular shaft or mine), the only one, the only institution that has responded, has been Los Alamos. And as you can see, it has responded fairly well. I want to acknowledge that publically; in Willy's terms, I think Los Alamos deserves high marks for having, what in my view is, a certain foresight and the courage to step forward and put it's money and it's time and it's energy into doing this. And it hasn't only been the engineers and the geologists. It has been encouraged by the physics division and by the theoretical division.

Nevertheless, I hope that if there are other contenders for a site such as this, that they will step forward, that they will prepare a serious proposal such as has been done at Los Alamos, and that they will go into competition with this present proposal so that we get the best of all possible facilities in the shortest possible time - because that's what we all want.

I'm finished.

WILLIAM FOWLER: Any comments?

BOB SHARP: I have one.

WILLIAM FOWLER: Yes?

BOB SHARP: In connection with what Al just said, if you have decided you really don't want to "compete," we would still like all the help we can get on the one we are working on if you have any ideas.

WILLIAM FOWLER: Well, I'm sure you all want to express some expression of gratitude to our host the Los Alamos National Laboratory. So let me thank all the people who really did the work.

With that, I DECLARE THIS WORKSHOP ADJOURNED!

AIP Conference Proceedings

		L.C. Number	ISBN
No. 1	Feedback and Dynamic Control of Plasmas	70-141596	0-88318-100-2
No. 2	Particles and Fields - 1971 (Rochester)	71-184662	0-88318-101-0
No. 3	Thermal Expansion - 1971 (Corning)	72-76970	0-88318-102-9
No. 4	Superconductivity in d-and f-Band Metals (Rochester, 1971)	74-18879	0-88318-103-7
No. 5	Magnetism and Magnetic Materials - 1971 (2 parts) (Chicago)	59-2468	0-88318-104-5
No. 6	Particle Physics (Irvine, 1971)	72-81239	0-88318-105-3
No. 7	Exploring the History of Nuclear Physics	72-81883	0-88318-106-1
No. 8	Experimental Meson Spectroscopy - 1972	72-88226	0-88318-107-X
No. 9	Cyclotrons - 1972 (Vancouver)	72-92798	0-88318-108-8
No. 10	Magnetism and Magnetic Materials - 1972	72-623469	0-88318-109-6
No. 11	Transport Phenomena - 1973 (Brown University Conference)	73-80682	0-88318-110-X
No. 12	Experiments on High Energy Particle Collisions - 1973 (Vanderbilt Conference)	73-81705	0-88318-111-8
No. 13	π-π Scattering - 1973 (Tallahassee Conference)	73-81704	0-88318-112-6
No. 14	Particles and Fields - 1973 (APS/DPF Berkeley)	73-91923	0-88318-113-4
No. 15	High Energy Collisions - 1973 (Stony Brook)	73-92324	0-88318-114-2
No. 16	Causality and Physical Theories (Wayne State University, 1973)	73-93420	0-88318-115-0
No. 17	Thermal Expansion - 1973 (lake of the Ozarks)	73-94415	0-88318-116-9
No. 18	Magnetism and Magnetic Materials - 1973 (2 parts) (Boston)	59-2468	0-88318-117-7
No. 19	Physics and the Energy Problem - 1974 (APS Chicago)	73-94416	0-88318-118-5
No. 20	Tetrahedrally Bonded Amorphous Semiconductors (Yorktown Heights, 1974)	74-80145	0-88318-119-3
No. 21	Experimental Meson Spectroscopy - 1974 (Boston)	74-82628	0-88318-120-7
No. 22	Neutrinos - 1974 (Philadelphia)	74-82413	0-88318-121-5
No. 23	Particles and Fields - 1974 (APS/DPF Williamsburg)	74-27575	0-88318-122-3
No. 24	Magnetism and Magnetic Materials - 1974 (20th Annual Conference, San Francisco)	75-2647	0-88318-123-1
No. 25	Efficient Use of Energy (The APS Studies on the Technical Aspects of the More Efficient Use of Energy)	75-18227	0-88318-124-X

No. 26	High-Energy Physics and Nuclear Structure - 1975 (Santa Fe and Los Alamos)	75-26411	0-88318-125-8
No. 27	Topics in Statistical Mechanics and Biophysics: A Memorial to Julius L. Jackson (Wayne State University, 1975)	75-36309	0-88318-126-6
No. 28	Physics and Our World: A Symposium in Honor of Victor F. Weisskopf (M.I.T., 1974)	76-7207	0-88318-127-4
No. 29	Magnetism and Magnetic Materials - 1975 (21st Annual Conference, Philadelphia)	76-10931	0-88318-128-2
No. 30	Particle Searches and Discoveries - 1976 (Vanderbilt Conference)	76-19949	0-88318-129-0
No. 31	Structure and Excitations of Amorphous Solids (Williamsburg, VA., 1976)	76-22279	0-88318-130-4
No. 32	Materials Technology - 1976 (APS New York Meeting)	76-27967	0-88318-131-2
No. 33	Meson-Nuclear Physics - 1976 (Carnegie-Mellon Conference)	76-26811	0-88318-132-0
No. 34	Magnetism and Magnetic Materials - 1976 (Joint MMM-Intermag Conference, Pittsburgh)	76-47106	0-88318-133-9
No. 35	High Energy Physics with Polarized Beams and Targets (Argonne, 1976)	76-50181	0-88318-134-7
No. 36	Momentum Wave Functions - 1976 (Indiana University)	77-82145	0-88318-135-5
No. 37	Weak Interaction Physics - 1977 (Indiana University)	77-83344	0-88318-136-3
No. 38	Workshop on New Directions in Mossbauer Spectroscopy (Argonne, 1977)	77-90635	0-88318-137-1
No. 39	Physics Careers, Employment and Education (Penn State, 1977)	77-94053	0-88318-138-X
No. 40	Electrical Transport and Optical Properties of Inhomogeneous Media (Ohio State University, 1977)	78-54319	0-88318-139-8
No. 41	Nucleon-Nucleon Interactions - 1977 (Vancouver)	78-54249	0-88318-140-1
No. 42	Higher Energy Polarized Proton Beams (Ann Arbor, 1977)	78-55682	0-88318-141-X
No. 43	Particles and Fields - 1977 (APS/DPF, Argonne)	78-55683	0-88318-142-8
No. 44	Future Trends in Superconductive Electronics (Charlottesville, 1978)	77-9240	0-88318-143-6
No. 45	New Results in High Energy Physics - 1978 (Vanderbilt Conference)	78-67196	0-88318-144-4
No. 46	Topics in Nonlinear Dynamics (La Jolla Institute)	78-057870	0-88318-145-2
No. 47	Clustering Aspects of Nuclear Structure and Nuclear Reactions (Winnepeg, 1978)	78-64942	0-88318-146-0
No. 48	Current Trends in the Theory of Fields (Tallahassee, 1978)	78-72948	0-88318-147-9
No. 49	Cosmic Rays and Particle Physics - 1978 (Bartol Conference)	79-50489	0-88318-148-7

No. 50	Laser-Solid Interactions and Laser Processing - 1978 (Boston)	79-51564	0-88318-149-5
No. 51	High Energy Physics with Polarized Beams and Polarized Targets (Argonne, 1978)	79-64565	0-88318-150-9
No. 52	Long-Distance Neutrino Detection - 1978 (C.L. Cowan Memorial Symposium)	79-52078	0-88318-151-7
No. 53	Modulated Structures - 1979 (Kailua Kona, Hawaii)	79-53846	0-88318-152-5
No. 54	Meson-Nuclear Physics - 1979 (Houston)	79-53978	0-88318-153-3
No. 55	Quantum Chromodynamics (La Jolla, 1978)	79-54969	0-88318-154-1
No. 56	Particle Acceleration Mechanisms in Astrophysics (La Jolla, 1979)	79-55844	0-88318-155-X
No. 57	Nonlinear Dynamics and the Beam-Beam Interaction (Brookhaven, 1979)	79-57341	0-88318-156-8
No. 58	Inhomogeneous Superconductors - 1979 (Berkeley Springs, W.V.)	79-57620	0-88318-157-6
No. 59	Particles and Fields - 1979 (APS/DPF Montreal)	80-66631	0-88318-158-4
No. 60	History of the ZGS (Argonne, 1979)	80-67694	0-88318-159-2
No. 61	Aspects of the Kinetics and Dynamics of Surface Reactions (La Jolla Institute, 1979)	80-68004	0-88318-160-6
No. 62	High Energy e^+e^- Interactions (Vanderbilt, 1980)	80-53377	0-88318-161-4
No. 63	Supernovae Spectra (La Jolla, 1980)	80-70019	0-88318-162-2
No. 64	Laboratory EXAFS Facilities - 1980 (Univ. of Washington)	80-70579	0-88318-163-0
No. 65	Optics in Four Dimensions - 1980 (ICO, Ensenada)	80-70771	0-88318-164-9
No. 66	Physics in the Automotive Industry - 1980 (APS/AAPT Topical Conference)	80-70987	0-88318-165-7
No. 67	Experimental Meson Spectroscopy - 1980 (Sixth International Conference, Brookhaven)	80-71123	0-88318-166-5
No. 68	High Energy Physics - 1980 (XX International Conference, Madison)	81-65032	0-88318-167-3
No. 69	Polarization Phenomena in Nuclear Physics - 1980 (Fifth International Symposium, Santa Fe)	81-65107	0-88318-168-1
No. 70	Chemistry and Physics of Coal Utilization - 1980 (APS, Morgantown)	81-65106	0-88318-169-X
No. 71	Group Theory and its Applications in Physics - 1980 (Latin American School of Physics, Mexico City)	81-66132	0-88318-170-3
No. 72	Weak Interactions as a Probe of Unification (Virginia Polytechnic Institute - 1980)	81-67184	0-88318-171-1
No. 73	Tetrahedrally Bonded Amorphous Semiconductors (Carefree, Arizona, 1981)	81-67419	0-88318-172-X
No. 74	Perturbative Quantum Chromodynamics (Tallahassee, 1981)	81-70372	0-88318-173-8

No.	Title		
No. 75	Low Energy X-ray Diagnostics-1981 (Monterey)	81-69841	0-88318-174-6
No. 76	Nonlinear Properties of Internal Waves (La Jolla Institute, 1981)	81-71062	0-88318-175-4
No. 77	Gamma Ray Transients and Related Astrophysical Phenomena (La Jolla Institute, 1981)	81-71543	0-88318-176-2
No. 78	Shock Waves in Condensed Matter - 1981 (Menlo Park)	82-70014	0-88318-177-0
No. 79	Pion Production and Absorption in Nuclei - 1981 (Indiana University Cyclotron Facility)	82-70678	0-88318-178-9
No. 80	Polarized Proton Ion Sources (Ann Arbor, 1981)	82-71025	0-88318-179-7
No. 81	Particles and Fields - 1981: Testing the Standard Model (APS/DPF, Santa Cruz)	82-71156	0-88318-180-0
No. 82	Interpretation of Climate and Photochemical Models, Ozone and Temperature Measurements (La Jolla Institute, 1981)	82-071345	0-88318-181-9
No. 83	The Galactic Center (Cal. Inst. of Tech., 1982)	82-071635	0-88318-182-7
No. 84	Physics in the Steel Industry (APS.AISI, Lehigh University, 1981)	82-072033	0-88318-183-5
No. 85	Proton-Antiproton Collider Physics - 1981 (Madison, Wisconsin)	82-072141	0-88318-184-3
No. 86	Momentum Wave Functions - 1982 (Adelaide, Australia)	82-072375	0-88318-185-1
No. 87	Physics of High Energy Particle Accelerators (Fermilab Summer School, 1981)	82-072421	0-88318-186-X
No. 88	Mathematical Methods in Hydrodynamics and Integrability in Dynamical Systems (La Jolla Institute, 1981)	82-072462	0-88318-187-8
No. 89	Neutron Scattering - 1981 (Argonne National Laboratory)	82-073094	0-88318-188-6
No. 90	Laser Techniques for Extreme Ultraviolt Spectroscopy (Boulder, 1982)	82-073205	0-88318-189-4
No. 91	Laser Acceleration of Particles (Los Alamos, 1982)	82-073361	0-88318-190-8
No. 92	The State of Particle Accelerators and High Energy Physics (Fermilab, 1981)	82-073861	0-88318-191-6
No. 93	Novel Results in Particle Physics (Vanderbilt, 1982)	82-73954	0-88318-192-4
No. 94	X-Ray and Atomic Inner-Shell Physics-1982 (International Conference, U. of Oregon)	82-74075	0-88318-193-2
No. 95	High Energy Spin Physics - 1982 (Brookhaven National Laboratory)	83-70154	0-88318-194-0
No. 96	Science Underground (Los Alamos, 1982)	83-70377	0-88318-195-9

RAYMOND H. FOGLER LIBRARY
DATE DUE

BOOKS ARE SUBJECT TO
RECALL AFTER TWO WEEKS